Lecture Notes in Computer Science 1261
Edited by G. Goos, J. Hartmanis and J. van Leeuwen

Springer
Berlin
Heidelberg
New York
Barcelona
Budapest
Hong Kong
London
Milan
Paris
Santa Clara
Singapore
Tokyo

Jan Mycielski Grzegorz Rozenberg
Arto Salomaa (Eds.)

Structures in Logic and Computer Science

A Selection of Essays in Honor of A. Ehrenfeucht

Springer

Series Editors

Gerhard Goos, Karlsruhe University, Germany

Juris Hartmanis, Cornell University, NY, USA

Jan van Leeuwen, Utrecht University, The Netherlands

Volume Editors

Jan Mycielski
University of Colorado at Boulder, Department of Mathematics
Campus Box 395, Boulder, CO 80309, USA
E-mail: jmyciel@euclid.colorado.edu

Grzegorz Rozenberg
Leiden University, Department of Computer Science
P.O. Box 9512, 2300 RA Leiden, The Netherlands
E-mail: rozenber@wi.leidenuniv.nl
and
University of Colorado at Boulder, Department of Computer Science

Arto Salomaa
Academy of Finland and Turku Centre for Computer Science
Lemminkäisenkatu 14A, FIN-20520 Turku, Finland
E-mail: asalomaa@sara.cc.utu.fi

Cataloging-in-Publication data applied for

Die Deutsche Bibliothek - CIP-Einheitsaufnahme

Structures in logic and computer science : a selection of essays / Jan Mycielski ... (ed.). - Berlin ; Heidelberg ; New York ; Barcelona ; Budapest ; Hong Kong ; London ; Milan ; Paris ; Santa Clara ; Singapore ; Tokyo : Springer, 1997
(Lecture notes in computer science ; Vol. 1261)
ISBN 3-540-63246-8

CR Subject Classification (1991): F

ISSN 0302-9743
ISBN 3-540-63246-8 Springer-Verlag Berlin Heidelberg New York

Typesetting: Camera-ready by author
SPIN 10548945 06/3142 – 5 4 3 2 1 0 Printed on acid-free paper

Preface

This book is dedicated to Andrzej Ehrenfeucht on the occasion of his 65th birthday on August 8, 1997. The editors feel priviledged to take part in a project honoring a scientist of such an exceptional stature. Ehrenfeucht's work is of seminal character both in mathematical logic and in theoretical computer science. The structure of this book reflects the two-fold nature of his work.

The contributors were invited by the editors. The editors would like to thank the authors and the referees for good and timely cooperation. Thanks are also due to the publisher, Springer-Verlag, and to Dr. Hendrik Jan Hoogeboom for his work on this book.

Each of the three editors has a different relation to Andrzej Ehrenfeucht.

For Jan Mycielski, Andrzej Ehrenfeucht is a collegue and friend with whom he has discussed mathematical questions since their student days in Poland.

For Grzegorz Rozenberg, Andrzej is his teacher, his collaborator, and his brother.

For Arto Salomaa, Andrzej Ehrenfeucht is a scientist whom he greatly admires but has never met, although he has joint papers with him.

Therefore, it is appropriate that this preface be concluded with a separate passage by each of the editors.

(1) It is an honor to be one of the editors of this volume dedicated to Andrzej Ehrenfeucht. I was invited to participate in this project by his energetic and prolific collaborator Grzegorz Rozenberg (because of my long association with Andrzej which will be described in a while). Arto, Grzegorz, and I started to think about prospective authors. Now we realize how many others also could and should have been invited to write, but it is too late, and we can only offer apologies for our thoughtless omissions.

Andrzej and I met long ago in Poland. We were born less than a year apart, and we studied in the 1950s, Andrzej at the University of Warsaw and I at the University of Wrocław. We were both fascinated by logic and foundations of mathematics. The fifties was a period of intense research in model theory. There was a lively exchange of ideas and papers between Berkeley (the school of Tarski) and Warsaw (the school of Mostowski). Andrzej and I met often at a seminar organized by J. Łoś in Toruń, where I learned from him about the latest developments. It was clear that Andrzej was the most talented young logician in Poland; time and again his deep insights and quick penetrating intelligence commanded the admiration of those who understood his work. Several basic theorems and methods of model theory of those years are due to him (and are presented in the well-known monographs of C.C. Chang and H.J. Keisler and that of W. Hodges). Later we both came to live in the United States, where Andrzej continued his brilliant research (mostly in other areas of mathematics and computer science). In the seventies we both came to settle in Boulder, Colorado, and since then we have writen a number of papers together.

Jan Mycielski

(2) I met Andrzej when I was working on my Master's thesis at the Electronics Department of the Technical University of Warsaw, Poland. Although I was in love with electronic hardware, I chose to write my thesis on the theory of algorithms albeit from an engineering point of view. The work on that thesis brought me in contact with Andrzej and very soon my excitement with electronics metamorphosed into a love of the mathematical theory of computation. Through Andrzej I discovered the beauty of mathematical logic, and became a member of the logic group at the Mathematical Institute of the Polish Academy of Sciences headed by A. Mostowski, where Andrzej was my mentor. Andrzej left Poland a couple of years before I did. We met again in the US in 1971 and resumed our collaboration.

Andrzej is for me a personification of science: he lives for and by science. He has made tremendous contributions to mathematical logic and theoretical computer science, but the breadth of his knowledge and creativity is amazing. Together with his life partner Patricia Baggett he is doing pioneering work in the teaching of mathematics in primary schools. An expert in the history of mathematics, dinosaurs, butterflies, spiders, trees, linguistics, physics, genetics, astronomy, ... Andrzej has also been an outstanding mountain climber – there is even a ridge in Yosemite National Park named by him. He is the most stimulating research partner – we have written over 100 papers together, and each new one is an exciting adventure. Andrzej's greatest gift is the ability to see elegant structures in the problems that he works on – this explains the title of this book.

Dear Andrzej,

I may be for you the magician of the playing cards, but you are for everybody, *the* Wizard of Science.

Grzegorz Rozenberg

(3) Dear Professor Ehrenfeucht,

I accepted with pleasure and enthusiasm the invitation to become an editor of this volume. I have followed and admired your work over the years. Although broad in scope, it is of impressive depth. We are about the same physical age; still I have always viewed you as a senior scientist. The reason for this is that some of your results were already well-known when I started my career; your work was referred to by my teacher in Berkeley, Alfred Tarski. I regret that we have never met. Such a meeting almost took place during the L systems conference in Noordwijkerhout in 1975, but your coming there was canceled at the last minute. I take this opportunity to wish you further success in science and all the best in general during the years to come.

Arto Salomaa

Curriculum Vitae of Andrzej Ehrenfeucht

Born: August 8, 1932, in Wilno, Poland.

Degrees:

M.Sc. in mathematics, University of Warsaw, Warsaw, Poland, 1955.

Ph.D. in Mathematics, Mathematical Institute of the Polish Academy of Sciences, Warsaw, Poland, 1961.

Positions:

1960-61, Adjunct, Mathematical Insitute, Polish Academy of Sciences, Warsaw, Poland.

1961-62, Assistant Professor, Mathematical Insitute, Polish Academy of Sciences, Warsaw, Poland.

1962-63, Assistant Professor, Department of Mathematics, University of California, Berkeley, USA.

1963-65, Assistant Professor, Mathematical Insitute, Polish Academy of Sciences, Warsaw, Poland.

1965-68, Assistant Professor, Department of Philosophy, Stanford University, Stanford, California, USA.

1968-71, Associate Professor, Department of Mathematics & Department of Computer Sciences, University of Southern California, Los Angeles, California, USA.

1971-74, Associate Professor, Department of Computer Science, University of Colorado, Boulder, Colorado, USA.

1974- , Professor, Department of Computer Science, University of Colorado, Boulder, Colorado, USA.

Publications:

Has published over two hundred papers and two books in mathematics, computer science, linguistics, psychology, and education.

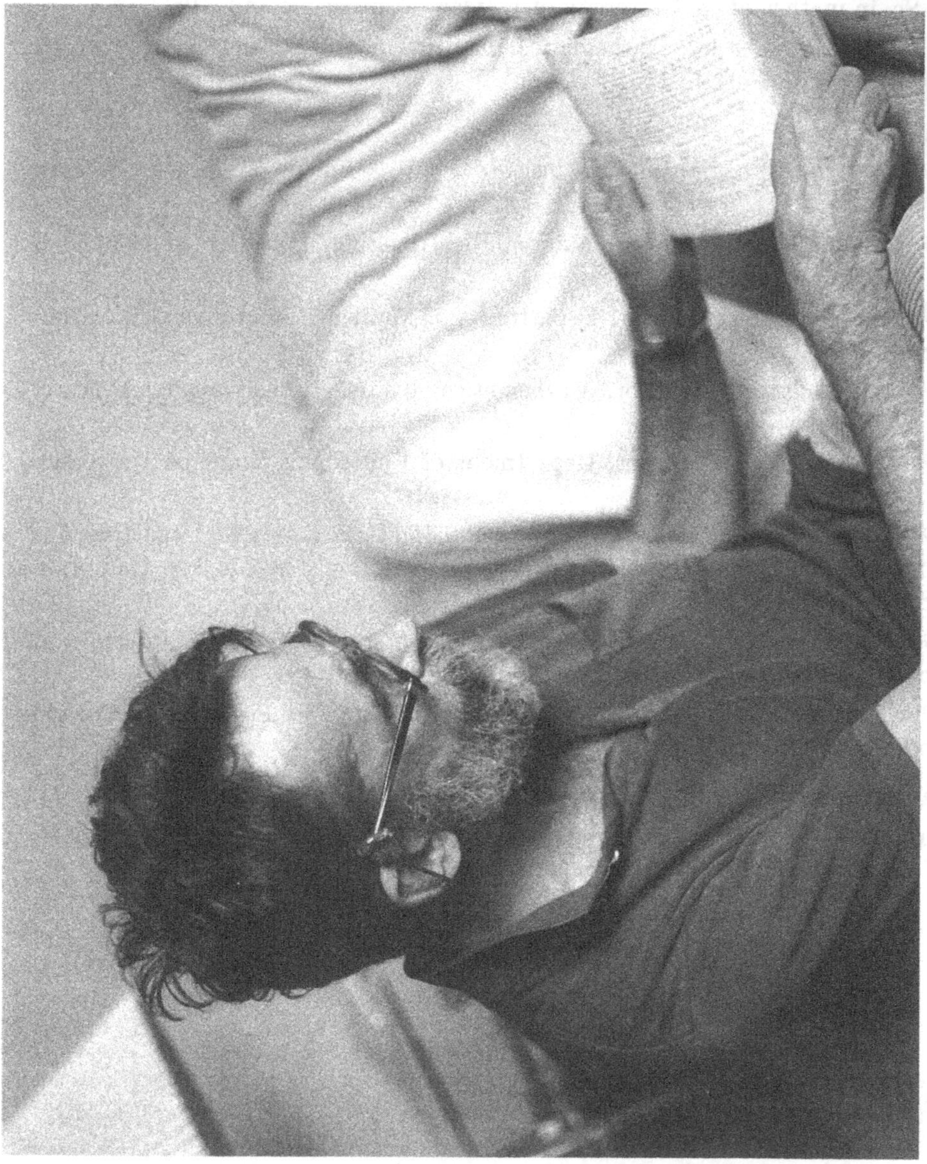

Contents

Combinatorics of Words

Algebra of Languages

Formal Language Theory

Computational Molecular Biology

On the Work of Andrzej Ehrenfeucht in Model Theory

Robert Vaught

Department of Mathematics
University of California
Berkeley, California 94720–3840

Abstract. This is a paper about a certain period of the develpoment of model theory upon which the work of A. Ehrenfeucht made an indelible mark. We will pay a special attention to his results about the theories of (Ord,<), (Ord,<,+), and (Ord,<,+,·). Also some of the history of the applications of Ramsey's theorem in model theory will be discussed.

§1. An outline of Ehrenfeucht's papers in model theory. I am much indebted to John Addison for helpful discussions and suggestions. Addison also introduced me to the superb, 1987 *Ω-Bibliography of Mathematical Logic*. It comes in six volumes (= topics), and Ehrenfeucht has contributions in all six!

At any rate, I learned in the Ω-Bibliography that Ehrenfeucht's first paper was a joint paper with Jerzy Loś, on the border of algebra and universal algebra. His next publication was in model theory, a very important paper written with Andrzej Mostowski in [1956]. It was the first paper in which Ramsey's Theorem (stated later) was applied to model theory. In [1957b], Ehrenfeucht expanded the work of [1956] and applied it for the first time to an important problem in model theory, in fact to the study of theories categorical in power. Then in [1958] he wrote his first *important abstract* (a concept implying in particular that there was no further publication) — giving two more very valuable applications of what is now called the Ehrenfeucht-Mostowski method. I shall call the three papers "the Ramsey papers". They will be discussed more later in this section.

Also in 1957 in [1957a], Ehrenfeucht announced his beautiful work which was presented in full in [1961a], in which he established a conjecture of Alfred Tarski concerning the theories of (Ord,<), (Ord,+), and (Ord,+,·) (Ord being the class of all ordinals) and the ordinals ω^ω, ω^{ω^ω} and $\omega^{\omega^{\omega^\omega}}$, respectively. The work in [1961a] will be the subject of the whole §2. The three Ramsey papers and the [1961a] paper are perhaps Ehrenfeucht's two greatest contributions to model theory. [1961a] and its important predecessor Fraïssé ([1952] and [1955b]) (discussed at the beginning of §2) depend on essentially *nothing* from previous model theory or perhaps even logic. Hence a large variety of people can read §2 and actually read an example of Ehrenfeucht at his best!

In [1959a,b], two important abstracts of Ehrenfeucht appeared on the decidability of the first-order theory of one function (unary operation) and the decidability of the first-order theory of one linear order. Ehrenfeucht never published full proofs of these results (although the abstracts did contain some hints).

Some time in 1956 or 1957, Ehrenfeucht had also stated to friends that he had a decision method for the theory of (Ord,+). In the late fifties and most of the sixties, there was an explosion of work on decision methods, done by many people. 'Finally', in [1969], M. Rabin obtained a "master" decision method, which absorbed all three Ehrenfeucht had discussed and many, many more. The first proof of the decidability of (Ord,+) appeared in Büchi [1965]. The first full proof of the decidability of one linear order appeared in a paper by H. Laüchli and J. Leonard [1966]. Finally, the decidability of the first-order theory of one function was first fully proved in J. Le Tourneau [1968]. There continued to be valuable papers in model theory (often with distinguished coauthors) up to 1987, which I have no time to describe. But somewhere in the sixties Andrzej turned to (theoretical) computer science as his primary interest.

Certainly the years 1956–1959 contained an outpouring of beautiful and deep results from A. Ehrenfeucht!

The three "Ramsey papers" all depended on the famous combinatory theorem of F. P. Ramsey [1929]:

Theorem (Ramsey). *For any set X, let $X^{(n)}$ be the set of all n-element subsets of X. If X is infinite and $X^{(n)} = C_1 \cup \cdots \cup C_k$ is a partition of X (say, into pairwise disjoint sets), then there is an infinite subset Y of X such that $Y^{(n)}$ is entirely included in some one C_j.*

Using Ramsey's Theorem, Ehrenfeucht and Mostowski created the "Ehrenfeucht-Mostowski method", and its first important applications were by Ehrenfeucht in [1957a] and [1958], mostly dealing with the study of theories categorical in some infinite power. I shall not even repeat what the theorems of these three papers say (this work can be found in the Model Theory books of Keisler [1965] and of Wilfred Hodges.) Instead, I will say something about how this subject has evolved since the "Ramsey papers" came out.

In 1965, Michael Morley's famous paper "Categoricity in powers" [1965a] appeared. It is certainly one of the greatest papers in model theory to date. In the paper, Morley is perhaps more dependent on the three "Ramsey papers" than on any other just previous papers. The "Ramsey papers" are acknowledged repeatedly in §3, "Results depending on Ramsey's theorem". Morley's paper, including §3, was the basis of an explosion of deep results obtained by many people and, above all, by Saharon Shelah — continuing up to the present time.

Another important follow-up to the three Ramsey papers was initiated by M. Morley in [1968]. In [1956], P. Erdös and T. Rado had proven the "Erdös-Rado theorem", which is definitely a Ramsey theorem for some of Cantor's infinite cardinals. It states:

Theorem (Erdös-Rado). *If $\overline{\overline{X}} \geq 2^{\kappa}_{n}$ $(= 2^{2^{\cdots^{2^{\kappa}}}}$ n times), $X^{(n)} = \bigcup(C_i : i \in I)$, and $\overline{\overline{I}} \leq \kappa$, then there is a subset Y of X with $\overline{\overline{Y}} > \kappa$ such that $Y^{(n)}$ is entirely included in some one C_i.*

Morley, in [1965b], proved the following important theorem using the Erdös-Rado theorem.

Theorem (Morley). *Let Σ be a set of formulas (with one free variable). We say the structure $\mathfrak{M}$ omits Σ if it has no element x such that for all ϕ in Σ, x satisfies ϕ in $\mathfrak{M}$. If for every $\alpha < \omega_1$, there is a model of T of power $\geq \beth_\alpha$ which omits Σ, then there is in every infinite power a model of T which omits Σ.*

Some readers will recognize that this is about something like a *Hanf number* for omitting types.

That is enough of vague remarks. Section 2 will present, in full, some of the beautiful mathematics in Ehrenfeucht [1961].

§2. Mostowski, Tarski, Fraïssé, and Ehrenfeucht. All notation used here will be defined in the first long paragraph of §2(a) below. That paragraph can be read or referred to now, if you wish.

A. Mostowski and A. Tarski announced in [1949] some work they had done just before World War II: they had carried out an elimination of quantifiers for the theory of well-orders (even of schematically-well-orders). With its help they inferred that for any ordinals α, β

$$(\alpha, <) \equiv (\beta, <) \quad \text{iff} \quad \alpha \underset{\omega}{\sim_\omega} \beta\ ; \tag{1}$$

so it follows that

$$(\mathrm{Ord}, <) \equiv (\omega^\omega, <) \tag{2}$$

Here $\alpha \underset{\omega}{\sim_\omega} \beta$ means that α and β are equivalent mod ω^ω (in the usual sense) and either $\alpha, \beta \geq \omega^\omega$ or $\alpha, \beta < \omega^\omega$). Ord is the class of all ordinals. $\mathfrak{A} \equiv \mathfrak{B}$ means structures $\mathfrak{A}$ and $\mathfrak{B}$ have the same true first-order sentences. It was a rather long and difficult elimination of quantifiers, at least by the standards of the time! (The reader will never *have* to know what an elimination of quantifiers is.)

Tarski conjectured sometime in the interval ([1939, 1950]) that

$$(\mathrm{Ord}, <, +) \equiv \omega^{\omega^\omega} \quad \text{and} \quad (\mathrm{Ord}, <, +, \cdot) \equiv \omega^{\omega^{\omega^\omega}}\ ! \tag{3}$$

In the remarkable paper [1957],[1961], which we will discuss in §2(b), Ehrenfeucht proved Tarski's conjectures (3). Notice that no ordinary elimination of quantifiers can be done for the theory of $(\mathrm{Ord}, <, +, \cdot)$, since it would prove that $(\omega, +, \cdot)$ has a decidable theory, contrary to Gödel in 1931! Most deep results *about special structures* have been attained or have been at least attainable by an elimination of quantifers. This state of affairs make Ehrenfeucht's verification of Tarski's $\omega^{\omega^{\omega^\omega}}$ conjecture seem all the more remarkable.

In [1952] and [1956] (and perhaps earlier), Roland Fraïssé had introduced a certain important group of definitions and theorems in general model theory (discussed in §2(a) below). With their help, he was able to give a new proof of (1) and (2) above, not involving any elimination of quantifiers, and quite short. Ehrenfeucht [1961] introduced a new general method (much like Fraïssé's and also very different) and using it, proved Tarski's conjectures (3). Both Fraïssé's work and that of Ehrenfeucht seemed to come out of the blue. Speaking of which, Fraïssé was working in Algiers and Ehrenfeucht in Warsaw — the latter then behind the deepest Iron Curtain. I remember very well, as a student of Tarski, his great interest in Fraïssé's new method and its proof of (1) and (2); then in 1956 or 1957, Mostowski wrote Tarski of Ehrenfeucht's proof of Tarski's conjecture (3) and Tarski was very excited about this.

Aside: In the summer of 1957, the National Science Foundation (at its richest ever) convened a six-week Summer Institute in Logic at Cornell. Many people had heard of Ehrenfeucht's work, so Sol Feferman, who had corresponded with Ehrenfeucht about it, agreed to give a summary of Ehrenfeucht's work (see Feferman [1961]). Often in §2 I am indebted to Feferman for some fine new notation and for some more serious things such as his version of Ehrenfeucht's work with no mention of games.

§2(a). Fraïssé's work. Read this paragraph only when you need to. The letters from i to r, possibly with subscripts or superscripts, range over $\omega = \{0, 1, 2, \ldots\}$. ℓ ranges over $\{1, 2\}$. The length r of a finite sequence $s = (a_1, \ldots, a_r)$ is denoted by $\ln(s)$. $\breve{A}$ (shorter than A^ω) means the set of all finite sequences of elements of A. $\mathfrak{M}$ is a *structure* if it is of the form $((M, (R_t : t \in T)), s) = (\mathfrak{M}, s)$ where $M \neq \emptyset$ and each R_t is a v_t-ary relation over M. Henceforth, $\mathfrak{M}, \mathfrak{N}, \mathfrak{M}_1$ always range over structures. The *similarity type* of the particular $\mathfrak{M}$ above is $((v_t : t \in T), \ln(s))$. Structures $\mathfrak{M}_1$ and $\mathfrak{M}_2$ are called *similar* if they have the same similarity type. $s \frown t$ is the *concatenation* of s and t, i.e. $(s_1, \ldots, s_{\ln(s)}, t_1, \ldots, t_{\ln(t)})$. For the $\mathfrak{M}$ above, we write $|\mathfrak{M}|$ for M, $\mathfrak{M}^R$ for $(M, (R_t : t \in T))$, the *relational part of* $\mathfrak{M}$, and $u^{\mathfrak{M}}$ for s. A *relational structure* is one having no individual constants. If $t \in \breve{M}$, write $(\mathfrak{M}, t)$ for $(\mathfrak{M}^R, u^{\mathfrak{M}} \frown t)$. It follows that in general $((\mathfrak{M}, s), t) = (\mathfrak{M}, s \frown t)$. Of course, $\mathfrak{M}_1 \underset{f}{\cong} \mathfrak{M}_2$ means f is an isomorphism of $\mathfrak{M}_1$ onto $\mathfrak{M}_2$. Corresponding to a similarity type $((\nu_t : t \in T), p)$ is a certain first order language L. L has equality as a logical 2-ary relation symbol; it has the non-logical relation symbols S_t of arity ν_t, for $t \in T$, and it has individual constants $c_1, \ldots, c_p$. L has infinitely many individual variables. Finally L has parentheses (and) . The *formulas* of L are all those formed from the symbols in the obvious way. For example, $\forall x\ (x < y \to \exists z(x < z \land z < y))$ is a formula of L (if $<$ is a symbol of L). A *sentence* (of L) is a formula with no free variables. We write $\mathfrak{M} \models \sigma$ if σ is true in $\mathfrak{M}$. A formula is prenex if it is of the form $Q_1 v_1 Q_2 v_2, \ldots, Q_k v_k \phi$ where the v_i's are variables, each Q is $\forall$ or $\exists$, and the formula ϕ contains no quantifiers. We write $\mathfrak{M}_1 \underset{n}{\equiv} \mathfrak{M}_2$ if $\mathfrak{M}_1$ and $\mathfrak{M}_2$ (are similar and) have the same true prenex sentences with n initial quantifers. We write $\mathfrak{M}_1 \equiv \mathfrak{M}_2$ (read $\mathfrak{M}_1$ and $\mathfrak{M}_2$ are elementarily equivalent) if for all n, $\mathfrak{M}_1 \equiv_n \mathfrak{M}_2$, i.e., if $\mathfrak{M}_1$ and $\mathfrak{M}_2$ have the same true sentences. α, β, γ always denote ordinals.

Definition 1, Proposition 2, and Theorems 3 and 4 below are all due to Fraïssé. Proofs can be found in Fraïssé [1955b], Feferman [1960], and Ehrenfeucht [1961].

Definition 1. *Let $\mathfrak{N}_1$ and $\mathfrak{N}_2$ be fixed, similar, purely relational structures. By recursion we put (for all $\mathfrak{M}_1$ with $\mathfrak{M}_1^R = \mathfrak{N}_1$, and all $\mathfrak{M}_2$ with $\mathfrak{M}_2^R = \mathfrak{N}_2$)*

$$\mathfrak{M}_1 \underset{0}{\approx} \mathfrak{M}_2 \quad \textit{iff} \quad \mathfrak{M}_1 \underset{0}{\equiv} \mathfrak{M}_2$$

$$\begin{aligned} \mathfrak{M}_1 \underset{n+1}{\approx} \mathfrak{M}_2 \quad & \textit{iff for all } x \in \mathfrak{M}_1 \textit{ there exists } y \in \mathfrak{M}_2 \\ & \textit{such that } (\mathfrak{M}_1, x) \underset{n}{\approx} (\mathfrak{M}_2, y) \textit{ and for all } y \in \mathfrak{M}_2 \\ & \textit{there exists } x \in \mathfrak{M}_1 \textit{ such that } (\mathfrak{M}_1, x) \underset{n}{\approx} (\mathfrak{M}_2, y) \end{aligned}$$

This defines $\mathfrak{M}_1 \underset{n}{\approx} \mathfrak{M}_2$ for *all* $\mathfrak{M}_1$ and $\mathfrak{M}_2$ and the two conditions above now hold for all $\mathfrak{M}_1, \mathfrak{M}_2$. Finally, we write $\mathfrak{M}_1 \approx \mathfrak{M}_2$ if for any n, $\mathfrak{M}_1 \underset{n}{\approx} \mathfrak{M}_2$.

It is easy to prove (using induction, of course, but I don't say just what statement should be literally proved by induction on n).

Proposition 2. (a) $\mathfrak{M} \underset{n}{\approx} \mathfrak{M}$

(b) *If* $\mathfrak{M}_1 \underset{n}{\approx} \mathfrak{M}_2$, *then* $\mathfrak{M}_2 \underset{n}{\approx} \mathfrak{M}_1$.

(c) *If* $\mathfrak{M}_1 \underset{n}{\approx} \mathfrak{M}_2$ *and* $\mathfrak{M}_2 \underset{n}{\approx} \mathfrak{M}_3$, *then* $\mathfrak{M}_1 \underset{n}{\approx} \mathfrak{M}_3$.

(d) *If* $\mathfrak{M}_1 \underset{m+n}{\approx} \mathfrak{M}_2$, *then* $\mathfrak{M}_1 \underset{m}{\approx} \mathfrak{M}_2$.

(e) (a) *through* (c) *hold for* $\approx$ *in place of* $\underset{n}{\approx}$.

Fraïssé's main results are in Theorem 3.

Theorem 3(a). *If* $\mathfrak{M}_1 \underset{n}{\approx} \mathfrak{M}_2$, *then* $\mathfrak{M}_1 \underset{n}{\equiv} \mathfrak{M}_2$.

Corollary 3(b). *If* $\mathfrak{M}_1 \approx \mathfrak{M}_2$, *then* $\mathfrak{M}_1 \equiv \mathfrak{M}_2$.

(Theorem 3(a) is proved by induction just as easily as 2(a)-(d).)

Theorem 3(c), below, will not be needed in §2(b), so will only be mentioned here briefly.

Theorem 3(c). *The converse of* (3b) *is true for structures* $\mathfrak{M}_1$ *and* $\mathfrak{M}_2$ *having finitely many relations.*

Fraïssé (in [1955b]) gave several applications of Theorems 3(a),3(b) to give short proofs of many results obtained using elimination of quantifiers. The most interesting one (already mentioned in the Introduction to §2) was a new proof of Theorem 4 below. Put $\alpha \underset{\omega^\gamma}{\sim} \beta$ if $\alpha \equiv \beta \pmod{\omega^\gamma}$ (in the usual sense) *and* either $\alpha, \beta < \omega^\gamma$ or $\alpha, \beta \geq \omega^\gamma$.

Theorem 4. (a) *If* $\alpha \underset{\omega^n}{\sim} \beta$ *then* $(\alpha, <) \underset{n}{\approx} (\beta, <)$. *Hence*

(b) *If* $\alpha \underset{\omega^\omega}{\sim} \beta$ *then* $(\alpha, <) \equiv (\beta, <)$

Theorem 4(a) is proved by induction on n, but it is not quite as simple as all the previous proofs.

§2(b). Ehrenfeucht's work. Henceforth we always assume that $|\mathfrak{M}_1|$ and $|\mathfrak{M}_2|$ are sets of urelements. If Q is any set of urelements, put $Q^* =$ the $\subseteq$-smallest set X such that for any k, and any $x_1, \ldots, x_k \in X$, $\{x_1, \ldots, x_k\} \in X$. One easily proves

Proposition (−1). (a) *Q^* exists.*
(b) *If $Q_1 \subseteq Q_2$ then $Q_1^* \subseteq Q_2^*$.*
(c) $Q^* = \bigcap\{K^* : K$ *is finite and* $K \subseteq Q\}$.

Let $\mathfrak{M}$ be a structure, say $\mathfrak{M} = ((M, (R_t : t \in T)), s)$, put $\mathfrak{M}^* = (M^*, \varepsilon, M, (R_t : t \in T), s)$ which we also write as $(M^*, \varepsilon, \mathfrak{M})$. *$\mathfrak{M}^*$ is still a structure!* (We just reindex all the R_t's plus ε and M.) Ehrenfeucht says that $\mathfrak{M}_1$ and $\mathfrak{M}_2$ are *indiscernible by means of finite sets,* if $\mathfrak{M}_1^* \equiv \mathfrak{M}_2^*$.

Definition 0. *By recursion, we put*

$$\mathfrak{M}_1 \underset{0}{\approx}^* \mathfrak{M}_2 \text{ iff } \mathfrak{M}_1 \underset{0}{\equiv} \mathfrak{M}_2 \quad (\text{iff } \mathfrak{M}_1 \underset{0}{\approx} \mathfrak{M}_2)$$

$$\mathfrak{M}_1 \underset{n+1}{\approx}^* \mathfrak{M}_2 \text{ iff } (\forall s \in \breve{M}_1)(\exists t \in \breve{M}_2)\ ((\mathfrak{M}_1, s) \underset{n}{\approx} (\mathfrak{M}_2, t) \quad (so\ \ \ell(s) = \ell(t))$$

$$and\ (\forall t \in \breve{M}_2)(\exists s \in \breve{M}_1)\ ((M_1, s) \underset{n}{\approx} (\mathfrak{M}_2, t)).$$

In defining $\mathfrak{M}_1 \underset{n+1}{\approx}^* \mathfrak{M}_2$ we only need to know $\underset{n}{\approx}^*$ for a certain *set* of previous $\mathfrak{M}_\ell$'s, which justifies this definition. We write $\mathfrak{M}_1 \approx^* \mathfrak{M}_2$ for: For any n, $\mathfrak{M}_1 \underset{n}{\approx}^* \mathfrak{M}_2$.

After extending Ehrenfeucht's [1961] definition of his game $H_n(\mathfrak{M}_1^*, \mathfrak{M}_2^*)$ to arbitrary structures (not just relational), one can easily prove for any similar structures $\mathfrak{M}_1$ and $\mathfrak{M}_2$:

$$\mathfrak{M}_1 \underset{n}{\approx}^* \mathfrak{M}_2 \text{ iff Player II has a winning strategy in the game } H_n(\mathfrak{M}_1^*, \mathfrak{M}_2^*) \tag{4}$$

(Indeed, if one *defines* $\mathfrak{M}_1 \underset{n}{\approx}^* \mathfrak{M}_2$ as just above, it is easy to prove that the conditions in Definition 0 hold.)

Proposition 0. (a) $\mathfrak{M}_1 \underset{n}{\approx}^* \mathfrak{M}_1$.
(b) *If* $\mathfrak{M}_1 \underset{n}{\approx}^* \mathfrak{M}_2$ *then* $\mathfrak{M}_2 \approx^* \mathfrak{M}_1$.
(c) *If* $\mathfrak{M}_1 \underset{n}{\approx}^* \mathfrak{M}_2$ *and* $\mathfrak{M}_2 \underset{n}{\approx}^* \mathfrak{M}_3$ *then* $\mathfrak{M}_1 \underset{n}{\approx}^* \mathfrak{M}_3$.
(d) *If* $\mathfrak{M}_1 \underset{m+n}{\approx}^* \mathfrak{M}_2$ *then* $\mathfrak{M}_1 \underset{m}{\approx}^* \mathfrak{M}_2$.

Each of (a)–(d) is very easily proved using induction.

Theorem 1. *Let $k \le n$. Let $s_1^1, \ldots, s_k^1 \in M_1 = |\mathfrak{M}_1|$ and $s_k^2, \ldots, s_k^2 \in M_2$ and $\ln(s_j^1) = \ln(s_j^2)$ for $1 \le j \le n$. Suppose*

$$(\mathfrak{M}_1, s_1^1 \frown \cdots \frown s_k^1) \underset{n-k}{\approx}{}^* (\mathfrak{M}_2, s_1^2, \frown \cdots \frown, s_k^2) \tag{0}$$

Put $t^\ell = u^{\mathfrak{M}_\ell} \frown s_1^\ell \frown \cdots \frown s_k^\ell$ $(\ell = 1,2)$; $e = \{(t^1(j), t^2(j)) : 1 \le j \le \ln t^1\}$ and $Q_\ell = Rng\ t^\ell$ $(\ell = 1,2)$. By (0) *above, and Proposition* 0(d), *$(\mathfrak{M}_1, s_1^1 \frown \cdots \frown s_k^1) \underset{0}{\approx}$ (i.e. $\underset{0}{\equiv}$) $(\mathfrak{M}_2, s_1^2 \frown \cdots \frown s_k^2)$, or, what is the same, $(\mathfrak{M}_1|Q_1, s_1^1 \frown \cdots \frown s_k^1) \underset{e}{\cong} (\mathfrak{M}_2|Q_2, s_1^2 \frown \cdots \frown s_k^2)$. The reader will be able to show that there is exactly one function f (extending e) such that $(Q_1^*, \varepsilon, Q_1) \underset{f}{\cong} (Q_2^*, \varepsilon, Q_2)$. Hence, clearly,*

$$(\mathfrak{M}_1^*|Q_1, s_1^1, \frown \cdots \frown s_k^1) \underset{f}{\cong} (\mathfrak{M}_2^*|Q_2, s_1^2 \frown \cdots \frown s_k^2) \tag{1}$$

or what is the same,

$$(\mathfrak{M}_1^*, s_1^1 \frown \cdots \frown s_k^1) \underset{0}{\equiv} (\mathfrak{M}_2^*, s_1^2 \frown \cdots \frown s_k^2) \text{ and}$$
$$f(s_j^1) = s_j^2 \text{ whenever } 1 \le j \le k. \tag{1'}$$

Now, suppose also that

$$x_1^1, \ldots, x_k^1 \in (Rng\ t_1)^* \text{ and } x_1^2, \ldots, x_k^2 \in (Rng\ t_2)^*, \text{ and} \tag{2}$$

$$f(x_j^1) = x_j^2 \text{ whenever } 1 \le j \le k. \tag{3}$$

Then $(\mathfrak{M}_1^*, (x_1^1, \ldots, x_k^1)) \underset{n-k}{\approx} (\mathfrak{M}_2^*, (x_1^2 \frown \cdots \frown x_k^2))$. (Note: this is the $\approx$ of §2(a).)

Proof. Let us write Theorem 1^- for the theorem in which only the last line in Theorem 1 is changed, and it is changed to

Then $(\forall x \in M_1^*)(\exists y \in M_2^*)((\mathfrak{M}_1^*, x_1^1, \ldots, x_k^1, x) \underset{n-k-1}{\approx} (\mathfrak{M}_2^*, x_1^2, \ldots, x_k^2, y))$.

First we will prove Theorem 1^-, with n fixed, by induction on k from $k = n$ down to $k = 0$.

It is easy to prove for $k = n$, in fact the last line of Theorem 1 holds, so certainly that of Theorem 1^- by Definition 0.

Suppose the hypothesis of Theorem 1^- (everything up to the last line of Theorem 1) holds for $k = n$. Then by (3), $f(x_j^1) = x_j^2$ $(1 \le j \le k)$, so by (1), $(\mathfrak{M}_1^*|Q_1, (x_1^1, \ldots, x_k^1)) \underset{f}{\cong} (\mathfrak{M}_2^*|Q_2, (x_1^2, \ldots, x_k^2))$. Hence $(\mathfrak{M}_1^*, (x_1^1, \ldots, x_k^1)) \underset{0}{\equiv} (\mathfrak{M}_2^*, (x_1^2, \ldots, x_k^2))$, so in fact Theorem 1 holds for $k = n$.

Now assume Theorem 1^- holds for $k+1$ $(k+1 \leq n)$, and assume that for k, the hypothesis of Theorem 1 (with its t's, e, Q's and f) holds. Suppose $x^1_{k+1} \in M_1$. By Proposition −1(c) for $Q = M_1$, there exists $s^1_{k+1} \in M_1$ such that

$$x^1_{k+1} \in (Rng(s^1_{k+1}))^* \tag{4}$$

Hence by (0) and Definition 1, there exists $s^2_{k+1} \in M_2$ such that

$$(\mathfrak{M}_1, t^1 \frown s^1_{k+1}) \underset{n-(k+1)}{\approx} (\mathfrak{M}_2, t^2 \frown s^2_{k+1})\ . \tag{5}$$

By (5), which is (0) for $k+1$, we can carry out for $k+1$ the part of Theorem 1 from just below (0) to just below (2). In this way we get for $k+1$, $t'^{\ell} = t^{\ell} \frown s^{\ell}_{k+1}$, $e' = \{(t'^1_j, t'^2_j),\ 1 \leq j \leq k+1\}$, $Q'_{\ell} = Rng\ t'^{\ell} = Rng(t^{\ell} \frown s^{\ell}_{k+1})$, and f' such that

$$(\mathfrak{M}^*_1 \mid Q'_1, s^1_1 \frown \cdots \frown s^1_{k+1}) \underset{f}{\cong} (\mathfrak{M}^*_2 \mid Q'_2, s^2_1 \frown \cdots \frown s^2_{k+1})\ . \tag{1'}$$

Clearly $f' \restriction Q_1$ fills all the conditions determining f, so $f' \restriction Q = f$. Now by Proposition 0(b), $(Rng(t^{\ell} \frown s^{\ell}_{k+1}))^* \supseteq Rng(t^{\ell})^*, (Rng(s^{\ell}_{k+1})^*$ for $\ell = 1,2$. Hence by hypothesis (2) for k, $x^{\ell}_1, \ldots, x^{\ell}_k \in Rng(t^{\ell} \frown s^{\ell}_{k+1})$ for $\ell = 1,2$, and by (4)

$$x^1_{k+1} \in (Rng(t^1 \frown s^1_{k+1}))^* \tag{6}$$

Put $x^2_{k+1} = f'(x'_{k+1})$. Since $x'_{k+1} \in (Rng(s^1_{k+1}))^*$ it follows that $f(x^1_{k+1}) \in Rng(f'(s^1_{k+1}))$. (Indeed, since $(a,b) = \{\{a\},\{a,b\}\}$ and $q = \{0,1,\ldots,q-1\}$, etc., f' preserves the notion "range of".) Now by applying Theorem 1 for $k+1$ we get at once that $(\mathfrak{M}^*_1, x^1_1, \ldots, x^1_{k+1}) \underset{n-(k+1)}{\approx} (\mathfrak{M}^*_2, x^2_1, \ldots, x^2_{k+1})$. Thus we have shown

$$\begin{gathered}(\forall x^1_{k+1} \in M^*_1)(\exists x^2_{k+1} \in M^*_2) \\ ((\mathfrak{M}^*_1, x^1_1, \ldots, x^1_k, x^1_{k+1}) \underset{n-(k+1)}{\approx} (\mathfrak{M}^*_2, x^2_1, \ldots x^2_k, x^2_{k+1}))\ .\end{gathered} \tag{7}$$

Now by Proposition 0(b) and (0) (in Theorem 1^-) we see that

$$(\mathfrak{M}_2, s^2_1 \frown \cdots \frown s^2_k) \underset{n-k}{\approx} (\mathfrak{M}_1, s^1 \frown \cdots \frown s^1_k)\ . \tag{8}$$

Thus hypothesis (0) of Theorem 1^- holds with $\mathfrak{M}_1$ and $\mathfrak{M}_2$ interchanged and s^1_j and s^2_j interchanged $(j = 1,2)$. [It is late at night and this manuscript must be emailed from Berkeley to the Netherlands before the absolute deadline tomorrow afternoon, so I'm going to leave a lot to the reader in the rest of this proof.] Following out the part of Theorem 1 from below (0) to (1′), we introduce $t^{\ell\wedge}$'s, $e^\wedge$, $Q^\wedge_\ell$'s and $f^\wedge$. Evaluating these as we go along, we find that $t^{\ell^\wedge} = t^{3-\ell}$ $(\ell = 1,2)$, and $f^\wedge = f^{-1}$. Now, perhaps skipping some things the reader will supply, we see that (3^) holds, since $f^{-1}(x^2_j) = x^1_j$ $(1 \leq j \leq k)$; and (2^) holds for our old $x_1^2, \ldots, x_k^2 \in (Rng\ t_2^\wedge)^*$ and $x_1^1, \ldots, x_k^1 \in (Rng\ t_1^\wedge)^*$. So by Theorem 1^- (for $\mathfrak{M}_2, \mathfrak{M}_1$, etc.) and perhaps also §2(a), Proposition 2(b) of Fraïssé)

$$(\forall y \in M_2^*)(\exists x \in M_1^*)((\mathfrak{M}_1^*, x_1^1, \ldots, x_k^1, x) \underset{n-(k+1)}{\approx} (\mathfrak{M}_2^*, x_1^2, \ldots x_k^2, x_k^2)) \ . \tag{9}$$

By (9), (7), and Definition 1, the last line of Theorem 1 and so also Theorem 1 holds.

Corollary 2. *If* $\mathfrak{M}_1 \underset{n}{\approx}^* \mathfrak{M}_2$ *then* $\mathfrak{M}_1^* \underset{n}{\approx}^* \mathfrak{M}_2^*$.

It is enough to put $k = 0$ in Theorem 1. Corollary $^2 3$ follows at once:

Corollary $^2 3$. *If* $\mathfrak{M}_1 \approx^* \mathfrak{M}_2$ *then* $\mathfrak{M}_1^* \approx \mathfrak{M}_2^*$.

Note 1: Corollary 2 and especially Corollary $^2 3$ were all that was needed for the applications, e.g. to (Ord,+,·). I think that the statement of Theorem 1 and its proof were hard to find. The part of its proof, up to (7), is very elegant, but the trickiest thing was to prove by induction these odd things about x_j^ℓ's $\in (Rng\ t_\ell)^*$, which would seem never to get to Corollary 2, until suddenly they disappear when $k = 0$.

Note 2: Ehrenfeucht [1961], has all of Theorem 1, Corollary 2, and Corollary $^2 3$, but with $\underset{n}{\equiv}$ or $\equiv$ in the conclusion, not $\underset{n}{\approx}$ or $\underset{n}{\equiv}$, respectively. If there are finitely many R_t's, as in $(\omega, <)$, $(\omega, +)$, and $(\omega, +, \cdot)$ below, then by Fraïssé's Theorem 3(c) of §2(a), $\mathfrak{M}_1^* \equiv \mathfrak{M}_2^*$ implies $\mathfrak{M}_1^* \approx \mathfrak{M}_2^*$, so Corollary $^2 3$ is the same as Ehrenfeucht's result, and Ehrenfeucht's Theorems 4,5 below would also not be improved by using $\approx$ in place of $\equiv$.

Turning now to Ehrenfeucht's application of these general theorems to the theories (Ord, <), (Ord, +), and (Ord, +, ·), the main thing is

Theorem 4. (a′) *If* $\alpha \underset{\omega^n}{\sim} \beta$, *then* $(\alpha, <)^* \underset{n}{\equiv} (\beta, <)^*$.

(a) *If* $\alpha \underset{\omega^\omega}{\sim} \beta$, *then* $(\alpha, <)^* \equiv (\beta, <)^*$.

(b) *If* $\alpha \underset{\omega^\omega}{\sim} \beta$, *then* $(\omega^\alpha, +)^* \equiv (\omega^\beta, +)^*$.

(c) *If* $\alpha \underset{\omega^\omega}{\sim} \beta$, *then* $(\omega^{\omega^\alpha}, +, \cdot)^* \equiv (\omega^{\omega^\beta}, +, \cdot)^*$.

For proofs see Ehrenfeucht [1961], Theorems 13–17, and Ehrenfeucht's earlier general theorems given here in Corollaries 2 and 3 above. The proof of Theorem 4 is also sketched very nicely in Feferman [1960]. In (a), Ehrenfeucht has improved Fraïssé's 4(b) in §2(a). This is used together with Cantor's canonical form $\gamma = \omega^{\alpha_1} n_1 + \cdots + \omega^{\alpha_k} n_k$ $(\alpha_1 > \alpha_2 > \ldots)$ and *more* to get (b) and (c), all short, but beautiful.

Corollary 5. (a) $(\text{Ord}, <)^* \equiv (\omega^\omega, <)^*$.

(b) $(\text{Ord}, +)^* \equiv (\omega^{\omega^\omega}, +)$.

(c) $(\text{Ord}, +, \cdot)^* \equiv (\omega^{\omega^{\omega}}, +, \cdot)$.

Corollary 5 has consequences concerning definable elements and elementary extensions (which, I think, were also all conjectured by Tarski) which we now give.

Let us agree to write $\mathrm{Df}(\mathfrak{A})$ for the set of elements $a \in \mathfrak{A}$ definable in $\mathfrak{A}$ (i.e., such that for some formula $\phi(v)$, we have for any $b \in \mathfrak{A}$, $(\mathfrak{A}, b) \models \phi$ iff $b = a$). We write α_i $(i = 1,2,3)$ for ω^ω, ω^{ω^ω}, or $\omega^{\omega^{\omega^\omega}}$, respectively, and $\underline{\alpha}_1$ for $(\alpha_1, <)$, $\underline{\alpha}_2$ for $(\alpha_2, +)$ and $\underline{\alpha}_3$ for $(\alpha_3, +, \cdot)$. $\underline{\mathrm{Ord}}_i$ (for $i = 1,2,3$) is defined similarly. We also write $\underline{\mathrm{Df}}(\alpha_i)$ for $\underline{\alpha}_i \upharpoonright \mathrm{Df}(\underline{\alpha}_i)$ and $\underline{\mathrm{Df}}(\mathrm{Ord}_i)$ analogously. Before making conjectures (3) in §2, Tarski showed (as you can do with effort) *something like* this:

$$\alpha_i \subseteq \mathrm{Df}(\underline{\alpha}_i), \mathrm{Df}(\underline{\mathrm{Ord}}_i) \quad \text{for} \quad \alpha = 1,2,3. \tag{10}$$

We write $\mathfrak{A} \prec \mathfrak{B}$ if $\mathfrak{A}$ is an elementary substructure of $\mathfrak{B}$, i.e., for any k, and $a_1, \ldots, a_k \in \mathfrak{A}$, and any $\phi(v_1, \ldots, v_k)$, we have $(\mathfrak{A}, a_1, \ldots, a_k) \models \phi(c_1, \ldots, c_k)$ iff $(\mathfrak{B}, a_1, \ldots, a_k) \models \phi(c_1, \ldots, c_k)$. For any structure $\mathfrak{A} = (\alpha, <, \text{anything else})$, it is easy to see that $\underline{\mathrm{Df}}(\mathfrak{A}) \prec \mathfrak{A}$, by Tarski's criterion for $\mathfrak{A} \prec \mathfrak{B}$ (see Chang-Keisler [1965]). Next, it is clear that for any $\mathfrak{A}, \mathfrak{B}$, if $\mathfrak{A} \prec \mathfrak{B}$ then $\mathrm{Df}(\mathfrak{A}) = \mathrm{Df}(\mathfrak{B})$.

Theorem 6. $\alpha_i = \mathrm{Df}(\underline{\mathrm{Ord}}_i)$ *(so* $\underline{\alpha}_i = \underline{\mathrm{Df}}(\underline{\mathrm{Ord}}_i)$.*)*

Proof. We only need to prove $\supseteq$. Suppose $\beta \in \mathrm{Df}(\underline{\mathrm{Ord}}_i)$. Then for some formula ψ (we now write sloppily but clearly) $\underline{\mathrm{Ord}}_i \models \forall v(\psi(v) \leftrightarrow v = \beta)$. Thus $\underline{\mathrm{Ord}}_i \vdash \exists! v\psi(v)$. Hence, since $\underline{\mathrm{Ord}}_i \equiv \underline{\alpha}_i$ (by Corollary [2]3), we have $\underline{\alpha}_i \models \exists! v\psi(v)$. We choose $\beta' \in \alpha_i$ such that $\underline{\alpha}_i \models \psi[\beta']$. By (10), we also have, for some θ, $\underline{\mathrm{Ord}}_i \models \forall v(\theta(v) \leftrightarrow v = \beta')$. Thus $\underline{\alpha}_i \models \forall v(\theta(v) \leftrightarrow \psi(v))$. So also, $\underline{\mathrm{Ord}}_i \models \forall v(\theta(v) \leftrightarrow \psi(v))$. Hence $\beta = \beta'$. So $\alpha_i = \mathrm{Df}(\underline{\mathrm{Ord}}_i)$.

It follows easily that

Theorem 7. $\underline{\alpha}_i \prec \underline{\mathrm{Ord}}_i$.

(Tarski knew that Theorems 6 and 7 would follow from $\underline{\alpha}_i \equiv \underline{\mathrm{Ord}}_i$. His conjectures had some basis!)

Feferman [1960] gave a different, correct proof of 7 and 8. In Ehrenfeucht [1961] please correct misprint $\prec$ in 18 to ε. In order to make his lemma and proof correct, he should, I think, have added after $(*)$, that also $x \in \mathrm{Df}(\underline{\mathrm{Ord}}_i)$, and put something similar in the Lemma's hypothesis.

Our discussion of Ehrenfeucht [1961] would not be complete without mentioning the remarkable work of John Doner in [1969]. Everyone knows roughly how to extend successor, +, $\cdot$, exponentiation to hyperexponentiation, and on for $6, 7, \ldots, n, \ldots$. In Doner and Tarski [1969] these $\mathcal{O}_n$'s are studied carefully and some arithmetic facts are developed which Doner needed for [1969]. In [1969] Doner was able to extend more or less the entire work of Ehrenfeucht [1961] to all these $\mathcal{O}_n$'s!

References

Papers by A.Ehrenfeucht

[1957a] Application of games to some problems of mathematical logic. *Bull. Acad. Polon. Sci., Sér. Sci Math. Astronom. Phys.* **5** (1957), 35–37.

[1957b] On theories categorical in power. *Fund. Math.* **44** (1957), 241–248.

[1957c] Two theories with axioms built by means of pleonasms. *J. Symb. Logic* **22** (1957), 36–38.

[1958] Theories having at least continuum many non-isomorphic models in each infinite power. *Amer. Math. Soc. Notices* **5** (1958), 680–681.

[1959a] Decidability at the theory of one function. *Amer. Math. Soc. Notices* **6** (1959), p.268.

[1959b] Decidability of the theory of linear ordering relation. *Amer. Math. Soc. Notices* . **6** (1959), 268–269.

[1959c] A decidable theory which has exactly one decidable complete extension. *Amer. Math. Soc. Notices* **6** (1959), p.269.

[1961a] An application of games to the completeness problem for formalized theories. *Fund. Math.* **49** (1961), 129–141.

[1961b] Separable theories. *Bull. Acad. Polon. Sci., Sér. Sci. Math. Astronom. Phys.* **9** (1961), 17–19.

[1965] Elementary theories with models without automorphisms. *The Theory of Models*, Proceedings of the 1963 International Symposium at Berkeley, (North-Holland, Amsterdam, 1965), 70–76.

[1972] *Bull. Acad. Polon. Sci., Sér. Sci. Math. Astronom. Phys.* **20** (1972), 425–427.

[1973] Discernible elements in models for Peano arithmetic. *J. Symb. Logic* **38** (1973), 291–292. There are continuum many ω_0-categorical theories.

A.Ehrenfeucht and G.Fuhrken

[1971a] A finitely axiomatizable complete theory with atomless $F_1(T)$. *Arch. Math. Logik und Grundlagenforsch.* **14** (1971), 325–328.

[1971b] On models with undefinable elements. *Math. Scand.* **28** (1971), 325–328.

A.Ehrenfeucht and G.Kreisel

[1966] Strong models of arithmetic. *Bull. Acad. Polon. Sci., Ser. Sci. Math. Astronom. Phys.* **14** (1966), 107–110.

A.Ehrenfeucht and H.Läuchli

[1962] Rigid theories. *J. Symb. Logic* **27** (1962), 475–476.

A.Ehrenfeucht and J.Loś

[1954] Sur les produits cartesiens des groups cycliques infinis. *Bull. Acad. Polon. Sci., Sér. Sci. Math. Astronom. Phys.* **2** (1954), 261–263.

A.Ehrenfeucht and A.Mostowski

[1956] Models of axiomatic theories admitting automorphisms. *Fund. Math.* **43** (1956), 50–68.

[1961] A compact space of models of first order theories. *Bull. Acad. Polon. Sci., Sér. Sci. Math. Astronom. Phys.* **9** (1961), 369–373.

Other References

Büchi — Büchi, J. Richard. Transfinite automata recursions and weak second order theory of ordinals. *Proceedings of the 1964 International Congress of Logic, Methodology, and Philosophy of Science* held in Hebrew University, Jerusalem, (North-Holland, Amsterdam, 1965), 3–23.

Chang & Keisler — Chang, C. C. and H. J. Keisler. *Model Theory*, (North-Holland, Amsterdam, 1973), xii+550 pp.

Doner — Doner, John E. An Extended Arithmetic of Ordinal Numbers and its Metamathematics. Doctoral Dissertation, University of California, Berkeley, 1969, vi+312 (Also published by System Development Corporation.)

Doner & Tarski — Doner, John E. and Alfred Tarski. An extended arithmetic of ordinal numbers. *Fund. Math.* **65** (1969), 95–127.

Erdös & Rado — Erdös, P. and T. Rado. A partition calculus in set theory, *Bull. Amer. Math. Soc.* **62** (1956), 427–489.

Feferman — Feferman, Solomon. Some recent work of Fraisse and Ehrenfeucht. *Summaries of Talks Presented at the Summer Institute for Symbolic Logic Cornell University 1957*, Second Edition, Communications Research Division, Institute for Defense Analyses, 1960, 201–209.

Fraïssé — Fraïssé, Roland. Sur les rapports entre la théorie des relations et le sémantique au sens A. Tarski. Communication au Colloque de logique mathématique, Paris, 1952, 1 p.

Fraïssé — Fraïssé, Roland. Sur quelques classifications des systèmes de relations. *Publ. Sci. Univ. Alger Ser. A*, **1** (1954), 35–182.

Fraïssé — Fraïssé, Roland. Sur quelques classification des systèmes de relations. Thèses présentees a las Faculté des Sciences de l'Université de Paris, Imprimerie Durand, Chartres, 1955, 1–154.

Fraïssé — Fraïssé, Roland. Sur quelques classifications des relations, basees sur des isomorphismes restreints. I. Étude generale. II. Application aux relations d'ordres. *Alger-Mathematiques* **2** (1955), 16–60, 273–295.

Laüchli & Leonard — Laüchli, H. and J. Leonard, On the elementary theory of linear order, *Fund. Math.* **59** (1966), 109–116.

LeTourneau — LeTourneau, John Joseph. Decision Problems Related to the Concept of Operation, Doctoral Dissertation, University of California, Berkeley, 1968, vi+122

Morley — Morley, Michael. Categoricity in power. *Trans. Amer. Math. Soc.* **114**, pp. 514–538 (1965).

Morley — Morley, Michael. Omitting classes of elements, *The Theory of Models*, Proceedings of the 1963 International Symposium at Berkeley, (North-Holland, Amsterdam, 1965), 265–273.

Mostowski & Tarski Mostowski, Andrzej and Alfred Tarski. Arithmetical classes and types of well-ordered systems. *Bull. Amer. Math. Soc.* **55** (1949–50), p.65.

Rabin Rabin, Michael O. Decidability of second-order theories and automata on infinite trees. *Trans. Amer. Math. Soc.* **141**, 1–35 (1969).

Ramsey Ramsey, Frank P. On a problem of formal logic, *Proc. London Math. Soc.* **30** (1929), 264–286.

Syntax vs. Semantics on Finite Structures

Natasha Alechina[1] and Yuri Gurevich*[2]

[1] University of Birmingham
[2] University of Michigan

Abstract. Logic preservation theorems often have the form of a syntax/semantics correspondence. For example, the Łos-Tarski theorem asserts that a first-order sentence is preserved by extensions if and only if it is equivalent to an existential sentence. Many of these correspondences break when one restricts attention to finite models. In such a case, one may attempt to find a new semantical characterization of the old syntactical property or a new syntactical characterization of the old semantical property. The goal of this paper is to provoke such a study.

1 Introduction

It is well known that famous theorems about first-order logic fail in the case when only finite structures are allowed (see, for example, [8]). A more careful examination shows that it is wrong to lump all these failing theorems together. On one side we have theorems like completeness or compactness where the failure is really and truly hopeless. On the other side there are theorems like the Łos-Tarski theorem, which we prefer to formulate in the following form:

Theorem 1 (Łos and Tarski). *A first order formula is preserved by extensions iff it is equivalent to an existential formula.*

Theorems of this form are known as preservation theorems. Classical preservation theorems fail not only when the class of models is restricted to allow only finite models, but also when the language is modified (for example, only up to k many variables allowed in a formula). Recently a lot of interesting work was done on rescuing preservation theorems in non-classical contexts. Rosen and Weinstein in [12] start with analyzing the failure of the Łos-Tarski theorem on finite models and come up with a generalized notion of a preservation theorem: $L \cap EXT \subseteq L'$, where EXT is the set of formulas preserved by extensions on finite models, L is some first order quantifier prefix class, and L' is the existential fragment of $L^{\omega}_{\infty\omega}$ or positive Datalog. They also show that the analogue of the Łos - Tarski theorem fails for $L^{\omega}_{\infty\omega}$. Barwise and van Benthem in [4] rescue preservation theorems in $L^{\omega}_{\infty\omega}$ and its fragments (on arbitrary models) using a generalized notion of 'consequence along some model relation'.

* Partially supported by NSF grant CCR 95-04375. During the work on this paper (1995-96 academic year), this coauthor was with CNRS in Paris, France

We view the Łos-Tarski theorem not so much as a preservation theorem, but rather as a theorem relating syntax and semantics. On the one hand, it can be seen as the semantical characterization of existential first-order formulas. It is natural to ask if there is an alternative characterization of such formulas in the case of finite structures[3]. It turns out that there is a natural characterization of that sort; see Theorem 5.

On the other hand, the Łos-Tarski theorem can be seen as a syntactical characterization of the semantical property 'being preserved by extensions' in the context of first-order logic. We say that a property $\mathcal{P}$ has a *syntactical characterization in the context of a logic* L if there is a recursive class F of formulas of L, such that

1. Every formula in F has the property $\mathcal{P}$.
2. Every L-formula which has the property $\mathcal{P}$ is L-equivalent to a formula in F.

A natural question arises whether such a characterization for the formulas preserved by extensions exists in the case of finite structures. We know that the classical characterization fails (Gurevich - Shelah, Tait), but this does not rule out the existence of another characterization.

The Łos-Tarski theorem is not the only syntax/semantics theorem.

Some other theorems of the same kind are:

Theorem 2 (Lyndon). *A first order formula is monotone in a predicate P iff it is equivalent to a formula positive in P*[4].

Theorem 3. *A first order formula is preserved by homomorphisms iff it is equivalent to a positive existential formula.*

Theorem 4. *A first order universal formula is preserved by finite direct products iff it is equivalent to a universal Horn formula.*

In this paper we concentrate on syntax-to-semantics characterization in the context of finite structures. We also make some preliminary remarks on the problem of semantics-to-syntax characterization. This is only the beginning; our investigation provides many questions and few answers.

The rest of the paper is organized as follows. In section 2, we give a semantical characterization of existential first order formulas on finite structures. In section 3 we use similar techniques to characterize positive existential, existential Horn and universal Horn formulas on finite structures. Section 4 deals with the problem of characterizing semantical properties in the context of first order logic. Here, we only have negative results which state that the class of formulas

[3] Indeed this question has been asked by Johan van Benthem (in September 1995) and by H. Jerome Keisler (in March 1996) when one of the authors lectured on finite model theory.

[4] A couple of years ago Jörg Flum asked explicitly if there is an alternative characterization of monotonicity over finite structures.

preserved under a given construction cannot be itself the desired characterization, since it is not recursive. In section 5 some syntactical characterizations are proposed in the context of extensions of first order logic. We conclude with stating some open problems.

Conventions. We assume that all structures are over a finite vocabulary which does not contain functional symbols of positive arity, unless explicitly stated otherwise. We assume that classes of structures are closed under isomorphisms, and 'there is a unique structure' or 'there are finitely many structures' means 'up to isomorphism'.

The notation is usually fairly standard. We use $M \preceq N$, where M, N are structures, for 'M is a substructure of N', that is: the universe of M is a subset of the universe of N, and the interpretation of all predicates and constants in M and N is the same on the universe of M. $Mod(T)$, where T is a theory, denotes the class of finite models of T.

For the sake of brevity, we speak about sentences and classes of structures rather than formulas and global relations (for the definition of global relations, see [8]).

2 Existential Formulas on Finite Structures

As we have already mentioned, the Łos-Tarski's characterization of existential first order formulas fails for finite structures (see Theorem 10). A question arises whether there is any natural characterization of existential formulas on finite structures. It turns out that such a characterization exists.

Define a *minimal* structure in a class K to be a structure in K with no proper substructures in K.

Let $\min(K) = \{M \in K : M \text{ is minimal in } K\}$.

Theorem 5. *Let K be an arbitrary class of finite structures over the same finite vocabulary which does not contain functional symbols of positive arity.*

The following are equivalent:

1. *K is closed under extensions and $min(K)$ is finite.*
2. *There is an existential first-order sentence φ such that K is the collection of finite models of φ.*

Proof. $1 \longrightarrow 2$. Let $A_1, .., A_j$ be the minimal models of K of cardinalities $n_1, .., n_j$ respectively. The desired φ has the form

$$\varphi_1 \vee \ldots \vee \varphi_j,$$

where φ_i states that there are elements $x_1, \ldots, x_{n_i}$ which form a structure isomorphic to A_i.

$2 \longrightarrow 1$. Fix an appropriate φ and let n be the number of quantifiers plus the number of constants in φ. By the classical theorem, K is closed under extensions.

Since there are only finitely many structures in K of cardinality $\leq n$, it suffices to prove that every A in K has a substructure of cardinality $\leq n$ satisfying φ. But this is obvious. The desired substructure is formed by a set of at most n elements witnessing that A satisfies φ. □

Remark. The same holds for any global relation K (not only for a class of structures, i.e. a 0-ary global relation).

In one aspect, the semantical characterization of existential formulas in Theorem 5 is even preferable to the classical Łos-Tarski characterization: K is not supposed to be finitely axiomatizable in first-order logic. It may seem that Theorem 5 and its proof survive in the case of arbitrary structures. Flum pointed out that this is wrong ([7]). Counterexamples include (*i*) the class of extensions of a given (up to isomorphism) infinite structure, and (*ii*) the class of non-well-founded orders. In both cases, the class in question satisfies 1 (in case (*ii*) this happens because there are no minimal models in the class) but is not definable by an existential sentence. To adapt Theorem 5 to the case of infinite structures, the condition that $\min(K)$ is finite may be replaced by the following stronger condition: There exists a finite class $K_0 \subseteq K$ of finite structures such that every K-structure extends some K_0-structure.

Obviously, the Łos-Tarski theorem cannot be generalized as 'A class of arbitrary structures is closed under extensions if, and only if, it is axiomatizable by an existential formula'. Here is an example (building on the Gurevich-Shelah counter-example, cf. Theorem 10) of a class which is closed under extensions on arbitrary structures but is not finitely axiomatizable. Consider the class K_0 of finite linear orders with the successor relation, a minimal element a and a maximal element b, and close this class under extensions. Let us call the resulting class K. Assume by contradiction that K is definable by a first order formula. Then, by the Łos-Tarski's theorem, K is definable by an existential sentence. By the analog of Theorem 5 for arbitrary structures, $\min(K)$ is finite. But it is easy to see that $\min(K) = K_0$; in particular, if you remove the successor of some element c in a structure $M \in K_0$, then the remaining structure does not belong to K_0 because the element c does not have a successor there. Thus $\min(K)$ is infinite, which gives the desired contradiction.

We say that Theorem 5 *survives the restriction to a theory T* if for every class K of models of T, the following are equivalent:

1. K is closed under T-extensions (that is, if $M \in K$, $N \models T$ and $M \preceq N$ then $N \in K$) and $\min(K)$ is finite.
2. There is an existential sentence φ such that $K = \{M \in Mod(T) : M \models \varphi\}$.

Proposition 6. *If T is axiomatized by an $\exists^*\forall^*$ sentence, then Theorem 5 survives the restriction to T.*

Proof. $1 \longrightarrow 2$: as before.
$2 \longrightarrow 1$. Suppose that T is given by a sentence $\alpha = \exists x_1 \ldots \exists x_k \forall \bar{y} \psi$ and $K = \{M \in Mod(T) : M \models \exists z_1 \ldots \exists z_n \chi\}$, where ψ and χ are quantifier-free.

It is obvious that K is closed under T-extensions. Let $M \in K$. There are k elements satisfying $\forall \bar{y}\psi(x_1, \dots, x_k)$ and n elements satisfying χ in M. The substructure generated by these $k+n$ elements still satisfies α and φ. Therefore the minimal structures in K are of size less or equal to $k+n+$ the number of constants. The rest of the proof is as above. □

Observe that the same proof works in case T is a universal theory. However,

Proposition 7. *Theorem 5 may not survive the restriction to a theory given by an infinite set of existential axioms.*

Proof. Let φ be $\exists x(x = x)$. The following existential theory T has infinitely many minimal models:

$$T = \exists x_1 x_2 (x_1 R x_2), \exists x_1 x_2 x_3 (x_1 R x_2 R x_3), \exists x_1 x_2 x_3 x_4 (x_1 R x_2 R x_3 R x_4), \dots$$

Notice that every cycle $a_1 R a_2 R \dots R a_1$ (without any additional edges) is a model of T. It is also a minimal model, for, if one of the elements is deleted, and the longest chain of the form $a_1 R a_2 \dots R a_n$ has length n, then

$$\exists x_1 \dots x_{n+1}(x_1 R x_2 \dots R x_{n+1})$$

is no longer satisfied. □

Proposition 8. *Theorem 5 may not survive the restriction to a theory given by a $\forall\exists$ sentence α.*

Proof. Let α be $\forall x \exists y R(x, y)$ and φ be $\exists x(x = x)$. Every cycle $a_1 R a_2 R \dots R a_1$ is a minimal model of $\alpha \wedge \varphi$. □

Proposition 9. *Theorem 5 fails if functions are allowed.*

Proof. Let φ be $\exists x(f(x) = f(x))$. It has infinitely many minimal models, for example with $f(a_1) = a_2, f(a_2) = a_3, \dots, f(a_n) = a_1$. (If any element is deleted, the resulting substructure is not closed under functional application.) □

Theorem 5 allows us to simplify the counterexample to the Łos-Tarski theorem on finite structures given by Gurevich and Shelah in [8].

Theorem 10 (Tait 1959, Gurevich - Shelah 1984). *There exists a first order formula which is preserved by extensions on finite structures but is not equivalent to any existential formula on finite structures.*

Proof. Consider the first order language with equality containing two constants a and b, and two binary predicates $<$ and S (which is going to denote the successor relation) in addition to equality.

Let χ be $\chi_1 \wedge \chi_2 \wedge \chi_3 \rightarrow \chi_4$, where:

1. χ_1 says that $<$ is a linear order, that is, χ_1 is the conjunction of

$\chi_{11} = \forall x \forall y (x < y \vee y < x \vee x = y)$,
$\chi_{12} = \forall x \neg (x < x)$,
$\chi_{13} = \forall x \forall y \forall z (x < y \wedge y < z \rightarrow x < z)$;

2. χ_2 says that S is consistent with the successor relation in the following sense:
$$\chi_2 = \forall x \forall y (S(x,y) \rightarrow x < y \wedge \neg \exists z (x < z < y));$$
3. χ_3 says that a is the least element and b the greatest element, and a is not equal to b, that is, $\chi_3 = \chi_{31} \wedge \chi_{32} \wedge a < b$, where $\chi_{31} = \forall x (a \leq x)$ and $\chi_{32} = \forall x (x \leq b)$ (where $\leq$ is defined as usual);
4. χ_4 says $\forall x (x < b \rightarrow \exists y S(x,y))$ (every element except for b has a successor). Together with χ_1 - χ_3, χ_4 implies that S is in fact the successor relation.

Notice that χ is preserved by extensions. For, let $M \models \chi$ and N be a proper extension of M. Suppose $M \not\models \chi_1 \wedge \chi_2 \wedge \chi_3$. Since $\neg(\chi_1 \wedge \chi_2 \wedge \chi_3)$ is existential, it is preserved by extensions. Therefore $N \models \chi$. Suppose $M \models \chi_1 \wedge \chi_2 \wedge \chi_3$. It suffices to check that N fails to satisfy $\chi_1 \wedge \chi_2 \wedge \chi_3$. Consider an element c of N which does not belong to M. If it does not fit in the linear order between a and b, χ_1 or χ_3 are false in N. Since M is finite, c has to fit between a and the successor of a, or between the successor of a and its successor, et cetera, or between the predecessor of b and b. In all these cases χ_2 is violated.

This formula has infinitely many minimal models, because every initial segment of natural numbers with a interpreted as 0, b as the greatest element and the predicates having their standard interpretation is a minimal model of χ. Indeed, let M be such a model (an initial segment of natural numbers), and N a proper substructure of M. Since χ_1, χ_2, χ_3 are universal sentences satisfied in M, they are satisfied in N, but χ_4 necessarily fails in N.

It follows from Theorem 5 that χ is not equivalent to an existential formula. □

Observe that χ is not preserved by extensions on infinite structures. Let ω^R be the order type of the reversed ω: $\ldots, 3, 2, 1, 0$. Consider a structure of order type $\omega + \omega^R$:
$$a = 0, 1, \ldots, \ldots, -2, -1 = b$$
with the standard successor relation. It is a model of χ. If we extend this structure by putting an element in the middle, then χ is not true any more.

If a formula with only unary predicates, constants and equality is preserved by extensions on finite models, it is equivalent to an existential formula. This is easy to check. However, one binary relation is sufficient for a counter-example.

Theorem 11. *Let L be a language with only one binary relation and without constants or equality. There is an L-sentence preserved by extensions that is not equivalent to any existential sentence on finite structures.*

We only sketch a proof of Theorem 11. The proof consists in step-by-step replacing the formula χ above with a formula which does the same job but contains only one binary predicate.

As a first step, we get rid of the individual constants and equality. The first auxiliary language contains $<$, S and two more unary predicates, F ('first') and L ('last'). Let $x \equiv y =_{df} \forall z([(x < z) \leftrightarrow (y < z)] \wedge [(z < x) \leftrightarrow (z < y)] \wedge [S(x, z) \leftrightarrow S(y, z)] \wedge [S(z, y) \leftrightarrow S(z, y)] \wedge [F(x) \leftrightarrow F(y)] \wedge [L(x) \leftrightarrow L(y)])$.

Our first modification of χ is

$\chi' = \chi'_1 \wedge \chi'_2 \wedge \chi'_3 \rightarrow \chi'_4$, where χ'_1 is like in the proof of Theorem 10, but with $=$ replaced by $\equiv$, χ'_2 is like χ_2, χ'_3 is the conjunction of

$$\forall x \forall y (F(x) \rightarrow x < y \vee x \equiv y);$$

$$\forall x \forall y (L(x) \rightarrow y < x \vee x \equiv y);$$

$$\forall x \forall y (F(x) \wedge L(y) \rightarrow x < y),$$

and χ'_4 is the conjunction of $\forall x(\neg L(x) \rightarrow \exists y S(x, y))$ and $\exists x F(x) \wedge \exists x L(x)$.

We leave it for the reader to check that this formula is preserved by extensions on finite structures and has infinitely many minimal models. (Notice that if $M \models \chi'$ and N is a proper extension of M, N may satisfy the antecedent of χ' provided that every new element fits into some $\equiv$-equivalence class in M. But then every element in N still has a successor.)

Now we translate χ' into the language containing just one binary predicate R. We are going to code old elements with clusters consisting of three points, and code the relations $<$, S, F and L between the clusters in terms of relation R among their components.

We use the following abreviations: $M(x)$ ('x is a main element') for $R(x, x)$; $A(x)$ ('x is an auxiliary element') for $\neg R(x, x)$; and $Reg(x, x_1, x_2)$ ('x, x_1, x_2 is a regular triple') for the conjunction of

- $M(x)$;
- $A(x_1) \wedge A(x_2)$;
- $R(x_1, x) \wedge R(x_2, x)$ ('x_1, x_2 are auxiliary to x');
- $R(x, x_1) \wedge \neg R(x, x_2)$ ('x_1 is the first auxiliary, x_2 is the second auxiliary').

Let $Reg(x, x_1, x_2)$, $Reg(y, y_1, y_2)$. The relations between the main elements are coded as follows:

- $x < y \Leftrightarrow x_1 R y_1$;
- $S(x, y) \Leftrightarrow x_2 R y_2$;
- $F(x) \Leftrightarrow x_1 R x_2$;
- $L(x) \Leftrightarrow x_2 R x_1$.

Abbreviate $\forall z(R(x, z) \leftrightarrow R(y, z)) \wedge (R(z, x) \leftrightarrow R(z, y))$ as $x \equiv y$.

Translate $\chi'_1 - \chi'_4$ introduced above using this coding. For example, χ'_{11} becomes

$$\forall x x_1 x_2 y y_1 y_2 (Reg(x, x_1, x_2) \wedge Reg(y, y_1, y_2) \rightarrow x_1 R y_1 \vee y_1 R x_1 \vee x \equiv y).$$

Denote the resulting translations $\chi_1'' - \chi_4''$.

To guarantee that the coding is well defined, we need the following conditions. Let χ_0'' be the conjunction of the universal closure of

$M(x) \wedge M(y) \wedge R(x,y) \rightarrow x \equiv y$ (main elements are not connected or they are indistinguishable);

$A(y) \wedge A(z) \wedge R(x,y) \wedge R(y,x) \wedge R(x,z) \wedge R(z,x) \rightarrow y \equiv z$ (every main element has only one first auxiliary element up to $\equiv$);

$M(x) \wedge A(y) \wedge A(z) \wedge R(y,x) \wedge \neg R(x,y) \wedge R(z,x) \wedge \neg R(x,z) \rightarrow y \equiv z$ (every main element has only one second auxiliary);

$A(x) \wedge M(y) \wedge M(z) \wedge R(x,y) \wedge R(x,z) \rightarrow y \equiv z$) (every auxiliary element has only one 'master').

Again we leave it to the reader as an exercise to verify that $\chi_0'' \wedge \ldots \wedge \chi_3'' \rightarrow \chi_4''$ is preserved by extensions on finite structures and has infinitely many minimal models.

3 More Examples

In this section we look at some other well known preservation theorems from classical logic and give some characterizations for the finite case.

The following characterizations of universal Horn and existential Horn formulas are known in classical model theory (cf. [5], p.337, Propositions 6.2.8 and 6.2.9).

Fact 12. *A first order formula is preserved by substructures and finite direct products iff it is equivalent to a universal Horn formula.*

Fact 13. *A first order formula is preserved by extensions and finite direct products iff it is equivalent to an existential Horn formula.*

We do not know if the first statement fails on finite structures. The second one is false:

Proposition 14. *There exists a formula which is preserved by extensions and finite direct products on finite structures but is not equivalent to an existential Horn formula.*

Proof. Consider the formula $\exists x \exists y (x \neq y) \wedge \chi$, where χ is defined as in the proof of Theorem 10. This formula is preserved by extensions on finite structures (since both conjuncts are) and is not equivalent to an existential formula (since it has infinitely many minimal models). It is easy to show that it is preserved by direct products. Let M_1 and M_2 satisfy the formula. Obviously if M_1 and M_2 contain at least two elements, so does their product. With respect to χ, we have two cases. First, assume that for all elements e of M_1 and M_2 $e < e$ holds. Then all elements of $M_1 \times M_2$ are reflexive as well, and the antecedent of χ which states that the order is strict, fails. Therefore χ is true on $M_1 \times M_2$. Now suppose that one of the structures (let it be M_2) has a non-reflexive element e. Since M_1 has at least two elements e_1 and e_2, $M_1 \times M_2$ contains (e_1, e) and (e_2, e) which are incomparable. Hence the antecedent of χ which states the existence of a linear order, fails, and χ itself is true on $M_1 \times M_2$. □

A characterization of universal Horn and existential Horn formulas on finite structures can be easily obtained from Theorem 5.

Theorem 15. *Let K be an arbitrary class of finite structures over the same finite vocabulary which does not contain functional symbols of positive arity.*
The following are equivalent:

1. *K is closed under extensions and direct products, and min(K) is finite.*
2. *K is definable by an existential Horn formula.*

Proof. 1 $\longrightarrow$ 2. By Theorem 5, K is definable by an existential formula φ. We also know that φ is preserved by finite direct products of finite structures. It remains to show that φ is preserved by finite direct products of arbitrary structures.

By contradiction, assume that this is not the case. Then there are possibly infinite structures $M_1, \ldots, M_k$, each satisfying φ, such that their product does not satisfy φ. Since φ is existential, there are finite substructures of structures M_i satisfying φ. Then φ is true on the product of these substructures, since it is preserved by products of finite structures. But this product is a substructure of the big product, so φ is true in the big product, which contradicts the assumption.

The direction from 2 to 1 is easy. □

Analogously it is possible to characterize universal Horn sentences.

We say that a structure M is a *minimal counterexample* to the class K if M is not in K and every proper substructure of M is in K.

Theorem 16. *Let K be an arbitrary class of finite structures over the same finite vocabulary which does not contain functional symbols of positive arity.*
The following are equivalent:

1. *K is closed under substructures and direct products, and K has finitely many (up to isomorphism) minimal counterexamples.*
2. *K is definable by a universal Horn sentence.*

Proof. 1 $\longrightarrow$ 2. By Theorem 5, K is definable by a universal statement. Classical argument (see [5], Proposition 6.2.8) shows how to construct a finite counterexample to preservation by direct products for any universal sentence which is not equivalent to a Horn sentence. Since K is closed under direct products, K is definable by a Horn sentence.

2 $\longrightarrow$ 1. Let φ be a universal Horn sentence. Then the set of models of φ is closed wrt direct products and substructures. Since φ is a universal formula, $\neg\varphi$ is existential and has finitely many minimal models. This means that φ has finitely many minimal counterexamples. □

Remark. Observe that the same reasoning again applies to any global relation (not only to a class of structures).

Observe that on arbitrary structures the restriction on minimal models or minimal counterexamples cannot be thrown away. For example, take a class K of (finite) graphs which do not contain cycles. It is closed under substructures and direct products but is not first-order definable. The class of (finite) graphs which do contain cycles is preserved by extensions and direct products as well, but it is not first order definable hence not definable by an existential Horn formula.

Given two structures M and N in the same relational vocabulary, a mapping $h : M \longrightarrow N$ is a *homomorphism* (or a *homomorphism into*) if for every n-ary relation R in the vocabulary of M and N, and for all n-tuples of elements $x_1, \ldots, x_n$ in M, if $R_M(x_1, \ldots, x_n)$, then $R_N(h(x_1), \ldots, h(x_n))$. Observe that if a class K of structures K is closed under homomorphisms, then K is closed under extensions and under adding more tuples in the relations. The latter means that if $M \in K$ and M' is like M except for $R_M \subset R_{M'}$, then $M' \in K$. Let M and N be two structures in the same vocabulary. Call N a *weak substructure* of M if N can be obtained from M by deleting elements or decreasing relations in M. Let $\min_h(K)$ be a collection of $M \in K$ such that no proper weak substructure of M belongs to K.

It is a known fact that if a first order formula is preserved by homomorphisms on arbitrary structures, then it is equivalent to a positive existential formula. It is unknown whether this result remains true on finite structures. We have the following characterization of positive existential first order sentences on finite structures:

Theorem 17. *Let K be a class of finite structures of a given finite vocabulary. The following are equivalent:*

1. *K is closed under homomorphisms and $\min_h(K)$ is finite.*
2. *K is definable by a positive existential sentence.*

Proof. The direction from 2 to 1 is easy. Assume 1. Then $\min_h(K)$ has finitely many elements, $M_1, \ldots, M_n$. As in the proof of Theorem 5, we describe each of them by an existential formula, but now we describe only the positive diagram of each M_i, $1 \leq i \leq n$. More precisely, let $\Delta_i(\bar{a})$ be the positive diagram of M_i, and let φ_i be $\exists \bar{x} \Delta_i(\bar{x})$. Let φ be the disjunction of φ_i. Obviously, φ is a positive existential formula. It remains to show that φ defines K.

Suppose $M \in K$. Then for some M_i there is a homomorphism $h : M_i \longrightarrow M$. By the classical result, $M \models \varphi_i$, hence $M \models \varphi$.

Suppose $M \models \varphi$. Then for some i, $M \models \varphi_i$. This gives a homomorphism $h : M_i \longrightarrow M$, hence $M \in K$. □

We do not have a semantic characterization of the property 'being positive in a given predicate P' on finite structures (cf. Lyndon's theorem stated in the Introduction). It is straightforward to characterize *existential* sentences positive (negative) in a given predicate P analogously to the theorem above. Given the characterization of existential sentences positive (negative) in a given predicate, the characterization of universal sentences negative (positive) in a given predicate follows. We leave the following theorem as an exercise for the reader.

Theorem 18. *Let φ be an existential or a universal sentence. If φ is increasing (resp. descreasing) in P on finite structures, then it is equivalent, on finite structures, to an existential sentence φ' positive (resp. negative in P).*

4 Undecidability of Semantical Properties

In this section we show that the easiest answer to the question whether a given semantical property has a syntactic characterization (in the sense defined in the Introduction), namely, taking the class of formulas having the property itself as a characterization, often does not work. We prove undecidability results, both for finite and arbitrary models, for a large class of semantical properties. We do it by using Turing machine encodings and recursive inseparability. There are much easier ways of proving undecidability for particular properties. For example, Johan van Benthem pointed to us that monotonicity is undecidable since a formula φ is valid iff $\varphi \vee \neg q$, for a fresh propositional variable q, is upward monotone in q.

For a formal definition of a Turing machine, we refer the reader to e.g. [6]. Here we give a brief and informal description of the kind of Turing machines appropriate to our purposes.

A Turing machine T has a 'head' and a linear tape, divided into cells, which is bounded on the left and unbounded on the right. The head can move one cell at a time to the left or to the right, read and write in a cell a symbol from some finite alphabet $\mathcal{A}$, or erase the current symbol in the cell; let us say that an empty cell contains the letter '*blank*'.

For technical reasons, we introduce an additional symbol α which is not in $\mathcal{A}$ and which occurs in the leftmost cell of the tape. T has finitely many states $q_1, \ldots, q_n$. The instructions of T are of the form: 'If the state is q and you are reading a, replace a with b, move the head one cell to the right (or to the left, or stay in the same cell) and change the state to q''.

Without loss of generality we restrict attention to Turing machines which make $\geq max(3, s_T)$ steps, where s_T is the number of states of T.

An *n-th configuration* of T is the description of T after n steps of its computation.

It is well known that the problem whether a Turing machine with a given set of instructions stops after a finite number of steps (the halting problem) is undecidable (see, for example, [11]).

To formalize Turing machine computations we use several predicates.

Denote the configurations of a given Turing machine T by C_0 (initial configuration), $C_1, \ldots, C_t, \ldots$. Let

$Q(q, t)$ mean that in configuration C_t the state of the machine is q;
$H(i, t)$ mean that in configuration C_t the head is reading cell i;

$L_a(i,t)$ mean that in configuration C_t the letter a is in cell i.

We will use the following formulas in the proofs to follow:

ψ_1 is $\chi_1 \wedge \ldots \wedge \chi_4$ from the proof of Theorem 10 (with a replaced by 0),

ψ_2 says that there are at least $max(3, s_T)$ elements, where s_T is the number of states of T; that for each t, there is precisely one state q and cell i with $Q(q,t)$ and $H(i,t)$; for every pair (i,t), precisely one of $\{L_a(i,t) : a \in \mathcal{A}\cup\{\alpha, blank\}\}$ holds.

ψ_3 is $Q(0,0) \wedge H(0,0) \wedge L_\alpha(0,0) \wedge \forall x(x \neq 0 \to L_{blank}(x,0))$ (this is the description of C_0);

ψ_4 is conjunction of formulas describing instructions of T, e.g.:

$$\forall y \forall t(Q(q,t) \wedge H(y,t) \wedge L_a(y,t) \to \exists y' \exists t'(S(t,t') \wedge S(y,y') \wedge Q(q',t') \wedge$$
$$\wedge H(y',t') \wedge L_b(y,t') \wedge \forall v(v \neq y \to \bigwedge_{a \in \mathcal{A}\cup\{\alpha, blank\}} (L_a(v,t) \to L_a(v,t'))))$$

Similarly for the case when the head has to move left (replace $S(y,y')$ by $S(y',y)$) or not move (replace $S(y,y')$ by $y = y'$). It is assumed that the head does not move to the left when it reads α.

Recall that two sets X and Y, say of strings in a given alphabet, are called ***recursively inseparable*** (in short, r.i.) if there is no recursive set R with $X \subseteq R$ and $R \cap Y = \emptyset$ (see, for example, [11]).

We will use the following well known fact:

Lemma 19 (Reduction Lemma). *Suppose (X_1, X_2) are r.i., and there is a recursive function f such that*

$$\forall x(x \in X_1 \to f(x) \in Y_1)$$
$$\forall x(x \in X_2 \to f(x) \in Y_2)$$

Then (Y_1, Y_2) are r.i.

This gives us a method of proving that two sets are recursively inseparable by reducing to them some sets which are already known to be recursively inseparable.

Fix two members q_1, q_2 of the alphabet of states. Then $\mathcal{M}_1$ and $\mathcal{M}_2$ are r.i., where

$\mathcal{M}_1$ is the set of Turing machines which halt in state q_1.
$\mathcal{M}_2$ is the set of Turing machines which halt in state q_2.

Recursive inseparability of $\mathcal{M}_1$ and $\mathcal{M}_2$ follows from the existence of two r.e. sets of natural numbers which are recursively inseparable (Theorem XII in [11]). Let X_1 and X_2 be two such sets. Then for any natural number n, let T_n be a Turing machine which is computing X_1 and X_2 and halting in q_1 if $n \in X_1$ and in q_2 if $n \in X_2$. The function $f(n) = T_n$ is the required reduction.

Assume that we have proved that sets X and Y are r.i.; then any $X' \supseteq X$ and $Y' \supseteq Y$ are recursively inseparable (otherwise the set separating X' and Y' would also separate X and Y).

Theorem 20. *The following are r.i.:*

The set F_1 of logically false first order formulas.
The set F_2 of first order formulas which have only one finite model of size > 1.

Proof. Given a Turing machine T, we are going to write a formula β_T, such that β_T is logically false if T halts in q_1 and has exactly one finite model if it halts in q_2. In other words,

$$\forall T(T \in \mathcal{M}_1 \to \beta_T \in F_1)$$
$$\forall T(T \in \mathcal{M}_2 \to \beta_T \in F_2)$$

First we give an informal description of β_T. It describes the 'standard model' of the computation of T, that is, it says that the model is linearly ordered, the first element corresponds to the initial configuration, every element has a successor satisfying the requirements imposed by instructions, the last element corresponds to a configuration where the state is q_2 and there is only one such configuration, and there is no configuration where the state is q_1.

More formally (we use the notation introduced above)

$$\beta_T = \psi_1 \wedge \psi_2 \wedge \psi_3 \wedge \psi_4 \wedge Q(q_2, b) \wedge \forall t(t < b \to \neg Q(q_2, t) \wedge \neg Q(q_1, t))$$

If T halts in q_1, then β_T has neither finite nor infinite model. Indeed, assume that every non-halting configuration has a successor, that is, the linear order is long enough to correspond to the whole computation of T. Then there is a q_1-configuration, or one of the instructions has been violated. If the order is not long enough, again one of the instructions is violated, since the instructions imply the existence of successor.

If T halts in q_2, then β_T has precisely one model, which is finite and has the size equal to the length of the computation of T. □

Observe that this result cannot be strengthened by restricting the cardinality of the model in F_2 by some n. In that case the recursive set of formulas not having a model of cardinality less or equal to n would separate F_1 and F_2.

This gives us a whole range of recursive inseparability / undecidability results for semantical preservation properties. Namely, logically false formulas are usually preserved by semantical constructions. For semantical constructions of interest for us in this paper, it holds that if a formula has precisely one finite model of size > 1, then it is not preserved by this construction.

Corollary 21. *Each of the following pairs are r.i.:*

1. *The set of formulas preserved by extensions on arbitrary models (EPA) and the set of formulas* not *preserved by extensions on finite models (not-EPF);*
2. – *The set of formulas preserved by direct products on arbitrary structures.*
   – *The set of formulas which are not preserved by direct products on finite structures.*
3. – *The set of formulas preserved by direct products and substructures on arbitrary structures.*

- *The set of formulas which are not preserved by direct products and substructures on finite structures.*

4. – *The set of formulas preserved by direct products and extensions on arbitrary structures.*
   – *The set of formulas which are not on preserved by direct products and substructures on finite structures.*
5. – *The set of formulas preserved by homomorphisms on arbitrary structures.*
   – *The set of formulas which are not preserved by homomorphisms on finite struc tures.*
6. – *The set of formulas monotone in a given predicate P on arbitrary structures.*
   – *The set of formulas which are not monotone in P on finite structures.*

Proof. For (1), notice that $F_1 \subseteq EPA$, $F_2 \subseteq \overline{EPF}$. The proofs of (2), (3), (4) and (5) are analogous.

For (6), replace β_T from the previous theorem by

$$\beta'_T = \beta_T \wedge \forall t \neg P(t).$$

If T halts in q_1, β_T has no model, hence β'_T does not. Therefore β'_T is trivially monotone in P. If T halts in q_2, β'_T has exactly one model where P is empty. If P is extended, β'_T fails. Hence in this case β'_T is not monotone in P. □

Corollary 22. *The following sets of formulas are not recursive:*

- *The set of formulas preserved by substructures.*
- *The set of formulas preserved by extensions.*
- *The set of formulas preserved by direct products.*
- *The set of formulas preserved by direct products and substructures.*
- *The set of formulas preserved by direct products and extensions.*
- *The set of formulas preserved by homomorphisms.*
- *The set of formulas monotone in a given predicate P.*

Proof. If any of those classes were recursive, it would separate a pair which we proved to be recursively inseparable. □

Corollary 23. *Corollary 22 remains true if we restrict attention to finite structures.*

5 From Semantics to Syntax

5.1 Łos-Tarski as a Normal Form Theorem

So far, we do not have any syntactic characterization of semantical properties in the context of first-order logic restricted to finite models. Here we give characterizations of EPF and monotonicity in the context of extensions of first-order logic where the notion of a substructure is definable (monadic second order logic, fixed point logic).

But before going into that, we would like to reformulate the Łos-Tarski Theorem as a *normal form theorem* for formulas preserved by extensions.

Proposition 24. *There is a partial recursive function which reduces every EPA formula to an equivalent existential formula.*

Proof. Define f as follows. Take a Turing machine which derives all logically true formulas in the vocabulary of φ; if φ is EPA, then after finitely many steps a tautology of the form $\varphi \leftrightarrow \psi$, where ψ is existential, will appear. Take ψ be $f(\varphi)$. □

Given the results above, there is no total recursive function f which assigns every formula an equivalent formula so that $f(\varphi)$ is existential iff φ is EPA - since that would make EPA decidable. Still the question remains whether there exists a total recursive function f, such that $f(\varphi)$ is an existential formula equivalent to φ in case φ is EPA (and arbitrary otherwise). We are going to show that this is not the case.

We use this opportunity to answer negatively a question of Andréka, van Benthem and Németi in [3]: is there a recursive function f which gives an upper bound on the number of variables needed to write an existential equivalent of an EPA formula, given the number of variables of this formula? In other words, is there a recursive bound on the number of quantifiers in the existential equivalent?[5]

We use the technique from [8], where an analogous result for finite structures is proved. Let $\sharp(\varphi)$ denote the number of quantifiers in φ.

Theorem 25. *There is no total recursive function f such that for every first order EPA formula φ, $f(\varphi)$ is greater or equal to the number of quantifiers needed to write the shortest existential equivalent of φ.*

Proof. By contradiction, assume that there is such a function f.

If ψ is an existential sentence, then every minimal model of ψ has at most $\sharp(\psi)$ elements. So f would also give a bound on the size of the minimal models of EPA sentences.

For every Turing machine T, we show how to write a formula φ_T which corresponds to a computation of T in the following way. If T halts, then φ_T is EPA and has a minimal model which has as many elements as there are steps in the computation of T. If φ_T is EPA, then the size of any minimal model of φ_T is less or equal to $f(\varphi_T)$. Therefore f can be used to solve the halting problem by letting T run for $f(\varphi_T)$ steps: we know that if T halts at all, it halts after at most $f(\varphi_T)$ steps. This shows that such a function f cannot exist.

Now we write down the formula in question. Given a Turing machine T with the halting state 1, let φ_T be as follows:

$$\varphi_T =_{df} \chi_1 \wedge \chi_2 \wedge \chi_3 \rightarrow \chi_4 \wedge \psi_2 \wedge \psi_3 \wedge \psi_4 \wedge Q(1,b) \wedge \forall t(t < b \rightarrow \neg Q(1,t)).$$

Claim 26. *If T halts, then φ_T is preserved by extensions on arbitrary structures. Moreover, φ_T has a minimal model of the size equal to the number of steps in the computation of T.*

[5] In the meantime, we have found out that this question is also answered in [12].

Proof of the claim. Let $M \models \varphi_T$. If the universal antecedent is false in M, then it will remain false in any extension of M, therefore φ_T is true on all extensions of M.

Assume that $M \models \chi_1 \wedge \chi_2 \wedge \chi_3 \wedge \chi_4 \wedge \psi_2 \wedge \psi_3 \wedge \psi_4 \wedge Q(1,b) \wedge \forall t(t < b \to \neg Q(1,t))$. Then M is a linear order where the initial point corresponds to the initial configuration of T, and each next point corresponds to the next step in the computation of T. Since T halts, on a finite distance from 0 there will be a point t with $Q(1,t)$. But the last conjunct says that this point has to be b. This gives us a finite rigid linear order as in Theorem 10, but on arbitrary structures. In this case, any extension satisfies the formula because in any proper extension one of $\chi_1 - \chi_3$ is violated. No proper submodel of this model satisfies the formula, because χ_4 is violated, so it is a minimal model.

Now the theorem follows. □

Corollary 27. *There is no total recursive function f such that, for every first order EPA formula φ, $f(\varphi)$ is an existential first order formula equivalent to φ.*

Proof. Assume that such function existed; then it would give a bound on the number of quantifiers of an existential equivalent. But this is impossible. □

5.2 Extensions of First-Order Logic

So far, we have no syntactic characterization of EPF and monotonicity in the context of first order logic. In this section, we show that such a characterization exists in the context of logics where the notion of a substructure is definable, namely (extensions of) monadic second order logic. On ordered finite structures, a characterization can be given in $FO^< + PFP$ (partial fixed point). The following weaker semantical property may be of interest on ordered finite structures: preservation by end extensions. We show that the problem of syntactical characterization of this weaker property has a positive solution in the context of e.g. $FO^< + IFP$ (inflationary fixed point). What is more, the reduction to normal form in all these cases is effective, unlike for first-order logic.

Theorem 28. *Let L be monadic second order logic. There is a linear time algorithm A such that for every formula φ of L, (i) $A(\varphi)$ is EPF and (ii) $A(\varphi)$ is equivalent to φ iff φ is EPF.*

Proof. Consider a formula φ of L. $A(\varphi)$ will be a formula saying 'there is a substructure satisfying φ'. Let X be a monadic second order variable not occurring in φ, and φ_X be the formula obtained from φ by restricting all quantifiers in φ to X, that is, replacing $\forall y$ by $\forall y \in X$, $\exists y$ by $\exists y \in X$, and $\forall Y$ and $\exists Y$ by $\forall Y \subseteq X$ and $\exists Y \subseteq X$, respectively. Then $A(\varphi)$ is $\exists X(\varphi_X \wedge X \neq \emptyset)$.

Obviously, $A(\varphi)$ is EPF. We check that if φ is EPF, then φ is equivalent to $A(\varphi)$. Let φ be EPF. Obviously, any formula φ implies $A(\varphi)$, just take X to be the whole universe. Assume that a structure M satisfies $A(\varphi)$. This means that there is a substructure of M which satisfies φ. Since φ is EPF, $M \models \varphi$.

Finally, if φ is not EPF, then φ is not equivalent to $A(\varphi)$ (which is always EPF). □

Remark. Theorem 28 generalizes to extensions of monadic second order logic.

We denote by $FO^<$ the restriction of first-order logic to the case of structures where $<$ is linear order. In the rest of this section we consider extensions of $FO^<$.

The next normal form uses a partial fixed point operator, i.e. it is about the language $FO^< + PFP$.

Recall that given any $\varphi(X, y)$, the partial fixed point operator generates a sequence $F_0, F_1, \ldots$, such that

$F_0 = \emptyset$ and
$F_n = \{y : \varphi(F_{n-1}, y)\}$.

Set $F_\infty = F_i$, if $F_i = F_{i+1}$ for some i, and $\emptyset$ otherwise. The partial fixed point $\mathrm{PFP}_{X,y}\varphi(X, y)$ of $\varphi(X, y)$ with respect to X, y denotes F_∞.

The following theorem is of no surprise.

Theorem 29. *Monadic second order quantification on ordered structures is expressible in $FO^< + PFP$.*

Proof. First we define a formula $\alpha(X, y)$, such that the fixed point operator applied to this formula with respect to X, y generates the list of all substructures of a given structure in a lexicographic order, repeated cyclically infinitely many times.

Let M be a structure with n elements, $1, \ldots, n$. There is a natural ordering on the substructures of M since they correspond to sequences of 0's and 1's of length n:

$$\begin{array}{cccc} 1\,2 \ldots & n-1 & n \\ 0\,0 \ldots & 0 & 0 \\ 0\,0 \ldots & 0 & 1 \\ 0\,0 \ldots & 1 & 0 \\ 0\,0 \ldots & 1 & 1 \\ & \ldots & \\ 1\,1 \ldots & 1 & 1 \end{array}$$

Let us call the subset of M corresponding to the ith number in this ordering F_i, with $F_0 = \emptyset$. Let us call F_{i+1} the successor of F_i, and F_0 the successor of F_{2^n-1}. Observe that given F_i for any $0 \leq i < 2^n$, we can describe the elements in its successor by a first order formula $\alpha(F_i, y)$. Namely, the maximal element of the complement of F_i is in F_{i+1}, and all elements such that they are in F_i and they are less than some element which is not in F_i, should be in F_{i+1}:

$$\alpha(F_i, y) =_{df} (\neg F_i(y) \wedge \forall z(\neg F_i(z) \rightarrow z \leq y)) \vee (F_i(y) \wedge \exists z(y \leq z \wedge \neg F_i(z)))$$

Now it is easy to check that $\mathrm{PFP}_{X,y}\alpha(X, y)$ generates the sequence described above.

Using the formula defined above, we can translate any expression of monadic second order logic into an $FO^< + PFP$ expression equivalent to it on ordered structures. Let $\exists X\varphi(X)$ be a monadic second order formula. Suppose we know how to translate $\varphi(X)$ into a $FO^{<} + PFP$ formula $\varphi'(X)$. Consider $\mathrm{PFP}_{X,y}\theta(X, y)$, where

$$\theta(X, y) = (\varphi'(X) \wedge X(y)) \vee (\neg\varphi'(X) \wedge \alpha(X, y)).$$

The operator starts with the empty X and checks whether $\varphi'(X)$ is true. If it is, then $F_\infty = \emptyset$. If not, the next substructure is checked. Eventually either a nonempty substructure X satisfying $\varphi(X)$ is found, or the fixed point is empty. The translation of $\exists X\varphi(X)$ is therefore $\varphi'(\emptyset) \vee \exists x(\mathrm{PFP}_{X,y}\theta(X, y))(x)$. □

Theorem 30. *Let L be $FO^{<} + PFP$. Then there is a linear time algorithm A, such that for every formula φ of L, (i) $A(\varphi)$ is EPF and (ii) $A(\varphi)$ is equivalent to φ iff φ is EPF.*

Proof. From the theorem above and the existence of the algorithm for monadic second order logic. □

It is impossible to check all substructures using the inflationary fixed point operator,[6] where the sequence generated by the operator is always increasing, i.e. $F_i \subseteq F_{i+1}$. If we restrict attention to the structures in the language of $<$, S and $=$ satisfying χ_1, χ_2, χ_{31} and χ_4, then the only legitimate substructures are those given by initial segments. We denote the first order part of such languages by $FO^{<,S}$. In this case, monadic second order quantification over substructures can be defined in a weaker fixed point logic, namely in $FO^{<,S} + IFP$. We leave this as an exercise to the reader.

Theorem 31. *Let L be $FO^{<,S} + IFP$. Then there is a linear time algorithm A, such that for every formula φ of L, (i) $A(\varphi)$ is EPF and (ii) $A(\varphi)$ is equivalent to φ iff φ is EPF.*

Proof. Analogously to Theorem 30. □

As before, the theorem also holds for extensions of $FO^{<,S} + IFP$.

It is known that Lyndon's theorem fails on finite structures; there is a first order sentence monotone in a given predicate P which is not equivalent to any formula positive in P ([1]). A question arises, is there any alternative characterization of monotonicity on finite structures?

We do not know the answer in the context of first-order logic; here we give a characterization in monadic second order logic and partial fixed point logic on ordered structures using the same trick as above.

Theorem 32. *Let L be monadic second order logic. There is a linear time algorithm A such that for every formula φ of L, (i) $A(\varphi)$ is positive in a predicate P and (ii) $A(\varphi)$ is equivalent to φ iff φ is monotone in P.*

---

[6] For a definition of inflationary fixed point operator, see [9].

Proof. Define $A(\varphi)$ to be $\exists X \subseteq P(X \neq \emptyset \wedge \varphi[P/X])$, where X is a new variable and $\varphi[P/X]$ denotes the result of replacing all occurrences of P in φ by X. The rest is as in Theorem 28. □

Theorem 33. *Let L be any extension of $FO^< + PFP$. Then there is a linear time algorithm A, such that for every formula φ of L, (i) $A(\varphi)$ is positive in a predicate P and (ii) $A(\varphi)$ is equivalent to φ iff φ is monotone in P.*

Proof. Follows from the theorem above and the fact that monadic second order quantification is definable in $FO^< + PFP$. □

6 Some Open Questions

Many open problems are implicit in the text above. Let us formulate some of them explicitly.

Problem 1. Is it true that, for every first order sentence φ, the following are equivalent:

1. φ is preserved by substructures and direct products on finite structures.
2. φ is equivalent, on finite structures, to a universal Horn formula?

Problem 2. Is it true that the following are equivalent:

1. φ is preserved by homomorphisms on finite structures.
2. φ is equivalent, on finite structures, to a positive existential first order sentence? [7]

Remark. The positive solution for this problem announced by Gurevich and Shelah in [10], has collapsed. Among the survivors is their construction, also announced in [10], of a series of formulas φ_n preserved by homomorphisms, such that the size of the smallest model of φ_n is a tower $2^{2^{2^{\cdot^{\cdot}}}}$ of n twos.

Problem 3. Does there exist a recursive class F of first order formulas such that

1. Every F-sentence has the property $\mathcal{P}$ on finite structures.
2. Every first order sentence having $\mathcal{P}$ is equivalent, on finite structures, to some F-sentence,

where $\mathcal{P}$ is one of: preserved by extensions, preserved by products, monotone in a given predicate.

Problem 4. Give semantical characterization, on finite structures, of the following classes of first order sentences: Horn sentences; sentences positive in a given predicate.

Acknowledgements. We are grateful to Johan van Benthem, Kees Doets and tho anonymous referee for their comments.

---

[7] This question is due to Phokion Kolaitis.

References

1. M. Ajtai and Y. Gurevich. Monotone versus positive. *Journal of ACM*, 34:1004 – 1015, 1987.
2. M. Ajtai and Y. Gurevich. DATALOG vs. first-order logic. In *Proceedings 30th IEEE Symposium on Foundations of Computer Science*, pages 142 – 146, 1989.
3. H. Andréka, J. van Benthem, and I. Németi. Submodel preservation theorems in finite variable fragments. In M. de Rijke A. Ponse and Y. Venema, editors, *Modal Logic and Process Algebra*. Cambridge University Press, 1994.
4. J. Barwise and J. van Benthem. Interpolation, preservation, and pebble games. Technical Report ML-96-12, ILLC, University of Amsterdam, 1996.
5. C. C. Chang and H. J. Keisler. *Model theory*. North Holland, 1973.
6. H. - D. Ebbinghaus and J. Flum. *Finite model theory*. Springer, 1995.
7. J. Flum. Private communication.
8. Y. Gurevich. Toward logic tailored for computational complexity. In M. M. Richter et al., editor, *Computation and Proof Theory*. Springer lecture Notes in Mathematics 1104, 1984.
9. Y. Gurevich. Logic and the challenge of computer science. In E. Börger, editor, *Current Trends in Theoretical Computer Science*, pages 1 – 57. Computer Science Press, 1988.
10. Y. Gurevich. On finite model theory. In S. R. Buss and P. J. Scott, editor, *Feasible Mathematics*, pages 211 – 219. Birkhauser, 1990.
11. H. Rogers. *Theory of recursive functions and effective computability*. MIT Press, 1992.
12. E. Rosen and S. Weinstein. Preservation theorems in finite model theory. In *Logic and Computational Complexity*, volume 960 of *Lecture Notes in Computer Science*, pages 480–502, 1995.
13. W. Tait. A counterexample to a conjecture of Scott and Suppes. *Journal of Symbolic Logic*, 24:15 – 16, 1959.

Expressive Power of Unary Counters

Michael Benedikt[1] H. Jerome Keisler[2]

[1] Bell Laboratories, 1000 East Warrenville Rd., Naperville, IL 60566, USA,
Email: benedikt@bell-labs.com
[2] University of Wisconsin, Madison Wisconsin 53706,
Email: keisler@math.wisc.edu

Abstract. We compare the expressive power on finite models of two extensions of first order logic L with equality. $L(Ct)$ is formed by adding an operator $count\{x : \varphi\}$, which builds a term of sort $\mathbf{N}$ that counts the number of elements of the finite model satisfying a formula φ. Our main result shows that the stronger operator $count\{t(x) : \varphi\}$, where $t(x)$ is a term of sort $\mathbf{N}$, cannot be expressed in $L(Ct)$. That is, being able to count elements does not allow one to count terms.
This paper also continues our interest in new proof techniques in database theory. The proof of the unary counter combines a number of model-theoretic techniques that give powerful tools for expressivity bounds: in particular, we discuss here the use of indiscernibles, the Paris-Harrington form of Ramsey's theorem, and nonstandard models of arithmetic.

1 Introduction

Most database query languages are based on some version of first-order logic. However, practical query languages such as SQL generally supplement their pure first-order component with certain primitives, among them the ability to count over the database. An active line of research in database theory has been to model the impact of counting on a database language by studying extensions of first-order logic by *counting quantifiers* [17] [11]. The work of Väänänen and Kolaitis [21] is similar in spirit. They study the impact of adding certain kinds of generalized quantifiers. A particular class of such quantifiers, the generalized unary quantifiers, basically permit the ability to test whether the number of elements satisfying a property is in a certain set of integers. In each case, the goal is to characterize the expressive power of various counting languages, and to identify those with and without good analytical and computational properties.

In [17] logics with n-ary counters $FO + C^n$ are introduced. A method of proving upper bounds on the complexity of these languages is introduced, relying on an Ehrenfeucht-Fraïssé game construction for counting (see also [19]). This technique is used to prove a hierarchy theorem for this sequence of logics, in the case of unnested counters. The games are also exploited in [10] and [11] to get complexity bounds for languages allowing counting quantifiers of the form $\exists i$.

As opposed to the works cited above, we will consider languages with counting constructs that can interact with arbitrary expressions over the integers.

We shall consider three extensions of first order logic formed by adding term-building operators which count the number of elements satisfying a formula. The smallest of these is a first-order logic with unary counters, like the language $FO + C^1$ considered in [17], but with arbitrary integer predicates applicable to the counters. The expressive power of sentences of this language corresponds, in the terminology of [21], to first-order logic supplemented by every ***generalized unary quantifier***. As shown in [21], this language is extremely large. However, we will show some interesting limits on its expressive power by displaying two counting languages with more expressive power. In particular, we will show that it is impossible to count the number of equivalence classes of a binary relation in the language of unary counters, and it is impossible to count the number of connected components of a graph. Our results also serve to show that the ability to count the number of elements satisfying a property does not suffice to count the number of terms.

In the process, we will introduce modifications to the Ehrenfeucht-Fraïssé argument that we think are interesting in their own right. We will make use of two model-theoretic techniques: nonstandard universes and indiscernibles. Indiscernibles and nonstandard models have shown up either implicitly or explicitly in several recent works in database theory [30] [4]. These techniques were recently used to settle a number of other problems concerning the expressive power of query languages [3], and they can be used to simplify the bookkeeping involved in many Ehrenfeucht-Fraïssé arguments. We believe these techniques can be particularly helpful for proving expressivity bounds for languages involving aggregates.

As mentioned before, one of the principal reasons for studying the expressive power of query languages is to give insight into the design of languages with desirable properties. At the end of this paper we will apply our main theorem to show that a particular desirable property—the weakest precondition property—fails for a natural database language with unary counters.

Organization: Section 2 gives the definition of the languages we will deal with in this paper. Section 3 describes the nonstandard framework we use to analyze expressivity of queries, and gives some introductory examples of its usefulness. Section 4 outlines the proof of the main result. Section 5 gives a version of this result for a language similar to the tuple relational calculus with range-restriction, and gives an application of these results to the weakest precondition problem for this language. We also discuss the connection of this paper with [21], whose results are similar in spirit but technically incomparable.

2 Preliminaries

We give the notation for first-order logic and several counting languages.

Let L be a first-order language with equality and finitely many relation, function and constant symbols. Let $\mathbb{U}$ be an infinite set. When we talk about *finite L-structures*, we will mean structures whose domain is a finite subset of $\mathbb{U}$. Let Λ be a countable set of relations and functions on the set $\mathbf{N}$ of natural numbers which contains at least equality and a constant for each $n \in \mathbf{N}$, and remains fixed throughout our discussion.

Our first extension of L, denoted by $L(Ct)$, adds to L the term building operator

$$count\{x : \varphi(x, \ldots)\}$$

which bounds the variable x, and has a symbol for every element of Λ. The models for $L(Ct)$ are finite structures $\mathcal{A} = \langle A, \ldots \rangle$ with vocabulary L, and the count operator is interpreted in $\mathcal{A}$ as the cardinality of the set of all elements $x \in A$ which satisfy the formula φ in $\mathcal{A}$. Similarly, the term $\tau(\mathbf{y}) = count\{x : \varphi(x, \mathbf{y})\}$ defines a function that for each $\mathbf{y}$ returns the cardinality of the set of all elements x in $\mathcal{A}$ satisfying $\varphi(x, \mathbf{y})$.

More precisely, the language $L(Ct)$ has terms of the two sorts U and $\mathbf{N}$. The terms of sort U are the same as for first order logic; in particular, the variables are of sort U. $L(Ct)$ has the usual first order rules for building formulas, plus

- For each formula φ and variable x, $count\{x : \varphi\}$ is a term of sort $\mathbf{N}$.
- If $t_1, \ldots, t_n$ are terms of sort $\mathbf{N}$, then $f(t_1, \ldots, t_n)$ is a term of sort $\mathbf{N}$, and $r(t_1, \ldots, t_n)$ is a formula for each n-ary function $f \in \Lambda$ (relation $r \in \Lambda$).

Our second extension of L, denoted by $L(Tm)$, adds to $L(Ct)$ the more general term-building operator

$$count\{t(x, \ldots) : \varphi(x, \ldots)\}$$

where φ is a formula and $t(x, \ldots)$ is a term of sort $\mathbf{N}$. In a model $\mathcal{A}$, it counts the number of distinct values of $t(x, \ldots)$ such that x satisfies φ in $\mathcal{A}$. Thus, we always have $count\{t(x, \ldots) : \varphi(x, \ldots)\} \leq count\{x : \varphi(x, \ldots)\}$.

Our third extension, $L(Ct, \mathbf{N})$, adds to $L(Ct)$ variables of sort $\mathbf{N}$ and quantifiers over variables of sort $\mathbf{N}$. The language $L(Ct, \mathbf{N})$ was considered in [17]. The main object of study in that paper was a language like $L(Ct, \mathbf{N})$ which had counts over n-tuples of variables of sort U, but did not allow nesting of counts.

We note that $L(Ct)$ is already quite big. It includes such quantifiers as the Hartog quantifier and Rescher quantifier, and can express arbitrarily computationally complex sentences, as well as noncomputable ones. In expressive power for sentences (as opposed to terms) $L(Ct)$ corresponds exactly to the language of all generalized unary quantifiers. For every S that is in the powerset of N^n for some n,this language allows a new formula constructor Q_S which acts on sequences of n-fomulae $\phi_1, \ldots, \phi_n$. $Q_S(\phi_1(x), \ldots, \phi_n(x))$ binds x and holds if and only if $c_1, \ldots, c_n \in S$, where c_i is the cardinality of $\{x : \phi_i(x)\}$. Generalized quantifiers are of enormous interest both in infinitary model theory (dating back to [24]) and in finite model theory (see the overview in [21]). Included in this line of work are a number of results limiting the expressive power of particular generalized quantifiers or of finite collections of quantifiers [18]. In contrast, [21] shows a limitation on the expressive power of a first-order logic supplemented with *all* simple generalized unary quantifiers. Simple generalized unary quantifiers are a restricted class of unary quantifiers corresponding to restricting $L(Ct)$ to allowing only unary predicates on the integer sort. In [21] it is shown that the language of simple unary counters is a proper subset of the language $L(Tm)$ above.

In this paper we shall prove an extension of this last result, by showing that $L(Tm)$ and $L(Ct, \mathbf{N})$ are proper extensions of $L(Ct)$. That is, even with all arithmetic extensions and arithmetic relations of arbitrary arity available, counting elements does not allow one to count terms.

We remark that quantifiers of sort U can be eliminated from each of the languages $L(Ct)$, $L(Tm)$, and $L(Ct, \mathbf{N})$. An existential quantifier can be eliminated by replacing a formula $\exists x \varphi(x, \ldots)$ by $\neg count\{x : \varphi(x, \ldots)\} = 0$.

In a vocabulary L with function symbols, one might also consider the term builder $count\{s(x, \ldots) : \varphi(x, \ldots)\}$, where the term s is of sort U. This term builder is already definable in $L(Ct)$, because the equation

$$count\{s(x, \ldots) : \varphi(x, \ldots)\} = count\{z : \exists x(z = s(x, \ldots) \wedge \varphi(x, \ldots))\}$$

holds in all finite models.

3 Model-theoretic techniques and expressive bounds

We discuss here some model-theoretic techniques that are helpful for giving expressive bounds, and which will be used in the main result of Section 4.

3.1 Nonstandard Models

The work in [3] made use of nonstandard models and indiscernibles as techniques for analyzing expressivity bounds. Here we discuss them in more detail.

The naive approach to showing that a property Q is not expressible in some language $\mathcal{L}$ is to get two models that agree on all sentences of $\mathcal{L}$, but disagree on Q. The problem immediately encountered in applying this technique in finite-model theory is the following: Any two finite models which satisfy the same sentences of a first order language L are isomorphic, and thus satisfy the same sentences of any reasonable logic, including $L(Ct)$, $L(Tm)$, and $L(Ct, \mathbf{N})$. The standard technique for circumventing this problem is via Ehrenfeucht-Fraïssé games ([8], [7]). One decomposes the sentences of the logic into countably many fragments $\mathcal{L}_n$, and then constructs for each n two finite models M_n and N_n agreeing on fragment $\mathcal{L}_n$ but disagreeing on Q.

Here, we give an alternative to this construction. Inexpressibility bounds are obtained by finding two hyperfinite (meaning, informally for now, "infinitely large finite") models M and N agreeing on all queries in $\mathcal{L}$, but disagreeing on Q. The first virtue of this technique is as a way of abstracting away from the bookkeeping involved in Ehrenfeucht-Fraïssé constructions. For example, if one is interested in showing the inexpressibility of connectivity within pure first-order logic, one need only look at the two hyperfinite graphs G_1 and G_2, where G_1 is a single hyperfinite cycle, while G_2 is the union of two hyperfinite cycles. A single game argument shows these two to be elementarily equivalent in first-order logic, but only one is connected, hence connectivity is not first-order definable.

The above example may appear to make the technique of nonstandard models useful more as a convenience than as an essential tool. However, the technique becomes particularly useful when dealing with expressivity results for higher-order logics. We will defer a more detailed discussion of the use of nonstandard models in higher-order languages to the full paper. The use of nonstandard universes (as opposed to elementary extensions) allows one to work with hyperfinite extensions of traditional aggregate constructs such as Counts, Sums, and set formers. For example, consider the query Q over complex objects (see [5],[29],) asking whether a given set of sets A contains two sets of differing parity. One can show this query to be inexpressible in the nested relational algebra [5], a higher-order analog of the relational algebra, by considering a natural counterexample: two hyperfinite structures S_1 and S_2, the first consisting of two hyperfinite sets of differing parity, the second consisting of two hyperfinite sets of the same parity. This argument can be formalized in a straightforward way using the definitions below, and can be easily generalized to higher-order queries. Arguments such as this are difficult to prove using other techniques (see the discussion in [23]) and can't be formalized using the 'flat' ultraproduct or elementary extension constructions. On the other hand, direct constructions using ultraproducts, when available, are often more concrete and more accessible in terms of exposition than the use of a nonstandard universe (compare, for example [3] and [28]).

In the previous paragraph, we spoke informally of hyperfinite structures. We now give some formal definitions, following the exposition in [3]:

For any set S, the **superstructure** $V(S)$ over S is defined as $V(S) = \bigcup_{n<\omega} V_n(S)$ where $V_1(S) = S$, and $V_{n+1}(S) = V_n(S) \cup \{X \mid X \subset V_n(S)\}$.

We will work with the structure $\langle V(S), \in \rangle$ considered as a structure for the first-order language for the epsilon relation. A **bounded-quantifier formula** in this language is a formula built up from atomic formulas by the logical connectives and the quantifications: $\forall X \in Y$, $\exists X \in Y$, where X and Y are variables.

A **nonstandard universe** consists of a pair of superstructures $V(S)$ and $V(Y)$ and a mapping $* : V(S) \to V(Y)$ which is the identity when restricted to S (i.e. ${}^*x = x$ for each x in S) and which satisfies

1. $Y = {}^*S$.
2. *(Transfer Principle)* For any bounded quantifier formula $\phi(v_1, \ldots, v_n)$ and any list $a_1, \ldots, a_n$ of elements from $V(S)$, $\phi(a_1, \ldots, a_n)$ is true in $V(S)$ if and only if $\phi({}^*a_1, \ldots, {}^*a_n)$ is true in $V(Y)$.

(For technical reasons one also assumes that $\emptyset \notin S$ and $x \cap V(S) = \emptyset$ for all $x \in S$, and similarly for Y). An element of $V(Y)$ is **standard** if it is in the image of the $*$-map. An element of $V(Y)$ is **internal** if it is an element of some standard set. Elements of $V(Y)$ that are not internal are called **external**. An internal map is a map whose graph is an internal set.

We will assume that our universe also satisfies the following

3. *(Countable Saturation Principle)* For every standard set A, every countable collection $\Sigma(x, \mathbf{v})$ of bounded-quantifier formulas, and every vector $\mathbf{c}$

of internal sets, if each finite subset of $\Sigma(x, \mathbf{c}/\mathbf{v})$ is satisfied in $V(Y)$ by an element of A, then $\Sigma(x, \mathbf{c}/\mathbf{v})$ is satisfied in $V(Y)$ by an element of A.

We often omit the $*$ when convenient: for example, if $<$ is an ordering on a set A, and $x_1, x_2 \in {}^*A$, then we will write $x_1 < x_2$ rather than $x_1 {}^*\!< x_2$. If Q is a query on schema SC, and M is a *-database (an element of the set $^*DB$, where DB is the set of all SC-databases), then we will refer to $Q(M)$ rather than $^*Q(M)$.

For our proofs we take the base set S of $V(S)$ to be the disjoint union of the domain $\mathbb{U}$ from which our finite structures are taken and $\mathbb{N}$.

By a ***-finite set** we mean any set in the *-image of the collection of finite subsets of some standard set. Equivalently, an internal set B is *-finite if there is an internal bijection of B onto an initial segment of $^*\mathbb{N}$. In particular, we can talk about *-finite structures, which will have their underlying domains being contained in the *-image of the finite powerset of $\mathbb{U}$. By the transfer principle, such sets B have a well-defined cardinality, which is a (possibly nonstandard) positive integer, as well as a well-defined parity, sum, etc. By a **hyperfinite set**, we mean a *-finite set whose cardinality is not a standard integer. We similarly talk about hyperfinite structures, orderings etc. to mean those whose cardinality is a nonstandard integer.

We can now talk formally about hyperfinite structures. We cannot assume in general that the semantic function for a given logic will agree with the semantics obtained by considering the structure "externally". However, for first-order logic, we have the following result:

Proposition 1. *[3] Let L be a language for which each symbol is internal, and let M be an L-structure such that the domain of M is internal and the interpretation of each symbol in L is internal. Let $\phi(\mathbf{x})$ be a formula of L that has standard finite cardinality (i.e. number of symbols). Then the internal satisfaction predicate $^*\models$ agrees with the external satisfaction predicate $\models$ on ϕ. That is, if $\mathbf{c}$ is a finite sequence of parameters from M, then $M \,^*\!\models \phi(\mathbf{c})$ iff $M \models \phi(\mathbf{c})$.*

Given the above proposition we will not distinguish the two kinds of satisfaction predicates when we are dealing with finitary first-order ϕ's.

The nonstandard technique is now based on the following simple proposition:

Proposition 2. *[3] The following are equivalent for any boolean query Q on finite structures:*

- *There are two hyperfinite structures that agree on every query in the first-order language L but disagree on $^*Q$.*
- *Q is not expressible in L.*

It is now easy to formalize the inexpressibility arguments mentioned in the beginning of the section. We can form, for example, a graph consisting of two hyperfinite chains (such a graph exists by transfer plus saturation), and show, by a single Ehrenfeucht-Fraïssé game, that this graph, considered as an infinite structure, is elementarily equivalent to any graph consisting of a single hyperfinite cycle. The inexpressibility result now follows from Proposition 2.

3.2 Indiscernibles

Although the use of nonstandard models eliminates the need to construct countably many counterexample models, and relieves some of the combinatorial burden in a game argument, it may still be difficult to reason about elementary equivalence in an arbitrary hyperfinite structure.

To simplify the analysis of elementary equivalence, we will often want to restrict our attention to models whose algebraic structure is "as simple as possible". Indiscernibility is a method for capturing the intuition that the domain of our structures should have no unnecessary algebraic dependencies among its elements. Indiscernible sets were introduced by Ehrenfeucht and Mostowski [9], and are implicit in many of the Ramsey-theoretic constructions used in Ehrenfeucht-Fraïssé constructions [25],[30].

We now define indiscernibles formally. Let I be any ordered set. A sequence $B = \langle B_i \rangle_{i \in I}$, whose elements come from an infinite L-structure M is **indiscernible** if for every formula $\phi(\mathbf{x})$, ϕ is satisfied in M by either every increasing (in the order on I) subsequence of B or by no increasing subsequence of B. Indiscernibles are discussed at length in [6].

Within an indiscernible set, the logical structure of the model reduces to a simple ordering. Indiscernibles were used in algorithms for eliminating constraints from constraint queries in [3] [30] and [25].

An infinite set of indiscernibles need not exist in an arbitrary infinite structure. For example, considering the structure $M = \langle N, +, < \rangle$, we easily see that there can be no indiscernible set of size bigger than 1! However, it is easy to show, using saturation, that for any infinite structure, there is an infinite set of indiscernibles in the nonstandard extension $^*M$: this makes the use of indiscernibles particularly powerful in conjunction with nonstandard methods.

In this work, we will consider the use of indiscernibles in collapsing logics with counting quantifiers to first-order logics. The construction in the next section will give an example of the use of indiscernibles to reduce every formula in the language $L(Ct)$ to a first-order formula. Since several previous expressibility bounds on aggregates make use of some sort of "Count Elimination" [22], we hope to investigate this phenomenon in more generality in forthcoming work.

4 Nonstandard Models and Unary Counters

Consider the logics $L(Ct)$, $L(Tm)$, and $L(Ct, \mathbf{N})$ defined in Section 2. We are interested in investigating the relative expressive powers of these languages, using the techniques mentioned above. In particular, we wish to show that $L(Ct)$ cannot express important properties expressible in $L(Tm)$, and $L(Ct, \mathbf{N})$. Our plan, of course, will be to find two nonstandard models which satisfy the same sentences of $L(Ct)$ but do not satisfy the same sentences of $L(Tm)$ or of $L(Ct, \mathbf{N})$.

Our first result will show that the language of unary counters $L(Ct)$ cannot count the number of equivalence classes in an equivalence relation, while $L(Ct, \mathbf{N})$ can express this. We now fix a particular first order language L^0.

Definition 3. Let the language L^0 have one unary predicate symbol S and one binary predicate symbol E. Let θ be the following sentence of $L^0(Tm)$:

$$count\{count\{y : E(x,y)\} : x = x\} = count\{z : S(z)\}$$

If E is an equivalence relation, θ says that the number of distinct sizes of equivalence classes of E is equal to the number of elements of S.

Definition 4. Let θ^+ be the following sentence of $L^0(Ct, \mathbf{N})$:

$$\forall i[\forall j(Bit(j,i) \Leftrightarrow \exists x\; count\{y : E(x,y)\} = j) \Rightarrow Setcard(i) = count\{z : S(z)\}],$$

where i, j are variables of sort $\mathbf{N}$, $Bit(j,i)$ is true exactly when the jth bit of the binary representation of i is set, and $Setcard(i)$ is the cardinality of the set coded by the binary representation of i.

Since our arithmetic permits arbitrary functions on the integers, the predicates *Bit* and *Setcard* are certainly expressible. Note that θ^+ expresses the same property as θ, that is, for every finite model $\mathcal{C}$ for L^0, $\mathcal{C}$ satisfies θ if and only if it satisfies θ^+. The sentence θ^+ does not have nested counts.

We shall prove the following theorem, which shows that neither θ nor θ^+ is expressible in $L^0(Ct)$, so that both $L(Tm)$ and $L(Ct, \mathbf{N})$ are proper extensions of $L(Ct)$.

Theorem 5. *For every sentence φ of $L^0(Ct)$ there is a finite model $\mathcal{C}$ in which θ is not equivalent to φ: i.e., the unary counter language cannot express θ.*

As mentioned above, our plan will be to find two nonstandard models which satisfy the same sentences of $L^0(Ct)$ but do not satisfy the same sentences of $L^0(Tm)$ or of $L^0(Ct, \mathbf{N})$.

We let $\mathcal{N}$ be the structure with universe set $\mathbf{N}$ and a symbol for $+$, $\times$, and every relation and function in the set Λ. For each *finite model $\mathcal{A}$ for L, each term t and formula φ of $L(Tm)$ is interpreted in the natural way, using the functions and relations on the extension $^*\mathcal{N}$. Terms of sort $\mathbf{N}$ are interpreted as functions with values in $^*\mathbf{N}$.

Our goal is to prove the following.

Theorem 6. *There exist *finite models $\mathcal{A}$ and $\mathcal{B}$ which satisfy the same sentences of $L^0(Ct)$ such that θ is false in $\mathcal{A}$ but true in $\mathcal{B}$.*

Given this theorem, the inexpressibility of θ in $L^0(Ct)$ now follows from the general results of the previous section.

The proof of this theorem will give a canonical example of the use of indiscernibles and nonstandard universes together. We will first show that a special set of indiscernibles exists, and then make use of them to prove Theorem 6.

We use the usual notation for intervals in $^*\mathbf{N}$; for example, $(J, K] = \{x \in {}^*\mathbf{N} : J < x \text{ and } x \leq K\}$. By a *finite sequence in $^*\mathbf{N}$ we mean an an element of the star-image of the finite sequences in $\mathbf{N}$. By the transfer principle, each *finite sequence in $^*\mathbf{N}$ is a function $\mathbf{d} = \langle d_1, \ldots, d_H \rangle$ from the interval $(0, H]$ into $^*\mathbf{N}$ for some hyperinteger $H \in {}^*\mathbf{N}$.

Lemma 7. *There is a strictly increasing *finite sequence* $\mathbf{d} = \langle d_1, \ldots, d_H \rangle$ *in* $^*\mathbf{N}$ *such that* $0 < d_1 < H$ *and* $\mathbf{d}$ *is indiscernible in* $^*\mathcal{N}$*; that is, any two finite increasing subsequences of* $\mathbf{d}$ *satisfy the same first order formulas in* $^*\mathcal{N}$.
Proof. By the Paris-Harrington form of Ramsey's theorem, for every $e, r \in \mathbf{N}$ there exists $m \in \mathbf{N}$ such that for every partition $P : [m]^e \to r$ there is a subset $Q \subseteq m$ which is homogeneous for P and has size $|Q| \geq \min(Q)$. Then for each finite set φ of formulas in the language of $\mathcal{N}$ there is a strictly increasing finite sequence $\langle d_1, \ldots, d_h \rangle$ in $\mathbf{N}$ such that $0 < d_1 < h$ and $\mathbf{d}$ is indiscernible for φ. Using saturation once again on $^*\mathcal{N}$, it follows that there is a strictly increasing *finite sequence $\mathbf{d} = \langle d_1, \ldots, d_H \rangle$ in $^*\mathbf{N}$ such that $0 < d_1 < H$ and $\mathbf{d}$ is indiscernible in $^*\mathcal{N}$. □

Hereafter we let $\mathbf{d} = \langle d_1, \ldots, d_H \rangle$ be as in Lemma 7, and put $d_0 = 0$.

Corollary 8. *H and d_1 are infinite, and $\frac{d_J}{d_{J-1}}$ is infinite for each $J \in (1, H]$ (i.e. for each standard integer n, $d_J > n \cdot d_{J-1}$) for each $J \in (1, H]$).*
The proofs are in the appendix.

We now define the *finite structures $\mathcal{A}$ and $\mathcal{B}$ for L^0.

Definition 9. Let $\mathcal{A} = \langle A, E, S \rangle$ where $A = (0, d_H]$, E is the equivalence relation on A with equivalence classes $(d_{J-1}, d_J], J \in (0, H]$, and $S = (0, d_1]$. Let $K = d_1$ and $\mathcal{B} = \langle B, F, S \rangle$ where $B = (0, d_K]$, $F = E \cap B \times B$, and $S = (0, d_1]$.

Lemma 10. *The sentence θ is false in $\mathcal{A}$ and true in $\mathcal{B}$.*

Lemma 11. *$\mathcal{A}$ and $\mathcal{B}$ satisfy the same sentences of $L^0(Ct)$.*

Lemma 10 is proved straightforwardly by insepection; the proof of Lemma 11 is quite involved, and can be found in the appendix.

5 Applications to weakest preconditions

We now outline a version of the previous results for a language L' whose concrete syntax is closer to existing database languages. The description we give below is based on the tuple relational calculus with range restriction [1], and on our analysis of the PRL constraint language [15] [16].

We have a signature $\{R_1, \ldots, R_n\}$, and for each $i \leq n$ a finite set of **attribute symbols** A_i. The **arity** of R_i is the cardinality of A_i. For each R_i we also fix a countably infinite set U_i. We now define our language L'.

We have variables of sort i for each $i \leq n$, and an integer sort. A formula will be built out of atomic formulas of the form:

- $x.a = y.b$ where x and y are variables of the same sort i, and a and b are symbols in A_i, or of the form
- $P(\tau_1, \ldots, \tau_n)$, where P is an integer predicate and the τ_j are terms of type integer.

Formulae are built up by the logical connectives and quantifications $\forall x \in R_i\ \phi(x)$, and $\exists x \in R_i\ \phi(x)$, where x has sort i.

Terms are built up via composition from atomic terms, which are of one of the forms:

1. $f(\mathbf{x})$, where f is a symbol for some function from tuples of integers to integers,and $\mathbf{x}$ is a tuple of variables of integer sort. The corresponding language $L'(Ct)$ adds the ability to form terms from formulas via the rule:
2. $Count\{x \in R_i : \phi(x)\}$, where x has sort i and ϕ is a formula.

The language $L'(Tm)$ further supplements this by permitting

3. $Count\{\tau(x) : x \in R_i \wedge \phi(x)\}$, where τ is a term.

A structure for L' consists of an assignment to each relational variable R_i a finite collection of elements of the set of functions $U_i^{A_i}$ (i.e. tuples). Satisfaction relative to an assignment of variables is defined exactly as in the language $L(Ct)$.

Let our signature have two relations R and S, and let R have attributes a and b, while S has attributes c and d.

Theorem 12. *The $L'(Tm)$ sentence θ' below is not expressible in $L'(Ct)$:*

$$\theta' \equiv count\{x.a : x \in R \wedge x = x\} = count\{z \in S : z = z\}.$$

The proof is found in the appendix.

5.1 Preconditions and Definable transactions

Let $\mathcal{L}$ be any of the standard logical languages (first-order logic, infinitary logic, etc.). We let $s(\mathcal{L})$ denote the sentences of $\mathcal{L}$. We will talk about a database (that is, a finite structure as in the previous section) satisfying a sentence or open formula of $\mathcal{L}$: we mean this in the usual sense. By a **transaction** on databases, we mean simply any function mapping databases to databases. In the following discussion, we will let D range over databases for a particular signature, and $\mathcal{T}$ denote the set of database transactions for this signature. $TERM(\mathcal{L})$ denotes the set of terms of $\mathcal{L}$ and for Γ a collection of terms and D a database, we let $\Gamma(D)$ denote all elements obtained from applying a vector of terms in Γ to elements in the underlying set of D. That is:

$$\Gamma(D) = \{\boldsymbol{\tau}(\mathbf{y}) : \mathbf{y} \subset D \text{ and } \boldsymbol{\tau} \subset \Gamma\}$$

We now discuss two classes of transactions on finite models associated with $\mathcal{L}$. The definitions are taken from [2]. The class $\mathcal{WPC}(\mathcal{L})$ (transactions with *weakest preconditions* with respect to $\mathcal{L}$) is defined as

$$\{T \in \mathcal{T} \mid \exists \text{ recursive } f : s(\mathcal{L}) \rightarrow s(\mathcal{L}) : \forall D\ \forall \alpha \in s(\mathcal{L}) : T(D) \models \alpha \Leftrightarrow D \models f(\alpha)\}$$

A transaction T has weakest preconditions for $\mathcal{L}$ if we can statically determine whether the database resulting from T will satisfy a constraint in $\mathcal{L}$.

The set of *L-definable* transactions is the collection

$$\mathcal{DEF}(\mathcal{L}) = \{T \mid \forall R\, \exists\, \mathcal{L} \text{ formula } \beta_R(\mathbf{x})\ \exists\, \Gamma \subset TERM(\mathcal{L}) \text{ such that}$$
$$\forall D\, \forall \mathbf{t} : T(D) \models R(\mathbf{t}) \Leftrightarrow \mathbf{t} \subset \Gamma(D) \wedge D \models \beta_R(\mathbf{t})\}$$

The class of definable transactions are those that can be expressed using a finite set of $\mathcal{L}$ terms and $\mathcal{L}$ formulae.

In [2] it is observed that for first-order languages, we have containment, $\mathcal{DEF}(\mathcal{L}) \subset \mathcal{WPC}(\mathcal{L})$. That is, definable transactions all admit weakest preconditions. This is a desirable closure property for a query language to possess. [2] investigates weakest preconditions over a number of logics, and over a number of transaction languages. In [16],[2], and [15] applications of weakest precondition closure to database integrity maintenance are discussed.

Given the importance of weakest precondition closure for database query languages, it is important to see if natural extensions of the relational calculus, such as $L'(Ct)$, posess this closure property. It follows from the main result of this paper that containment does *not* hold for the language $L'(Ct)$.

Corollary 13. *There are $L'(Ct)$-definable transactions that do not posess weakest preconditions over $L'(Ct)$.*

6 Conclusions and future work

Languages with aggregate constructs are not nearly so well understood as the relational calculus. In particular, issues of optimization, complexity, safety, and expressiveness remain open for many models of aggregation. We are interested in developing a usable set of rewrite-rules for simplifying languages such as $L(Ct)$ and $L(Tm)$, and getting semantic characterizations of the definable transactions that are available (along the lines of Gaifman's locality theorem [12] or the bounded degree property of [22]).

We are interested in studying the relationship of the language $L(Tm)$ to various other counting languages (those discussed, for example, in [17]). In particular, it would be helpful to know whether languages with binary counters can express all sentences of $L(Tm)$, and similarly for n-ary counters.

Techniques for proving expressiveness bounds on query languages are hard to come by. We've presented here one technique for analyzing expressivity of query languages, based on the use of indiscernibles and nonstandard models, that we believe can be useful outside of the context of aggregates. In particular, we hope to investigate the interaction of these techniques with the use of Ehrenfeucht-Fraïssé games for logics with counting [17] [13].

Questions in this work were motivated by considering the closure under weakest preconditions of various extensions of first-order logic. The closure of a specification language under weakest-preconditions of definable transactions is helpful for integrity constraint maintenance [2],[26],[27]. It is therefore important to find logics that include aggregate operators that have this closure property, are of manageable complexity, and allow for optimization and analysis. The results of this paper show that $L(Ct)$ is not closed under weakest-preconditions, and we suspect that the same is true for $L(Tm)$. We would like to discover natural weakest-precondition closed logics containing these languages.

In [21] it was shown that the sentence saying that the number of classes of an equivalence relation is even, cannot be expressed in first-order logic supplemented with every simple generalized unary quantifier. The same proof given in this paper suffices to show that this sentence cannot be expressed in the language $L(Ct)$, and hence cannot be expressed even with arbitrary generalized unary quantifiers. However, [21] also shows the inexpressibility of the sentence above for the finite-variable infinitary logic L^{∞}_{ω} supplemented by all simple unary quantifiers. The sentences here and the one mentioned above in [21] *are* expressible in infinitary logic over $L(Ct)$, so it also remains to separate $L(Tm)$ from infinitary logic over $L(Ct)$.

Acknowledgements: The authors wish to thank Jouko Väänänen for several helpful discussions, Leonid Libkin for numerous improvements to the text, and Bell Laboratories for sponsoring the visit by the second author during which this research was done.

References

1. S. Abiteboul, R. Hull, V. Vianu. *Foundations of Databases.* Addison-Wesley, 1995.
2. M. Benedikt, T. Griffin, and L. Libkin Verifiable properties of database transactions. In *Proceedings of 15th ACM Symposium on Principles of Database Systems*, pages 117–128, Montreal Canada, June 1996.
3. M. Benedikt, G. Dong, L. Libkin, L. Wong. Relational expressive power of constraint query languages. In *Proceedings of 15th ACM Symposium on Principles of Database Systems*, pages 5–17, Montreal Canada, June 1996.
4. M. Benedikt and L. Libkin. On the structure of queries in constraint query languages. In *Proceedings of 11th IEEE Symposium on Logic in Computer Science, New Brunswick, New Jersey* 1996.
5. P. Buneman, S. Naqvi, V. Tannen, L. Wong. Principles of programming with complex objects and collection types. *Theoretical Computer Science*, 149(1):3–48, 1995.
6. C. C. Chang and H. Jerome Keisler. *Model Theory.* North-Holland Elsevier 1990.
7. H.-D. Ebbinghaus and J. Flum. *Finite Model Theory.* Springer Verlag, 1995.
8. A. Ehrenfeucht. An application of games to the completeness problem for formalized theories. *Fundamentae Mathematicae*, 49:129–141, 1961.
9. A. Ehrenfeucht and A. Mostowski. Models of axiomatic theories admitting automorphisms. *Fundamentae Mathematicae*, 43:50–68, 1956.
10. K. Etessami. Counting quantifiers, successor relations, and logarithmic space. In *Proceedings of 10th IEEE Conference on Structure in Complexity Theory*, May 1995, pages 2–11.
11. K. Etassami and N. Immerman. Tree canonization and transitive closure. Tenth Annual IEEE Symposium on Logic in Computer Science. 1995.
12. H. Gaifman. On local and nonlocal properties. In J. Stern, editor, *Logic Colloquium '81*, pages 105–135. North -Holland, 1982.
13. E. Gradel and M. Otto Inductive definability with counting on finite structures. In E. Borger (ed.), *Computer Science Logic*, LNCS 702 Springer (1993), 231-247.

14. Erich Gradel and Yuri Gurevich. Metafinite model theory. In *Logic and Computational Complexity, International Workshop LCC'94* Indianapolis, IN,313-366.
15. T. Griffin and H. Trickey. Integrity Maintenance in a Telecommunications Switch. In IEEE Data Engineering Bulletin V. 17, No. 2., June 1994.
16. T. Griffin, H. Trickey, and C. Tuckey. Update constraints for relational databases. Technical Memorandum AT&T Bell Laboratories, 1992
17. S. Grumbach and C. Tollu. On the expressive power of counting. Fourth International Conference on Database Theory. 1992.
18. L. Hella Logical hierarchies in PTIME. Proc. 7th IEEE Symp. on Logic in Computer Science Pages 360-368,1992
19. N. Immerman, E.S. Lander. Describing graphs: a first-order approach to graph canonization. In *Complexity Theory Retrospective*, pp. 59-81. Springer, 1990.
20. C. Karp. Finite quantifier equivalence. In *The Theory of Models*, edited by J. Addison, L. Henkin, and A. Tarski, North-Holland 1965, 407-412.
21. P. Kolaitis and J.Väänänen. Generalized Quantifiers and Pebble Games on Finite Structures. Annals of Pure and Applied Logic 74 (1995) 23-75.
22. L. Libkin and L. Wong. New techniques for studying set languages, bag languages and aggregate functions. In *Proceedings of the 13th Conference on Principles of Database Systems,* Minneapolis MN, May 1994, pages 155–166.
23. L. Libkin and L. Wong. On representation and querying incomplete information in databases with bags. Information Processing Letters 56 (1995), 209-214
24. P. Lindström. First order predicate logic with generalized quantifiers. Theoria, 32:186-195,1966
25. J. Paredaens, J. Van den Bussche, and D. Van Gucht. First-order queries on finite structures over the reals. In *Proceedings of 10th IEEE Symposium on Logic in Computer Science, San Diego, California*, pages 79–87, 1995.
26. X. Qian. An effective method for integrity constraint simplification. In *Fourth International Conference on Data Engineering*, 1988.
27. X. Qian. *The Deductive Synthesis of Database Transactions.* PhD thesis, Stanford University, 1989.
28. O.Belagradek, A. Stolboushkin, M. Tsaitlin. On order-generic queries. Manuscript. To appear.
29. V. Tannen, Tutorial: Languages for collection types, *in* "Proceedings of 13th Symposium on Principles of Database Systems," Minneapolis, May 1994.
30. J. Van Den Bussche and M. Otto. First-order queries on databases embedded in an infinite. structure. Information Processing Letters, to appear.

A Proofs

We will prove the remaining lemmas used to prove Theorem 6. Hereafter we let $\mathbf{d} = \langle d_1, \ldots, d_H \rangle$ be as in Lemma 7, and put $d_0 = 0$.

Proof of corollary 8. By indiscernibility, d_1 is infinite, and thus H is infinite. Since $\mathbf{d}$ is strictly increasing and $d_1 < H$, we have $2d_1 \leq d_H$. By indiscernibility, $2d_{J-1} \leq d_J$ for each $J \subset (1, H]$, and by indiscernibility again, $2^n d_{J-1} \leq d_J$ for each J and each finite n. □

Proof of lemma 11. We have $count\{x : S(x)\} = d_1$ in both $\mathcal{A}$ and $\mathcal{B}$. In $\mathcal{A}$, $count\{count\{y : E(x,y)\} : x = x\} = H > d_1$, so θ fails in $\mathcal{A}$. In $\mathcal{B}$, $count\{count\{y : E(x,y)\} : x = x\} = d_1$, so θ holds in $\mathcal{B}$. □

It remains to show that $\mathcal{A}$ and $\mathcal{B}$ satisfy the same sentences of $L^0(Ct)$. In order to do this we introduce an auxiliary first order vocabulary L^1 and corresponding models $\mathcal{A}^1$ and $\mathcal{B}^1$.

Definition 14. Let $f : A \to (0, H]$ be the function such that $f(x) = J$ whenever $J \in (0, H]$ and $x \in (d_{J-1}, d_J]$. Let L^1 be a first order vocabulary with countably many unary relations $\min_n(x), \max_n(x)$ and countably many binary relations $x \preceq_n y$ for $n \in \mathbf{N}$. Let $\mathcal{A}^1$ be the model for L^1 with universe A such that

- $\mathcal{A}^1 \models \min_n(x)$ iff $f(x) \leq n$,
- $\mathcal{A}^1 \models \max_n(x)$ iff $H - n \leq f(x)$,
- $\mathcal{A}^1 \models x \preceq_n y$ iff $f(x) + n \leq f(y)$.

Let $\mathcal{B}^1$ be the model for L^1 with universe B such that

- $\mathcal{B}^1 \models \min_n(x)$ iff $f(x) \leq n$,
- $\mathcal{B}^1 \models \max_n(x)$ iff $K - n \leq f(x)$,
- $\mathcal{B}^1 \models x \preceq_n y$ iff $f(x) + n \leq f(y)$.

Lemma 15. *For every formula $\varphi(\mathbf{x})$ of $L^0(Ct)$ there is a formula $\varphi^1(\mathbf{x})$ of $L^1(Ct)$ such that for all $\mathbf{a}$ in A, $\mathcal{A} \models \varphi[\mathbf{a}]$ if and only if $\mathcal{A}^1 \models \varphi^1[\mathbf{a}]$, and for all $\mathbf{b}$ in B, $\mathcal{B} \models \varphi[\mathbf{b}]$ if and only if $\mathcal{B}^1 \models \varphi^1[\mathbf{b}]$.*
Proof: Put $(x = y)^1 = (x = y)$, $E(x,y)^1 = (x \preceq_0 y \wedge y \preceq_0 x)$, and $S(x)^1 = min_1(x)$. Then use the formation rules in the obvious way to define φ^1 for arbitrary φ. □

Definition 16. Let $\mathbf{a} \equiv_0 \mathbf{b}$ mean that $|\mathbf{a}| = |\mathbf{b}|$ and $(\mathcal{A}^1, \mathbf{a})$ and $(\mathcal{B}^1, \mathbf{b})$ satisfy the same atomic formulas of L^1. We use a similar notation for pairs of tuples which are both in A or in B.

Lemma 17. *$\mathcal{A}^1$ and $\mathcal{B}^1$ satisfy the same sentences of L^1. In fact, $\mathbf{a} \equiv_0 \mathbf{b}$ if and only if $(\mathcal{A}^1, \mathbf{a})$ and $(\mathcal{B}^1, \mathbf{b})$ satisfy the same formulas of L^1.*
Proof: Using the fact that we have $\preceq_n$ as an atomic relation in the language for each n, and the fact that $\mathcal{A}^1$ and $\mathcal{B}^1$ are countably saturated, we derive that the relation $\mathbf{a} \equiv_0 \mathbf{b}$ has the back and forth property. It follows that whenever $\mathbf{a} \equiv_0 \mathbf{b}$, player $\exists$ has a winning strategy in the Ehrenfeucht-Fraïssé game with ω moves between $(\mathcal{A}^1, \mathbf{a})$ and $(\mathcal{B}^1, \mathbf{b})$. In addition, we have that the empty sequences from each model are $\equiv_0$-equivalent, since there are no atomic sentences in L^1. Thus $(\mathcal{A}^1, \mathbf{a})$ and $(\mathcal{B}^1, \mathbf{b})$ are elementarily equivalent by Karp's theorem in [20]. □

Lemma 18. *Let $\Gamma(\mathbf{x})$ be a set of quantifier-free formulas of L^1 maximal consistent with the theory of $\mathcal{A}^1$ and let $\psi(\mathbf{x}, y)$ be a quantifier-free formula in L^1. Let $s = domain(\mathbf{x}) \cup \{\min, \max\}$. There exist $\alpha_1, \ldots, \alpha_m \in s, \beta_1, \ldots, \beta_m \in \{-1, 1\}, \gamma_1, \ldots, \gamma_m \in \mathbf{Z}, \delta \in \mathbf{Z}$ such that whenever $(\mathcal{A}^1, \mathbf{a})$ satisfies $\Gamma(\mathbf{x})$,*

$$count\{y : \psi(\mathbf{a}, y)\} = \delta + \sum_{j=1}^{m} \beta_j d_{f(a_{\alpha_j}) + \gamma_j},$$

and whenever $(\mathcal{B}^1, \mathbf{b})$ *satisfies* $\Gamma(\mathbf{x})$,

$$count\{y : \psi(\mathbf{b}, y)\} = \delta + \sum_{j=1}^{m} \beta_j d_{f(b_{\alpha_j})+\gamma_j},$$

with the convention that $f(a_{\min}) = f(b_{\min}) = 0$, $f(a_{\max}) = H$, *and* $f(b_{\max}) = K$.
Proof: Given $\Gamma(\mathbf{x})$, any quantifier-free formula $\psi(\mathbf{x}, y)$ of L^1 says that either y belongs to a subset of $\{x_i : i < |\mathbf{x}|\}$, or $y \notin \{x_i : i < |\mathbf{x}|\}$ and $f(y)$ belongs to a finite union of disjoint "intervals" with endpoints at a finite distance from elements of

$$\{f(x_i) : i \in s\}.$$

In the model $(\mathcal{A}, \mathbf{a})$, such an interval will have the form $(u + \gamma, u' + \gamma']$ where $\gamma, \gamma' \in \mathbf{Z}$ and u, u' belong to $\{f(x_i) : i \in s\}$. The number of elements y of A such that $f(y)$ belongs to such an interval is equal to the difference $d_{u'+\gamma'} - d_{u+\gamma}$. A similar computation holds for $(\mathcal{B}, \mathbf{b})$. □

We now prove our main lemma, which shows that in the models $\mathcal{A}^1$ and $\mathcal{B}^1$ both quantifiers and counts can be eliminated.

Lemma 19. *For each formula* φ *of* $L^1(Ct)$ *there is a quantifier-free formula* φ^1 *of* L^1 *such that* $\mathcal{A}^1 \models \varphi \Leftrightarrow \varphi^1$ *and* $\mathcal{B}^1 \models \varphi \Leftrightarrow \varphi^1$.
Proof: We argue by induction on the complexity of φ. As remarked earlier, we may assume without loss of generality that φ has no quantifiers, because the existential quantifier $\exists x \psi(x, \ldots)$ may be replaced by $\neg count\{x : \psi(x, \ldots)\} = 0$. The hard case of the induction is the case where φ has the form

$$r(count\{y : \psi_i(\mathbf{x}, y)\} : i \leq j)$$

for some j-ary relation r on $\mathbf{N}$. For simplicity we let $j = 1$, so that

$$\varphi = r(count\{y : \psi(\mathbf{x}, y)\}).$$

By inductive hypothesis we have a quantifier-free formula $\psi^1(\mathbf{x}, y)$ of L^1 which is equivalent to ψ in both models.
Claim 1. Suppose $\mathbf{a} \equiv_0 \mathbf{b}$. Then $(\mathcal{A}^1, \mathbf{a}) \models \varphi$ if and only if $(\mathcal{B}^1, \mathbf{b}) \models \varphi$.
Proof of Claim 1: By the preceding lemma, $(\mathcal{A}^1, \mathbf{a})$ satisfies equation (18) and $(\mathcal{B}^1, \mathbf{b})$ satisfies the corresponding equation (18). The claim now follows by the indiscernibility of the sequence $\mathbf{d}$ in ${}^*\mathcal{N}$.
Claim 2. There is a quantifier-free formula $\varphi^1(\mathbf{x})$ of L^1 such that $\mathcal{A}^1 \models \varphi \Leftrightarrow \varphi^1$.
Proof of Claim 2: Let $\Sigma(\mathbf{x})$ be the set of all quantifier-free formulas $\sigma(\mathbf{x})$ of L^1 such that $\mathcal{A}^1 \models \varphi \Rightarrow \sigma$. Suppose $\mathcal{A}^1 \models \Sigma[\mathbf{a}]$. Then the set of formulas

$$\{\varphi(\mathbf{x})\} \cup \{\eta(\mathbf{x}) : \eta \text{ is quantifier-free in } L^1 \text{ and } \mathcal{A}^1 \models \eta[\mathbf{a}]\}$$

is finitely satisfiable in $\mathcal{A}^1$. Since ${}^*\mathcal{N}$ is saturated, this set of formulas is satisfiable in $\mathcal{A}^1$ by some tuple $\mathbf{c}$. Then $\mathcal{A}^1 \models \varphi[\mathbf{c}]$ and $\mathbf{a} \equiv_0 \mathbf{c}$. By Lemma 17 and the saturation of ${}^*\mathcal{N}$, there exists $\mathbf{b}$ in B such that $\mathbf{a} \equiv_0 \mathbf{b} \equiv_0 \mathbf{c}$. By Claim 1,

$(\mathcal{B}^1, \mathbf{b}) \models \varphi$ and $(\mathcal{A}^1, \mathbf{a}) \models \varphi$. Thus every tuple which satisfies Σ in $\mathcal{A}^1$ satisfies φ. Since $^*\mathcal{N}$ is saturated, there is a finite conjunction $\varphi^1(\mathbf{x})$ of formulas in $\Sigma(\mathbf{x})$ such that $\mathcal{A}^1 \models \varphi \Leftrightarrow \varphi^1$, and the claim is proved.

Now let $\mathbf{b}$ be a tuple in B. By Lemma 17 and the saturation of $^*\mathcal{N}$, there exists $\mathbf{a}$ in A such that $\mathbf{a} \equiv_0 \mathbf{b}$. By Claim 2, $\mathcal{A}^1 \models (\varphi \Leftrightarrow \varphi^1)[\mathbf{a}]$. By Claim 1, $\mathcal{B}^1 \models (\varphi \Leftrightarrow \varphi^1)[\mathbf{b}]$. □

Corollary 20. *$\mathcal{A}^1$ and $\mathcal{B}^1$ satisfy the same sentences of $L^1(Ct)$. Moreover, $\mathcal{A}$ and $\mathcal{B}$ satisfy the same sentences of $L^0(Ct)$.*

This completes the proof of Lemma 11, and hence Theorem 6.

Proof of Theorem 12. Fix some bijection K from the universe for tuples of S to the universe for tuples of R. Let $\mathcal{M}$ be the class of L' models of the form $M' = \langle R, S \rangle$ which satisfy

- the sets $\{x.a : x \in R\}$, $\{x.b : x \in R\}$, $\{x.c : x \in S\}$, are pairwise disjoint.
- $x.c = x.d$ for each $x \in S$, – $x.c \neq y.c$ for distinct x, y in S.
- For each x in S, $K(x.c)$ is in R.

We define a mapping F that maps $\mathcal{M}$ to models for the language L^0 defined previously. $F(M')$ is defined to have domain equal to R, the predicate S is interpreted by the K-image of the attributes of the relation S in M' , and the interpretation of the binary predicate E is defined by $xEy \leftrightarrow x.a = y.a$.

We also define a mapping G from formulae and terms of $L(Ct)$ to formulae and terms (respectively) of $L'(Ct)$ as follows:
$G(x.a = y.a) = xEy$, where x and y are any two variables of sort R, (not necessarily distinct).
$G(x.c = y.c) = G(x.d = y.d) = G(x.c = y.d) = G(x.d = y.c) = x = y$, where x, y have sort S.
$G(x.att1 = y.att2) = false$, for all attribute/sort combinations not listed above.
$G(P(\tau_1, \ldots, \tau_n)) = P(G(\tau_1), \ldots, G(\tau_n))$, $G(\forall x \in R\ \phi(x)) = \forall x\ G(\phi(x))$
$G(\exists x \in R\ \phi(x)) = \exists x\ G(\phi(x))$, $G(\forall x \in S\ \phi(x)) = \forall x\ S(x) \wedge G(\phi(x))$
$G(\exists x \in S\ \phi(x)) = \exists x\ S(x) \wedge G(\phi(x))$, $G(f(\mathbf{x})) = f(\mathbf{x})$,
$G(Count\{x \in R : \phi(x)\}) = Count\{x : G(\phi(x))\}$.

Proposition 21. *1) The maps F and G are surjections.*
2) $M' \models \phi(\mathbf{x}) \Leftrightarrow F(M') \models G(\phi(\mathbf{x}))$.
1) is proved by inspection and 2) by straightforward induction on complexity.

Theorem 12 now follows, since if there were a formula of $L'(Ct)$ expressing exactly those models that satisfy θ', then by applying the inverse of G to this formula we would get a sentence of $L(Ct)$ expressing exactly those models satisfying θ , contradicting the main theorem of the previous section.

Proof of Corollary 13. Consider the transaction T on structures for the signature SC defined above as follows: $R \Leftarrow \Pi_a(R)$, $S \Leftarrow S$, where $\Pi_a(R)$ is defined via $x \in \Pi_a(R) \Leftrightarrow (\exists\, y \in R\; x.a = y.a \wedge x.b = y.a)$. Then T is clearly $L(Ct)$-definable. The precondition of the $L'(Ct)$-sentence

$$count\{y \in R : y = y\} = count\{z \in S : z = z\}$$

is exactly the sentence θ' of Theorem 12 cited above, which is not expressible in $L'(Ct)$.

Some Strange Quantifiers

Wilfrid Hodges

Queen Mary and Westfield College
University of London

Abstract. We report a recent Tarski-style semantics for a language which includes the branching quantifiers on which Andrzej Ehrenfeucht made the first breakthrough in 1958. The semantics is equivalent to Henkin's game-theoretic semantics, but unlike Henkin's it is compositional. We use second-order formulas to give a new (and with any luck, more manageable) description of this Tarski-style semantics. Finally we apply the new description to present a compositional and fully abstract semantics for the slightly more limited syntax of Hintikka and Sandu, answering a question of Sandu.

1 Introduction

In 1967 I owned a cat called Andrzej. Its mother had dropped it behind a bush at the swimming pool at UCLA, and my wife and I responded to an appeal from the swimming pool staff for someone to take the kitten off their hands. Since I was in the middle of writing a DPhil thesis on Ehrenfeucht-Mostowski models, Andrzej seemed the obvious name. Andrzej was black all over, and he could jump fantastically high. It took him precisely three jumps to get from the yard outside to the window-ledge of our upstairs apartment in West LA. The next year we left Los Angeles. John Perry in the Philosophy Department kindly adopted Andrzej; he had him neutered and renamed him Black Black Panther.

I only once met Andrzej Ehrenfeucht, very briefly in Jerry Malitz's room at Boulder in 1980. If I remember, he was trying out on Jerry an idea about recognising finite sequences in DNA molecules. But since my days at UCLA I have come to respect Ehrenfeucht more and more for his incisive contributions in practically every branch of model theory, from back-and-forth games, through decidability results on linear orderings, to the beginnings of stability and non-structure theory. This essay develops some path-breaking work which he did on quantifiers in the 1950s. As far as I know, he never came back to it later.

2 Dependences between quantifiers

Consider this, from Serge Lang's book 'Diophantine geometry' [9], p. 67:

> Let V be a non-singular projective variety, and X a hyperplane section. Given an integer d, there exists a positive integer e depending only on V and d such that for any positive divisor Y on V of degree d, the divisors $Y + eX$ and $-Y + eX$ are ample.

A standard first-order symbolisation of this would run along the lines

$$\forall V\ \forall d\ \exists e\ \forall X\ \forall Y\ \phi(V, d, e, X, Y).$$

We note that Lang introduced X too early, and he had to cancel the error by adding the meta-level remark that e depends only on V and d.

That's the classical first-order approach. In the 1950s Leon Henkin challenged it. He asked what would happen if we incorporated a version of Lang's metatheoretic device into the object language, so that we could say explicitly that something exists independently of something else introduced earlier. I shall use the following notation, which is based on a later suggestion of Hintikka and Sandu [7]:

$$(\exists x/y_1 \dots y_n)$$

means 'There is x independent of $y_1, \dots, y_n$'. The dual universal quantifier is

$$(\forall x/y_1 \dots y_n).$$

As we shall see, there is no problem about giving a formal semantics for this universal quantifier, though it seems not to correspond to any easy device in English. In this notation, Lang's claim becomes

$$\forall V\ \forall X\ \forall d\ (\exists e/X)\ \forall Y\ \phi(V, d, e, X, Y),$$

with the quantifiers in the same order that Lang put them in.

Henkin asked whether these new slash quantifiers, or at least the special case

$$\forall x \exists v \forall y (\exists w/xv)$$

(which he called $Q_{2,2,d_2}$), can always be reduced to standard first-order expressions. He records that

> It seemed to me, intuitively, that the answer ... was negative

but he saw no proof of this. In September 1958 he mentioned the question to Andrzej Ehrenfeucht, and Ehrenfeucht brought him back a strong negative answer within two days. To quote Henkin again,

> THEOREM (Ehrenfeucht). *In a first-order predicate calculus enriched with the quantifier $Q_{2,2,d_2}$ one can define the quantifier R such that $(Rx)\phi(x)$ is true if and only if there are infinitely many values of x for which $\phi(x)$ is true.*

The compactness theorem implies that first-order logic has no such device. (Cf. Henkin [4] pp. 181f.)

Several people have claimed that a language with devices like the slash can't have a 'compositional' or Tarski-style semantics. More precisely, people doubted whether there could be any way of interpreting formulas in a structure, so that the interpretation of a quantified formula $Q\phi$ is a function of the quantifier Q and the interpretation of ϕ. It's clear at once that there is a problem with slash operators, the semantic scope of a quantifier need not be the same as its syntactic scope. In a recent paper [8] I showed how to get around this problem. This note will paraphrase the Tarski-style semantics of [8] from a new angle.

Before we go to details, the reader may wonder how I can be sure that the semantics of [8] agrees with the intended meanings of the slash quantifiers. The answer is that we have a game semantics, proposed by Henkin in [4], which everybody accepts as correct. For the parts of the language where this game semantics applies, one easily shows that it agrees with the semantics of [8]. (I shall not repeat the argument here. It should be added that Hintikka and Sandu also propose putting slashes on disjunctions and conjunctions. My account in [8] covers these, but for simplicity I ignore them here.)

Henkin's game semantics is not compositional. As we build up a sentence ψ by adding quantifiers, we simultaneously build up the definition of a game $G(\psi)$. For example when ψ is Lang's sentence

$$\forall V\ \forall X\ \forall d\ (\exists e/X)\ \forall Y\ \phi(V,d,e,X,Y),$$

the game $G(\psi)$ runs as follows:

> Player $\forall$ chooses an element V', then he chooses an element X', then an element d', after which player $\exists$ chooses an element e' without knowing what the element X' was; finally player $\forall$ chooses an element Y'. Player $\exists$ wins the game if and only if the sentence
>
> $$\phi(V',d',e',X',Y')$$
>
> is true.

Note that the slash in $(\exists e/X)$ means that the choice for X is hidden from player $\exists$ when she chooses for e. The sentence ψ is defined to be true if and only if player $\exists$ has a winning strategy for $G(\psi)$. Thus the semantics of the sentence ψ as a whole is defined in terms of the game. But no semantics is on offer for the subformulas of ψ, and so the semantics is not compositional.

Henkin's semantics gives us a game for which $\exists$ has a winning strategy if and only if ψ is true. But in general there is no reason to expect a game for which $\exists$ has a winning strategy if and only if ψ is *not* true; so there is no natural game semantics for negation. The $\neg$ in $\neg\psi$ has to be read classically, as follows: $\neg\psi$ is true if and only if ψ is not true.

Hintikka [5] pointed out that in the special case of classical first-order logic, where the slashes are empty, the games $G(\psi)$ are always determined (since they are games of perfect information and their length is finite). So for first-order logic we can represent negation by transposing the players. Barwise and Etchemendy [1] make good use of this idea in their teaching software 'Tarski's world'.

Unfortunately the games used by Henkin are not of perfect information. Following Hintikka we can introduce a symbol $\sim$ and interpret it as a sign that the players must change places; but in general it no longer expresses negation.

For example neither player has a winning strategy for the game $G(\psi)$ when ψ is the sentence

$$\forall x (\exists y/x)\ x = y$$

interpreted in a structure with more than one element. So $\sim \psi$ is false, though $\neg\phi$ is true.

3 Trumps

In this section and the next, we consider how to interpret formulas in a fixed structure A. All formulas are assumed to come from a language L of suitable signature for A. The language is first-order with connectives $\neg, \wedge, \vee$ and slashed quantifiers as above. An occurrence of a variable after the slash in a quantifier is free unless it is bound by another quantifier further out.

For motivation, consider the sentence ψ, viz.:

$$\forall x (\exists y/x)\ P(x,y),$$

Thinking classically, one would say that an element a of A 'satisfies' the formula $(\exists y/x)P(x,y)$ if we get a true sentence by interpreting the variable x as a name of a. This won't work straightforwardly, because we don't know what the quantifier $(\exists y/x)$ is supposed to mean when x is a name of a. If one tries to fix this up, one realises quickly that the truth of ψ in A depends not on what happens separately at each element a, but on whether there is a single value for y that works simultaneously for all values of x.

This suggests that instead of looking for a condition on *elements* of A, we should seek a condition on *sets of elements* of A. Namely, a set X of elements of A should 'satisfy' $(\exists y/x)P(x,y)$ if this formula is true for all elements of X *in a uniform way*. Instead of saying that X 'satisfies' the formula, I shall say that X is a ***trump*** of the formula. (Roughly speaking, a trump is a set of possible positions in a game which are all winning for a player, for a uniform reason.)

This idea works. If $\phi(v_0, \ldots, v_{n-1})$ is a formula with n free variables as shown, a trump of ϕ will be a nonempty set of ordered n-tuples of elements of A, all of which make ϕ true in a uniform way. We can give a definition for the set of trumps of a formula ϕ in terms of the sets of trumps of the immediate subformulas of ϕ. There are two sets of 0-tuples of elements of A, namely $\emptyset$ and $1 = \{\emptyset\}$. A sentence ϕ is true if and only if ϕ has a trump (which must then be 1).

The definition of trumps makes sense for classical logic too, but it becomes trivial. Classical formulas $\phi(v_0, \ldots, v_{n-1})$ have the property:

> A nonempty set X of n-tuples is a trump of ϕ if and only if every tuple in X satisfies ϕ.

Let us say that formulas with this property are ***flat***. If ϕ is flat, then there is a trump of ϕ if and only if there is a tuple which satisfies ϕ.

The rules of syntax allow $\neg$ to appear between a slashed variable and the quantifier which binds it. Take for example

$$\forall x \neg (\forall y/x)\, R(x,y).$$

How should we interpret

$$\neg (\forall y/x)\ R(x,y)?$$

Classically we would say that an element a satisfies it if and only if a doesn't satisfy $(\forall y/x)\ R(x,y)$. But in general the formula $(\forall y/x)\ R(x,y)$ is not flat, so the notion of satisfaction doesn't apply. Any solution to this problem is bound to be ad hoc, but the following idea seems to work smoothly.

We introduce an operator $\downarrow$ which flattens the formula that follows it. Thus a trump X of $\downarrow \phi$ is a nonempty set such that for each $\bar{a}$ in X, $\{\bar{a}\}$ is a trump of ϕ. To ensure that $\neg$ is only applied to flat formulas, we introduce Hintikka's game negation $\sim$ and define $\neg$ as an abbreviation for $\sim\downarrow$. This seems to give the right results in those cases where intuition is any guide. (In the case of the example above, where I don't have much intuition, our semantics makes it equivalent to $\forall x \neg \forall y\ R(x,y)$; see the end of section 6.)

In [8] I spelt out the recursive definition of the set of trumps of a formula. Here I shall repeat it in a new style. A trump X of a formula $\phi(v_0, \ldots, v_{n-1})$ is a nonempty n-ary relation on the structure A satisfying certain conditions. We can write these conditions as a sentence $\tau_\phi(R)$ to be satisfied by the expanded structure (A, X), with R a predicate symbol to stand for X. The sentence τ_ϕ is completely classical and slash-free, but in general it is higher-order. We read $\tau_\phi(R)$ as 'R is either empty or a trump of ϕ'. One can also use $\tau_\phi(R)$ informally as a statement of the metalanguage; in which case it's often convenient to write it '$R \models \phi$'.

The problems with negation suggest that we need some guiding principles. In [8] I proposed that our semantics should be ***fully abstract***, to borrow a term from the computer scientists. In our context this means the following:

> If ψ and ψ' are two formulas with the same free variables, then the following are equivalent:
> 1. In every structure, the interpretation of ψ is the same as that of ψ'.
> 2. For every sentence χ (possibly containing symbols not in ψ or ψ') and every structure A, the truth-value of χ in A never changes if we replace an occurrence of ψ as a subformula of χ by an occurrence of ψ'.

The implication from left to right is compositionality. The implication from right to left ensures that the interpretations carry no extra baggage.

4 First level: Negation-normal formulas

We begin with the simplest formulas of the language, namely the negation-normal formulas. These formulas never have the negation sign $\neg$ except immediately in front of atomic formulas. The game semantics of Henkin and Hintikka is completely adequate for this level of the language.

Base clause. For an atomic or negated atomic formula $\phi(v_0, \ldots, v_{n-1})$, a trump of ϕ is a nonempty n-ary relation T on the domain $|A|$ of A, such that the sentence $\tau_\phi(R)$, viz.:

$$\forall \bar{v}\ (R(\bar{v}) \to \phi(\bar{v}))$$

is true of (A, T).

For the remaining clauses I simply state what τ is and leave the reader to translate this into a definition of trumps.

Clause for $\wedge$. Given formulas $\phi(v_0, \ldots, v_{n-1})$ and $\psi(v_0, \ldots, v_{n-1})$, the formula $\tau_{\phi\wedge\psi}(R)$ is

$$\tau_\phi(R) \wedge \tau_\psi(R).$$

Clause for $\vee$. Given formulas $\phi(v_0, \ldots, v_{n-1})$ and $\psi(v_0, \ldots, v_{n-1})$, the formula $\tau_{\phi\vee\psi}(R)$ is

$$\exists S \exists T\ ((R \subseteq S \cup T) \wedge \tau_\phi(S) \wedge \tau_\psi(T)).$$

The clauses for quantifiers will use $\bar{v}$ as an abbreviation for $v_0 \ldots v_{n-1}$, and $\forall \bar{v}$ as an abbreviation for $\forall v_0 \ldots \forall v_{n-1}$. If W is a subset of $\{v_0, \ldots, v_{n-1}\}$, we write $\bar{x} \simeq_W \bar{y}$ to mean the conjunction

$$\bigwedge\{(x_i = y_i) : v_i \notin W\}.$$

Note that W plays no role in the clause for $(\forall v_n/W)$. In game-theoretic terms, this is because W expresses a restriction on the strategies of player $\forall$, and for negation-normal sentences these strategies are irrelevant to truth.

Clause for $(\forall v_n/W)$. Given a formula $\phi(v_0, \ldots, v_n)$ and a subset W of $\{v_0, \ldots, v_{n-1}\}$, the formula $\tau_{(\forall v_n/W)\phi}(R)$ is

$$\exists S\ (\tau_\phi(S) \wedge \forall \bar{v}(R(\bar{v}) \to \forall v_n\ S(\bar{v}, v_n))).$$

Clause for $(\exists v_n/W)$. Given a formula $\phi(v_0, \ldots, v_n)$ and a subset W of $\{v_0, \ldots, v_{n-1}\}$, the formula $\tau_{(\exists v_n/W)\phi}(R)$ is

$$\exists S\ (\tau_\phi(S) \wedge \forall \bar{v} \exists v_n \forall \bar{x}\ (R(\bar{x}) \wedge \bar{v} \simeq_W \bar{x} \to S(\bar{x}, v_n))).$$

Note that when W is empty, $\simeq_W$ is equality and the formula reduces to

$$\exists S\ (\tau_\phi(S) \wedge \forall \bar{v}\ (R(\bar{v}) \to \exists v_n\ S(\bar{v}, v_n))).$$

To apply the clauses for $\wedge$ and $\vee$, we need to be able to add redundant variables to a formula. In particular the interpretation of the formula with the added variable and the interpretation of the formula without it should determine each other. In classical logic this is completely trivial, but here it needs a little care. The following *clause for redundant variables* is proved by induction on the complexity of formulas:

Proposition 1. *Given a formula* $\phi(v_0, \ldots, v_{n-1})$, *where this formula is also written as* $\psi(v_0, \ldots, v_n)$, *the formula* $\tau_\psi(R)$ *is*

$$\forall v_n \exists S \, (\tau_\phi(S) \wedge \forall \bar{v} \, (R(\bar{v}, v_n) \rightarrow S(\bar{v}))).$$

Thanks to Proposition 1, we can say that two formulas $\phi(v_0, \ldots, v_{n-1})$ and $\psi(v_0, \ldots, v_{n-1})$ are *logically equivalent* if in every structure they have the same trumps.

Note that $\tau_\phi(\emptyset)$ holds for any ϕ. Note also that every occurrence of the variable R in these formulas $\tau_\phi(R)$ is negative. In fact induction on the complexity of formulas shows the following, which I shall refer to as the *downwards property*:

Proposition 2. *If* $(A, X) \models \tau_\phi(R)$ *and* $Y \subseteq X$, *then* $(A, X) \models \tau_\phi(R)$. *(Or in our briefer notation, if* $X \models \phi$ *and* $Y \subseteq X$ *then* $Y \models \phi$.*)*

The downwards property allows us to simplify some of the clauses above:

Proposition 3. $R \models (\forall v_n / W)\phi \; \Leftrightarrow R \times |A| \models \phi$.

Write R^π for the projection of the relation R along the last coordinate; in other words, $R^\pi(a_1, \ldots, a_{n-1})$ if and only if for some a_n, $R(a_1, \ldots, a_n)$.

Proposition 4. *When* W *is empty,* $R \models (\exists v_n)\phi \Leftrightarrow R^\pi \models \phi$.

Write R_b for the set of tuples $\bar{a}$ such that $R(\bar{a}, b)$.

Proposition 5. *If* $\psi(v_0, \ldots, v_n)$ *is* $\phi(v_0, \ldots, v_{n-1})$, *then* $R \models \psi \Leftrightarrow$ *for all* b, $R_b \models \phi$.

Every formula τ_ϕ is a $\exists^1_1$ formula, i.e. it can be brought to prenex form so that it consists of a string of existential second-order quantifiers followed by a first-order formula. A result of Enderton [2] and Walkoe [10] states that the converse is true for sentences: for every $\exists^1_1$ sentence σ there is a sentence ϕ in the language discussed in this section, in fact a sentence using just positive occurrences of the quantifier $Q_{2,2,d_2}$, for which $\tau_\phi(R)$ is logically equivalent to

$$R \rightarrow \sigma.$$

(This is just as easy to prove using the game semantics as it is using ours.)

For formulas with free variables, it's not clear what kind of counterpart to the Enderton-Walkoe result we might hope for. If ϕ is flat, then $\tau_\phi(R)$ can always be written as

$$\forall \bar{v} \, (R(\bar{v}) \rightarrow \sigma(\bar{v}))$$

where σ is a $\exists^1_1$ formula; one can show that up to logical equivalence, any $\exists^1_1$ formula may occur as σ. But when ϕ is not flat, things become much more complicated and I hesitate to make any conjecture.

5 Example

This example illustrates how we test the truth of a sentence χ in a structure A with domain $|A|$, by an induction on the complexity of χ. The sentence χ is

$$\forall v_0 \forall v_1 \forall v_2 \, (\neg P(v_0, v_1, v_2) \vee \exists v_3 \, (v_3 = v_1 \wedge \theta(v_0, v_3, v_2))),$$

where P is a relation symbol and θ is a formula. (The reason for choosing this example will emerge in section 7.)

We write P^A for the interpretation of the symbol P in A. The following clauses are all equivalent. Several of the equivalences use the downwards property.

1. χ is true in A.
2. $1 \models \chi$.
3. (By clause for $\forall$:)

$$|A|^3 \models \neg P(v_0, v_1, v_2) \vee \exists v_3 \, (v_3 = v_1 \wedge \theta(v_0, v_3, v_2)).$$

4. (By clause for $\vee$:)

$$\exists S \subseteq |A|^3 \, (((|A|^3 \setminus S) \models \neg P(v_0, v_1, v_2)) \wedge$$
$$S \models \exists v_3 \, (v_3 = v_1 \wedge \theta(v_0, v_3, v_2))).$$

5. (By base clause and clause for $\exists$, writing $v_3 = v_1$ as $=^\star (v_0, \ldots, v_3)$ and $\theta(v_0, v_3, v_2)$ as $\theta^\star(v_0, \ldots, v_3)$:)

$$\exists S \subseteq |A|^3 \, (P^A \subseteq S \wedge$$
$$\exists T \subseteq |A|^4 \, (T \models (=^\star (v_0, \ldots, v_3) \wedge \theta^\star(v_0, \ldots, v_3)) \wedge$$
$$\forall v_0 \forall v_1 \forall v_2 \, (S(v_0, v_1, v_2) \to \exists v_3 \, T(v_0, \ldots, v_3)))).$$

6. (By clause for $\wedge$ and base clause:)

$$\exists T \subseteq |A|^4 \, (\forall v_0 \ldots \forall v_3 \, (T(v_0, \ldots, v_3) \to v_3 = v_1) \wedge$$
$$T \models \theta^\star(v_0, \ldots, v_3) \wedge$$
$$\forall v_0 \forall v_1 \forall v_2 \, (P(v_0, v_1, v_2) \to \exists v_3 \, T(v_0, \ldots, v_3))).$$

7. (By simplifying:)

$$\{(a, b, c, b) : P(a, b, c)\} \models \theta^\star(v_0, \ldots, v_3).$$

8. (By clause for redundant variables:)

$$\forall b \, \{(a, b, c) : P(a, b, c)\} \models \theta(v_0, v_3, v_2).$$

In short, χ is true in A if and only if for every element b the set $\{(a, b, c) : (a, b, c) \in P^A\}$ is either empty or a trump of $\theta(v_0, v_3, v_2)$.

6 Second level: Adding negation

This section shows how to add $\sim$ and $\downarrow$ to the language. If we introduce game negation $\sim$, we need to know not only those sets of tuples which uniformly make a formula true, but also those sets which uniformly make it false. So we need to define the *cotrumps* of a formula, by simultaneous induction with the trumps. As τ stands for Trump, so κ will stand for Cotrump. I give some typical cases.

Clause for $\wedge$ continued. Given formulas $\phi(v_0, \ldots, v_{n-1})$ and $\psi(v_0, \ldots, v_{n-1})$, the formula $\kappa_{\phi\wedge\psi}(R)$ is

$$\exists S \exists T\ ((R \subseteq S \cup T) \wedge \kappa_\phi(S) \wedge \kappa_\psi(T)).$$

Clause for $(\forall v_n/W)$ continued. Given a formula $\phi(v_0, \ldots, v_n)$ and a subset W of $\{v_0, \ldots, v_{n-1}\}$, the formula $\kappa_{(\forall v_n/W)\phi}(R)$ is

$$\exists S\ (\kappa_\phi(S) \wedge \forall \bar{v} \exists v_n \forall \bar{x}\ (R(\bar{x}) \wedge \bar{v} \simeq_W \bar{x} \to S(\bar{x}, v_n))).$$

Clause for $\sim$. Given a formula $\phi(v_0, \ldots, v_{n-1})$, the formula $\tau_{\sim\phi}(R)$ is

$$\kappa_\phi(R)$$

and the formula $\kappa_{\sim\phi}(R)$ is

$$\tau_\phi(R).$$

Clause for $\downarrow$. Given a formula $\phi(v_0, \ldots, v_{n-1})$, the formula $\tau_{\downarrow\phi}(R)$ is

$$\forall \bar{v}\ (R(\bar{v}) \to \exists S\ (\tau_\phi(S) \wedge S(\bar{v}))),$$

and the formula $\kappa_{\downarrow\phi}(R)$ is

$$\forall \bar{v}\ (R(\bar{v}) \to \exists S\ (\neg\tau_\phi(S) \wedge \forall \bar{x}(S(x) \leftrightarrow \bar{x} = \bar{v}))).$$

Then as mentioned earlier, we can define $\neg$ to be $\sim\downarrow$. To check that this agrees with the base case, we have the following chain of equivalent statements, where as before we interpret in a structure A, and $X \not\models \phi$ means $\kappa_\phi(X)$:

$$R \models \sim\downarrow P(v_0, \ldots, v_{n-1}).$$

$$R \not\models \downarrow P(v_0, \ldots, v_{n-1}).$$

$$\forall \bar{v}\ (R(\bar{v}) \to \bar{v} \notin P^A).$$

The downwards property still holds, both for trumps and for cotrumps. The clause for redundant variables holds both for trumps and for cotrumps; we duly revise the definition of logical equivalence, to say that in any structure the two formulas have the same trumps and the same cotrumps. Also the semantics is fully abstract, as shown in [8].

So far so good. But otherwise the addition of $\sim$ and $\downarrow$ does undeniably complicate the story. We can ease the pressure a little by defining $\uparrow$ to mean $\sim\downarrow\sim$ ('flattening up' as opposed to 'flattening down'). The semantics of $\uparrow$ is straightforward:

Proposition 6.

$$\tau_{\downarrow\phi}(R) \Leftrightarrow \forall\bar{v}\ (R(\bar{v}) \to \tau_\phi(\{\bar{v}\}));$$

$$\tau_{\uparrow\phi}(R) \Leftrightarrow \forall\bar{v}\ (R(\bar{v}) \to \neg\kappa_\phi(\{\bar{v}\})).$$

(And dual clauses for $\kappa_{\downarrow\phi}$ and $\kappa_{\uparrow\phi}$.)

Then we can say that a formula ϕ is *generalised negation-normal* if $\sim$ never occurs in ϕ except immediately in front of atomic formulas or in $\uparrow$. Standard arguments show:

Proposition 7. *Every formula $\phi(v_0, \ldots, v_{n-1})$ is logically equivalent to a generalised negation-normal formula $\psi(v_0, \ldots, v_{n-1})$.*

For example the problematic sentence $\forall x \neg(\forall y/x)\ R(x,y)$ mentioned in section 3 becomes successively

$$\forall x\ \sim\downarrow (\forall y/x)\ R(x,y),$$

$$\forall x \uparrow\sim (\forall y/x)\ R(x,y),$$

$$\forall x \uparrow (\exists y/x)\ \sim R(x,y).$$

One can show that $\uparrow (\exists x/W)$ is equivalent to $\uparrow \exists x$. In game-theoretic terms the crucial point is that $\uparrow$ reduces both trumps and cotrumps to questions about winning strategies for player $\forall$, and slashes on existential quantifiers don't affect strategies of $\forall$. Hence the sentence is equivalent to

$$\forall x \uparrow \exists y\ \sim R(x,y).$$

The formula after $\uparrow$ is classical, so that $\uparrow$ is redundant and we are left with

$$\forall x \exists y\ \sim R(x,y)$$

which is the negation-normal form of the classical equivalent mentioned in section 3.

7 The mug's game

Jeroen Gronendieck, who is known among other things for having provided a compositional semantics for discourse representation theory [3], said the following to me. (I don't remember his exact words.)

> It's a mug's game. You give a compositional semantics, and then they change some feature of the syntax and your semantics doesn't work any more, so you have to start all over again.

And so it happened. After [8] was written, Hintikka's book [6] arrived, with a new suggested notation that replaces the slash. I shall refer to this new notation as Hintikka II. But in fact the notation used by Hintikka and Sandu [7]—let me call it Hintikka-Sandu I—was already different from that of [8] in a subtle way. Gabriel Sandu has rightly pressed me to give a compositional semantics that works explicitly for Hintikka-Sandu I. This I shall do in a moment. My own feeling, if I wanted to enter this particular battlefield, would be that the artificiality of the semantics needed to cope with these variants is fair evidence that the variants are unnatural.

In what follows I shall deal with Hintikka II and Hintikka-Sandu I only for generalised negation-normal formulas. This is purely to avoid some tiresome book-keeping; nothing of principle is missing. If we allowed negation to appear in other places, we would need to distinguish between positive and negative occurrences of quantifiers within subformulas. Also I give proofs only for trumps, since everything dualises to cotrumps.

We can paraphrase the notation of Hintikka-Sandu I as follows:

> An occurrence of $(\exists x/W)$ is read as $(\exists x/W \cup V)$ where V is the set of all variables v such that the occurrence lies within the scope of an existential quantifier $\exists v$ (possibly with its own slash) with no intervening quantifier $\forall v$.

This is easier to say in game terms: player $\exists$ is never allowed to remember her own preceding choices. There is a dual clause for the universal quantifiers.

Clearly everything that can be said in this notation can also be said in ours, so strength is no argument in its favour.

Imagine we are on the inside of a formula ϕ, sitting at a subformula $\theta(x, y, z)$ and looking out. Nothing in θ itself indicates whether the variables x, y, z are going to be bound by existential or universal quantifiers. So for a compositional semantics we have to take all cases into account. The formula $\theta(x, y, z)$ needs eight different interpretations, one for each subset W of the set $\{x, y, z\}$ of variables. As we build up a formula and the interpretations of its subformulas recursively, half the available possibilities drop out each time that we quantify one of the variables x, y, z.

This is complicated, but it is only an elaboration of what we did already in order to cope with game negation $\sim$. There we gave each formula two interpretations, one by trumps and the other by cotrumps. We find out which interpretation to use when we know whether the formula occurs positively or negatively in the sentence as a whole. Formally, instead of defining just $\tau_\phi(R)$ and $\kappa_\phi(R)$ for a formula $\phi(v_0,\ldots,v_{n-1})$, we now have to define $\tau^W_\phi(R)$ and $\kappa^W_\phi(R)$ for each subset W of $\{v_0,\ldots,v_{n-1}\}$. The corresponding sets are called W-trumps and W-cotrumps.

Here is a problematic situation. We have a formula $\phi(v_0,\ldots,v_{n-1})$ which is also written $\psi(v_0,\ldots,v_n)$ with a redundant variable v_n that doesn't occur in the formula. Now we put ψ into a larger formula in which the variable v_n gets quantified existentially, so that according to Hintikka-Sandu I it is implicitly slashed in ψ. It seems that the formula ψ is prepared for this; the variable v_n is taken into account in trumps of ψ. But how can ϕ understand what it means for player $\exists$ ***not to know what element is assigned to*** v_n, since v_n doesn't appear in ϕ at all? This confusing type of situation is very typical of slashed quantifiers.

Fortunately there is an answer. But I had to write it out carefully before I believed it. It runs as follows. Given a formula $\phi(v_0,\ldots,v_{n-1})$, we write $\phi^\star$ for ϕ written with added redundant variable w at the end. Much as before, we write S^π for $\{\bar{a} :$ for some $w, S(\bar{a},w)\}$, the projection of S along w.

Proposition 8. *If w is a slashed variable in each existential quantifier of ϕ but otherwise doesn't occur in ϕ, then $\tau_{\phi^\star}(R)$ is equivalent to $\tau_\phi(R^\pi)$.*

Proof. By induction on the complexity of ϕ.

Case ϕ atomic or negated atomic.

$$\begin{aligned} &\forall\bar{v}\forall w\ (R(\bar{v},w) \to \phi^\star(\bar{v},w)) \\ &\Leftrightarrow\ \forall\bar{v}\ (R^\pi(\bar{v}) \to \phi(\bar{v})). \end{aligned}$$

Case $\wedge$. Here $(\phi\wedge\psi)^\star = \phi^\star \wedge \psi^\star$, so

$$\begin{aligned} \tau_{(\phi\wedge\psi)^\star}(R) &\Leftrightarrow\ \tau_{\phi^\star}(R)\ \wedge\ \tau_{\psi^\star}(R) \\ &\Leftrightarrow\ \tau_\phi(R^\pi)\wedge\tau_\psi(R^\pi) \\ &\Leftrightarrow\ \tau_{\phi\wedge\psi}(R^\pi). \end{aligned}$$

Case $\vee$. Here $(\phi\vee\psi)^\star = \phi^\star\vee\psi^\star$. So

$$\begin{aligned} &\tau_{(\phi\vee\psi)^\star}(R) \\ &\Leftrightarrow\ \exists S\exists T\ ((R\subseteq S\cup T)\ \wedge\ \tau_{\phi^\star}(R)\ \wedge\ \tau_{\psi^\star}(R)) \\ &\Leftrightarrow\ \exists S\exists T\ ((R\subseteq S\cup T)\ \wedge\tau_\phi(R^\pi)\ \wedge\ \tau_\psi(R^\pi)). \end{aligned}$$

Now $R\subseteq S\cup T$ implies $R^\pi\subseteq S^\pi\cup T^\pi$. The converse fails, but if $R^\pi\subseteq S^\pi\cup T^\pi$ then there are S',T' with $S'^\pi = S^\pi$ and $T'^\pi = T^\pi$, such that $R\subseteq S\cup T$ (take $S' = \{(\bar{a},b) : S^\pi(\bar{a})\}$ and likewise with T). So this last condition is equivalent to

$$\tau_{\phi\vee\psi}(R^\pi).$$

Case $\forall$.

$$\tau_{((\forall v_n/W)\phi)^*}(R) \Leftrightarrow \exists S\ (\tau_{\phi^*}(S) \wedge \forall\bar{v}\forall w\ (R(\bar{v},w) \to \forall v_n\ S(\bar{v},v_n,w)))$$

$$\Leftrightarrow \exists S\ (\tau_\phi(S^\pi) \wedge \forall\bar{v}\forall w\ (R(\bar{v},w) \to \forall v_n\ S(\bar{v},v_n,w))).$$

Again if

$$(1)\quad \forall\bar{v}\forall w\ (R(\bar{v},w) \to \forall v_n\ S(\bar{v},v_n,w))$$

then

$$(2)\quad \forall\bar{v}\ (R^\pi(\bar{v}) \to \forall v_n\ S^\pi(\bar{v},v_n)).$$

The converse fails in general, but if (2) holds then there is S' with $S'^\pi = S^\pi$ such that (1) holds with S' for S. (Argue as in the previous case.) So the condition reduces to

$$\exists S\ (\tau_\phi(S^\pi) \wedge \forall\bar{v}\ (R^\pi(\bar{v}) \to \forall v_n\ S^\pi(\bar{v},v_n)))$$

which is clearly equivalent to $\tau_{(\forall v_n/W)}(R^\pi)$.

Case $\exists$.

$$\tau_{((\exists v_n/W\cup\{w\})\phi)^*}(R)$$

$$\Leftrightarrow \exists S\ (\tau_{\phi^*}(S) \wedge \forall\bar{v}\forall w\exists v_n\forall\bar{x}\forall y\ (R(\bar{x},y)\ \wedge \bar{v} \simeq \bar{x} \to\ S(\bar{x},v_n,y))),$$

where the condition $w = y$ drops out because w is after the slash. By the same argument as in the previous case, we continue

$$\Leftrightarrow \exists S\ (\tau_\phi(S^\pi) \wedge \forall\bar{v}\exists v_n\forall\bar{x}\ (R^\pi(\bar{x}) \wedge \bar{v} \simeq \bar{x} \to\ S^\pi(\bar{x},v_n)))$$

which is equivalent to

$$\tau_{(\exists v_n/W\cup\{w\})\phi}(R^\pi).$$

Case $\downarrow$.

$$\tau_{(\downarrow\phi)^*}(R) \Leftrightarrow \forall\bar{v}\forall w\ (R(\bar{v},w) \to \tau_{\phi^*}(\{\bar{v}w\}))$$

$$\Leftrightarrow \forall\bar{v}\ (R^\pi(\bar{v}) \to \tau_\phi(\{\bar{v}\}))$$

$$\Leftrightarrow \tau_{\downarrow\phi}(R^\pi).$$

Case $\uparrow$.

$$\tau_{(\uparrow\phi)^*}(R) \Leftrightarrow \forall\bar{v}\forall w\ (R(\bar{v},w) \to \neg\kappa_{\phi^*}(\{\bar{v}w\}))$$

$$\Leftrightarrow \forall\bar{v}\ (R^\pi(\bar{v}) \to \neg\kappa_\phi(\{\bar{v}\}))$$

$$\Leftrightarrow \tau_{\uparrow\phi}(R^\pi).$$

Given Proposition 8, it is now easy to prove:

Proposition 9. *The semantics described for Hintikka-Sandu I is fully abstract.*

Proof. One direction follows from the fact that the semantics is compositional. In the other direction, we have to show that if ψ and ψ' have the same variables but there is a structure B in which some set T is a W-trump for ψ but not a W-trump for ψ', then we can change the truth-value of some sentence χ in some structure A by replacing an occurrence of ψ in χ by an occurrence of ψ'. For a typical example, suppose that $\psi(v_0, v_1, v_2)$ and $\psi'(v_0, v_1, v_2)$ both have the free variables shown, and W is the set $\{v_1\}$. Let A be the structure B expanded by adding a relation symbol P whose interpretation P^A is T. Let χ be the sentence

$$\forall v_0 \forall v_1 \forall v_2\ (\neg P(v_0, v_1, v_2) \vee \exists v_3\ (v_3 = v_1 \wedge \psi(v_0, v_3, v_2))),$$

of the Example in section 5 above, with ψ for θ.

Following the example, we write $\psi(v_0, v_3, v_2)$ as $\psi^\star(v_0, \ldots, v_3)$. The variable v_3 in ψ is bound in χ by an existential quantifier, so $\{v_3\}$-trumps are in force. By the argument of the example, down to the last step but one, χ is true in A if and only if

$$\{(a, b, c, b) : P(a, b, c)\} \models \psi^\star(v_0, \ldots, v_3).$$

The final step of the example fails because the variable v_3 is no longer redundant; it appears in slashes of ψ, even if only implicitly. Instead we have to use Proposition 8, which tells us that

$$\{(a, b, c, b) : P(a, b, c)\} \models \psi^\star(v_0, \ldots, v_3) \Leftrightarrow$$

$$P \models \psi(v_0, v_3, v_2)$$

where we have projected along the added variable v_1. But this last condition is equivalent to

$$T \models \psi(v_0, v_1, v_2).$$

This is true, so χ is true in A.

Now let χ' be χ with ψ' in place of ψ. Exactly the same argument goes through, and we have that χ' is true in A if and only if $T \models \psi'(v_0, v_1, v_2)$. This is false, and hence χ' is false in A.

We turn to Hintikka II. In this notation, $(\forall x//W)$ means that the choice of element for x is hidden from all later quantifiers over variables in W. Thus our previous

$$\forall x \exists y (\exists z/x)\ \phi$$

becomes

$$(\forall x//z) \exists y \exists z\ \phi.$$

The problem with this notation is essentially the same as with Hintikka-Sandu I: when we are interpreting the formula $\exists z \phi$, we don't know what information is going to be hidden by a choice made further out in the formula. The remedy is the same; we have to carry all possibilities forward with us until they can be eliminated.

For this we need to carry up definitions of ω-trump and ω-cotrump of a formula $\phi(v_0, \ldots, v_{n-1})$. This time, ω must be a function that assigns to each variable v_i $(i < n)$ a finite set of variables, which may include variables not in the set $\{v_0, \ldots, v_{n-1}\}$. Together with these trumps and cotrumps, we assign to each formula ϕ its set $\beta(\phi)$ of bound variables. When we bind a variable, say by a quantifier $(\forall v_j//W)$, we pass upwards only those ω where $\omega(v_j) = W \cap \beta(\phi)$. This gives a compositional semantics, but I hold out no hopes of making it fully abstract.

8 Brief remarks

A question: What sets of subsets of an infinite set X are the sets of trumps for a formula $\phi(v_0)$ in the language of equality, interpreted with X as domain?

After spending a year and a half with these slash quantifiers, I still find them immensely confusing. That said, I agree with Hintikka and Sandu that the idea behind them is very natural. The semantics above should fall into place as part of a more general theory of imperfect information, whose shape I see only dimly.

Nearly forty years on, Andrzej Ehrenfeucht's result on these quantifiers is probably still the most revealing in the literature.

References

1. Barwise, J., Etchemendy, J.: Tarski's world 3.0. CSLI/SRI International, Menlo Park CA, 1991.
2. Enderton, H. B.: Finite partially ordered quantifiers. Zeitschrift für Math. Logic und Grundlagen der Math. **16** (1970) 393–397.
3. Gronendieck, J., Stokhof, M.: Dynamic predicate logic. Linguistics and Philosophy **14** (1991) 39–100.
4. Henkin, L.: Some remarks on infinitely long formulas. Infinitistic methods, Pergamon Press, Oxford and PAN, Warsaw 1961, pp. 167–183.
5. Hintikka, J.: Logic, language-games and information. Clarendon Press, Oxford 1973.
6. Hintikka, J.: The principles of mathematics revisited. Cambridge University Press, Cambridge 1996.
7. Hintikka, J., Sandu, G.: Game-theoretical semantics. Handbook of logic and language, ed. van Benthem, J., ter Meulen, A., Elsevier 1996, pp. 361–410.
8. Hodges, W.: Compositional semantics for a language of imperfect information. Logic Journal of the IGPL (1997), to appear.
9. Lang, S.: Diophantine geometry. John Wiley and Sons, New York 1962.
10. Walkoe, W. J.: Finite partially-ordered quantification. Journal of Symbolic Logic **35** (1970) 535–555.

Pebble Games in Model Theory

James F. Lynch

Department of Mathematics and Computer Science,
Box 5815, Clarkson University,
Potsdam, NY 13699-5815, USA

Abstract. This article describes the Ehrenfeucht game and its applications in two areas of model theory: definability theory and random model theory. Most of the examples are taken from finite model theory, where the Ehrenfeucht game is one of the most useful tools. Extensions of the Ehrenfeucht game to logics more expressive than first-order logic are described, and applications to definability and random models in these logics are outlined.

1 Introduction

We will discuss several combinatorial games, collectively known as pebble games. The original pebble game was invented by A. Ehrenfeucht [10] as a method for proving elementary equivalence of models. In this article, we will show how this game and its extensions have been used to solve problems concerned with the expressive power of logical languages, with particular attention to expressive power over finite models.

We begin with a notion due to R. Fraïssé [15], which is essentially a finer analysis of the notion of elementary equivalence. Let $\mathcal{L}$ be some fixed first-order language with equality and a finite number of relational and constant symbols, but no function symbols. (This restriction is not essential. We can, of course, interpret functions as relations, at the cost of making our notation and definitions more cumbersome.) The *depth*, or *quantifier rank*, of a formula in $\mathcal{L}$ is its deepest nesting of quantifiers. More precisely, atomic formulas have depth 0. Inductively, assume the formulas α and β have depth d_1 and d_2 respectively. Then $\neg\alpha$ has depth d_1, $\alpha \vee \beta$ has depth $\max(d_1, d_2)$, and $\exists x(\alpha)$ has depth $d_1 + 1$.

Let $\mathfrak{A}$ and $\mathfrak{B}$ be two $\mathcal{L}$-structures and $d \in \omega$. We put $\mathfrak{A} \equiv^d \mathfrak{B}$ if $\mathfrak{A}$ and $\mathfrak{B}$ agree on all sentences in $\mathcal{L}$ of depth at most d. Clearly $\equiv^d$ is an equivalence relation. Further, as Fraïssé showed, $\equiv^d$ has finite rank, and every $\equiv^d$ class is definable by a sentence of depth d.

Ehrenfeucht invented a game that gives a combinatorial characterization of $\equiv^d$. Again, let $\mathfrak{A}$ and $\mathfrak{B}$ be two $\mathcal{L}$-structures, say

$$\mathfrak{A} = \langle A, R_{0,1}, \ldots, R_{0,i}, c_{0,1}, \ldots, c_{0,j} \rangle \text{ and}$$
$$\mathfrak{B} = \langle B, R_{1,1}, \ldots, R_{1,i}, c_{1,1}, \ldots, c_{1,j} \rangle.$$

We imagine two players, referred to as I and II in the original paper. Some recent papers have used the more colorful appellations Spoiler and Duplicator, but we will keep the original names. There are two sets of d pebbles, the pebbles in each set numbered $1, \ldots, d$. The game consists of d rounds, also numbered $1, \ldots, d$. In each round r, player I places one of the pebbles numbered r on some element of either structure. Then player II places the other pebble r on some element of the other structure. Let a_r be the element in $\mathfrak{A}$ that was pebbled in round r (by either player) and b_r be the element in $\mathfrak{B}$ that was pebbled in round r. At the conclusion of the game, player II has won if the mapping $a_r \mapsto b_r$, $1 \leq r \leq d$, induces a *partial isomorphism* from $\mathfrak{A}$ to $\mathfrak{B}$. More specifically,

- For $1 \leq r, s \leq d$, $a_r = a_s$ if and only if $b_r = b_s$.
- For every relational symbol R in $\mathcal{L}$ (say R is m-ary) and every m-tuple $(r_1, \ldots, r_m) \in \{1, \ldots, d\}^m$,

$$\mathfrak{A} \models R(a_{r_1}, \ldots, a_{r_m}) \text{ if and only if } \mathfrak{B} \models R(b_{r_1}, \ldots, b_{r_m}) \ .$$

- For every constant symbol c in $\mathcal{L}$ and $1 \leq r \leq d$,

$$\mathfrak{A} \models a_r = c \text{ if and only if } \mathfrak{B} \models b_r = c \ .$$

If the mapping $a_r \mapsto b_r$ is not a partial isomorphism, then player I has won. We say that player II has a winning strategy if, no matter how player I chooses, player II can always respond in such a way that he will eventually win. We put $\mathfrak{A} \sim^d \mathfrak{B}$ if player II has a winning strategy for the d-round game on $\mathfrak{A}$ and $\mathfrak{B}$. It is not difficult to see that $\sim^d$ is an equivalence relation on $\mathcal{L}$-structures. The central result, due to Ehrenfeucht, is that $\sim^d$ is precisely the same as $\equiv^d$:

Theorem 1 (Ehrenfeucht [10]). *For any two $\mathcal{L}$-structures $\mathfrak{A}$ and $\mathfrak{B}$ and $d \in \omega$, $\mathfrak{A} \equiv^d \mathfrak{B}$ if and only if $\mathfrak{A} \sim^d \mathfrak{B}$.*

Most applications of this theorem use the implication, if $\mathfrak{A} \sim^d \mathfrak{B}$ then $\mathfrak{A} \equiv^d \mathfrak{B}$. Neither direction is difficult to prove, but we shall not do it here. Instead, we will give examples of the game and its extensions, some trivial and some not, to illustrate the extremely broad range of applications of pebble games.

Our first example is intended only to provide some intuitive feeling for pebble games. Consider the two undirected graphs $\mathfrak{A}$ and $\mathfrak{B}$ in Fig. 1. It is easily seen that player II has a winning strategy for the one and two round games. But player I can always win the three round game. There are a number of ways he can ensure this. One way is to place his first pebble on an arbitrary vertex of $\mathfrak{A}$. Then, after player II has responded in $\mathfrak{B}$, player I places his second pebble on the vertex of $\mathfrak{B}$ that is "opposite" b_1, i.e., a distance of 3 from b_1. Player II cannot match this move; the best he can do is place his second pebble on one of the two vertices in $\mathfrak{A}$ that are a distance of 2 from a_1. Any other choice leads to an immediate loss. Assuming player II has made such a choice (as shown in Fig. 1), in the third round player I places his third pebble so that it is adjacent to a_1 and a_2. Player II cannot respond successfully to this choice.

Although we are not going to prove Theorem 1, it is worth noting that the proof of the implication, if player I wins then there exists a sentence of depth

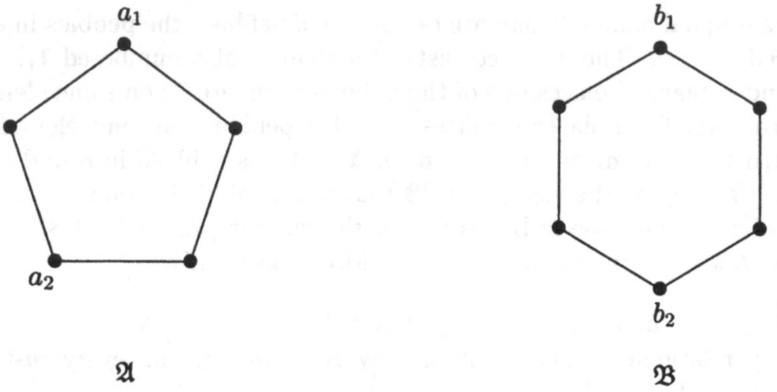

Fig. 1. Example of a Three-Pebble Game

d that distinguishes 𝔄 from 𝔅, essentially shows how the winning strategy of player I yields such a sentence. Thus in our example, the sentence

$$\exists x_1 \forall x_2 [x_1 = x_2 \lor E(x_1, x_2) \lor \exists x_3 (E(x_1, x_3) \land E(x_3, x_2))]$$

is true for 𝔄 but false for 𝔅.

In the next two sections, we will apply the Ehrenfeucht game to definability theory and convergence laws for random models. Some of the theorems we will discuss were first proven by other methods, e.g., quantifier elimination, but the Ehrenfeucht game gives shorter, more insightful proofs. In other instances, the Ehrenfeucht game is the only known approach. This is often the case in finite model theory. We will also describe extensions of the game that characterize higher-order logics. This is not intended to be an exhaustive survey; rather, we hope to convey a feeling for the scope and limitations of pebble games. The references will guide interested readers to further material.

2 Definability Theory

Let $\mathcal{L}$ be a logical language and $\mathcal{C}$ be a class of $\mathcal{L}$-structures. A property $\mathcal{P}$ is simply a subclass of $\mathcal{C}$. $\mathcal{P}$ is *definable* or *expressible* in $\mathcal{L}$ over $\mathcal{C}$ if there is a sentence σ in $\mathcal{L}$ such that for all $\mathfrak{A} \in \mathcal{C}$,

$$\mathfrak{A} \in \mathcal{P} \Leftrightarrow \mathfrak{A} \models \sigma \ .$$

Definability questions have long been central to mathematical logic. The connections between computational complexity, database theory, and definability have provided additional impetus to its study. We first describe the use of the Ehrenfeucht game in first-order definablity theory. Then we will show how extensions of the game can be applied to more powerful logics.

2.1 First-Order Logic

A method for proving the undefinability of $\mathcal{P}$ in a first-order logic $\mathcal{L}$ is the following. Given an arbitrary $d \in \omega$, show that there exist $\mathfrak{A} \in \mathcal{P}$ and $\mathfrak{B} \in \mathcal{C} - \mathcal{P}$ such that $\mathfrak{A} \sim^d \mathfrak{B}$. This method is sound and complete. Soundness means that to prove $\mathcal{P}$ is not definable in $\mathcal{L}$, it suffices to show that for all $d \in \omega$, there exist $\mathfrak{A}$ and $\mathfrak{B}$ as above. Completeness means that if $\mathcal{P}$ is not definable in $\mathcal{L}$, then this method is guaranteed to succeed. Of course, in practice, it can be difficult to find $\mathfrak{A}$ and $\mathfrak{B}$, or even to show their existence.

We will give two well-known examples of this method. The first is an illustration of the general principle that first-order logic cannot define properties about the size of structures. Let $\mathcal{L}$ be any first-order relational language and $\mathcal{C}$ be the class of all $\mathcal{L}$-structures, or the class of all finite $\mathcal{L}$-structures. Then no sentence in $\mathcal{L}$ can define the property that the size of the structure is finite and even. To show this, for any $d \in \omega$, take $\mathfrak{A}$ and $\mathfrak{B}$ to be any finite structures of size at least d, but of different parity, such that all their relations are empty. In other words, they are just sets of elements. Player II has a simple winning strategy: in each round r, he seeks to maintain the condition

$$\bigwedge_{1 \le i,j \le r} (a_i = a_j \leftrightarrow b_i = b_j) \; . \tag{1}$$

It is easily seen by induction on r that he can do this. Then at the conclusion of the game, (1) implies that player II has won.

The next example is more interesting. We show that connectivity is not definable in first-order logic. This property can be formulated for arbitrary relational structures, but for clarity, we will consider only the usual notion of connectivity on undirected graphs. Such a graph is connected if and only if, for every pair of vertices, there is a path between them. As pointed out by H. Gaifman [16], when $\mathcal{C}$ is the class of all graphs, compactness arguments or ultraproduct constructions easily prove that connectivity is not definable in first-order logic, but when $\mathcal{C}$ is the class of finite graphs, these arguments do not work. However, the Ehrenfeucht game method still succeeds. R. Fagin [13] gave the first proof that connectedness over finite graphs is undefinable in first-order logic. In fact, he proved it is undefinable in the more powerful existential monadic second-order logic, which we will describe later. The proof is quite difficult. The following proof that connectivity over finite graphs is undefinable in first-order logic is much simpler, but is still a good example of a nontrivial application of the Ehrenfeucht game.

We will use the notion of distance between vertices of a graph. Given a graph $\mathfrak{A} = \langle V, E \rangle$ and vertices $u, v \in V$, $\delta(u,v)$ is the length of the shortest path between u and v, or it is ∞ if there is no path between them. Then δ is obviously a metric on V. For any $v \in V$ and $\varepsilon \in [0, \infty]$, the ε-neighborhood of v is $N(v, \varepsilon) = \{ u \in V : \delta(v,u) \le \varepsilon \}$.

Let $1 \le d \in \omega$, $\mathfrak{A}$ be a graph consisting of a single cycle with at least $2 \cdot 3^{d-1} + 2$ vertices, and $\mathfrak{B}$ be two disjoint copies of $\mathfrak{A}$. Clearly $\mathfrak{A}$ is connected while $\mathfrak{B}$ is not. Next, we show that player II has a winning strategy for the d-round Ehrenfeucht game on $\mathfrak{A}$ and $\mathfrak{B}$. He moves so as to maintain the following condition after each round r:

$$\mathfrak{A} \restriction \left(\bigcup_{i=1}^{r} N(a_i, 3^{d-r}) \right) \cong \mathfrak{B} \restriction \left(\bigcup_{i=1}^{r} N(b_i, 3^{d-r}) \right) \tag{2}$$

via an isomorphism that maps each a_i to b_i. After round 1, (2) holds because $\mathfrak{A} \restriction N(a_1, 3^{d-1})$ and $\mathfrak{B} \restriction N(b_1, 3^{d-1})$ are just path graphs of length $2 \cdot 3^{d-1}$ with a_1 (resp. b_1) as their midpoint.

Now assuming (2) holds after round r, we will show that player II can choose so that it holds after round $r+1$. Let player I choose a_{r+1} in $\mathfrak{A}$. (Choosing in $\mathfrak{B}$ is similar.) If $a_{r+1} \notin \bigcup_{i=1}^{r} N(a_i, 2 \cdot 3^{d-r-1})$, then player II chooses any $b_{r+1} \notin \bigcup_{i=1}^{r} N(b_i, 2 \cdot 3^{d-r-1})$. This is possible because $|\bigcup_{i=1}^{r} N(b_i, 2 \cdot 3^{d-r-1})| \leq r(4 \cdot 3^{d-r-1} + 1) < r \cdot 5 \cdot 3^{d-r-1} < 2 \cdot 3^{d-1}$. We claim that $N(a_{r+1}, 3^{d-r-1}) \cap N(a_i, 3^{d-r-1}) = \emptyset$ for all $i = 1, \ldots, r$, and similarly for b_{r+1}. If not, let $c \in N(a_{r+1}, 3^{d-r-1}) \cap N(a_i, 3^{d-r-1})$ for some $i = 1, \ldots, r$. Then by the triangle inequality, $a_{r+1} \in N(a_i, 2 \cdot 3^{d-r-1})$, a contradiction. Thus (2) is maintained.

On the other hand, suppose player I has chosen $a_{r+1} \in N(a_i, 2 \cdot 3^{d-r-1})$ for some $i = 1, \ldots, r$. Then, using the isomorphism given by (2), player II chooses the b_{r+1} such that $a_{r+1} \mapsto b_{r+1}$. In this case $N(a_{r+1}, 3^{d-r-1}) \subseteq N(a_i, 3^{d-r})$, and similarly for b_{r+1}. Thus (2) still holds.

At the conclusion of the game, (2) implies that player II has won.

Combinatorial distances and neighborhoods can be generalized to arbitrary relational structures. W. Hanf [18] used these notions to give a sufficient condition for elementary equivalence of structures. Later, Gaifman [16] applied them to characterize the expressive power of first-order logic.

Although first-order logic is the foundation of logic, it plays a limited role in computer science. Many of the simplest computational problems exceed the expressive power of first-order logic, as illustrated by our result on evenness. Complexity and database theorists have resorted to various extensions of first-order logic in order to formalize their notions of computation. We shall describe three ways in which first-order logic has been augmented, and give examples of undefinablity results that have been proven using the Ehrenfeucht game or extensions of it.

2.2 Fixed Interpretations

In our first augmentation, $\mathcal{L}$ is still a first-order logic, but some of its relational symbols have fixed interpretations on the structures in $\mathcal{C}$. A widely studied case is when $\mathcal{L}$ has a binary relational symbol (usually written as $\leq$) which is interpreted as a linear order on the universe of each structure in $\mathcal{C}$. For example, let $\mathcal{L}$ be the first-order logic of one monadic symbol R and $\leq$. Then the class of all finite $\mathcal{L}$-structures (with $\leq$ interpreted as a linear order, and all isomorphic structures identified) can be regarded as the set of finite sequences of 0's and 1's. Again the property of evenness of the size of the structure cannot be expressed in this $\mathcal{L}$ over this $\mathcal{C}$, but the proof is not so easy as the one given earlier for evenness on all structures. As before, $R = \emptyset$ on both $\mathfrak{A}$ and $\mathfrak{B}$, but in order for player II to win the d-round game, it is not sufficient to take structures of size greater than or equal to d. They must be larger than 2^d. Let a_{-1} and a_0 be the least and largest, respectively, elements of $\mathfrak{A}$, using the $\leq$ ordering, and let b_{-1} and b_0 be the analogous elements of $\mathfrak{B}$. Player II's strategy is to maintain the condition that after each round r, for $-1 \leq i, j \leq r$, either

$$\begin{aligned} a_i - a_j &= b_i - b_j, \\ a_i - a_j, b_i - b_j &\leq -2^{d-r}, \text{ or} \\ a_i - a_j, b_i - b_j &\geq 2^{d-r} \ . \end{aligned}$$

As before, we use induction on r to show that this condition can be maintained.

A slight modification of the proof shows that the property "$|R|$ is even" is undefinable—replace R by its complement in $\mathfrak{A}$ and $\mathfrak{B}$. A stronger result is possible. Instead of $\leq$, let $\mathcal{L}$ contain a ternary relational symbol P, and let $\mathcal{C}$ be the set of all finite structures $\langle \{0, \ldots, n-1\}, P_n, R \rangle$ where $n \in \omega$, $P_n = \{ (x, y, z) \in \{0, \ldots, n-1\}^3 : x + y = z \}$, and $R \subseteq \{0, \ldots, n-1\}$. That is, $\mathcal{P}$ is interpreted as the ternary addition relation on $\{0, \ldots, n-1\}$. Here, evenness of the size of the structure is definable by $\neg \exists x \exists y (x + x + y \wedge \forall w \exists z (w + z = y))$. However, evenness of $|R|$ is not definable in $\mathcal{L}$. This result was inspired by a problem posed by R. McKenzie, J. Mycielski, and D. Thompson [35]. In this paper, they proved that a form of connectivity on certain infinite structures was not definable in first-order logic. Their structures were of the form $\langle \omega, +, Q \rangle$, where $Q \subseteq \omega^2$. Q was connected if, regarding ω^2 as the set of nonnegative integer-valued points in the cartesian plane, a chess king could visit all points in Q without leaving Q. The proof showed that if this kind of connectivity was definable, then the relation $D = \{ (x, y) \in \omega^2 : x \text{ divides } y \}$ could be defined by a first-order formula in $\langle \omega, + \rangle$. However, the theory of $\langle \omega, +, D \rangle$ was shown to be undecidable by J. Robinson [36], while M. Presburger (see J. Monk [34], Chapter 13) proved that the theory of $\langle \omega, + \rangle$ is decidable. Thus, the proof was ultimately a diagonalization. The authors posed the problem of whether the property that $R \subseteq \omega$ was finite and even was definable in first-order logic over the class of structures $\langle \omega, +, R \rangle$. This property seemed too weak to define the number-theoretic concepts required by their proof. However, its undefinability was shown by J. Lynch [28] using an Ehrenfeucht game.

The proof is rather technical, but the key idea is the construction of a set $A_d \subseteq \omega$ for every $d \in \omega$. It was shown that for every such d, there is an integer k_d such that for all finite sets $R_0, R_1 \subseteq A_d$, if $|R_0| = |R_1|$ or $k_d < |R_0|, |R_1|$, then $\langle \omega, +, R_0 \rangle \sim^d \langle \omega, +, R_0 \rangle$. Further, for any $n \in \omega$, and any sets $R_0, R_1 \subseteq A_d \cap \{0, \ldots, n-1\}$, if $|R_0| = |R_1|$ or $k_d < |R_0|, |R_1|$, then $\langle \{0, \ldots, n-1\}, P_n, R_0 \rangle \sim^d \langle \{0, \ldots, n-1\}, P_n, R_1 \rangle$. This implies the undefinablity result for evenness mentioned earlier.

2.3 Second-Order Logics

Our second augmentation is a genuine strengthening of the logic itself, rather than the way it is interpreted. Here, we introduce second-order variables and quantifiers. A second-order variable is a symbol of the form Q_i^r for some $i, r \in \omega$. The r indicates that the symbol represents an r-ary relation, and the i is an index. Such variables can be quantified just like first-order variables. The meaning of $\exists Q_i^r$ is "there exists an r-ary relation Q_i^r," and similarly for $\forall Q_i^r$.

Second-order definability has attracted great interest in computer science because some of the fundamental complexity classes correspond to properties definable in second-order logic. One of the best known of these results is Fagin's characterization of NP by existential second-order logic [12]. This is the fragment of second-order logic restricted to sentences of the form $\exists Q_{i_1}^{r_1} \ldots \exists Q_{i_k}^{r_k} \varphi$, where φ is a first-order sentence. (The relational symbols occurring in φ will, in general, include $Q_{i_1}^{r_1}, \ldots, Q_{i_k}^{r_k}$ and others.) Fagin's theorem essentially says that a property of a class of finite models is recognized by some nondeterministic polynomial time-bounded Turing machine if and only if it is definable by an existential second-order sentence. As should be expected, definablity questions for second-order logic are quite difficult. But there have been some striking results on undefinablity by existential monadic second-order sentences. These are existential second-order sentences where all of the second-order variables are monadic, or unary, relational symbols. Fagin [13] proved the undefinablity of connectivity by such sentences. The complement of connectivity is definable by the existential monadic second-order sentence $\exists Q_1^1[\exists x \exists y(Q_1^1(x) \wedge \neg Q_1^1(y)) \wedge \forall x \forall y(Q_1^1(x) \wedge E(x,y) \rightarrow Q_1^1(y))]$. Consequently, this logic is not closed under complement. If it could be shown that the class of properties definable by the full existential second-order logic is not closed under complement, then by Fagin's theorem, it would follow that NP is not closed under complement, and this implies that P does not equal NP.

Fagin's proof of the undefinablity of connectivity used an extension of the Ehrenfeucht game. The parameters of this game are k and d. Here, player I begins by choosing unary relations $A_1, \ldots, A_k$ on the universe of the *first* structure $\mathfrak{A}$. Player II then chooses unary relations $B_1, \ldots, B_k$ on the universe of the second structure $\mathfrak{B}$. Then a d-round Ehrenfeucht game is played. Player II wins if the pebbled vertices induce not only a partial isomorphism between $\mathfrak{A}$ and $\mathfrak{B}$, but also between the structures expanded with the new relations $A_1, \ldots, A_k$ and $B_1, \ldots, B_k$. We put $\mathfrak{A} \rightarrow^{k,d} \mathfrak{B}$ if player II has a winning strategy for this game. Note the asymmetry in this condition reflecting the condition that player I must choose $A_1, \ldots, A_k$ in $\mathfrak{A}$. This is also seen in the theorem that relates Fagin's game to definablity.

Theorem 2 (Fagin [13]). *Let $k, d \in \omega$ and $\mathcal{L}$ be the fragment of existential monadic second-order logic consisting of all sentences of the form $\exists Q_1^1 \ldots Q_k^1 \varphi$, where φ is a sentence of depth d in some fixed first-order logic that includes the relational symbols $Q_1^1, \ldots, Q_k^1$. For any two $\mathcal{L}$-structures $\mathfrak{A}$ and $\mathfrak{B}$, $\mathfrak{A} \rightarrow^{k,d} \mathfrak{B}$ if and only if, for every sentence $\sigma \in \mathcal{L}$, $\mathfrak{A} \models \sigma$ implies $\mathfrak{B} \models \sigma$.*

Again, this theorem provides a sound and complete method for proving that a property $\mathcal{P}$ is not definable in existential monadic second-order logic: for every k and d, find $\mathfrak{A} \in \mathcal{P}$ and $\mathfrak{B} \in \mathcal{C} - \mathcal{P}$ such that $\mathfrak{A} \rightarrow^{k,d} \mathfrak{B}$.

M. de Rougemont [8] used this game to prove a stronger version of Fagin's result on undefinablity of connectivity. He showed that connectivity on finite graphs was still undefinable in existential monadic second-order logic, even when the logic is augmented with a binary relational symbol interpreted as a successor relation on the vertices.

More recently, M. Ajtai and Fagin [1] proved that there is no existential monadic second-order sentence that expresses whether two distinguished vertices of a directed graph have a directed path between them. This result is interesting for several reasons. It reveals a difference between definability on undirected and directed graphs. As P. Kanellakis [20] pointed out, there is an existential monadic second-order sentence that holds if and only if two distinguished vertices in an undirected graph have a path between them. It also shows that the two seemingly similar properties of connectivity and reachability are quite different, because Fagin's proof of undefinability of connectivity on undirected graphs easily extends to directed graphs.

In addition, the proof uses a modification of Fagin's game that may have numerous applications to definability theory. It is based on the observation that starting the game with $\mathfrak{A}$ and $\mathfrak{B}$ already chosen gives a potential advantage to player I, since he can choose $A_1, \ldots, A_k$ so that player II's choice of $B_1, \ldots, B_k$ is difficult. At the very least, it makes the description of player II's strategy more complicated. Instead, let the game begin with player I selecting $\mathfrak{A} \in \mathcal{P}$. Then player I chooses $A_1, \ldots, A_k$ (without knowledge of $\mathfrak{B}$). Next, player II chooses $\mathfrak{B} \in \mathcal{C} - \mathcal{P}$ and $B_1, \ldots, B_k$. The game then proceeds as before. Note that this is a game played on the classes of structures $\mathcal{C}$ and $\mathcal{P}$ rather than on individual structures. Thus we put $\mathcal{P} \Rightarrow^{k,d} \mathcal{C} - \mathcal{P}$ if player II has a winning strategy for this game.

Theorem 3 (Ajtai and Fagin [1]). *Let $k, d \in \omega$ and $\mathcal{L}$ be the fragment of existential monadic second-order logic consisting of all sentences of the form $\exists Q_1^1 \ldots Q_k^1 \varphi$, where φ is a sentence of depth d in some fixed first-order logic that includes the relational symbols $Q_1^1, \ldots, Q_k^1$. For any class $\mathcal{C}$ of $\mathcal{L}$-structures and any property $\mathcal{P} \subseteq \mathcal{C}$, $\mathcal{P} \Rightarrow^{k,d} \mathcal{C} - \mathcal{P}$ if and only if $\mathcal{P}$ is not definable by a sentence in $\mathcal{L}$.*

One more point of interest is that in applying this theorem, Ajtai and Fagin never gave an explicit description of $\mathfrak{A}$ and $\mathfrak{B}$. They showed that if player II chooses $\mathfrak{A}$ and $\mathfrak{B}$ randomly, then with high probability, he can ultimately win the game by simply "copying" player I's choice of $A_1, \ldots, A_k$.

The Ajtai-Fagin game has had several other successes. T. Schwentick [37] used it to show that connectivity is undefinable in existential monadic second-order logic with a relation $\leq$ interpreted as linear order, thus improving the result of de Rougemont. His proof, which is an explicit construction of $\mathfrak{A}$ and $\mathfrak{B}$, is very involved, and uses a sophisticated result of D. Coppersmith [7] about permutation groups.

2.4 Infinitary Logics

A third way of augmenting first-order logic is to introduce operators that perform recursion on first-order definable properties. These include transitive closure logic, least fixed-point logic, and partial fixed-point logic. Their full definitions are somewhat technical, and we refer the reader to the book of H.-D. Ebbinghaus and J. Flum [9]. They are, however, included in an even more powerful logic whose syntax is a simple extension of first-order logic. This is the so-called infinitary logic $L^{\omega}_{\infty\omega}$. For $d \in \omega$, let $L^{d}_{\infty\omega}$ be the extension of first-order logic obtained by closure under conjunction and disjunction of arbitrary sets of formulas, provided only the variables $x_1, \ldots, x_d$ occur among them. Then

$$L^{\omega}_{\infty\omega} = \bigcup_{d \in \omega} L^{d}_{\infty\omega} .$$

Since sentences in $L^{\omega}_{\infty\omega}$ can have infinite depth, it would not be useful to categorize them according to their quantifier depth. A more meaningful characterization is simply the number of distinct variables they contain. We can then define an equivalence relation on models with respect to sentences in $L^{d}_{\infty\omega}$, for each $d \in \omega$. Let $\mathfrak{A}$ and $\mathfrak{B}$ be models of the same type. Then $\mathfrak{A} \equiv^{d}_{\infty} \mathfrak{B}$ if and only if they agree on all sentences in $L^{d}_{\infty\omega}$.

There is a game for the infinitary logic $L^{\omega}_{\infty\omega}$ which does not seem to have a widely accepted name, but it is sometimes called the eternal game. Again, it is played by two players on two models $\mathfrak{A}$ and $\mathfrak{B}$, and there are two sets of pebbles numbered $1, \ldots, d$. The first d rounds are played exactly like the d-round Ehrenfeucht game, but after that, it may continue indefinitely. In succeeding rounds, player I chooses one of the models and moves a pebble to some (possibly the same) element in the same model. Then player II responds by moving the other pebble with the same number to some element in the other model. The game ends and player I wins if, after any round, the mapping $a_i \mapsto b_i$ for $i = 1, \ldots, d$ is not a partial isomorphism. We say that player II has a winning strategy if, no matter how player I moves, player II can always respond so that the game continues indefinitely. If this is the case, then we put $\mathfrak{A} \sim^{d}_{\infty} \mathfrak{B}$. Again, $\sim^{d}_{\infty}$ and $\equiv^{d}_{\infty}$ are the same equivalence relations. This was implicit in an article of J. Barwise [2].

Theorem 4 (Barwise [2]). *Let $d \in \omega$ and $\mathcal{L}$ be an $L^{d}_{\infty\omega}$ logic. For any two $\mathcal{L}$-structures $\mathfrak{A}$ and $\mathfrak{B}$, $\mathfrak{A} \equiv^{d}_{\infty} \mathfrak{B}$ if and only if $\mathfrak{A} \sim^{d}_{\infty} \mathfrak{B}$.*

An easy application of this theorem is that evenness of the size of a structure is undefinable in $L^{\omega}_{\infty\omega}$. In fact, the proof is exactly the same as the proof for first-order logic, except that the finite Ehrenfeucht game is replaced by the eternal game. The winning strategy remains the same. It is well-known (see e.g. [9]) that connectivity is definable in $L^{\omega}_{\infty\omega}$, but there are nontrivial properties that are not. One such is Hamiltonicity, the existence of a Hamilton cycle in a graph. De Rougemont proved that Hamiltonicity is not definable in inductive logic (a weaker variant of infinitary logic). His argument uses a pebble game and generalizes to $L^{\omega}_{\infty\omega}$. Other examples may be found in [9].

3 Convergence Laws

In this section, we proceed in a manner parallel to the previous section. We will define convergence laws and outline a general approach for proving them via pebble games. We will illustrate the approach by proving one of the first and best known convergence laws. Then we will describe convergence laws for each of the three augmented logics and briefly sketch their pebble games.

As before, $\mathcal{L}$ will denote a logic, and $\mathcal{C}$ will be a class of $\mathcal{L}$-structures. For every $n \in \omega$, $\mathcal{C}_n$ will be the set of structures in $\mathcal{C}$ with universe $\{0, \ldots, n-1\}$, and pr_n will be a probability measure on $\mathcal{C}_n$. For a sentence $\sigma \in \mathcal{L}$, we put $\mathrm{pr}(\sigma, n)$ for

$$\mathrm{pr}_n(\{\mathfrak{A} \models \sigma : \mathfrak{A} \in \mathcal{C}_n\}) .$$

For example, if $\mathcal{L}$ has a single r-ary relational symbol, and all structures in each $\mathcal{C}_n$ are equally likely, then

$$\mathrm{pr}(\sigma, n) = \frac{|\{\mathfrak{A} \models \sigma : \mathfrak{A} \in \mathcal{C}_n\}|}{2^{n^r}} .$$

A 0-1 law for $\mathcal{L}$ and $\mathcal{C}$ states that for every sentence $\sigma \in \mathcal{L}$, $\lim_{n\to\infty} \mathrm{pr}(\sigma, n) =$ 0 or 1. A convergence law merely states that this limit exists for all sentences σ.

3.1 First-Order Logic

Although the first theorems in this area were not proven with the Ehrenfeucht game, it has become a standard technique for proving 0-1 and convergence laws. The general idea is the following. Let $d \in \omega$ be fixed. Show that there are k (where k depends on d) $\equiv^d$ classes $\mathcal{D}_1, \ldots, \mathcal{D}_k$ such that

$$\lim_{n\to\infty} \mathrm{pr}\left(\mathfrak{A} \in \bigcup_{i=1}^{k} \mathcal{D}_i, n\right) = 1, \text{ and} \tag{3}$$

$$\lim_{n\to\infty} \mathrm{pr}(\mathfrak{A} \in \mathcal{D}_i, n) \text{ exists} \tag{4}$$

for each $i = 1, \ldots, k$. When $k = 1$, we have a 0-1 law. Usually, a combinatorial characterization of $\mathcal{D}_1, \ldots, \mathcal{D}_k$ is given, and the Ehrenfeucht game is used to show that they are $\equiv^d$ classes. Our first example is the classic 0-1 law due independently to Fagin [14] and Y. Glebskiĭ et al. [17]. We will describe it for undirected graphs, but it is easily seen to apply to all classes of structures of a given finite relational type with a uniform probability distribution on structures of a given size. That is, $\mathcal{L}$ is the first-order logic of graphs, and

$$\mathrm{pr}(\sigma, n) = \frac{|\{\mathfrak{A} \models \sigma : \mathfrak{A} \in \mathcal{C}_n\}|}{2^{n(n-1)/2}} .$$

Glebskiĭ et al. used quantifier elimination to prove the 0-1 law for this case, while Fagin used the completeness theorem. However, the proof based on the Ehrenfeucht game is a good example of the general method for proving 0-1 laws and convergence laws, and, as we shall see, a slight modification of the proof extends this 0-1 law to $L^{\omega}_{\infty\omega}$.

For a fixed $d \in \omega$, consider the following finite set of axioms.

$$\begin{aligned}
&\left\{\exists x_1 \ldots \exists x_{d-1}\left(\bigwedge_{1\leq i<j<d} x_i \neq x_j\right)\right\} \cup \\
&\left\{\forall x_1 \ldots \forall x_{d-1} \exists y\left[\left(\bigwedge_{1\leq i<j<d} x_i \neq x_j\right) \rightarrow\right.\right. \qquad (5)\\
&\left.\left.\left(\bigwedge_{1\leq i<d} x_i \neq y \wedge \bigwedge_{1\leq i\leq j} E(x_j,y) \wedge \bigwedge_{j<i<d} \neg E(x_j,y)\right)\right] : j = 0,\ldots,d-1\right\} .
\end{aligned}$$

These are called the extension axioms. They hold for a graph if it has at least $d-1$ vertices, and for any set of $d-1$ vertices, all possible 1-vertex extensions are realized. A graph that satisfies these axioms is said to be d-extendible. The proof of the 0-1 law consists in showing the following.

1. $\mathfrak{A} \sim^d \mathfrak{B}$ for any two d-extendible graphs.
2. Almost all graphs are d-extendible.

Both parts have elementary proofs. Part 1. follows because the extension axioms (5) state that player II can always match any move of player I. To prove Part 2., let θ_j, $j = 0,\ldots,d-1$, enumerate the extension axioms, where the j is as in (5). Then

$$\begin{aligned}
\mathrm{pr}\left(\neg\left(\bigwedge_{j=0}^{d-1}\theta_j\right), n\right) &= \mathrm{pr}\left(\bigvee_{j=0}^{d-1}\neg\theta_j, n\right)\\
&\leq \sum_{j=0}^{d-1} n^{d-1} \times [(\tfrac{1}{2})^j(\tfrac{1}{2})^{d-j-1}]^{n-d+1}\\
&= dn^{d-1}(\tfrac{1}{2})^{(d-1)(n-d+1)}\\
&\rightarrow 0 \text{ as } n \rightarrow \infty \ .
\end{aligned}$$

3.2 Fixed Interpretations

The following examples are taken from Lynch [27]. Let the vocabulary of $\mathcal{L}$ consist of an r-ary relational symbol R and the binary relational symbol S. R will be interpreted as a random relation on the universe. (We consider one symbol R for clarity only; all our results easily extend to any number of relational symbols $R_1,\ldots,R_m$.) S will be interpreted as a successor relation in two ways. For $n \in \omega$,

1. $\mathcal{C}_n = \{\langle\{0,\ldots,n-1\}, R, S_n\rangle : R \subseteq \{0,\ldots,n-1\}^r \text{ and } S_n = \{(i,i+1) : i < n-1\}$, and

2. $C'_n = \{\langle\{0,\ldots,n-1\}, R, S'_n\rangle : R \subseteq \{0,\ldots,n-1\}^r$ and $S'_n = \{(i, i+1 \pmod n)) : i < n\}$. We will call S'_n a cyclic successor to distinguish it from the usual successor in Case 1.

For $\sigma \in \mathcal{L}$, S is interpreted as S_n in Case 1. and S'_n in Case 2. We write $y = x + 1$ for $S(x, y)$. We use the uniform probability distribution on $\{R \subseteq \{0,\ldots,n-1\}^r\}$ and put pr_n (resp. pr'_n) for the probability of properties in C_n (resp. C'_n). For any structure $\langle A, R, \ldots\rangle$ and $B \subseteq A$, we abbreviate $\langle B, R \restriction B, \ldots\rangle$ by $\langle B, R, \ldots\rangle$.

It is easily seen that there is no 0-1 law for Case 1. Take the sentence

$$\sigma = \exists x \forall y (x \neq y + 1 \wedge R(x, \ldots, x)).$$

Then $\mathrm{pr}(\sigma, n) = 1/2$ for $n > 0$. In fact, for every rational in $[0,1]$ of the form $u/2^v$ where $u, v \in \omega$, there is $\sigma \in \mathcal{L}$ such that $\mathrm{pr}(\sigma, n) = u/2^v$ for sufficiently large n. However, there is a convergence law for Case 1., and the asymptotic probability is always of that form. The proof of this convergence law is a slight modification of the proof of the 0-1 law for Case 2. In the proof sketch that follows, we will try to show how the notion of d-extendibility generalizes to this class of structures.

Fix $d \in \omega$. We will show that there is some $\equiv^d$ class $\mathcal{D}$ of $\mathcal{L}$ structures with a cyclic successor such that $\mathrm{pr}'(\mathfrak{A} \in \mathcal{D}, n) \to 1$ as $n \to \infty$.

As before, this class of structures is defined by a winning strategy for player II in the d-round Ehrenfeucht game. However, the simple strategy of the previous case will not work because the extension axioms fail. For example, when $n > 1$, no structure in C'_n satisfies $\forall x \forall y \exists z (z = x + 1 \wedge y = z + 1)$. Player II must ***plan ahead***, i.e., look at the "neighborhood" around each point chosen by player I, and try to match that rather than just match the chosen points. This is made precise by the following:

Definition 5. 1. For $n \in \omega$ and $a, b \in \{0,\ldots,n-1\}$, let $\delta_n(a,b) = \min\{|m| : a = b + m \pmod n\}$. It is easily seen that δ_n is a metric.

2. For $n, m \in \omega$ and $a \in \{0,\ldots,n-1\}$, $N_n(a, m) = \{b \in \{0,\ldots,n-1\} : \delta_n(a,b) \leq m\}$.
3. For $i, j, n \in \omega$, $\mathfrak{A} = \langle\{0,\ldots,n-1\}, R, S'_n\rangle$, and $a_1, \ldots, a_i \in n$, the j-closure of $a_1, \ldots, a_i$ in $\mathfrak{A}$ is

$$\mathrm{Cl}^j(\mathfrak{A}; a_1, \ldots, a_i) = \langle \bigcup_{h=1}^{i} N_n(a_h, 3^j), R, S'_n, a_1, \ldots, a_i\rangle.$$

When the context is clear, we will also use $\mathrm{Cl}^j(\mathfrak{A}; a_1, \ldots, a_i)$ to denote the universe of that structure.

Let $\mathfrak{A}$ and $\mathfrak{B}$ be two structures with cyclic successor on which the d-round Ehrenfeucht game is to be played. Player II's strategy is to ensure that, after each round $r = 0, \ldots, d$,

$$\mathrm{Cl}^{d-r}(\mathfrak{A}; a_1, \ldots, a_i) \cong \mathrm{Cl}^{d-r}(\mathfrak{B}; b_1, \ldots, b_i) \ . \tag{6}$$

This condition is trivially true when $r = 0$, and when $r = d$, it implies that player II has won. If the following statement is true for $\mathfrak{A}$ and the analogous statement is true for $\mathfrak{B}$, then player II can follow this strategy.

For any $r < d$, any $\mathcal{L}$-structure $\mathfrak{C}$ with cyclic successor, and any $c_1, \ldots, c_{r+1} \in \mathfrak{C}$ such that $\mathrm{Cl}^{d-r-1}(\mathfrak{C}; c_{r+1}) \cap \mathrm{Cl}^{d-r-1}(\mathfrak{C}; c_1, \ldots, c_r) = \emptyset$,

$$\forall x_1 \ldots \forall x_r (\mathrm{Cl}^{d-r-1}(\mathfrak{A}; x_1, \ldots, x_r) \cong \mathrm{Cl}^{d-r-1}(\mathfrak{C}; c_1, \ldots, c_r) \rightarrow$$
$$\exists x_{r+1} (\mathrm{Cl}^{d-r-1}(\mathfrak{A}; x_1, \ldots, x_{r+1}) \cong \mathrm{Cl}^{d-r-1}(\mathfrak{C}; c_1, \ldots, c_{r+1}))) \ .$$

Note the similarity to the extension axioms. Indeed, the closure operators are definable in $\mathcal{L}$, and the statement above can be expressed as a first-order axiom. The rest of the proof consists in showing that $\mathfrak{A} \sim^d \mathfrak{B}$ for any two structures satisfying this statement, and almost all structures satisfy it.

The proof of the convergence law for $\mathcal{L}$-structures with the usual successor is very similar. We add the constant 0 to $\mathcal{L}$ (with the obvious fixed interpretation in C_n'). For every $\sigma \in \mathcal{L}$, form σ' by replacing all formulas of the form $y = x + 1$ by $y = x + 1 \wedge y \neq 0$. Then, for every $n \in \omega$ and $R \subseteq \{0, \ldots, n-1\}^r$,

$$\langle \{0, \ldots, n-1\}, R, S_n \rangle \models \sigma \Leftrightarrow \langle \{0, \ldots, n-1\}, R, S_n', 0 \rangle \models \sigma' \ .$$

The proof is completed by showing that the convergence law holds for the expanded $\mathcal{L}$ and C_n'. The key idea is to show that if

$$\mathrm{Cl}^{d-r}(\mathfrak{A}; 0, a_1, \ldots, a_r) \cong \mathrm{Cl}^{d-r}(\mathfrak{B}; 0, b_1, \ldots, b_r)$$

then, with probability asymptotic to 1, player II can ensure that it holds for $r+1$. That is, the $\sim^d$ class of almost all $\mathfrak{A}$ is determined by $\mathrm{Cl}^d(\mathfrak{A}; 0)$. The probability that $\mathrm{Cl}^d(\mathfrak{A}; 0)$ is in a given isomorphism class is $u/2^v$ for some $u, v \in \omega$, for sufficiently large n.

There are countable versions of these convergence laws. The language $\mathcal{L}$ is the same, but C is either the class of structures $\langle \omega, R, S \rangle$, where S is the successor relation on ω, or $\langle \mathbb{Z}, R, S' \rangle$, where S' is the successor relation on $\mathbb{Z}$. Let pr be the standard product measure in the power set of all r-tuples on ω and pr$'$ be the standard product measure in the power set of all r-tuples on $\mathbb{Z}$. Then for every $\sigma \in \mathcal{L}$,

$$\mathrm{pr}'(\{R \subseteq \mathbb{Z}^r : \langle \mathbb{Z}, R, S' \rangle \models \sigma) = 0 \text{ or } 1, \text{ and}$$
$$\mathrm{pr}(\{R \subseteq \omega^r : \langle \omega, R, S \rangle \models \sigma) = u/2^v \text{ for some } u, v \in \omega.$$

Closure operators and extension axioms have also been used to prove a convergence law on arithmetic subsequences for structures with addition (mod n). For $n \in \omega$, let C_n be the class of models

$$\langle \{0, \ldots, n-1\}, R, + \pmod{n}, S_n \rangle$$

where $R \subseteq \{0, \ldots, n-1\}^r$, and $\mathcal{L}$ be the appropriate first-order language (with a symbol for $+ \pmod{n}$). For every $\sigma \in \mathcal{L}$, there is a positive $a \in \omega$ (dependent on the depth of σ) such that for all $b = 0, \ldots, a-1$, $\lim_{n\to\infty} \mathrm{pr}(\sigma, an+b) = u_b/2^{v_b}$ for some $u_b, v_b \in \omega$.

Again, there is an analogous result for countable structures. Let pr be the standard product measure in the power set of all r-tuples in $\mathbb{Z}$ and $\mathcal{L}$ be the first-order language of structures $\langle \mathbb{Z}, R, +, S\rangle$ where $R \subseteq \mathbb{Z}^r$, $+$ is the usual addition operator, and S is the successor relation. For every $\sigma \in \mathcal{L}$, $\mu(\sigma) = u/2^v$ for some $u, v \in \omega$.

Extension axioms played a major role in the proofs of the convergence laws for structures with addition, although the technical details are quite different and more involved than the proofs about structures with successor. They have also been used to prove 0-1 and convergence laws for a variety of other structures, such as unary functions (Lynch [29]) and random graphs with variable edge probabilities (Lynch [30], J. Spencer [39][1]).

Extension axioms are not the only way to establish the conditions (3) and (4). Another approach uses the fact that the $\equiv^d$ class of a structure is determined by the number (up to d) of components it has in each connected $\equiv^d$ class. In the following, $\mathfrak{C} \sqsubseteq \mathfrak{A}$ means that $\mathfrak{C}$ is a component of $\mathfrak{A}$.

Theorem 6. *Let $\mathfrak{A}$ and $\mathfrak{B}$ be two structures of the same type such that for every connected structure $\mathfrak{C}$ of that type, either*

$$|\{\mathfrak{D} \sqsubseteq \mathfrak{A} : \mathfrak{D} \equiv^d \mathfrak{C}\}|, |\{\mathfrak{D} \sqsubseteq \mathfrak{B} : \mathfrak{D} \equiv^d \mathfrak{C}\}| \geq d$$

or

$$|\{\mathfrak{D} \sqsubseteq \mathfrak{A} : \mathfrak{D} \equiv^d \mathfrak{C}\}| = |\{\mathfrak{D} \sqsubseteq \mathfrak{B} : \mathfrak{D} \equiv^d \mathfrak{C}\}| \ .$$

Then $\mathfrak{A} \equiv^d \mathfrak{B}$.

This theorem has been used by a number of authors. Its proof is an easy application of the Ehrenfeucht game. It was used by Lynch in [30] to prove a convergence law for certain classes of random graphs. K. Compton [5] also used it to derive a sufficient and necessary condition for 0-1 laws on slow growing classes of finite structures. This article is especially significant because it was the first to use nontrivial methods from enumeration theory, such as generating functions and Tauberian theorems.

3.3 Second-Order Logic

As described in Section 2.3, there has been some success in analyzing the definablity of graph properties in monadic second-order logic, particularly for existential monadic second-order logic. Although the expressive power of this logic is quite limited, it is strong enough to define graph properties that do not have asymptotic probabilities (even on arithmetic subsequences). M. Kaufmann and

[1] The 0-1 law in this paper was first proven by S. Shelah and Spencer [38], using extension axioms to describe a method of quantifier elimination. In [39], Spencer used the same extension axioms to give a proof based on the Ehrenfeucht game.

Shelah [22] proved this for the monadic second-order logic of several binary relations, and Kaufmann [21] later improved it to existential monadic second-order logic.

The monadic second-order logic of several unary relations is not so badly behaved—it has a 0-1 law. This is a rather uninteresting logic, and we do not have a citation for the proof. However, the same logic augmented with a binary relational symbol $\leq$ interpreted as linear order is quite interesting. As J. Büchi [4] and C. Elgot [11] showed, the class of properties definable in this logic corresponds in a very precise way to the class of languages accepted by finite state automata. Lynch [32] proved that it has a convergence law on arithmetic subsequences. To see that it does not have a convergence law, let R be a unary relational symbol. There is a first-order formula that defines the successor relation $y = x + 1$ on $\langle\{0, \ldots, n-1\}, \leq\rangle$ and first-order formulas that identify the smallest and largest elements in $\langle\{0, \ldots, n-1\}, \leq\rangle$. Informally, we will use the symbols 0 and $n - 1$. Then the property "$|R|$ is even" is definable by

$$\begin{aligned}\exists Q[&R(0) \leftrightarrow Q(0) \\ &\wedge \forall x \forall y (y = x + 1 \rightarrow ((Q(x) \wedge \neg R(y) \vee \neg Q(x) \wedge R(y)) \leftrightarrow Q(y)) \\ &\wedge \neg Q(n-1)]\ .\end{aligned}$$

In other words, $Q(x)$ indicates the parity of $R \cap \{0, \ldots, x\}$. There is a convergence law for the first-order logic of several unary relations with $\leq$, which appeared in [27], although the main idea behind the proof was due to Ehrenfeucht. It used the Ehrenfeucht game to characterize the $\equiv^d$ classes. A certain Markov chain whose states were these classes was constructed, and by basic results in the theory of finite Markov chains, the convergence law followed. The proof of the convergence law on arithmetic subsequences for the monadic second-order logic used a natural generalization of the Ehrenfeucht game to monadic second-order logic. Fagin's game (see Theorem 2) is a special case of it.

The monadic second-order game has also been used by R. Ladner [26] to study finite state automata, and by Compton [6] and A. Woods [41] to prove 0-1 and convergence laws.

3.4 Infinitary Logics

There is a 0-1 law for the $L^\omega_{\infty\omega}$ logic of random relational structures with a uniform probability distribution on structures of a given size (Ph. Kolaitis and M. Vardi [25]). It was the culmination of several earlier results for logics weaker than $L^\omega_{\infty\omega}$. A. Blass, Y. Gurevich, and D. Kozen [3] proved the 0-1 law for least fixed-point logic, and Kolaitis and Vardi [24] extended it to partial fixed-point logic. These proofs used the compactness theorem and a model theoretic characterization of $\aleph_0$-categorical theories. It is interesting to note that the $L^\omega_{\infty\omega}$ 0-1 law has an elementary proof that is essentially the same as the one given in Section 3.1 for the first-order 0-1 law. The only difference is that the eternal game (see Theorem 4) is used instead of the finite Ehrenfeucht game.

Other convergence laws rely on a generalization of the component counting technique provided by Theorem 6. Instead of counting $\equiv^d$ classes of components, we count $\equiv^d_\infty$ classes. Kolaitis [23] has used this approach to prove a 0-1 law for the $L^\omega_{\infty\omega}$ logic of random equivalence relations, and Lynch [33] used it to prove $L^\omega_{\infty\omega}$ convergence laws for some classes of random graphs.

4 Future Directions

A prime motivation for studying definability problems (and to a lesser extent, convergence laws) comes from computational complexity. As we have alluded to in Section 2.3, proving certain undefinablity results would settle some major open problems in complexity theory. This hope has been largely unrealized. Even though pebble games provide a sound and complete method for settling many undefinablity results, they have not been successful in proving any results that are directly relevant to computational complexity. There are indications that pebble games are the wrong tool for these kinds of problems.

For example, it is easy to generalize the Ehrenfeucht game to arbitrary second-order logics, which might seem like a promising approach to proving P $\neq$ NP, the holy grail of complexity theory. But a close analysis of the correspondence between NP and existential second-order logic suggests that pebble games may not be effective in proving undefinablity of properties in NP. It was shown by Lynch [31] that every property in NP is definable by an existential second-order sentence whose prefix is of the form $\exists Q_1 \ldots \exists Q_i \forall x_1 \ldots \forall x_j \exists y_1 \ldots y_k$, where $Q_1, \ldots, Q_i$ are second-order variables, and $x_1, \ldots, x_j, y_1, \ldots, y_k$ are first-order. Pebble games seem to be effective in proving undefinability results for logics where the expressive power of a sentence depends on the number of alternations of quantifiers. Here, much of the expressive power is contained in the existentially quantified second-order variables.

Another kind of problem arises in attempting to use pebble games for logics that characterize other complexity classes. Most of these logics have the $\leq$ relation interpreted as linear order. For example, P is characterized by least fixed-point logic with $\leq$ (N. Immerman [19] and Vardi [40]), and PSPACE is characterized by partial fixed-point logic with $\leq$ (Vardi [40]). Pebble games for least fixed-point and partial fixed-point logic can be formulated. They are similar to the eternal game, but with additional restrictions that make winning strategies quite difficult to discover. The $L^\omega_{\infty\omega}$ logic with $\leq$ and its pebble game are conceptually simpler, but it is too powerful for finite model theory. As shown in [9], any property of finite models with $\leq$ is definable in $L^2_{\infty\omega}$. For any finite model $\mathfrak{A}$ with $\leq$, there is a first-order sentence $\sigma_\mathfrak{A}$ with only two variables that holds for models $\mathfrak{B}$ if and only if $\mathfrak{A} \cong \mathfrak{B}$. Then, given any class $\mathcal{P}$ of finite models with $\leq$, $\mathcal{P}$ is definable by $\bigvee_{\mathfrak{A}\in\mathcal{P}} \sigma_\mathfrak{A}$.

All this is not meant to imply that there are no more frontiers that can be explored with pebble games. They have proven their utility in the model theory of first-order logic and infinitary logic without $\leq$. There are still many interesting problems in these areas, and pebble games will continue to play a central role in solving them.

References

1. Ajtai, M., Fagin, R.: Reachability is harder for directed than for undirected finite graphs. J. Symbolic Logic **55** (1990) 113–150
2. Barwise, J.: On Moschovakis closure ordinals. J. Symbolic Logic **42** (1977) 292–296
3. Blass, A., Gurevich, Y., Kozen, D.: A zero-one law for logic with a fixed point operator. Inform. and Control **67** (1985) 70–90
4. Büchi, J.: Weak second-order arithmetic and finite automata. Z. Math. Logik Grundlagen Math. **6** (1960) 66–92
5. Compton, K.: A logical approach to asymptotic combinatorics I. first order properties. Adv. Math. **65** (1987) 65–96
6. Compton, K.: A logical approach to asymptotic combinatorics II: monadic second-order properties. J. Comb. Theory, Ser. A **50** (1989) 110–131
7. Coppersmith, D.: A left coset composed of n-cycles. Research Report rc 19511, IBM (1994)
8. de Rougemont, M.: Second-order and inductive definability on finite structures. Z. Math. Logik Grundlagen Math. **33** (1987) 47–63
9. Ebbinghaus, H.-D., Flum, J.: Finite Model Theory. Springer-Verlag (1995)
10. Ehrenfeucht, A.: An application of games to the completeness problem for formalized theories. Fund. Math. **49** (1961) 129–141
11. Elgot, C.: Decision problems of finite-automata design and related arithmetics. Trans. AMS **98** (1961) 21–51
12. Fagin, R.: Generalized first-order spectra and polynomial-time recognizable sets. Complexity of Computation. R. Karp, ed. SIAM-AMS Proc. **7**, Am. Math. Soc., New York (1974) 43–73
13. Fagin, R.: Monadic generalized spectra. Z. Math. Logik Grundlagen Math. **21** (1975) 89–96
14. Fagin R.: Probabilities on finite models. J. Symbolic Logic **41** (1976) 50–58
15. Fraïssé, R.: Sur quelques classifications des systéms de relations. Pub. Sci. Univ. Alger Sér. A **1** (1954) 35–182
16. Gaifman, H.: On local and nonlocal properties. Logic Colloquium '81. J. Stern, ed. North-Holland (1982) 105–135
17. Glebskiĭ,Y., Kogan, D., Liogon'kiĭ, M., Talanov, V.: Range and degree of realizability of formulas in the restricted predicate calculus. Kibernetika (Kiev) **2** (1969) 17–28; English translation: Cybernetics **5** (1972) 142–154
18. Hanf, W.: Model-theoretic methods in the study of elementary logic. J. Addison, L. Henkin, A. Tarski, eds. The Theory of Models. North-Holland (1965) 132–145
19. Immerman, N.: Relational queries computable in polynomial time. Inform. and Control **68** (1986) 86–104
20. Kanellakis, P.: Elements of relational database theory. The Handbook of Theoretical Computer Science. A. Meyer, M. Nivat, M. Paterson, D. Perrin, J. van Leeuwen, eds. North-Holland (1990)
21. Kaufmann, M.: A counterexample to the 0-1 law for existential monadic second-order logic. Internal Note #032, Computational Logic Inc. (1988)
22. Kaufmann, M., Shelah, S.: On random models of finite power and monadic logic. Discrete Math. **54** (1985) 285–293

23. Kolaitis, Ph.: On asymptotic probabilities of inductive queries and their decision problem. R. Parikh, ed. Logics of Programs '85. Lecture Notes in Computer Science **193**, Springer-Verlag (1985) 153–166
24. Kolaitis, Ph., Vardi, M.: The decision problem for the probabilities of higher-order properties. Proc. 19th ACM Symp. on Theory of Computing (1987) 425–435
25. Kolaitis, Ph., Vardi, M.: Infinitary logics and 0-1 laws. Inform. and Computation **98** (1992) 258–294
26. Ladner, R.: Application of model theoretic games to discrete linear orders and finite automata. Inform. and Control **33** (1977) 281–303
27. Lynch, J.: Almost sure theories. Ann. Math. Logic **18** (1980) 91–135
28. Lynch, J.: On sets of relations definable by addition. J. Symbolic Logic **47** (1982) 659–668
29. Lynch, J.: Probabilities of first-order sentences about unary functions. Trans. AMS **287** (1985) 543–568
30. Lynch, J.: Probabilities of sentences about very sparse random graphs. Random Struct. Alg. **3** (1992) 33–53
31. Lynch, J.: The quantifier structure of sentences that characterize nondeterministic time complexity. Comput. Complexity **2** (1992) 40–66
32. Lynch, J.: Convergence laws for random words. Australasian J. of Combinatorics **7** (1993) 145–156
33. Lynch, J.: Infinitary logics and very sparse random graphs. Proc. Eighth Ann. IEEE Symp. on Logic in Computer Science (1993) 191–198; J. Symbolic Logic (to appear)
34. Monk, J.: Mathematical Logic. Springer-Verlag, New York (1976)
35. McKenzie, R., Mycielski, J., Thompson, D.: On boolean functions and connected sets. Math. Systems Theory **5** (1971) 259–270
36. Robinson, J.: Definablilty and decision problems in arithmetics. J. Symbolic Logic **14** (1949) 98–114
37. Schwentick, T.: On winning Ehrenfeucht games and monadic NP. Technical Report 3/95, Institut für Informatik, Johannes Gutenberg-Universität Mainz (1995)
38. Shelah, S., Spencer, J.: Zero-one laws for sparse random graphs. J. AMS **1** (1988) 97–115
39. Spencer, J.: Zero-one laws via the Ehrenfeucht game. Discrete Appl. Math. **30** (1991) 235–252
40. Vardi, M.: The complexity of relational query languages. Proc. 14th ACM Symp. on Theory of Computing (1982) 137–146
41. Woods, A.: Counting finite models. J. Symbolic Logic (to appear)

An Interpretive Isomorphism Between Binary and Ternary Relations*

Dale Myers

University of Hawaii, Honolulu, HI 96822 USA

Abstract. It is well-known that in first-order logic, the theory of a binary relation and the theory of a ternary relation are mutually interpretable, *i.e.*, each can be interpreted in the other. We establish the stronger result that they are interpretively isomorphic, *i.e.*, they are mutually interpretable by a pair of interpretations each of which is the inverse of the other.

Introduction

Our main tool is a Cantor–Bernstein type theorem which constructs an interpretive isomorphism from a suitable pair of interpretations. In order for the theorem to apply, one of the two theories must be repetitive which means roughly that the model space of the theory contains ω disjoint copies of itself. We show that the theory of a binary relation is repetitive and then use the Cantor–Bernstein result plus previously known mutual interpretations to show that any language with one or more relation symbols of two or more places is interpretively isomorphic with the theory of a binary relation.

Most previously known interpretative isomorphisms were rather simple, e.g., lattices with meet and join vs. lattices defined in terms of the partial order relation. Among the more complicated interpretive isomorphisms are the isomorphism between Euclidean geometry and the theory of real–closed fields and the isomorphism between Peano's arithmetic and ZFC with a class form of the axiom of foundation and with the axiom of infinity replaced by its negation. There seemed to be no such isomorphism between the theory of a binary relation and that of a ternary. That they are isomorphic seems surprising.

In contrast, we show that two trivial languages with only unary relations and constants are nonisomorphic if they have a different number of unary relations or a different number of number of constants. The theory of equality and the theory of the language with one constant, for example, are not isomorphic.

History

The prototype of our main result is a theorem of Hanf (Hanf 1962, 1974) that the Lindenbaum–Tarski algebra of the theory of a binary relation is isomorphic

* The author thanks Adam Gajdor, Wilfrid Hodges and Jan Mycielski for helpful comments and suggestions.

to the algebra of the theory of any finite language with an undecidable theory. In (Hanf and Myers 1983; Myers 1989) the author reformulated Hanf's results using elementary model maps. Here we show that with appropriate modifications, many of these elementary model map results apply to interpretations.

Interpretive isomorphism has not been extensively studied. It appears in (Manders 1980) under the name "dual interpretability". Mutual interpretability, also called "bilateral interpretability", has a more extensive history, see (Mycielski 1977), (Mycielski et al. 1990), (Szczerba 1977), (Szmielew and Tarski 1952), (Montague 1957), and (Bouvere 1965). Our Cantor–Bernstein type theorem shows that many previously known mutually interpretable pairs of theories are in fact interpretively isomorphic.

Composite–free Formulas

Definition. A formula is *composite–free* iff its atomic formulas are of one of the forms $R(\overline{u})$, $x = y$, $x = c$, $c = x$, $g(\overline{u}) = x$, or $x = g(\overline{u})$ where $\overline{u}$ is a sequence of variables and R, c, g are symbols for a relation, a constant and an operation.

Other atomic formulas such as $R(x, a)$, $a = b$, and $g(x, f(x)) = y$ are not allowed.

Lemma. *Every formula can be reduced to a logically equivalent composite–free formula.*

Proof. In each atomic formula with terms which need to be eliminated, replace the terms with new variables, conjoin the equations which define the new variables and then existentially quantify. For example, $R(a, g(x, f(x)))$ is equivalent to

$$\exists u_a, u_g, u_f R(u_a, u_g) \wedge u_a = a \wedge u_g = g(x, u_f) \wedge u_f = f(x) .$$

It is also equivalent to

$$\forall u_a, u_g, u_f [u_a = a \wedge u_g = g(x, u_f) \wedge u_f = f(x)] \Longrightarrow R(u_a, u_g) .$$

□

We will need partial operations. Typically we will have a composite structure $\langle C, \langle A, g, \text{---} \rangle, *** \rangle$ where g is an operation on A but only a partial operation on C. To handle this situation we identify partial n–ary operations with $n+1$–place relations which satisfy the ***structural axiom*** of being single valued and we identify n–place operations with single–valued and total $n+1$–place relations. For purposes of readability we will write $y = g(\overline{x})$ or $g(\overline{x}) = y$ instead of the relational form $g(\overline{x}, y)$. When g is partial, these are not regarded as equalities and $g(\overline{x})$ is not considered a term and may not be substituted into formulas. Thus we allow $z = \sqrt{x}$ and $\sqrt{x} = z$ but we don't allow $(\sqrt{x})^2 = x$ or $\sqrt{x} = \sqrt{x}$ and we may not substitute $\sqrt{x}$ into $z^2 = x$ or $z = z$. Informally, we will relax this restriction when the variables range over the operation's domain as in $\forall x \geq 0 (\sqrt{x})^2 = x$. Formally, we will be dealing with composite–free formulas where even full operations must appear in one of the forms $y = g(\overline{x})$ or $g(\overline{x}) = y$.

Model Classes

All sentences and theories will be in the finitary first–order logic $L_{\omega\omega}$ with equality. All languages will be countable. A *theory* is a consistent deductively closed set of sentences. For any set T of sentences, let $\mathrm{Mod}(T)$ = the class of models of T. $\mathcal{X}$ is an *axiomatizable* (or *elementary*) class of structures if $\mathcal{X} = \mathrm{Mod}(T)$ for some T. A *finitely axiomatizable subclass* of $\mathrm{Mod}(T)$ is a set of the form $\mathrm{Mod}(T \cup \{\sigma\})$ for some sentence σ.

To avoid foundational difficulties, assume all structures lie in some ZFC–like subuniverse, which is a set in the full universe. The subuniverse of sets of rank $< \omega + \omega$ suffices; it satisfies Zermelo's axioms. With this assumption, model classes are sets.

Composite Structures

Consider the identity map $\mathbf{id} \restriction \mathbf{Lin}$:$\mathbf{Lin} \longrightarrow \mathbf{Lin}$ which maps the class of linear orders to itself. Set theoretically, $\mathbf{id} \restriction \mathbf{Lin} = \{\langle\langle A,<\rangle, \langle A,<\rangle\rangle : \langle A,<\rangle \in \mathbf{Lin}\}$. We want to claim that $\mathbf{id} \restriction \mathbf{Lin}$ is an axiomatizable class. To make sense of this claim, we encode the pair $\langle\langle A,<\rangle, \langle A,<\rangle\rangle$ as a single composite model $\langle A, A, <, A, <\rangle$. For purposes of readability, we add superfluous brackets and write $\langle A, \langle A,<\rangle, \langle A,<\rangle\rangle$. An appropriate language for this composite model has two unary relation symbols A_1 and A_2 for the second and third occurrences of A and two binary relation symbols $<_1$ and $<_2$. The two occurrences of $\langle A,<\rangle$ are identified with structures of the language of linear orders in the obvious way.

In the following, ---, ***, and ... will represent lists of relations and/or operations (regard constants as 0–place operations). Thus, $\langle A,$ ---$\rangle$ could be $\langle A, +, \cdot, 0, 1, \leq, +\rangle$ and $(\exists C, \ldots)\psi$ could be $(\exists C, f, a, R)\psi$.

For any structures $\mathfrak{A} = \langle A,$ ---$\rangle$ and $\mathfrak{B} = \langle B,$ ***$\rangle$ of possibly different similarity types, for any C including A and B and for any additional relations or operations ..., let $\langle C, \mathfrak{A}, \mathfrak{B}, \ldots\rangle$ be the structure $\mathfrak{C} = \langle C, A$ ---, B, ***, ...$\rangle$ of a language with symbols to name the unary relations A and B and the relations and operations in the lists ---, *** and .... Adding superfluous $\langle\rangle$'s to group together the elements of $\mathfrak{A}$ and $\mathfrak{B}$ gives $\langle C, \langle A,$ ---$\rangle, \langle B,$ ***$\rangle, \ldots\rangle$.

In $\langle C, \mathfrak{A}, \ldots\rangle$, the operations of $\mathfrak{A} = \langle A,$ ---$\rangle$ have domain A. If $C \neq A$, these will be partial operations on C.

For each model class of structures of the form $\langle C, \langle A,$ ---$\rangle, \langle B,$ ***$\rangle, \ldots\rangle$ we have an associated set of *structural axioms* which state that A and B are nonempty and that A and B include the domains and ranges of the relations and operations of $\langle A,$ ---$\rangle$ and $\langle B,$ ***$\rangle$ respectively.

Let $C = A \uplus B$ mean that C is the disjoint union of A and B. Thus $\{\langle A \uplus B, \langle A,$ ---$\rangle, \langle B,$ ***$\rangle\rangle : \ldots\} = \{\langle C, \langle A,$ ---$\rangle, \langle B,$ ***$\rangle\rangle : C = A \uplus B$ and ... $\}$. The associated structural axiom states that A and B are disjoint and that the universe is $A \cup B$.

Let $X \setminus Y = \{x \in X : x \notin Y\}$. Let $|X|$ be the cardinality of X. For any function $f : X \longrightarrow Y$, the *kernel* of f is the equivalence relation on X which relates $a, b \in X$ iff $f(a) = f(b)$. The *ntuple* operation on a set A is the n-ary operation on A whose value $ntuple(x_1, \ldots, x_n)$ on n elements of A is the element $(x_1, \ldots, x_n)$ of A^n.

Gaifman Maps

In a structure $\langle C, \mathfrak{A}, \ldots \rangle$ with $\mathfrak{A} = \langle A, \text{---} \rangle$, an element c is *definable from parameters in* A iff there is a formula $\theta(\overline{a}, x)$ with parameters from A such that c is the unique element which satisfies $\theta(\overline{a}, x)$ in $\langle C, \mathfrak{A}, \ldots \rangle$.

Definition (Gaifman). H is a *Gaifman map* iff H is an axiomatizable class of models of the form $\langle C, \mathfrak{A}, \mathfrak{B}, \ldots \rangle$ such that

- each $\langle C, \mathfrak{A}, \mathfrak{B}, \ldots \rangle \in H$ is determined up to isomorphism by $\mathfrak{A}$ in the following strong sense: $\mathfrak{C} = \langle C, \mathfrak{A}, \mathfrak{B}, \ldots \rangle$ & $\mathfrak{C}' = \langle C', \mathfrak{A}', \mathfrak{B}', {*}{*}{*} \rangle \in H \Longrightarrow$ every isomorphism $g : \mathfrak{A} \cong \mathfrak{A}'$ extends to an isomorphism $\overline{g} : \mathfrak{C} \cong \mathfrak{C}'$ and
- each element of $C \setminus A$ is definable from parameters in A.

We write $H(\mathfrak{A}) = \mathfrak{B}$ iff $(\exists C, \text{---})(\langle C, \mathfrak{A}, \mathfrak{B}, \text{---} \rangle \in H)$. These maps determine functions in the usual sense if equality is interpreted as isomorphism. Note that the entire structure $\langle C, \mathfrak{A}, \mathfrak{B}, \ldots \rangle$, not just $\mathfrak{B}$, must be determined up to isomorphism.

These maps are special cases of the elementary model maps of (Hanf and Myers 1983) and (Myers 1989), see also (Pillay 1977). Gaifman (Gaifman 1979) conjectures that the definability condition follows from the "rigidity" condition of the next lemma.

Lemma (Rigidity Lemma). *If $\mathfrak{C} = \langle C, \mathfrak{A}, \mathfrak{B}, \ldots \rangle$ is a model of a Gaifman map H, then every automorphism g of $\mathfrak{A}$ extends uniquely to an automorphism $\overline{g}$ of $\mathfrak{C}$. Moreover, $g \mapsto \overline{g}$ is an isomorphism from the automorphism group of $\mathfrak{A}$ to the automorphism group of $\mathfrak{C}$.*

Proof. See (Hodges 1975). Suppose f and f' are two automorphisms of $\mathfrak{C}$ which extend g and suppose $c \in C$. Let $\theta(a_1, \ldots, a_n, x)$ be a formula with parameters in A which defines c. Then in $\mathfrak{C}$, $\theta(g(a_1), \ldots, g(a_n), f(c)) \Longleftrightarrow \theta(f(a_1), \ldots, f(a_n), f(c))$ $\Longleftrightarrow \theta(a_1, \ldots, a_n, c) \Longleftrightarrow \theta(f'(a_1), \ldots, f'(a_n), f'(c)) \Longleftrightarrow \theta(g(a_1), \ldots, g(a_n), f'(c))$. Thus $f(c) = f'(c)$ since they are both the unique element satisfying $\theta(g(a_1), \ldots, g(a_n), x)$. Uniqueness guarantees that the composition of two automorphisms extends to the composition of the two extensions. Thus $g \mapsto \overline{g}$ is a homomorphism of automorphism groups. This homomorphism is 1–1 since different g's extend to different $\overline{g}$'s. The map is onto since any automorphism of $\mathfrak{C}$ is an extension of its restriction to $\mathfrak{A}$. □

For model classes $\mathcal{X}$ and $\mathcal{Y}$, H is a Gaifman map *from* $\mathcal{X}$ to $\mathcal{Y}$, written $H : \mathcal{X} \longrightarrow \mathcal{Y}$, iff H is a Gaifman map and $H(\mathfrak{A}) = \mathfrak{B}$ implies $\mathfrak{A} \in \mathcal{X}$ and $\mathfrak{B} \in \mathcal{Y}$.

Lemma. *If H and J are definable as Gaifman maps, so is their composition.*

Proof. If $H : \mathcal{X} \longrightarrow \mathcal{Y}$ and $J : \mathcal{Y} \longrightarrow \mathcal{Z}$ are definable as Gaifman maps with models of the form $\langle D, \mathfrak{A}, \mathfrak{B}, \text{---} \rangle$ and $\langle E, \mathfrak{B}, \mathfrak{C}, \text{***} \rangle$ respectively, then $J \circ H : \mathcal{X} \longrightarrow \mathcal{Z}$ can be defined by the Gaifman map $\{\langle D \cup E, \mathfrak{A}, \mathfrak{C}, \mathfrak{B}, D, \text{---}, E, \text{***} \rangle : D \cap E = \text{the universe of } \mathfrak{B} \text{ and } \langle D, \mathfrak{A}, \mathfrak{B}, \text{---} \rangle \in H \text{ and } \langle E, \mathfrak{B}, \mathfrak{C}, \text{***} \rangle \in J\}$. □

Interpretations

In our interpretations, we allow relativization and cartesian products as well as definitional expansions. We also allow equality to be interpreted by a definable equivalence relation. To adequately handle theories with one–element structures, we add an operation which adds a point to a structure. For more on interpretations see (Mycielski 1990) and (Szczerba 1977).

An interpretation of a theory T into a theory S determines a sentence map h from sentences of T's language to sentences of S's language. It also determines a contravariant model map $H : \text{Mod}(S) \longrightarrow \text{Mod}(T)$. These maps satisfy the duality relationship: $\mathfrak{A} \models h(\sigma)$ iff $H(\mathfrak{A}) \models \sigma$.

A simple ***identity*** interpretation interprets a theory T into a stronger theory S with added axioms. The sentence map is the identity map $\sigma \longrightarrow \sigma$. The model map from Mod(S) into Mod(T) sends a model of S to itself.

The model map is always definable as a Gaifman map. Here are four examples of interpretations and their Gaifman maps.

1. There are two simple ways to interpret the theory of abelian groups with product * and unit e into the theory of fields with operations and constants $+, \cdot$, $0, 1$. The sentence map may interpret a group–theory sentence $\phi(*, e)$ additively as $\phi(+, 0)$ or multiplicatively as $\phi^U(\cdot, 1)$ where the universe and hence the range of the bound variables is restricted to the set U of nonzero elements. The two associated model maps send a field $\langle F, +, \cdot, 0, 1\rangle$ to the abelian grops $\langle F, +, 0\rangle$ and $\langle F \setminus \{0\}, \cdot, 1\rangle$ respectively.

The additive interpretation is definable with the Gaifman map $\{\langle F, \langle F, +, \cdot, 0, 1\rangle, \langle F, +, 0\rangle\rangle : \langle F, +, \cdot, 0, 1\rangle \text{ is a field}\}$. The multiplicative interpretation is definable by $\{\langle F, \langle F, +, \cdot, 0, 1\rangle, \langle F \setminus \{0\}, \cdot, 1\rangle\rangle : \langle F, +, \cdot, 0, 1\rangle \text{ is a field}\}$.

2. The theory of reflexive linear orders is interpretable into the theory of irreflexive linear orders with the model map which sends an irreflexive order $\langle X, <\rangle$ to the reflexive $\langle X, \leq\rangle$. The Gaifman map is $\{\langle X, \langle X <\rangle, \langle X, \leq\rangle\rangle : \langle X, <\rangle$ is a linear order and $\leq = (< \cup =)\}$. This interpretation has an inverse and is an interpretive isomorphism.

3. The theory of the complex numbers can be interpreted into the theory of the reals via the model map which sends a model $\langle R, +, \cdot, 0, 1\rangle$ of the reals to $\langle R \times R, +', \cdot', (0,0), (1,0)\rangle$ where (a, b) is identified with $a + \text{ib}$ and $(a, b) +' (c, d) = (a + c, b + d)$ and $(a, b) \cdot' (c, d) = (ac - bd, ad + bc)$.

The sentence map sends a sentence such as $\forall p\, p + 0 = p$ about the complexes first to the sentence $\forall a + \text{ib}(a + \text{ib}) + (0 + i0) = (a + \text{ib})$ then to the sentence $\forall a, b\; a + 0 = a \wedge b + 0 = b$ about the reals. Thus a single complex variable p is interpreted using two real variables a, b.

The model map can be defined as a Gaifman map consisting of structures isomorphic to $\langle R\,\dot\cup (R\times R),\ \langle R,+,\cdot,0,1\rangle,\ \langle R\times R,+',\cdot',(0,0),\ (1,0)\rangle,\ \pi_1,\pi_2\rangle$ where π_1 and π_2 are the projection functions from $R\times R$ to R and $+'$ and $\cdot'$ are defined as above, for example $+'$ is defined with the axiom $\forall x,\, y\in R\times R$ $x+'y=z \Longleftrightarrow \big[\pi_1(x)+\pi_1(y)=\pi_1(z)\wedge \pi_2(x)+\pi_2(y)=\pi_2(z)\big]$. An element (a,b) of $R\times R$ is uniquely definable by the formula $\pi_1(x)=a\wedge\pi_2(x)=b$.

4. The theory of groups can be interpreted into the theory of abelian groups by the identity interpretation with Gaifman map $\big\{\langle G,\langle G,\cdot\rangle,\langle G,\cdot\rangle\rangle : \langle G,\cdot\rangle$ is an abelian group$\big\}$.

Our theorems will be stated and proved in terms of model maps instead of sentence maps.

Definition. An *interpretation* is a model map H which can be constructed by applying some sequence $H_1,H_2,\ldots,H_n$ of the following standard interpretations, *i.e.*, $H(\mathfrak{A})=H_1(H_2(\ldots(H_n(\mathfrak{A}))\ldots))$.

THE STANDARD INTERPRETATIONS. We give a Gaifman map for each interpretation. The axiomatizability, isomorphism and definability conditions are easy to verify.

- *Identity interpretations* are maps of the form $\mathbf{id}\restriction\mathcal{X}:\mathcal{X}\longrightarrow\mathcal{Y}$ where $\mathcal{X}$ is an axiomatizable class and $(\mathbf{id}\restriction\mathcal{X})(\mathfrak{A})=\mathfrak{A}$. The Gaifman map is $\{\langle A,\mathfrak{A},\mathfrak{A}\rangle:\mathfrak{A}\in\mathcal{X}\}$. If $H:\mathcal{Y}\longrightarrow\mathcal{Z}$, and $\mathcal{X}\subseteq\mathcal{Y}$ then $H\circ(\mathbf{id}\restriction\mathcal{X})$ is the restriction of H to $\mathcal{X}$.
- *Reduct interpretations* send a structure of a language L to its reduct to a smaller language L'. The Gaifman map is $\big\{\langle A,\mathfrak{A},\mathfrak{A}'\rangle:\mathfrak{A}$ is an L–structure with universe A and $\mathfrak{A}'$ is its reduct to $L'\big\}$.
- *Definitional expansions* send a structure $\langle A,\ldots\rangle$ to a definitional expansion $\langle A,\ldots,\text{---}\rangle$. The Gaifman map is $\big\{\langle A,\langle A,\ldots\rangle,\langle A,\ldots,\text{---}\rangle\rangle :$ the relations and operations of the list --- are definable from those of ... by some fixed list of definitions$\big\}$.
- *Restrictions* (the sentence maps are relativizations) send a structure to a substructure whose universe is defined by some formula $\theta(x)$. We require $\theta(x)$ to define a nonempty subset U which is closed under the operations of the language. The Gaifman map is $\big\{\langle A,\mathfrak{A},\langle U,\text{---}\rangle\rangle : U$ is the subset of A defined by $\theta(x)$, and --- is the list of relations and operations of $\mathfrak{A}$ restricted to $U\big\}$.
- *Quotient interpretations* send a structure to the quotient of the structure modulo a congruence relation defined by some formula $\theta(x,y)$. The Gaifman map is $\big\{\langle A\,\dot\cup\, B,\mathfrak{A},\mathfrak{B},f\rangle : f:A\longrightarrow B$ is onto, $\mathfrak{B}$ is the quotient of $\mathfrak{A}$ by f, and the kernel of f is defined by $\theta(x,y)$, *i.e.*, $\forall x,y\in A\ f(x)=f(y)\Longleftrightarrow \mathfrak{A}\models\theta(x,y)\big\}$.
- *Cartesian power interpretations* send a structure $\mathfrak{A}=\langle A,\text{---}\rangle$ to $\mathfrak{A}^n$ along with the projections and a unary relation for A which is identified with the diagonal elements of A^n. The Gaifman map is $\big\{\langle A^n,\ \langle A,\text{---}\rangle,\ \langle A^n,\ \text{***},\ A,\pi_1,\pi_2,\ldots,\pi_n\rangle\rangle$: For each i, $\pi_i:A^n\longrightarrow A$ is an onto map and for

each n–tuple $(a_1, a_2, \ldots, a_n)$ of elements from A, there is a unique $b \in A^n$ such that $\pi_1(b) = a_1$, $\pi_2(b) = a_2, \ldots, \pi_n(b) = a_n$. A is identified with the diagonal elements $(a, a, \ldots, a)$ which are defined by the condition that for each i, $\pi_i(a) = a$. Finally, the relations and operations of *** are cartesian powers of the relations and operations of ---. Hence for a binary relation R of --- and the corresponding R' of ***, we would have $R'(x, y)$ iff for each i, $R(\pi_i(x), \pi_i(y))\}$. The given axioms guarantee that $\langle A^n, ***\rangle$ is isomorphic to $\langle A, \text{---}\rangle^n$. Here A^n and $\pi_1, \ldots, \pi_n$ are any set and functions which satisfy the given conditions. The conditions guarantee that they are isomorphic to the real A^n and real projections. The universe A^n can't actually be the real A^n since A isn't a subset of the real A^n.

- *Single–point extensions* add a new point to the universe. The Gaifman map is $\{\langle A \cup \{p\}, \langle A, \text{---}\rangle, \langle A \cup \{p\}, A, \text{---}\rangle\rangle : p \notin A\}$. The new point is definable as the unique element not in A.

Comments about the cartesian power interpretation. Our definition of cartesian power is isomorphic to the usual one except for the addition of the unary relation A and the projection operations. These are needed to formalize the typical informal definitions about n–tuples. The standard interpretation of the theory of complex numbers into the theory of real numbers requires these projection operations.

In the associated sentence map, a sentence about the cartesian power structure $\langle A^n, ***, A, \pi_1, \pi_2, \ldots, \pi_n\rangle$ can be interpreted as a sentence about $\langle A, \text{---}\rangle$ as follows. (1) Reduce the sentence to a composition–free sentence. (2) Replace each variable x by an n–tuple $(x_1, \ldots, x_n)$ and each constant c by $(c, c, \ldots, c)$. (3) Drop the ( )'s after the quantifiers and replace each atomic formula with a conjunction of n "pointwise" atomic formulas for $\mathfrak{A}$. For example,

$$\forall x\, R[x, c] \longrightarrow \forall (x_1, \ldots, x_n)\, R\big[(x_1, \ldots, x_n), (c, c, \ldots, c)\big] \longrightarrow \forall x_1, \ldots, x_n \; R[x_1, c] \wedge \cdots \wedge R[x_n, c]\,.$$

The other types of composite–free atomic sentences are interpreted similarly:

$$x = g[y] \longrightarrow (x_1, \ldots, x_n) = g\big[(y_1, \ldots, y_n)\big] \longrightarrow \big[x_1 = g(y_1) \wedge \cdots \wedge x_n = g(y_n)\big]\,.$$
$$x = \pi_i[y] \longrightarrow (x_1, \ldots, x_n) = \pi_i\big[(y_1, \ldots, y_n)\big] \longrightarrow x_1 = x_2 = \cdots = x_n = y_i\,.$$

Comments about the single–point extension. Adding a single–point is a trivial construction but it isn't on any standard list of interpretations. For theories whose models have two or more elements, the next lemma shows that the single–point extension is constructible from the other interpretations. But applying the usual interpretations to a one–element structure always yields another one–element structure. If we want to add points to one–element models, something like the single–point extension construction is required.

In the associated sentence map, a sentence about the single–point extension model $\langle A \cup \{p\}, A, \text{---}\rangle$ can be interpreted as a sentence about $\langle A, \text{---}\rangle$ as follows. (1) Reduce the sentence to a composite–free sentence. (2) Add a new constant symbol to name p and replace each quantified subformula $\exists x \phi(x)$ and $\forall x \phi(x)$

with $\exists x \in A\phi(x) \vee \phi(p)$ and $\forall x \in A\ \phi(x) \wedge \phi(p)$ respectively. (3) Replace an equality $p = p$ with true. The new point p is not in the domain or range of the operations of $\langle A, \text{---}\rangle$ and, after the first step, all variables range over A. Hence (4) replace any other equality involving p with false, e.g., $g(p) = x$, $p = g(x)$, and $x = p$ are false. (5) Replace any nonequality atomic formula involving p with false (the relations are on A).

Lemma. *Single–point extensions are constructible from the other interpretations for structures with two or more elements.*

Proof. Suppose $\langle A, \text{---}\rangle$ has two or more elements. Let $\langle A^2, A, \text{***}\rangle$ be the reduct of the cartesian power with the set A of diagonal elements moved to the front. Restricting the relations and operations of *** to A gives back the original relations and operations ---. This gives $\langle A^2, A, \text{---}\rangle$. Let $\sim$ be the equivalence relation on A^2 which is equality on A and which groups all other elements into one equivalence class p. Taking the quotient modulo $\sim$ gives $\langle A \uplus \{p\}, A, \text{---}\rangle$. □

Lemma. *The following extension maps are interpretations.*

- *Quotient extensions* are maps of the form $\langle A, U, \text{---}\rangle \mapsto \langle A \uplus B, \langle A, U, \text{---}\rangle, B, f\rangle$ where $f : U \longrightarrow B$ is an onto function whose kernel is defined by some fixed formula $\theta(x, y)$ which, in any $\langle A, U, \text{---}\rangle$, defines an equivalence relation on U. Thus $f(x) = f(y) \iff \langle A, U, \text{---}\rangle \models \theta(x, y)$.
- *Cartesian power extension* are maps of the form $\langle A, \text{---}\rangle \mapsto \langle A \uplus A^n, \langle A, \text{---}\rangle, \mathit{ntuple}\rangle$ where *ntuple* is an n–ary operation on A which maps $A \times A \times \cdots \times A$ 1–1 onto A^n. Here A^n can be the real cartesian product A^n or any other set which satisfies the given conditions and hence looks like A^n.

Proof. We prove the cartesian power case; the quotient case is similar. Given $\mathfrak{A} = \langle A, \text{---}\rangle$, add a single point (single–point extension plus definitional expansion) to get $\mathfrak{B} = \langle A \uplus \{p\}, \langle A, \text{---}\rangle, p\rangle$. Take a cartesian power to get $\mathfrak{B}^n = \langle (A \uplus \{p\})^{n+1}, \langle A, \text{---}\rangle^{n+1}, p, A \uplus \{p\}, \pi_1, \pi_2, \ldots, \pi_{n+1}\rangle\rangle$. A is definable from $A \uplus \{p\}$ and p. We identify A^n with the definable subset of $n+1$ tuples of the form $A^n \times \{p\}$. Then $\mathit{ntuple}(a_1, \ldots a_n)$ is definable as the unique element y of $A^n \times \{p\}$ such that $\pi_1(y) = a_1, \ldots, \pi_n(y) = a_n$. Hence $\langle A \uplus A^n, \langle A, \text{---}\rangle, \mathit{ntuple}\rangle$ is constructible from $\mathfrak{B}^n$ by a definitional expansion, reduct and restriction. □

Theorem (Barwise–Gaifman). *Let L be a language and Struct(L) the class of structures for L. Let T be a theory and Lang(T) the set of sentences of the language of T. For any interpretation of the language L into the theory T with sentence map $h : L \longrightarrow$ Lang(T) and model map $H : \mathrm{Mod}(T) \longrightarrow$ Struct(L) and for any sentence $\phi \in$ Lang(T), the following are equivalent:*

- $T \vdash [\phi \iff h(\theta)]$ *for some θ in L.*
- $H(\mathfrak{A}) = H(\mathfrak{B})$ *implies $\mathfrak{A} \models \phi$ iff $\mathfrak{B} \models \phi$ for all models $\mathfrak{A}$ and $\mathfrak{B}$ of T.*

Proof. This follows from Feferman's (Feferman 1968) extension of Craig's interpolation theorem to many–sorted logic. It was noted in 1971 by Gaifman (Gaifman 1974) and Barwise (Barwise 1973). □

Theorem (Gaifman–Beth Definability Theorem). *Suppose H is a Gaifman map with structures of the form $\langle C, \mathfrak{A}, \mathfrak{B}, \ldots\rangle$ with $\mathfrak{A} = \langle A, \text{---}\rangle$. Then any relation on A definable in $\langle C, \mathfrak{A}, \mathfrak{B}, \ldots\rangle$ is definable in $\mathfrak{A}$.*

Proof. Suppose $\phi(x, y)$ is a formula of the language of $\langle C, \mathfrak{A}, \mathfrak{B}, \ldots\rangle$ which defines a relation on A. Let a, b be new constants. Then $\phi(a, b)$ is a sentence about $\langle C, \langle\mathfrak{A}, a, b\rangle, \mathfrak{B}, \ldots\rangle$. If $\langle C, \langle\mathfrak{A}, a, b\rangle, \mathfrak{B}, \ldots\rangle$ and $\langle C', \langle\mathfrak{A}, a, b\rangle, \mathfrak{B}', \ldots'\rangle$ are models of H, then $\langle C, \langle\mathfrak{A}, a, b\rangle, \mathfrak{B}, \ldots\rangle \cong \langle C', \langle\mathfrak{A}, a, b\rangle, \mathfrak{B}', \ldots'\rangle$. Hence $\langle C, \langle\mathfrak{A}, a, b\rangle, \mathfrak{B}, \ldots\rangle \models \phi(a, b) \iff \langle C', \langle\mathfrak{A}, a, b\rangle, \mathfrak{B}', \ldots'\rangle \models \phi(a, b)$. Hence, by the previous theorem, $H \models [\phi(a, b) \iff \theta^A(a, b)]$ for some sentence $\theta(a, b)$ in the language of $\langle\mathfrak{A}, a, b\rangle$. The previous theorem is being applied to the interpretation of the language of $\langle\mathfrak{A}, a, b\rangle$ into the theory of H with model map $\langle C, \langle\mathfrak{A}, a, b\rangle, \mathfrak{B}, \ldots\rangle \mapsto \langle\mathfrak{A}, a, b\rangle$ and sentence map $\theta(a, b) \mapsto \theta^A(a, b)$ where $\theta^A(a, b)$ is $\theta(a, b)$ with quantifiers relativized to A. □

We can always add a point but we can't always pick a point.

Corollary. *The elementary model map $\langle A\rangle \mapsto \langle A, p\rangle$ which chooses a point from an equality structure is not definable as a Gaifman map, i.e., there is no Gaifman map of the form $\{\langle A, \langle A\rangle, \langle A, p\rangle, \ldots\rangle : p \in A\}$. Likewise, there is no Gaifman definable model map $\langle A, <\rangle \mapsto \langle A, <, p\rangle$ which picks a point from each linear order.*

Proof. If there were such a map for equality structures, then, by the theorem above, p would be definable in the equality structure $\langle A\rangle$. But no point is definable in an equality structure of two or more elements. Likewise, if there were such a map for linear orders, there would be a formula $\theta(x)$ of the theory of linear orders which picks a point from each linear order. This is impossible by (Myers 1976b). □

Lemma. *Suppose H is a Gaifman map with structures of the form $\langle C, \mathfrak{A}, \mathfrak{B}, g\rangle$ where $\mathfrak{A} = \langle A, \text{---}\rangle$ and $C \setminus A$ is the image of the partial operation g on A. Then H is the model map of an interpretation.*

Proof. Assume the hypothesis of the theorem and suppose g is n–ary. Let $A_1 = C \setminus A$. Thus $C = A \uplus A_1$. Since $A_1 = g[A^n]$, $\langle C, A, g\rangle$ is *rigid* over A, that is, the identity map id_C is the only automorphism which fixes A. If $\langle C, \mathfrak{A}, \mathfrak{B}, g\rangle$ and $\langle C, \mathfrak{A}, \mathfrak{B}', g\rangle \in H$, then, by the isomorphism condition, $\langle C, \mathfrak{A}, \mathfrak{B}, g\rangle \cong \langle C, \mathfrak{A}, \mathfrak{B}', g\rangle$ by an isomorphism which extends id_A. By rigidity, this isomorphism is id_C and hence $\mathfrak{B} = \mathfrak{B}'$. By Beth's theorem, B and all the relations and operations of $\mathfrak{B}$ are definable from A, g, and the relations and operations of $\mathfrak{A}$. Hence $\langle C, \mathfrak{A}, \mathfrak{B}, g\rangle$ is a definitional expansion of $\langle C, \mathfrak{A}, g\rangle = \langle A \uplus A_1, \mathfrak{A}, g\rangle$.

Let D_g be the n–ary relation $\{\overline{x} \in A^n : g(\overline{x}) \in C \setminus A\}$. Let $\equiv_g$ be the $2n$–ary relation on A defined by $\overline{x} \equiv_g \overline{y}$ iff $g(\overline{x}) = g(\overline{y})$. By the Gaifman–Beth

Definability Theorem, D_g and $\equiv_g$ are definable in $\mathfrak{A}$. $\langle A \uplus A_1, \mathfrak{A}, g\rangle$ is specified up to isomorphism as being the expansion and extension of $\mathfrak{A}$ such that $g : D_g \longrightarrow A_1$ is an onto map with kernel $\equiv_g$.

Now regard A^n as a set (instead of an n–ary relation on A). Thus the *ntuple* operation is the n–ary operation on A whose value $ntuple(x_1, \ldots, x_n)$ on n elements of A is the corresponding element of A^n. The image of this operation on the n–ary relation D_g is a corresponding unary relation U_g on the set A^n. The image of the $2n$–ary relation $\equiv_g$ on A is a corresponding equivalence relation E_g on U_g.

We now construct $\langle A \uplus A^n \uplus A_1, C, \mathfrak{A}, ntuple, D_g, \equiv_g, U_g, E_g, A_1, g', g\rangle$ from $\mathfrak{A}$ using interpretations. Given $\mathfrak{A}$, add (definitional expansion) a unary relation for A getting $\langle A, \mathfrak{A}\rangle$. Add (cartesian power extension) A^n and *ntuple* getting $\langle A \uplus A^n, \mathfrak{A}, ntuple\rangle$. Add (definitional expansion) D_g and $\equiv_g$. Add (definitional expansion) U_g and E_g. Add (quotient extension) A_1 as the quotient of U_g via a map $g' : U_g \longrightarrow A_1$ whose kernel is E_g. Define $g : D_g \longrightarrow A_1$ to be the n–ary partial operation on A with domain D_g such that for $(x_1, \ldots, x_n) \in D_g$, $g(x_1, \ldots, x_n) = g'(ntuple(x_1, \ldots, x_n))$. Let C be $A \uplus A_1$. Adding g and C gives $\langle A \uplus A^n \uplus A_1, C, \mathfrak{A}, ntuple, D_g, \equiv_g, U_g, E_g, A_1, g', g\rangle$. Taking a reduct and restriction gives $\langle C, \mathfrak{A}, g\rangle$. As noted above, $\langle C, \mathfrak{A}, \mathfrak{B}, g\rangle$ is a definitional expansion. A reduct and a restriction gives $\mathfrak{B}$. Since $\langle C, \mathfrak{A}, \mathfrak{B}, g\rangle \in H$, the $\mathfrak{B}$ we have constructed is $H(\mathfrak{A})$. □

Corollary. *H is the model map of an interpretation iff H is definable as a Gaifman map with structures of the form $\langle C, \mathfrak{A}, \mathfrak{B}, g_1, \ldots, g_n\rangle$ where $\mathfrak{A} = \langle A$, ---$\rangle$ and every element of $C \setminus A$ is a value of one of the partial operations g_i applied to elements of A.*

Proof. $\Longrightarrow$: In the cartesian power interpretation, the definable *ntuple* map does the job. In the single–point extension, the map which sends everything to the new point works. In quotient interpretations, the given function f works. In the other cases, $C \setminus A$ is empty. Finally, Gaifman maps with the described covering operations are closed under composition.

$\Longleftarrow$: First restrict the g_i's so that their ranges are disjoint and cover $C \setminus A$. Then apply the construction of g in the preceding lemma to each g_i. □

Gaifman conjectured (Gaifman 1974) and claimed (Gaifman 1979) a stronger more interesting version of the next theorem with the rigidity condition of the Rigidity Lemma replacing the definability condition in the definition of Gaifman maps. No proof was published.

Theorem (Gaifman's Interpretation Theorem). *The model maps defined by interpretations are exactly the model maps defined by Gaifman maps.*

Proof. By the preceding corollary, interpretations are Gaifman maps.

Suppose H is a Gaifman map with structures of the form $\langle C, \mathfrak{A}, \ldots\rangle$. Then in every such model, every element of $C \setminus A$ is definable by some formula $\theta(\overline{y}, x)$ and some list $\overline{a}$ of parameters from A. For each $\theta(\overline{y}, x)$, we may assume without loss of generality that $\vdash \forall \overline{y} \in A(\exists \text{ at most one } x)\theta(\overline{y}, x)$. If not, replace $\theta(\overline{y}, x)$ with $\theta(\overline{y}, x) \wedge \neg\exists x, x'[x \neq x' \wedge \theta(\overline{y}, x) \wedge \theta(\overline{y}, x')]$.

Only finitely many such formulas are needed. Otherwise we could add a new constant c and the theory $\{\forall \overline{y} \in A \neg\theta(\overline{y}, c) : \forall \overline{y} \in A(\exists \text{ at most one } x)\theta(\overline{y}, x)\}$ would be finitely consistent. By compactness, there would be a model with an undefinable element c. Hence only finitely many formulas are needed, say $\theta_1, \ldots, \theta_n$.

These formulas define n partial operations $g_1, \ldots, g_n$ which satisfy the conditions of the preceding corollary. □

Are there any other constructions which deserve to be called "interpretations"? We feel that any construction which is sufficiently first-order, sufficiently well-defined and sufficiently constructive to be an "interpretation" should be definable as a Gaifman map. Thus Gaifman's theorem implies that our list of standard interpretations is complete. It would not be complete if single-point extensions were omitted.

Embeddings and Isomorphisms

Definition. For any interpretations $H : \mathcal{X} \longrightarrow \mathcal{Y}$ and $J : \mathcal{X} \longrightarrow \mathcal{Y}$:

- A Gaifman map H is a *Gaifman embedding* iff (1) each $\langle C, \mathfrak{A}, \mathfrak{B}, \ldots \rangle \in H$ is determined up to isomorphism by $\mathfrak{B}$, *i.e.*, $\mathfrak{C} = \langle C, \mathfrak{A}, \mathfrak{B}, \ldots \rangle$ and $\mathfrak{C}' = \langle C', \mathfrak{A}', \mathfrak{B}', *** \rangle \in H \Longrightarrow$ every isomorphism $g : \mathfrak{B} \cong \mathfrak{B}'$ extends to an isomorphism $\overline{g} : \mathfrak{C} \cong \mathfrak{C}'$, and (2) each element of $C \setminus B$ is definable from parameters in B where B is the universe of $\mathfrak{B}$.
- H and J are *isomorphic*, written $H \cong J$, iff $(\forall \mathfrak{A}, \mathfrak{B}, \mathfrak{B}')(H(\mathfrak{A}) = \mathfrak{B}$ and $J(\mathfrak{A}) = \mathfrak{B}'$ implies $\mathfrak{B} \cong \mathfrak{B}')$.
- H is an *interpretive isomorphism* from $\mathcal{X}$ to $\mathcal{Y}$, written $H : \mathcal{X} \cong \mathcal{Y}$, iff for some interpretation $G : \mathcal{Y} \longrightarrow \mathcal{X}$, $G \circ H \cong \mathbf{id} \restriction \mathcal{X}$ and $H \circ G \cong \mathbf{id} \restriction \mathcal{Y}$.
- $\mathcal{X}$ and $\mathcal{Y}$ are *interpretively isomorphic*, written $\mathcal{X} \cong \mathcal{Y}$, iff $H : \mathcal{X} \cong \mathcal{Y}$ for some H.
- $H : \mathcal{X} \leq \mathcal{Y}$, "$\mathcal{X} \leq \mathcal{Y}$ *via* H", iff $H : \mathcal{X} \cong \mathcal{Z}$ for some finitely axiomatizable subclass $\mathcal{Z}$ of $\mathcal{Y}$.

Lemma. *If H is a Gaifman embedding of $\mathcal{X}$ onto $\mathcal{Y}$, then $H^{-1} = \{\langle C, \mathfrak{B}, \mathfrak{A}, \ldots \rangle : \langle C, \mathfrak{A}, \mathfrak{B}, \ldots \rangle \in H\}$ is a Gaifman embedding of $\mathcal{Y}$ onto $\mathcal{X}$. Hence H is a Gaifman isomorphism and $\mathcal{X}$ and $\mathcal{Y}$ are interpretively isomorphic.*

Proof. Clear. □

If $\mathcal{X}$ is a finitely axiomatizable subclass of $\mathcal{Y}$, then $\mathcal{X} \leq \mathcal{Y}$ via the identity interpretation $\mathbf{id} \restriction \mathcal{X}$. If the models of $\mathcal{Y}$ are definitional expansions of those of $\mathcal{X}$ via some fixed list of definitions, then $\mathcal{X} \cong \mathcal{Y}$ via a definitional expanison interpretation. Cartesian power interpretations, and cartesian, quotient and single point extensions are Gaifman embeddings. Reducts, restrictions, and quotient interpretations need not be embeddings.

Lemma (Automorphism Group Lemma). *If $H(\mathfrak{A}) = \mathfrak{B}$ where H is a Gaifman embedding, then $\mathfrak{A}$ and $\mathfrak{B}$ have isomorphic automorphism groups.*

Proof. Suppose H is a Gaifman embedding and $\mathfrak{C} = \langle C, \mathfrak{A}, \mathfrak{B}, \ldots \rangle \in H$. By the Rigidity Lemma, the automorphism groups of $\mathfrak{A}$ and $\mathfrak{B}$ are both isomorphic to the automorphism group of $\mathfrak{C}$. □

ω–like Model Classes

Let **Eq** be the class of equality structures, that is, structures (sets essentially) for the empty similarity type. In **Eq**, let **1**, **2**, ..., be the finitely axiomatizable subclasses of the one element, two element, ... models of **Eq**.

Let **FinLin**, "**Fin**itary **Lin**ear orders", be the class of discrete irreflexive linear orders $\langle B, < \rangle$ with first and last elements. "Discrete" means that every element after the first has an immediate predecessor and every element before the last has an immediate successor. In **FinLin**, let **1**, **2**, ..., be the finitely axiomatizable subclasses of one element, two element, ... linear orders. All infinite members of **FinLin** are elementarily equivalent to orders of type $\omega + \omega^*$.

FinSuc. Let **FinSuc**, "**Fin**itary **Suc**cessor", be the axiomatizable class of unary operation structures $\langle A, f \rangle$ such that:

- There is a unique "initial" point a which is not the value of any other point.
- There is a unique "terminal" point b such that $f(b) = b$.
- There are no other cycles: $f^n(x) \neq x$ if $x \neq b$ and $n > 0$.
- f maps $A \setminus \{b\}$ $1-1$ onto $A \setminus \{a\}$.
- For any n, if $f^n(a) = b$ then $A = \{a, f(a), f^2(a), \ldots, f^{n-1}(a), b\}$.

In **FinSuc**, let **1**, **2**, ..., be the finitely axiomatizable subclasses of the one element, two element, ... models of **FinSuc**. The structures of **1**, **2**, **3**, and **4** are as pictured with initial points at the left.

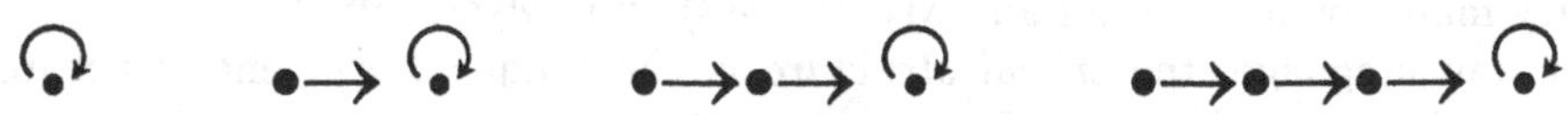

Eq, **FinLin** and **FinSuc** are the smallest axiomatizable classes containing their subclasses **1**, **2**, ... If elementarily equivalent structures are identified, each is homeomorphic to the one–point compactification of ω. **Eq** and **FinLin** are finitely axiomatizable; **FinSuc** is not.

Definition. A model is ***rigid*** iff the identity is the only automorphism.

In a finite rigid model all elements are definable.

Definition. An axiomatizable model class O is *ω–like* iff (1) it has a finitely axiomatizable subclass **1** which consists of the isomorphism type of a finite rigid model and (2) for some interpretive isomorphism **Succ**, **Succ**: $O \cong O \setminus$ **1**.

Lemma. FinLin *is* ω*-like.*

Proof. Let **1** be as in the definition of **FinLin**. Let **Succ**: **FinLin** $\cong$ **FinLin**$\setminus$ **1** be the map which adds a single new last point. Thus **Succ(1)** = **2**, **Succ(2)** = **3**, ...

To construct **Succ**($\langle A, <\rangle$) as an interpretation, add a new point (single–point extension) to get $\langle A \uplus \{p\}, A, <\rangle$. Define $<'$ to be $(<) \cup (A \times \{p\})$ and add it (definitional expansion) to get $\langle A \uplus \{p\}, \langle A, <\rangle, <'\rangle$. Then reduct to $\langle A \uplus \{p\}, <'\rangle$. The inverse is the restriction interpretation which deletes the last element. □

Lemma. FinSuc *is* ω*-like.*

Proof. Let **1** be as in the definition of **FinSuc**. Let **Succ**: **FinSuc** $\cong$ **FinSuc**$\setminus$**1** be the map which adds a predecessor to the initial point. This new point becomes the new initial point. **Succ** can be constructed as an interpretation as in the previous lemma. □

Theorem. Eq *is not* ω*-like.*

Proof. Suppose **Eq** were ω*-like.* Thus there is an interpretive isomorphism **Succ** and a finitely axiomatizable subclass **1** consisting of the isomorphism type of a finite rigid model such that **Succ**: **Eq** $\cong$ **Eq**$\setminus$**1**. The class **1** must be the isomorphism type of a 1–element equality structure since no other equality structure is rigid. By the Automorphism Group Lemma, **Succ(1)** must also be rigid which is impossible since **Succ(1)** $\in$ **Eq**$\setminus$**1**. □

Repetition and the Interpretive Cantor–Bernstein Theorem

In this section and the next, the definitions and theorems regarding interpretations are virtually identical to the analogous definitions and theorems for elementary maps found in (Hanf and Myers 1983) and (Myers 1989).

For any language, the ***trivial structure*** is the unique 1–element structure whose relations are all empty. Let *true* and *false* be the true and false truth values. We write "=" instead of "$\Longleftrightarrow$" between truth values. We have defined the disjoint union relation $C = A \uplus B$ for sets. For classes of models, we'll define an operation.

Definition. Suppose $\mathcal{X}, \mathcal{Y}, \mathcal{U}$, and $\mathcal{V}$ are axiomatizable classes and $H : \mathcal{X} \longrightarrow \mathcal{U}$ and $J : \mathcal{Y} \longrightarrow \mathcal{V}$ are Gaifman maps. Suppose the models of H and J are of the form $\langle D, \mathfrak{A}, \mathfrak{B}, D, \text{---}\rangle$ and $\langle E, \mathfrak{A}, \mathfrak{B}, E, \text{***}\rangle$ respectively where the symbols for --- and *** are disjoint. Let L be the language whose symbols are those of $\mathcal{X}$, those of $\mathcal{Y}$, a new constant a and a new propositional symbol p. For $\mathfrak{A}$ in $\mathcal{X}$ or $\mathcal{Y}$, let $\langle \mathfrak{A} \uplus \{a\}, a, p\rangle = \langle A \uplus \{a\}, A, \ldots, a, p\rangle$ be a structure for L such that the relations and operations (including constants) of "..." are interpreted as in $\mathfrak{A}$ if they are symbols of $\mathfrak{A}$ with total operations becoming partial. Any other symbols of L are interpreted as the empty relation or the constant operation whose value is that of a (this is the only reason for adding a).

Let $\mathcal{X} \uplus \mathcal{Y} = \{\langle \mathfrak{A} \uplus \{a\}, a, p\rangle : (\mathfrak{A} \in \mathcal{X} \wedge p = true) \vee (\mathfrak{A} \in \mathcal{Y} \wedge p = false)\}$.

Let $H \uplus J : \mathcal{X} \uplus \mathcal{Y} \longrightarrow \mathcal{U} \uplus \mathcal{V}$ be the Gaifman map $\{\langle D \uplus E \uplus \{a\}, \langle \mathfrak{A} \uplus \{a\}, a, p\rangle, \langle \mathfrak{B} \uplus \{a\}, a, p\rangle, D, ---, E, ***\rangle : D \cap E = \emptyset, \langle \mathfrak{A} \uplus \{a\}, a, p\rangle \in \mathcal{X} \uplus \mathcal{Y}, \langle \mathfrak{B} \uplus \{a\}, a, p\rangle \in \mathcal{U} \uplus \mathcal{V}$ and either (1) $p = true$, $\langle D, \mathfrak{A}, \mathfrak{B}, ---\rangle \in H$, and $\langle E, ***\rangle$ is the trivial structure or (2) $p = false$, $\langle E, \mathfrak{A}, \mathfrak{B}, ***\rangle \in J$, and $\langle D, ---\rangle$ is trivial$\}$. □

Lemma. *For any axiomatizable classes $\mathcal{X}$, $\mathcal{Y}$, $\mathcal{Z}$, $\mathcal{U}$, $\mathcal{V}$:*

(a) $\mathcal{X} \uplus \mathcal{Y} \cong \mathcal{U} \uplus \mathcal{V}$ if $\mathcal{X} \cong \mathcal{U}$ and $\mathcal{Y} \cong \mathcal{V}$.
(b) $\mathcal{X} \uplus \mathcal{Y} \cong \mathcal{Y} \uplus \mathcal{X}$.
(c) $\mathcal{X} \uplus (\mathcal{Y} \uplus \mathcal{Z}) \cong (\mathcal{X} \uplus \mathcal{Y}) \uplus \mathcal{Z}$.
(d) $\mathcal{X} \cong \mathcal{U} \uplus \mathcal{V}$ if $\mathcal{U}$ and $\mathcal{V}$ are complementary axiomatizable subclasses of $\mathcal{X}$.

Proof. (a) $H \uplus J : \mathcal{X} \uplus \mathcal{Y} \cong \mathcal{U} \uplus \mathcal{V}$ if $H : \mathcal{X} \cong \mathcal{U}$ and $J : \mathcal{Y} \cong \mathcal{V}$.
(b) Let $H : \mathcal{X} \uplus \mathcal{Y} \cong \mathcal{Y} \uplus \mathcal{X}$ be $\{\langle A \uplus \{a\}, \langle \mathfrak{A} \uplus \{a\}, a, p\rangle, \langle \mathfrak{A} \uplus \{a\}, a, \neg p\rangle\rangle : \langle \mathfrak{A} \uplus \{a\}, a, p\rangle \in \mathcal{X} \uplus \mathcal{Y}\}$.
(c) Send $\langle \mathfrak{A} \uplus \{a\}, a, true\rangle \in \mathcal{X} \uplus (\mathcal{Y} \uplus \mathcal{Z})$ ($\therefore \mathfrak{A} \in \mathcal{X}$) to $\langle \langle \mathfrak{A} \uplus \{a\}, a, true\rangle \uplus \{b\}, b, true\rangle \in (\mathcal{X} \uplus \mathcal{Y}) \uplus \mathcal{Z}$.
Send $\langle \langle \mathfrak{A} \uplus \{a\}, a, true\rangle \uplus \{b\}, b, false\rangle \in \mathcal{X} \uplus (\mathcal{Y} \uplus \mathcal{Z})$ ($\therefore \mathfrak{A} \in \mathcal{Y}$) to $\langle \langle \mathfrak{A} \uplus \{a\}, a, false\rangle \uplus \{b\}, b, true\rangle \in (\mathcal{X} \uplus \mathcal{Y}) \uplus \mathcal{Z}$.
Send $\langle \langle \mathfrak{A} \uplus \{a\}, a, false\rangle \uplus \{b\}, b, false\rangle \in \mathcal{X} \uplus (\mathcal{Y} \uplus \mathcal{Z})$ to $\langle \mathfrak{A} \uplus \{a\}, a, false\rangle \in (\mathcal{X} \uplus \mathcal{Y}) \uplus \mathcal{Z}$.
(d) Let $H : \mathcal{X} \longrightarrow \mathcal{U} \uplus \mathcal{V}$ be $\{\langle A, \mathfrak{A}, \langle \mathfrak{A} \uplus \{a\}, a, p\rangle\rangle : (\mathfrak{A} \in \mathcal{U} \wedge p = true) \vee (\mathfrak{A} \in \mathcal{V} \wedge p = false)\}$. □

Note that $\langle A \uplus B, \mathfrak{A}, \mathfrak{B}\rangle \cong \langle A' \uplus B', \mathfrak{A}', \mathfrak{B}'\rangle$ iff $\mathfrak{A} \cong \mathfrak{A}'$ and $\mathfrak{B} \cong \mathfrak{B}'$. Hence $\langle A \uplus B, \mathfrak{A}, \mathfrak{B}\rangle$ serves as an ordered pair.

Definition. Suppose $\mathcal{X}$ and $\mathcal{Y}$ are axiomatizable classes and $H : \mathcal{X} \longrightarrow \mathcal{U}$ and $J : \mathcal{Y} \longrightarrow \mathcal{V}$ are Gaifman maps.

Let $\mathcal{X} \times \mathcal{Y} = \{\langle A \uplus B, \mathfrak{A}, \mathfrak{B}\rangle : A \cap B = \emptyset, \mathfrak{A} = \langle A, ---\rangle \in \mathcal{X}$, and $\mathfrak{B} = \langle B, ***\rangle \in \mathcal{Y}\}$ where the symbols of *** are renamed if necessary in order to be disjoint from those of ---.

Let $H \times J : \mathcal{X} \times \mathcal{Y} \longrightarrow \mathcal{U} \times \mathcal{V}$ be the map $\{\langle F \uplus G, \langle A \uplus B, \mathfrak{A}, \mathfrak{B}\rangle, \langle C \uplus D, \mathfrak{C}, \mathfrak{D}\rangle, F, ---', G, ***'\rangle : F \cap G = \emptyset, \langle F, \mathfrak{A}, \mathfrak{C}, ---'\rangle \in H$, and $\langle G, \mathfrak{B}, \mathfrak{D}, ***'\rangle \in J\}$.

Lemma. *For any axiomatizable classes $\mathcal{X}$, $\mathcal{Y}$, $\mathcal{U}$, and $\mathcal{V}$, and any axiomatizable class $\mathbf{1}$ consisting of the isomorphism type of a finite rigid model:*

(a) $\mathcal{X} \times \mathcal{Y} \cong \mathcal{U} \times \mathcal{V}$ if $\mathcal{X} \cong \mathcal{U}$ and $\mathcal{Y} \cong \mathcal{V}$.
(b) $\mathcal{X} \times \mathbf{1} \cong \mathcal{X}$.

Proof. (a) $H \times J : \mathcal{X} \times \mathcal{Y} \cong \mathcal{U} \times \mathcal{V}$ if $H : \mathcal{X} \cong \mathcal{U}$ and $J : \mathcal{Y} \cong \mathcal{V}$.
(b) $H : \mathcal{X} \times \mathbf{1} \cong \mathcal{X}$ where $H = \{\langle A \uplus B, \langle A \uplus B, \mathfrak{A}, \mathfrak{B}\rangle, \mathfrak{A}\rangle : \mathfrak{A} = \langle A, \ldots\rangle \in \mathcal{X}$ and $\mathfrak{B} = \langle B, ***\rangle \in \mathbf{1}\}$. □

Note (Gajdor). If $\mathbf{1}$ and $\mathbf{1}'$ are finitely axiomatizable isomorphism types of two finite rigid models, then $\mathbf{1} \cong \mathbf{1}'$. By (d) above, $\mathbf{1} \cong \mathbf{1} \times \mathbf{1}' \cong \mathbf{1}'$.

Definition. For any axiomatizable classes $\mathcal{X}$ and $\mathcal{Y}$:

- $\mathcal{X} \leq \mathcal{Y}$ iff $\mathcal{X} \leq \mathcal{Y}$ via some interpretative isomorphism iff $\mathcal{X} \cong \mathcal{Z} \subseteq \mathcal{Y}$ for some finitely axiomatizable subclass $\mathcal{Z}$ of $\mathcal{Y}$.
- $\mathcal{X}$ *repeats* in $\mathcal{Y}$ iff $\mathcal{X} \times O \leq \mathcal{Y}$ for some ω–like axiomatizable class O. We say $\mathcal{X}$ *repeats in* $\mathcal{Y}$ *over* O when O is the ω–like space involved.
- $\mathcal{X}$ is *repetitive* iff $\mathcal{X}$ repeats in itself.

Lemma. *For axiomatizable classes $\mathcal{X}, \mathcal{Y}, \mathcal{U}, \mathcal{V}$:*

(a) $\leq$ is transitive.
(b) if $\mathcal{X} \leq \mathcal{Y}$, $\mathcal{Y}$ repeats in $\mathcal{U}$, and $\mathcal{U} \leq \mathcal{V}$, then $\mathcal{X}$ repeats in $\mathcal{V}$.
(c) if $\mathcal{X}$ repeats in $\mathcal{Y}$, then $\mathcal{Y} \cong \mathcal{X} \uplus \mathcal{Y}$.

Proof. (a) Clear.
(b) If $\mathcal{X} \leq \mathcal{Y}$, $\mathcal{Y}$ repeats in $\mathcal{U}$ over O, and $\mathcal{U} \leq \mathcal{V}$, then $\mathcal{X} \times O \leq \mathcal{Y} \times O \leq \mathcal{U} \leq \mathcal{V}$.
(c) Suppose $\mathcal{X}$ repeats in $\mathcal{Y}$ over O and $\mathbf{1}$ is the isomorphism type of the finite rigid structure such that $O \cong (O \backslash \mathbf{1})$. Then $H : \mathcal{X} \times O \cong \mathcal{X}_\omega$ for some H and some finitely axiomatizable subclass $\mathcal{X}_\omega$ of $\mathcal{Y}$. Let $\sim \mathcal{X}_\omega = \mathcal{Y} \setminus \mathcal{X}_\omega$. Let $\mathcal{X}_1$ and $\mathcal{X}_{\omega-1}$ be the images under H of the finitely axiomatizable subclasses $\mathcal{X} \times \mathbf{1}$ and $\mathcal{X} \times (O \backslash \mathbf{1})$. Then $\mathcal{X} \cong \mathcal{X} \times \mathbf{1} \cong \mathcal{X}_1$ and, since $O \cong (O \backslash \mathbf{1})$, $\mathcal{X}_\omega \cong \mathcal{X} \times O \cong \mathcal{X} \times (O \backslash \mathbf{1}) \cong \mathcal{X}_{\omega-1}$. Thus $\mathcal{Y} \cong \mathcal{X}_\omega \uplus (\sim \mathcal{X}_\omega) \cong (\mathcal{X}_1 \uplus \mathcal{X}_{\omega-1}) \uplus (\sim \mathcal{X}_\omega) \cong (\mathcal{X} \uplus \mathcal{X}_\omega) \uplus (\sim \mathcal{X}_\omega) \cong \mathcal{X} \uplus (\mathcal{X}_\omega \uplus \sim \mathcal{X}_\omega) \cong \mathcal{X} \uplus \mathcal{Y}$. □

Theorem (Interpretive Cantor–Bernstein Theorem). *For any axiomatizable classes $\mathcal{X}$ and $\mathcal{Y}$ such that either $\mathcal{X}$ or $\mathcal{Y}$ is repetitive, $\mathcal{X} \leq \mathcal{Y}$ and $\mathcal{Y} \leq \mathcal{X}$ implies $\mathcal{X} \cong \mathcal{Y}$.*

Proof. Suppose $\mathcal{X}$ repeats in itself, $\mathcal{X} \leq \mathcal{Y}$, and $\mathcal{Y} \leq \mathcal{X}$. We want $\mathcal{X} \cong \mathcal{Y}$.

$\mathcal{Y} \leq \mathcal{X}$ and $\mathcal{X}$ repeats in $\mathcal{X} \Longrightarrow$, by (b) of the previous lemma, $\mathcal{Y}$ repeats in $\mathcal{X} \Longrightarrow$, by (c) of the previous lemma, $\mathcal{X} \cong \mathcal{Y} \uplus \mathcal{X}$.

$\mathcal{X}$ repeats in $\mathcal{X}$ and $\mathcal{X} \leq \mathcal{Y} \Longrightarrow$, by (b) of the previous lemma, $\mathcal{X}$ repeats in $\mathcal{Y} \Longrightarrow$, by (c) of the previous lemma, $\mathcal{Y} \cong \mathcal{X} \uplus \mathcal{Y}$.

Thus $\mathcal{X} \cong \mathcal{Y} \uplus \mathcal{X} \cong \mathcal{X} \uplus \mathcal{Y} \cong \mathcal{Y}$. □

Repetitive Theories

We call a theory with no axioms a *language*. We will use the same name for a language and its class of structures and the same name for a theory and its class of models. In theorem statements, $\mathbf{Lang}_2$, will be the language of a binary relation and **FinLin** will be the theory of finite linear orders. In proofs, $\mathbf{Lang}_2$ and **FinLin** will be corresponding classes of structures.

Lemma. $\mathbf{Lang}_2$ *is repetitive, where* $\mathbf{Lang}_2$ *is the language of a binary relation.*

Proof. We want an H such that $H : \mathbf{Lang_2} \times \mathbf{FinLin} \leq \mathbf{Lang_2}$. For any binary relation S, let $U_S = \{x : \exists y\ y \neq x \wedge ySx\}$ and consider y a predecessor of x if ySx. For the relation S pictured below, the 3 points in the A box and the 2 points in the B box are in U_S. The points in A have one irreflexive predecessor not in U_S; those in B have one reflexive predecessor not in U_S.

Let $H = \{\langle C, \langle A \uplus B, \langle A, R\rangle, \langle B, <\rangle\rangle, \langle C, S\rangle\rangle : A \uplus B = U_S$, every element not in $A \uplus B$ is the unique predecessor of a unique element in $A \uplus B$, $A = \{x \in A \uplus B : x$ has an irreflexive predecessor not in $A \uplus B\}$, $B = \{x \in A \uplus B : x$ has a reflexive predecessor not in $A \uplus B\}$, $\langle A, R\rangle$, $\langle B, <\rangle$ are substructures of $\langle C, S\rangle$, $\langle B, <\rangle \in \mathbf{FinLin}\}$.

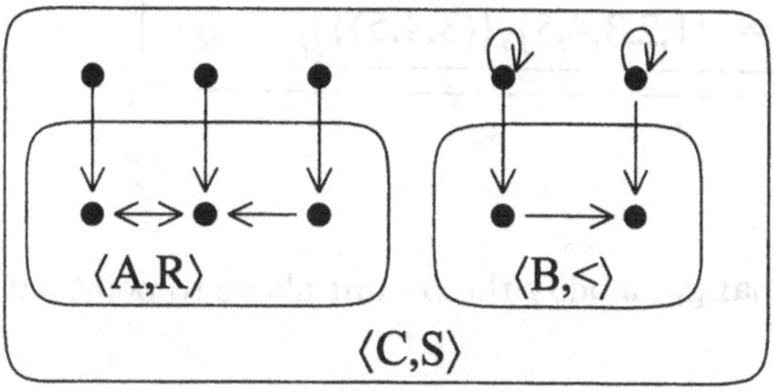

The range is the finitely axiomatizable subclass $\{\langle C, S\rangle :$ the conditions of H are satisfied with U_S, A, and B defined as above and with R and $<$ being S restricted to A and $B\}$. □

Interpretive Language Isomorphisms

Theorem. $\mathbf{Lang_2} \cong \mathbf{Lang_3} \cong \mathbf{Lang}_n$, *where* $\mathbf{Lang_2}$, $\mathbf{Lang_3}$ *and* $\mathbf{Lang}_n$ *are the languages of a binary, ternary, and n–ary relation respectively, with* $n \geq 2$.

Proof. First we show that $\mathbf{Lang_2} \cong \mathbf{Lang_3}$. By the repetitiveness of $\mathbf{Lang_2}$ and the Interpretation Cantor–Bernstein Theorem, it suffices to show that $\mathbf{Lang_2} \leq \mathbf{Lang_3}$ and $\mathbf{Lang_3} \leq \mathbf{Lang_2}$. Let R be a binary relation and S a ternary.

$\mathbf{Lang_2} \leq \mathbf{Lang_3}$ via the map $\{\langle A, \langle A, R\rangle, \langle A, S\rangle\rangle : (\forall xyz)(S(x,y,z) \Longleftrightarrow R(x,y))\}$. The range is the finitely axiomatizable subclass $\{\langle A, S\rangle : (\forall xyzz')(S(x,y,z) \Longleftrightarrow S(x,y,z'))\}$.

Now we wish to encode a ternary relation structure $\langle A, S\rangle$ with a binary relation structure $\langle B, R\rangle$. In a binary structure $\langle B, R\rangle$, an *S–unit* is a 6-tuple (x', y', z', x, y, z) such that $x'Ry'Rz', x'Rx, y'Ry, z'Rz$ and $x', \ldots, z$ are not R–related in any other way. The S–unit is said to *point out* the triple (x,y,z). The S–unit in the figure below points out the triple $(3,4,5)$.

$\mathbf{Lang_3} \leq \mathbf{Lang_2}$ via the map $\{\langle B, \langle A, S\rangle, \langle B, R\rangle\rangle : (B \setminus A) =$ domain of R, every element of $B \setminus A$ is in a unique S–unit, if (x,y,z) is pointed out by an

S–unit then $x, y, z \in A$ and $S(x,y,z)$, if $S(x,y,z)$ then (x,y,z) is pointed out by a unique S–unit, R relates no pairs other than those required above $\}$.

The range of G is the finitely axiomatizable subclass $\{\langle B, R\rangle$: each element in R's domain is in a unique S–unit, no two S–units point out the same triple, and R relates no pairs other than those in S–units$\}$.

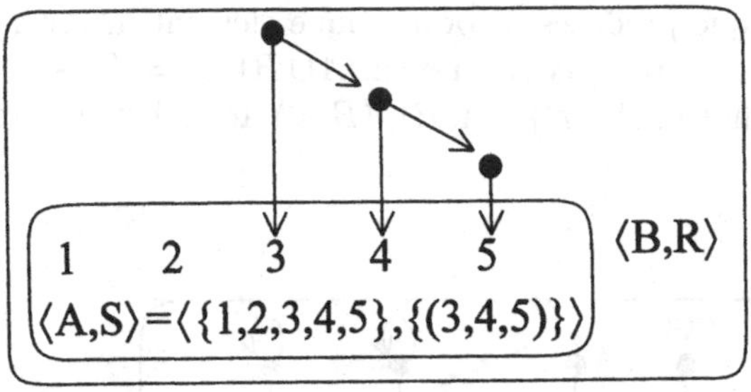

To show $\mathbf{Lang}_2 \cong \mathbf{Lang}_n$, modify the S–unit above to point out an n–tuple instead of a triple. □

Theorem. $\mathbf{Lang}_2 \cong \mathbf{Lang}_{1.5,1.5}$, *where* $\mathbf{Lang}_2$ *is the language of a binary relation and* $\mathbf{Lang}_{1.5,1.5}$ *is the language of two unary functions.*

Proof. By the above results, it suffices to show that $\mathbf{Lang}_2 \leq \mathbf{Lang}_{1.5,1.5}$ and $\mathbf{Lang}_{1.5,1.5} \leq \mathbf{Lang}_3$. Let R be a binary relation, S a ternary relation and f and g two unary functions.

$\mathbf{Lang}_{1.5,1.5} \leq \mathbf{Lang}_3$ via the map $\{\langle A, \langle A, f, g\rangle, \langle A, S\rangle\rangle : \forall xyz\; S(x,y,z) \iff f(x) = y \wedge g(x) = z\}$.

$\mathbf{Lang}_2 \leq \mathbf{Lang}_{1.5,1.5}$ via the map $\{\langle B, \langle A, R\rangle, \langle B, f, g\rangle\rangle : A = range(f) = range(g),\ \forall x \in A\; f(x) = g(x) = x,\ \forall x \in B \setminus A\; f(x) R g(x),\ \forall (x,y) \in R\; \exists! z \in B \setminus A\; f(z) = x \wedge g(z) = y\}$.

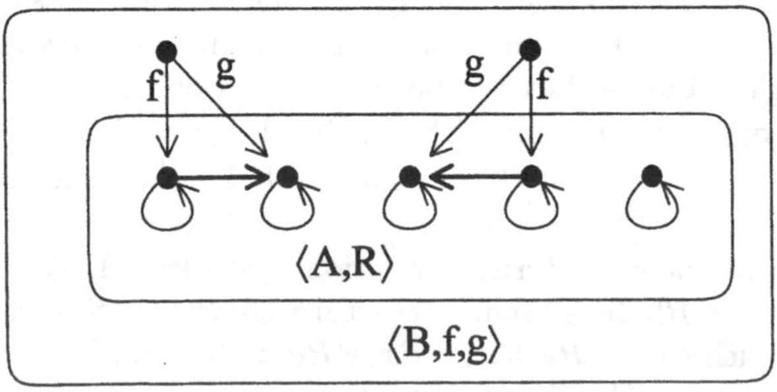

In the figure, the horizontal arrows are the pairs of R. □

Definition. A language is

- *finite* iff it has only finitely many symbols,
- *undecidable* iff it has ≥ 1 relation or operation symbols of ≥ 2 places or ≥ 2 unary operation symbols,
- *single-function* iff it has exactly one unary operation symbol and all other symbols are unary relations or constants,
- *trivial* iff each symbol is either a unary relation or a constant.

Theorem (Undecidable Language Isomorphism Theorem). *All finite undecidable languages are interpretively isomorphic to the theory* **Lang**$_2$ *of a binary relation.*

Proof. The theorem may be proved by checking that the elementary maps constructed for a similar theorem in (Hanf and Myers 1983) are Gaifman embeddings. For completeness we give a direct proof.

Let L be an undecidable language. Suppose, for example, $L = \mathbf{Lang}_{1,1.5,3}$, the language with symbols for a unary predicate U, unary function f, and ternary relation S. Then $\mathbf{Lang}_2 \cong L = \mathbf{Lang}_{1,1.5,3}$ follows from the facts below.

$\mathbf{Lang}_2 \leq \mathbf{Lang}_{1,1.5,3}$ via the map which encodes the binary R with the ternary S, adds a new point for the value of the operation f and makes the relation U empty.

$$\{\langle B, \langle A, R\rangle, \langle B, U, f, S\rangle, c\rangle : B = A \uplus \{c\}, U = \emptyset, \forall x f(x) = c,$$
$$\forall xyz\; xRy \iff S(x, y, z)\}.$$

$\mathbf{Lang}_{1,1.5,3} \leq \mathbf{Lang}_{1.5,1.5,3.5}$ via a map which changes n-ary relations to $(n+1)$-ary $0-1$ operations. $\{\langle B, \langle A, U, f, S\rangle, \langle B, u, f', s\rangle, 0, 1\rangle : B = A \uplus \{0, 1\}, 0 \neq 1, u(x) = 1$ if $x \in U$ else $0, f'(x) = f(x)$ if $x \in A$ else $x, s(x, y, z) = 1$ if $S(x, y, z)$ else $0\}$.

$\mathbf{Lang}_{1.5,1.5,3.5} \leq \mathbf{Lang}_7$ via a map which encodes a list of operations into one large relation: $\{\langle A, \langle A, f, g, h\rangle, \langle A, F\rangle\rangle : \forall xyz, rst\; F(x, y, z, r, s, t) \iff r = f(x) \wedge s = g(x) \wedge t = h(x, y, z)\}$.

$\mathbf{Lang}_7 \leq \mathbf{Lang}_2$ by the first theorem of this section. □

Open problem. Is the theory $\mathbf{Lang}_{1.5}$ of a unary function $\cong$ the theory $\mathbf{Lang}_{1.5,1}$ of a unary function and a unary relation? Are all finite single-function languages interpretively isomorphic?

An argument similar to the one for undecidable languages shows that all finite single-function languages are mutually Gaifman embeddable. If **FinSuc** were finitely axiomatizable, all finite single-function languages would be isomorphic.

Theorem. *The following theories are all isomorphic with* **Lang**$_2$*: the theory of a symmetric irreflexive relation (***SymIrr***), the theory of partial orders (***PartOrd***), the theory of lattices (***Latt***), the theory of semigroups (***Semigrp***).*

Proof. The listed theories are clearly embeddable in an undecidable language and hence in $\mathbf{Lang_2}$. We must show that $\mathbf{Lang_2}$ is embeddable in these theories.

$\mathbf{Lang_2} \leq \mathbf{SymIrr}$ via the Gaifman map of (Rabin 1965, page 62).

$\mathbf{SymIrr} \leq \mathbf{PartOrd}$ via the map $\{\langle B, \langle A, R\rangle, \langle B, <\rangle\rangle : \langle A, R\rangle \in \mathbf{SymIrr}$, $\langle B, <\rangle \in \mathbf{PartOrd}$, (1) $\forall xy\, x < y \Longrightarrow x \in A \wedge y \in B \setminus A$, (2) $\forall b \in (B \setminus A) \exists$ exactly two x's in A with $x < b$, (3) $\forall xy \in A\; x \neq y \Longrightarrow \exists$ at most one $b\; x < b \wedge y < b$, (4) $\forall xy \in A\; xRy \Longleftrightarrow \exists b \in (B \setminus A)\; x < b \wedge y < b\}$.

$\mathbf{SymIrr} \leq \mathbf{Latt}$ via the map which adds a top and bottom point to the partial order $\langle B, <\rangle$ of the previous paragraph. This results in a partial order in which every pair has a greatest lower bound and least upper bound. These bounds define the meet and join of a lattice.

$\mathbf{Latt} \leq \mathbf{Semigrp}$ via $\{\langle A, \langle A, \wedge, \vee\rangle, \langle A, \wedge\rangle\rangle : \langle A, <\rangle \in \mathbf{Latt}\}$. □

$\mathbf{Lang_2}$ is *universal* for finitely axiomatizable theories with finitely many nonlogical symbols in the sense that each such theory is $\leq \mathbf{Lang_2}$.

Nonisomorphism

The simplest way to establish that two theories are nonisomorphic with respect to interpretation is to show that they differ with respect to a property or function which is preserved under interpretative isomorphism. In this section, we assume that the languages of our theories involve only finitely many symbols.

A property is *preserved under interpretive isomorphism* iff whenver one of two isomorphic theories has the property, so does the other. A function on theories is *preserved* iff it always assigns the same value to isomorphic theories.

The following functions and properties are preserved under interpretative isomorphism – see (Szczerba 1977) and (Prazmowski and Szczerba 1976): consistency and completeness, axiomatizability and finite axiomatizability, the degree of decidability, having only finite models, the number of nonisomorphic models of a given infinite power, the isomorphism type of the theory's Lindenbaum–Tarski algebra, the isomorphism type of the Boolean algebra of formulas with at most $x_1, \ldots, x_n$ free, having an atomic model, having a countable saturated model, stability.

The following are not preserved: having a one–element model, being a variety (having an equational axiomatization), being closed with respect to substructures (having a universal axiomatization), being closed with respect to unions of chains.

Examples of groups of theories which are mutually nonisomorphic with respect to interpretations.

The theories	*The invariant property*
• a binary relation • an equivalence relation	degree of decidability
• a unary relation • a function • a permutation	Lindenbaum–Tarski algebra (Simons 1971)
• **Eq** • **FinSuc** • **FinLin**	number of nonisomorphic countable models

Recall that a language is trivial if it has only constants and unary relations. This includes the equality language **Eq** with no nonlogical symbols.

Theorem (Trivial Language Nonisomorphism Theorem). *Two finite trivial languages are isomorphic iff they have the same number of unary relations and the same number of constants.*

Proof. The proof uses the Automorphism Group Lemma: if $\mathfrak{A}$ is mapped to $\mathfrak{B}$ by an interpretive isomorphism, then $\mathfrak{A}$ and $\mathfrak{B}$ have isomorphic automorphism groups.

First, isomorphic finite trivial languages must have the same number of unary relations. In a structure for a trivial language, the orbit of a point under automorphisms is an atom of the Boolean algebra of the structure's definable sets. A finite trivial language with n unary relations has a structure in which 2^n orbits have exactly two points and no orbit has more than two points. The automorphism group for this structure is Z_{2^n}. No language with fewer unary relations has a structure with this automorphism group.

Finally, isomorphic finite trivial languages with the same number of unary relations must have the same number of constants. Suppose there are n unary relations. Consider structures whose automorphism group is isomorphic to Z_{2^n}. Each such structure can be partioned into two parts: the constant part (possibly empty) consisting of elements named by constants and the complementary nonconstant part of elements not named by constants. The constant part is rigid, *i.e.*, has no automorphisms. Hence the nonconstant part is isomorphic to the unique structure with n unary relations whose automorphism group is Z_{2^n}. Thus the number of nonisomorphic models with automorphism group Z_{2^n} equals the number of isomorphism types of the constant part. A language with more constants has more such isomorphism types. Hence two languages with the same number of unary relations but a different number of constants can not be isomorphic. □

Two theories have isomorphic cylindric algebras iff they are interpretively isomorphic with respect to interpretations constructed from identity interpretations, reduct interpretations and definitional expansions. These are the interpretations which preserve the universe. Pigozzi (Henkin et al. 1971, page 362) has shown that languages with nonequivalent finite similarity types have nonisomorphic cylindric algebras. For the case with countable similarity types, see

(Myers 1976a). For the classification of finite languages which have isomorphic Lindenbaum–Tarski sentence algebras, see (Hanf and Myers 1983). For a classification of languages which have isomorphic Boolean formula algebras, see (Faust 1982).

References

Barwise, J. A preservation theorem for interpretations. Cambridge Summer School in Mathematical Logic, Cambridge, 1971, Lecture Notes in Mathematics, Springer-Verlag, Berlin **337** (1973) 618–621

Bouvere, K.: Synonymous Theories. The Theory of Models (Addison, Henkin and Tarski, eds.), North-Holland, Amsterdam (1965) 402–406

Fause, D.: The Boolean algebra of formulas of first-order logic. Annals of Mathematical Logic **23** (1982) 27–53

Feferman, S.: Lectures on proof theory. Proceedings of the Summer School in Logic (Leeds, 1967), Lecture Notes in Mathematics, Berlin **70** (1968) 1–107

Gaifman, H.: Operations on relational structures, functors and classes. I. Proceedings of the Tarski Symposium, 1971, Proceedings of Symposia in Pure Mathematics **25**, American Mathematical Society, Providence, R.I., (first edition: 1974; second edition: 1979) 21–39

Hanf, W.: Primitive Boolean algebras. Proceedings of the Alfred Tarski Symposium, Proceedings of Symposia in Pure Mathematics, American Mathematical Society, Providence, R.I. **25** (1974) 75–90

Hanf, W.: The Boolean algebra of logic. Bulletin of the American Mathematics Society **81** (1975) 587–589

Hanf, W., Myers, D.: Boolean sentence algebras: isomorphism constructions. Journal of Symbolic Logic **48** (1983) 329–338

Henkin, L., Mong, D., Tarski, A.: Cylindric Algebras. North-Holland, Amsterdam, 1971

Hodges, W.: A normal form for algebraic constructions II. Logique et Analyse **18** (1975) 429–487

Manders, K.: First-order logical systems and set-theoretic definability. preprint 1980

Montague, R.: Contributions to the Axiomatic Foundations of Set Theory. Ph.D. Thesis, University of California, 1957

Mycielski, J.: A lattice of interpretability types of theories. Journal of Symbolic Logic **42** (1977) 297–305

Mycielski, J., Pudlak, P., Stern, A.: A lattice of chapters of mathematics (interpretations between theories [theorems]. Memoirs of the American Mathematical Society **84** (1990) 1–70

Myers D.: Cylindric algebras of first-order theories. Transactions of the American Mathematical Society **216** (1976a) 189–202

Myers, D.: Invariant uniformization. Fundamenta Mathematicae **91** (1976b) 65–72

Myers, D.: Lindenbaum–Tarski algebras. Handbook of Boolean Algebras (Monk, ed.), Elsevier Science Pulishers, B.V. (1989) 1168–1195

Pillay, A.: Gaifman operations, minimal models and the number of countable models. Ph.D. Thesis, University of London, 1977

Prazmowski, K., Szczerba, L.: Interpretability and categoricity. Bulletin de l'Academie Polonaise des Sciences **24** (1976) 309–312

Rabin, M.: A simple method for undecidability proofs and some applications. Logic, Methodology and Philosophy of Science, Proc. 1964 International Congress, (Bar-Hillel, ed.), North-Holland, Amsterdam, (1965), 58–68

Simons, R.: The Boolean algebra of sentences of the theory of a function. Ph.D. Thesis, Berkeley, 1971

Szczerba, L.: Interpretability of elementary theories. Logic, Foundations of Mathematics and Computability Theory (Butts and Hintikka, eds.), D. Reidel Publishing Co., Dordrecht-Holland (1977) 126–145

Szmielew, W., Tarski, A.: Mutual interpretability of some essentially undecidable theories. Proceedings of the International Congress of Mathematicians, Cambridge, Massachusetts, American Mathematical Society, Providence (1952) 734

Vagueness – a Rough Set View

Zdzislaw Pawlak

Institute of Theoretical and Applied Informatics
Polish Academy of Sciences
ul. Baltycka 5 Gliwice 44 000, Poland

Abstract. Vagueness for a long time has been studied by philosophers, logicians and linguists. Recently researchers interested in AI contributed essentially to this area.
In this paper we present a new approach to vagueness, called rough set theory. The starting of the theory theory is the assumption that fundamental mechanisms of human reasoning are based on the ability to classify object of interest, i.e. group objects into similarity classes, which form granules (basic concepts) of knowledge about the universe of discourse (e.g. color, height, weight etc.). Every union of basic concepts is called a precise (crisp) concept, otherwise the concept is called imprecise (rough). Thus rough concepts (sets) cannot be expressed in terms of elementary concepts (set). Therefore with each imprecise concept a pair of precise concepts, called its lower and upper approximation, is associated. Approximations are basic operations of rough set theory.
The paper contains basics of rough set theory, shows some of its applications, and the relationship to fuzzy sets, the theory of evidence, discriminant analysis and boolean reasoning methods are pointed out.

1 Introduction

Vagueness has been studied for many years by researchers interested in mathematics, philosophical logic and philosophy of language (see e.g. [1, 2, 6, 8, 12, 13, 30, 36, 47, 59, 69, 70, 72]). Recently, researchers interested in AI contributed essentially to this area of research. The most important contributions seemingly are fuzzy set theory (see [92]) and the theory of evidence (see [74]).

This paper presents another approach to vagueness based on rough set theory (see [60]).

Rough set theory bears on the assumption that we have initially some information (knowledge) about elements of the universe we are interested in. Evidently to some elements of the universe the same information can be associated and consequently the elements can be *similar* or *indiscernible*, in view of the available information. Similarity is assumed to be a reflexive and symmetric relation, whereas the indiscernibility relation - also transitive. Thus similarity is a tolerance relation and indiscernibility is an equivalence relation.

The concepts of similarity and indiscernibility attracted attention of philosophers and logicians for many years (see e.g. [88, 91]), nevertheless these concepts are still far of being understood fully.

2 The Boundary-line Approach to Vagueness

The idea of vagueness is usually connected with the so called "boundary-line" approach first formulated by Frege (see [21]), who writes:

"The concept must have a sharp boundary. To the concept without a sharp boundary there would correspond an area that had not a sharp boundary-line all around ".

Thus according to Frege "the concept without a sharp boundary", i.e. vague concept, must have boundary-line examples which cannot be classified, on the basis of available information, neither to the concept nor to its complement. For example the concept of an *odd (even) number* is precise, because every number is either odd or even - whereas the concept of a *beautiful women* is vague, because for some women we cannot decide whether they are beautiful or not (there are boundary-line cases).

In the rough set approach vagueness is due to the lack of information about some elements of the universe. If with some elements the same information is associated, in view of this information these elements are indiscernible. For example if some patients suffering from a certain disease display the same symptoms, they are indiscernible with respect to these symptoms. It turns out that the indiscernibility leads to the boundary-line cases, i.e. some elements cannot be classified neither to the concept nor to its complement, in view of the available information and thus form the boundary-line cases.

Now let us present these ideas more formally.

Suppose we are given a finite not empty set U called the *universe*, and let I be a binary relation on U. By $I(x)$ we mean the set of all $y \in U$ such that yIx. If I is reflexive and symmetric, i.e.

$$xIx, \text{ for every } x \in U,$$

$$xIy, \text{ implies } yIx \text{ for every } x, y \in U,$$

then I is a tolerance relation. If I is also transitive, i.e. xIy and yIz implies xIz, then I is an equivalence relation. In this case $I(x) = [x]_I$, i.e. $I(x)$ is an equivalence class of the relation I containing element x. If I is a tolerance relation and xIy, then x, y are called *similar* with respects to I (*I-similar*), whereas if I is an equivalence relation and xIy, then x, y are referred to as *indiscernible* with respect to I (*I-indiscernible*). For the sake of simplicity we will assume in this paper that I is an equivalence relation.

Let us define now two following operations on sets

$$I_*(X) = \{x \in U : I(x) \subseteq X\},$$

$$I^*(X) = \{x \in U : I(x) \cap X \neq \emptyset\},$$

assigning to every subset X of the universe U two sets $I_*(X)$ and $I^*(X)$ called the *I-lower* and the *I-upper approximation* of X respectively. The set

$$BN_I(X) = I^*(X) - I_*(X)$$

will be referred to as the *I-boundary region* of X.

If the boundary region of X is the empty set, i.e. $BN_I(X) = \emptyset$, then X will be called ***crisp*** (*exact*) with respect to I; in the opposite case, i.e. if $BN_I(X) \neq \emptyset$, X will be referred to as ***rough*** (*inexact*) with respect to I.

Thus rough sets seems to be a natural mathematical model of vague concepts.

One can easily show the following properties of approximations:

1) $I_*(X) \subseteq X \subseteq I^*(X)$,
2) $I_*(\emptyset) = I^*(\emptyset) = \emptyset, I_*(U) = I^*(U) = U$,
3) $I^*(X \cup Y) = I^*(X) \cup I^*(Y)$,
4) $I_*(X \cap Y) = I_*(X) \cap I_*(Y)$,
5) $X \subseteq Y$ implies $I_*(X) \subseteq I_*(Y)$ and $I^*(X) \subseteq I^*(Y)$,
6) $I_*(X \cup Y) \supseteq I_*(X) \cup I_*(Y)$,
7) $I^*(X \cap Y) \subseteq I^*(X) \cap I^*(Y)$,
8) $I_*(-X) = -I^*(X)$,
9) $I^*(-X) = -I_*(X)$,
10) $I_*(I_*(X)) = I^*(I_*(X)) = I_*(X)$,
11) $I^*(I^*(X)) = I_*(I^*(X)) = I^*(X)$.

It is easily seen that the lower and the upper approximation of a set are interior and closure operations in a topology generated by the indiscernibility relation. Thus vagueness is related to some topological properties of inexact concepts.

Vagueness can be also characterized numerically by defining the following coefficient, called the ***accuracy*** of ***approximation***

$$\alpha_I(X) = \frac{|I_*(X)|}{|I^*(X)|},$$

where $|X|$ denotes the cardinality of X.

Obviously $0 \leq \alpha_I(X) \leq 1$. If $\alpha_I(X) = 1$, X is ***crisp*** with respect to I (the concept X is ***precise*** with respect to I), and otherwise, if $\alpha_I(X) < 1$, X is ***rough*** with respect to I (the concept X is vague with respect to I).

3 Topological Classification of Vagueness

It turns out that the above considerations give rise to the following four basic classes of rough sets, i.e. four classes of vagueness:

a) $I_*(X) \neq \emptyset$ and $I^*(X) \neq U$, iff X is *roughly I-observable*,
b) $I_*(X) = \emptyset$ and $I^*(X) \neq U$, iff X is ***internally I-unobservable***,
c) $I_*(X) \neq \emptyset$ and $I_*(X) = U$, iff X is ***externally I-unobservable***,
d) $I_*(X) = \emptyset$ and $I^*(X) = U$, iff X is ***totally I-unobservable***.

The intuitive meaning of this classification is the following.

If X is roughly *I*-observable we are able to decide for some elements of U whether they belong to X or $-X$.

If X is internally I-unobservable we are able to decide whether some elements of U belong to $-X$, but we are unable to decide for any element of U whether it belongs to X or not.

If X is externally I-unobservable we are able to decide for some elements of U whether they belong to X, but we are unable to decide for any element of U whether it belongs to $-X$ or not.

If X is totally I-unobservable, we are unable to decide for any element of U whether it belongs to X or $-X$.

That means, that X is roughly observable if there are some elements in the universe which can be positively classified, to X or $-X$.

External I-unobservability of a set refers to a situation when positive classification is possible for some elements, but it is impossible to determine that an element does not belong to X.

4 An Example

In this section we will illustrate the above ideas intuitively, by means of an indiscernibility relation generated by data.

Data are often presented as a table, columns of which are labeled by *attributes*, rows by *objects* of interest and entries of the table are *attribute values*. For example, in a table containing information about patients suffering from a certain disease objects are *patients* (strictly specking their ID's), attributes can be, for example, *blood pressure, body temperature* etc., whereas the entry corresponding to object *Smiths* and the attribute *blood preasure* can be *normal.* Such tables are known as *information systems.*

Table 1. Example of an information system

Patient	Headache	Muscle-pain	Temperature	Flu
p1	no	yes	high	yes
p2	yes	no	high	yes
p3	yes	yes	very high	yes
p4	no	yes	normal	no
p5	yes	no	high	no
p6	no	yes	very high	yes

Each row of the table can be seen as information about specific patient. For example patient p2 is characterized in the table by the following attribute-value set

(Headache, yes), (Muscle-pain, no), (Temperature, high), (Flu, yes),

which form information about the patient.

Obviously cach subset of attributes defines an indiscernibility (equivalence) relation on the set of patients. Patients are indiscernible by a set of attributes if they have the same values of the attributes.

For example, patients p2, p3 and p5 are indiscernible with respect to the attribute Headache, patients p3 and p6 are indiscernible with respect to the attributes Muscle-pain and Flu, and patients p2 and p5 are indiscernible with respect to the attributes Headache, Muscle-pain and Temperature.

Patient p2 has flu, whereas patient p5 does not, and they are indiscernible with respect to the attributes Headache, Muscle-pain and Temperature, hence flu cannot be characterized in terms of the attributes Headache, Muscle-pain and Temperature. Thuse p2 and p5 are the boundary-line cases, which cannot be properly classified in view of the available knowledge. The remaining patients p1, p3 and p6 display symptoms which enable us to classify them with certainty as having flu, patients p2 and p5 cannot be excluded as having flu and patient p4 for sure does not have flu, in view of the displayed symptoms. Thus the lower approximation of the set of patients having flu is the set {p1, p3, p6} and the upper approximation of this set is the set {p1, p2, p3, p5, p6}, whereas the boundary-line cases are patients p2 and p5. Similarly p4 does not have flu and p2, p5 cannot be excludes as having flu, thus the lower approximation of this concept is the set {p4}, whereas – the upper approximation – is the set {p2, p4, p5} and the boundary region of the concept "not flu" is the set {p2, p5}, the same as in the previous case. Hence the accuracy of approximation of "flu", $\alpha(X_{flu}) = 3/5$ and $\alpha(X_{notflu}) = 1/3$.

5 Vagueness and Uncertainty

A vague concept has a boundary-line cases, i.e. elements which cannot be with *certainty* classified as elements of the concept i.e., we are uncertain whether the boundary-line cases belong to the concept or not. Hence *uncertainty* is related to the *membership* of elements to a set. Therefore in order to discuss the problem of uncertainty from the rough set perspective we have to define a *rough membership function*, and investigate its properties.

The rough membership function can be defined employing the relation I in the following way (see [63]):

$$\mu_X^I(x) = \frac{|X \cap I(x)|}{|I(x)|}.$$

Obviously $0 \le \mu_X^I(x) \le 1$.

The rough membership has a probabilistic flavour and can be interpreted as a conditional probability which expresses a degree to which an element belongs to a set. For example, patient p1 can be classified as having flu on the basis of his body temperature with probability 3/5.

The rough membership function can be used to define the approximations and the boundary region of a set, as shown below:

$$I_*(X) = \{x \in U : \mu_X^I(x) = 1\},$$

$$I^*(X) = \{x \in U : \mu_X^I(x) > 0\},$$
$$BN_I(X) = \{x \in U : 0 < \mu_X^I(x) < 1\}.$$

Thus there exists a strict connection between vagueness and uncertainty. As we mentioned above vagueness is related to sets (concepts), whereas uncertainty is related to elements of sets, and the rough set approach shows clear connection between the two concepts.

It can be shown (see [63]) that the rough membership function has the following properties:

a) $\mu_X^I(x) = 1$ iff $x \in I_*(X)$,
b) $\mu_X^I(x) = 0$ iff $x \in U - I^*(X)$,
c) $0 < \mu_X^I(x) < 1$ iff $x \in BN_I(X)$,
d) If $I = \{(x,x) : x \in U\}$, then $\mu_X^I(x)$ is the characteristic function of X,
e) If xIy, then $\mu_X^I(x) = \mu_X^I(y)$,
f) $\mu_{U-X}^I(x) = 1 - \mu_X^I(x)$ for any $x \in U$,
g) $\mu_{X \cup Y}(x) \geq \max(\mu_X^I(x), \mu_Y^I(x))$ for any $x \in U$,
h) $\mu_{X \cap Y}^I(x) \leq \min(\mu_X^I(x), \mu_Y^I(x))$ for any $x \in U$,
i) If $\mathbf{X}$ is a family of pair wise disjoint sets of U, then $\mu_{\cup \mathbf{X}}^I(x) = \sum_{X \in \mathbf{X}} \mu_X^I(x)$ for any $x \in U$.

The above properties show clearly the difference between fuzzy and rough memberships. In particular properties g) and h) show that the rough membership can be regarded as a generalization of of fuzzy membership.

6 Applications

Rough set theory has found many interesting applications. The rough set approach seems to be of fundamental importance to AI and cognitive sciences, especially in the areas of machine learning, knowledge acquisition, decision analysis, knowledge discovery from databases, expert systems, inductive reasoning and pattern recognition. It seems of particular importance to decision support systems.

The main advantage of rough set theory is that it does not need any preliminary or additional information about data – like probability in statistics, or basic probability assignment in Dempster-Shafer theory and grade of membership or the value of possibility in fuzzy set theory.

Rough set theory has been successfully applied in many real-life problems e.g., in medicine, pharmacology, engineering, banking, financial and market analysis and others. Some exemplary applications are listed below.

Medicine turned out to be a very interesting domain of application of rough sets (see e.g., [28, 82, 83, 84, 85, 87]). In pharmacology the analysis of relationships between the chemical structure and the antimicrobial activity of drugs (see [43, 44, 45, 46]) has been successfully investigated. Banking applications include evaluation of a bankruptcy risk (see [80, 81]) and market research (see [24, 95]). Very interesting results have been also obtained in speaker independent speech recognition (see [10, 14, 15, 16, 17]) and acoustics (see [40, 41]). The rough set approach seems also important for various engineering applications, like diagnosis of machines using vibroacoustics symptoms (noise, vibrations) (see [55, 56, 57]), material sciences (see [33]) and process control (see e.g., [53, 54, 67, 86, 96]). Application in linguistics (see e.g., [26, 27, 39, 51]) and environment (see [29]), databses (see e.g., [3, 4, 5, 73]) are other important domains, where rough set proved to be a valuable tool.

More about applications of rough set theory can be found in the references (see e.g., [48, 49, 78, 93]). Besides, many other fields of application, e.g., time series analysis, image processing and character recognition, are being extensively explored.

7 Conclusion

Rough set theory seems to be well suited as a mathematical model of vagueness and uncertainty. Vagueness is a property of sets (concepts) and is strictly related to the existence to the boundary region of a set, whereas uncertainty is a property of elements of sets and is related to the rough membership function. In the rough set approach both concepts are closely related and are due to the indiscernibility caused by insufficient information about the world.

Rough set theory overlaps to a certain degree many other mathematical theories. Particularly interesting is the relationship with fuzzy set theory and Dempster-Shafer theory of evidence. The concepts of rough set and fuzzy set are different since they refer to various aspects of imprecision (see [63]) whereas the connection with theory of evidence is more substantial (see [76]). Besides, rough set theory is related to discriminant analysis (see [42]), Boolean reasoning methods (see [77]) and others. The relationship between rough set theory and decision analysis is presented in (see [64, 79]). More details concerning these relationships can be found in the references. Nevertheless rough set theory can be viewed in its own rights as an independent discipline with considerable achievements to its credit.

References

1. Ballmer, T.T., Pinkal, M., (eds.).: Approaching Vagueness. North-Holland, Amsterdam (1983)
2. Ballweg, J.: Vagueness or context dependence? Supervaluation revisited in a semantics based on scales. Approaching Vagueness, edited by T.T. Ballmer and M. Pinkal. North-Holland, Amsterdam (1983)
3. Beaubouef, T., and Petry, F.E.: A rough set model for relational databases. In: W. Ziarko (ed.), *Rough Sets, Fuzzy Sets and Knowledge Discovery. Proceedings of the International Workshop on Rough Sets and Knowledge Discovery* (RSKD'93), Banff, Alberta, Canada, October 12–15, Springer-Verlag, Berlin (1993) 100–107
4. Beaubouef, T., and Petry, F.E.: Rough querying of crisp data in relational databases. In: T.Y. Lin and A.M. Wildberger (eds.), (1995), The Third International

Workshop on Rough Sets and Soft Computing Proceedings (RSSC'94), San Jose State University, San Jose, California, USA, November 10–12 (1995) 85–88

5. Beaubouef, T., Petry, F.E., and Buckles, B.P.: Extension of the relational Database and its algebra with rough set techniques. Computational Intelligence: An International Journal **11** (1995) 233–245
6. Black, M.: Vagueness. The Philosophy of Sciences **2** (1937) 427–455
7. Black, M.: Language and philosophy. Ithaca, New York (1949)
8. Black, M.: Vagueness - an exercise in logical analysis (1937). Reprinted in Language and Philosophy. Cornell University Press. Ithaca, New York (1954)
9. Black, M.: Reasoning with loose concepts. Dialog **2** (1963) 1–12
10. Brindle, D.: Speaker-independent speech recognition by rough sets analysis. In: T.Y. Lin (ed.), The Third International Workshop on Rough Sets and Soft Computing Proceedings (RSSC'94), San Jose State University, San Jose, California, USA, November 10–12 (994) 376–383
11. Burns, L.: Vagueness and coherence. Synthese **68** (1986)
12. Copilowish, I.M.: Borderline cases, vagueness and ambiguity. Philosophy of Science **6** (1937)
13. Chatterjee, A.: Undrestanding vagueness. Pragati Publications, Dehli (1994)
14. Czyżewski , A.: Speaker-Independent Recognition of Digits - Experiments with Neural Networks, fuzzy logic and rough sets. Journal of the Intelligent Automation and Soft Computing (1995) (to appear)
15. Czyżewski, A., and Kaczmarek, A.: Multilayer knowledge based system for speaker-independent recognition of isolated words. In: W. Ziarko (ed.), Rough Sets, Fuzzy Sets and Knowledge Discovery. Proceedings of the International Workshop on Rough Sets and Knowledge Discovery (RSKD'93), Banff, Alberta, Canada, October 12–15, Springer-Verlag, Berlin (1993) 387–394
16. Czyżewski, A., and Kaczmarek, A.: Speaker-independent recognition of isolated words using rough sets In: P.P. Wang (ed.), Second Annual Joint Conference on Information Sciences PROCEEDINGS, September 28 – October 1, Wrightsville Beach, North Carolina, USA (1995) 397–400
17. Czyżewski A., and Kaczmarek, A.: Speech recognition systems based on rough sets and neural networks. In: T.Y. Lin and A.M. Wildberger (eds.), The Third International Workshop on Rough Sets and Soft Computing Proceedings (RSSC'94), San Jose State University, San Jose, California, USA, November 10-12 (1995) 97–100
18. Czogała, E., Mrózek, A. and Pawlak, Z.: The Idea of Rough-Fuzzy Controller. International Journal of Fuzzy Sets and Systems, **72** (1995) 61–63
19. Evans, G.: Can there be vague objects? Analysis **38** (1978)
20. Fine, K.: Vagueness, truth and logic. Synthese **30** (1975) 265–300
21. Frege, G.: Grundgesetze der Arithmentik, 2, in Geach and Black (ed.) Selections from the Philosophical Writings of Gotlob Frege, Blackweil, Oxford (1903) 1970
22. Forbes, G.: Thisness and vagueness. Synthese **54** (1983)
23. Garrete, B.J.: Vagueness and identity. Analysis **48** (1988)
24. Golan, R., and Edwards, D.: Temporal rules discovery using datalogic/R+ with stock market data. In: W. Ziarko Rough Sets, Fuzzy Sets and Knowledge Discovery. Proceedings of the International Workshop on Rough Sets and Knowledge Discovery (RSKD'93), Banff, Alberta, Canada, October 12–15, Springer-Verlag, Berlin, (1993) 74–81
25. Grzymała-Busse, J.: Rough Sets. Advances in Imaging and Electrons Physics **94** (1995) to appear

26. Grzymała-Busse, J.W., Sedelow, S.Y., and Sedelow, W.A. Jr.: Machine learning & knowledge acquisition, rough sets, and the English semantic code. Proc. of the Workshop on Rough Sets and Database Mining, 23rd Annual ACM Computer Science Conference CSC'95, Nashville, TN, March 2 (1995) 91–109
27. Grzymała-Busse, J.W., and Than, S.: Data compression in machine learning applied to natural language. Behavior Research Methods, Instruments, & Computers **25** (1993) 318–321
28. Grzymała-Busse, and Woolerly, L.: Improving prediction of preterm birth using a new classification scheme and rule induction. Proc. of the 18-th Annual Symposium on Computer Applications in Medical Care (SCAMC), Washington D.C. November 5–9 (1994) 730–734
29. Gunn, J.D., and Grzymała-Busse, J.W.: Global temperature stability by rule induction: an interdisciplinary bridge. Human Ecology, **22** (1994) 59–81
30. Hemple, C.: Vagueness and logic. Philosophy of Science **6** (1939)
31. Hempel, C.G.: Fundamental of concept formation in empirical sciences. University of Chicago Press, Chicago (1952)
32. Hunt, E.B.: Concept formation. John Wiley and Sons, New York (1974)
33. Jackson, A.G., Ohmer, M., and Al-Kamhawi, H.: Rough sets analysis of chalcopyrite semiconductor band gap data. In: T.Y. Lin (ed.), The Third International Workshop on Rough Sets and Soft Computing Proceedings (RSSC'94), San Jose State University, San Jose, California, USA, November 10–12 (1994) 408–417
34. Jackson, A.G., LeClair, S.R., Ohmer, M.C., Ziarko. W., and Al-Kamhwi, H.: Rough sets applied to material data. Acta Metallurgica et Materialia (to appear)
35. Johnsen, B.: Is vague identity incoherent? Analysis **49** (1989)
36. Khatchadourian, H.: Vagueness. Philosophical Quarterly **12** (1962)
37. Kindt, W.: Two approaches to vagueness: theory of interaction and topology. Approaching Vagueness, edited by T.T. Ballmer and M. Pinkal. North-Holland, Amsterdam (1983)
38. Kreuse, R., Schwecke, E., Heinsohn, J.: Uncertainty and vagueness in knowledge based systems in numerical methods. Springer-Verlag, New York (1991)
39. Kobayashi, S., Yokomori, T.: Approximately learning regular languages with respect to reversible languages: A rough set based analysis. In: P.P. Wang (ed.), Second Annual Joint Conference on Information Sciences PROCEEDINGS, September 28 – October 1, Wrightsville Beach, North Carolina, USA (1995) 91–94
40. Kostek, B.: Statistical versus artificial intelligence based processing of subjective test results. 98th Convention of the Audio Engineering Society, Paris, February 25–28, Preprint **4018** (1995)
41. Kostek B.: Rough set and fuzzy set methods applied to acoustical analyses. Journal of the Intelligent Automation and Soft Computing (1995) (to appear)
42. Krusińska E., Słowiński R., and Stefanowski J.: Discriminant versus rough set approach to vague data analysis. Applied Stochastic Models and Data Analysis **8** (1992) 43–56
43. Krysiński, J.: Rough set approach to the analysis of structure - activity relationship of quaternary imidazolium compounds. Arzneimittel Forschung / Drug Research **40/II** (1990) 795–799
44. Krysiński, J.: Grob Mengen Theorie in der Analyse der Struktur Wirkungs Beziehungen von quartaren Pyridiniumverbindungen. Pharmazie **46/12** (1992) 878–881
45. Krysiński, J.: Analysis of structure - activity relationships of quaternary ammonium compounds. In: R. Słowiński (ed.), Intelligent Decision Support. Handbook

of Applications and Advances of the Rough Set Theory, Kluwer Academic Publishers, Dordrecht (1992) 119–136

46. Krysiński, J.: Application of the rough sets theory to the analysis of structure-activity-relationships of antimicrobial pyridinium compounds. Die Pharmazie, **50** (1995) 593–597
47. Kubiński, T.: Vague terms. Studia Logica. **7** (1958) 115–179 (in Polish)
48. Lin, T.Y., (ed.).: The Third International Workshop on Rough Sets and Soft Computing Proceedings (RSSC'94), San Jose State University, San Jose, California, USA, November 10–12 (1994)
49. Lin, T.Y., Cercone, N.: Rough sets and data mining, analysis of imperfect data. Kluwer Academic Publishers (1997)
50. Machina, K.F.: Truth, belif and vagueness. Journal of Philosophical Logic **5** (1976)
51. Moradi, H., Grzymała-Busse, J., and Roberts, J.: Entropy of English text: Experiments with humans nad machine learning system based on rough sets. In: P.P. Wang (ed.), Second Annual Joint Conference on Information Sciences PROCEEDINGS, September 28 – October 1, Wrightsville Beach, North Carolina, USA (1995) 87–88
52. Mrózek, A.: Rough sets and dependency analysis among attributes in computer implementations of expert's inference models. International Journal of Man-Machine Studies **30** (1989) 457–471
53. Mrózek, A.: Rough sets in computer implementation of rule-based control of industrial processes. In: R. Słowiński (ed.), Intelligent Decision Support. Handbook of Applications and Advances of the Rough Set Theory, Kluwer Academic Publishers, Dordrecht (1992) 19–31
54. Munakata, T.: Rough control: Basic ideas and applications. In: Wang, P.P., (ed.), Second Annual Joint Conference on Information Sciences PROCEEDINGS, September 28 – October 1, Wrightsville Beach, North Carolina, USA (1995) 340–343
55. Nowicki, R., Słowiński, R., and Stefanowski, J.: Rough sets analysis of diagnostic capacity of vibroacoustic symptoms. Journal of Computers and Mathematics with Applications **24/2** (1992) 109–123
56. Nowicki, R., Słowiński, R., and Stefanowski, J.: Evaluation of vibroacoustic diagnostic symptoms by means of the rough sets theory. Journal of Computers in Industry **20** (1992) 141–152
57. Nowicki, R., Słowiński, R., and Stefanowski, J.: Analysis of diagnostic symptoms in vibroacoustic diagnostics by means of the rough set theory. In: R. Słowiński (ed.), Intelligent Decision Support. Handbook of Applications and Advances of the Rough Set Theory, Kluwer Academic Publishers, Dordrecht (1992) 33–48
58. Myers, C.M.: Vagueness, arbitrariness and matching. International Logic Review **31** (1985)
59. Noonan, H.W.: Vague objects. Analysis **42** (1982)
60. Pawlak, Z.: Rough sets - theoretical aspects of reasoning about data. Kluwer Academic Publishers (1991)
61. Pawlak, Z.: Vaguenes and uncertainty - a rough set perspective. Computational Intelligence an Informational Journal (1995) 227–232
62. Pawlak Z., Grzymała-Busse J. W., Słowiński R., and Ziarko, W.: Rough sets. Communication of the ACM **38** (1995) 88–95
63. Pawlak, Z., Skowron, A.: Rough membership functions. In Yaeger, R.R., Fedrizzi, M., and Kacprzyk, J. (Eds.), *Advances in the Dempster Shafer Theory of Evidence,* John Wiley and Sons (1994) 251–271

64. Pawlak Z., and Słowiński R.: Rough set approach to multi-attribute decision analysis, Invited Review. European Journal of Operational Research **72** (1994) 443–459
65. Peirce, C.S.: Vague. Dictionary of Philosophy and Psychology, edited by J.M. Baldwin. London (1902)
66. Pelletier F.J.: Another argument against vague objects. The Journal of Philosophy **86** (1986)
67. Płonka, L., and Mrózek, A.: Rule-based stabilization of the inverted pendulum. Computational Intelligence: An International Journal **11** (1995) 348–356
68. Popper, K.: The logic of scientific discovery. London. Hutchinson (1959)
69. Rolf, B.: A theory of vagueness. Journal of Philosophical Logic **9** (1980)
70. Russell, B.: Vagueness. Australian Journal of Philosophy **1** (1923) 84–92
71. Russell, B.: An inquiry into meaning and truth. George Allen and Unwin, London (1950)
72. Sainsbury, R.M.: What is a vague object? Analysis **49** (1989)
73. Shenoi, S.: Rough sets in fuzzy databases. In: P.P. Wang (ed.), Second Annual Joint Conference on Information Sciences PROCEEDINGS, September 28 – October 1, Wrightsville Beach, North Carolina, USA (1995) 263–264
74. Shafer, G.: A Mathematical theory of evidence. Princeton, NJ. Princeton University Press (1976)
75. Skala H.J., Termini, S., Trillas, E.: Aspects of Vagueness. D. Reidel, Dordrecht, Holland (1984)
76. Skowron, A., Grzymała-Busse, J.: From the rough set theory to evidence theory. In Yaeger, R.R., Fedrizzi, M., and Kacprzyk, J. (eds.). Advances in the Dempster Shafer Theory of Evidence. John Wiley and Sons (1994) 193–236
77. Skowron A.,and Rauszer, C.: The discernibility matrices and functions in information systems. In: R. Słowiński (ed.), Intelligent Decision Support. Handbook of Applications and Advances of the Rough Set Theory, Kluwer Academic Publishers, Dordrecht (1992) 311–362
78. Słowiński, R., (ed.).: Intelligent Decision Support. Handbook of Applications and Advances of the Rough Set Theory, Kluwer Academic Publishers, Dordrecht (1992)
79. Słowiński, R.: Rough set learning of preferential attitude in multi-criteria decision making. In: J. Komorowski and Z.W. Raś (eds.), Methodologies for Intelligent Systems. Lecture Notes in Artificial Intelligence Vol. 689, Springer-Verlag, Berlin (1993) 642–651
80. Słowiński, R., and Zopounidis, C.: Applications of the rough set approach to evaluation of bankruptcy risk. Working Paper 93-08, Decision Support System Laboratory, Technical University of Crete, Chania, June (1993)
81. Słowiński, R., and Zopounidis, C.: Rough set sorting of firms according to bankruptcy risk. In: M. Paruccini (ed.), *Applying Multiple Criteria Aid for Decision to Environmental Management*, Kluwer, Dordrecht, Netherlands (1994) 339–357
82. Słowiński, K.: Rough classification of HSV patients. in: R. Słowiński (ed.), Intelligent Decision Support. Handbook of Applications and Advances of the Rough Set Theory, Kluwer Academic Publishers, Dordrecht (1992) 77–93
83. Słowiński, K., Słowiński, R., Stefanowski, J.: Rough sets approach to analysis of data from peritoneal lavage in acute pancreatitis. Medical Informatics **13/3** (1988) 143–159
84. Słowiński, K., and Sharif, E.S.: Rough sets approach to analysis of data of diatnostic peritoneal lavage applied for multiple injuries patients. In: W. Ziarko (ed.),

Rough Sets, Fuzzy Sets and Knowledge Discovery. Proceedings of the International Workshop on Rough Sets and Knowledge Discovery (RSKD'93), Banff, Alberta, Canada, October 12–15, Springer-Verlag, Berlin (1993) 420–425
85. Słowiński, K., Stefanowski, J., Antczak, A., Kwias, Z.: Rough sets approach to the verification of indications for treatment of urinary stones by extracorporeal shock wave lithotripsy (ESWL). In: T.Y. Lin and A.M. Wildberger (eds.), The Third International Workshop on Rough Sets and Soft Computing Proceedings (RSSC'94), San Jose State University, San Jose, California, USA, November 10–12 (1995) 93–96
86. Szladow, A. J., and Ziarko, W.: Knowledge-based process control using rough sets. In: R. Słowiński (ed.), Intelligent Decision Support. Handbook of Applications and Advances of the Rough Set Theory, Kluwer Academic Publishers, Dordrecht (1992) 49–60
87. Tanaka, H., Ishibuchi, H., and Shigenaga, T.: Fuzzy inference system based on rough sets and its application to medical diagnostic. In: R. Słowiński (ed.), ntelligent Decision Support. Handbook of Applications and Advances of the Rough Set Theory, Kluwer Academic Publishers, Dordrecht (1992) 111–117
88. Thomason, R.: Identity and vagueness. Philosophical Studies **42** (1982)
89. Travis, Ch.: Vagueness observation and sorites. Mind **94** (1985)
90. Tye, M.: Vague object. Mind **99** (1990)
91. Williamson, T.: Identity and discrimination. Basil Blackwell, Cambridge, Massachusetts (1990)
92. Zadeh, L.: (1965). Fuzzy sets. Information and Control **8** (1965) 338-353
93. Ziarko, W., (ed.).: Rough Sets, Fuzzy Sets and Knowledge Discovery. Proceedings of the International Workshop on Rough Sets and Knowledge Discovery (RSKD'93), Banff, Alberta, Canada, October 12–15, Springer-Verlag, Berlin (1993)
94. Ziarko, W.: Variable Precision Rough Set Model. Journal of Computer and System Sciences **40** (1993) 39–59
95. Ziarko, W., Golan, R., and Edwards, D.: An application of DATALOGIC/R knowledge discovery tool to identify strong predictive rules in stock market data. In: Proc. AAAI Workshop on Knowledge Discovery in Databases, Washington, DC(1993) 89–101
96. Ziarko, W., and Katzberg, J.: Control algorithms acquisition, analysis and reduction: machine learning approach. In: Knowledge-Based Systems Diagnosis, Supervision and Control, Plenum Press, Oxford (1989) 167–178

Ehrenfeucht Games, the Composition Method, and the Monadic Theory of Ordinal Words

Wolfgang Thomas

Institut für Informatik und Praktische Mathematik
Christian-Albrechts-Universität zu Kiel
D-24098 Kiel, Germany

Abstract. When Ehrenfeucht introduced his game theoretic characterization of elementary equivalence in 1961, the first application of these "Ehrenfeucht games" was to show that certain ordinals (considered as orderings) are indistinguishable in first-order logic and weak monadic second-order logic. Here we review Shelah's extension of the method, the "composition of monadic theories", explain it in the example of the monadic theory of the ordinal ordering $(\omega, <)$, and compare it with the automata theoretic approach due to Büchi. We also consider the expansion of ordinals by recursive unary predicates (which gives "recursive ordinal words"). It is shown that the monadic theory of a recursive ω^n-word belongs to the $2n$-th level of the arithmetical hierarchy, and that in general this bound cannot be improved.

1 Introduction

One of the most successful tools of mathematical logic, in particular of those parts of logic which are relevant to computer science, is the method of "Ehrenfeucht games", introduced by A. Ehrenfeucht in [Ehr61]. These games are a very clear and intuitive formulation of Fraïssé's characterization [Fr54] of elementary equivalence (indistinguishability of relational structures by first-order formulas). In [Ehr61] the method was extended to weak monadic second-order logic; later other logical systems were covered, such as infinitary logic, modal and temporal logics, transitive closure logic, and fixed point logics, to mention a few. Today one speaks of the "Ehrenfeucht-Fraïssé method". Already in [Ehr61], Ehrenfeucht provided first applications by showing that the first-order theory and the weak monadic second-order theory of an ordinal ordering $(\alpha, <)$ depends only on its "ω-tail". (If $\alpha = z + \omega^y + \omega^n c_n + \ldots + \omega^0 c_0$, where $+$ is ordinal addition, $y \geq \omega$, and $c_i < \omega$, then $\omega^n c_n + \ldots + \omega^0 c_0$ is the ω-tail of α.)

This work opened a series of hundreds of papers, especially in the theory of finite models, where the technique was applied to clarify the expressive power of logical systems, often connected with decidability proofs. The decidability of the first-order and weak monadic second-order theory of linear orderings were among the first results, shown by Ehrenfeucht [Ehr59] and Läuchli [Lä68], respectively.

In this paper, we take up the Ehrenfeucht-Fraïssé method in the context of full monadic second-order logic over orderings, explain its extension by Shelah (for the "composition of monadic theories"), and show some results on the monadic theory of "ordinal words" (expansions of ordinals by unary predicates).

At the time of the appearance of [Ehr61], Büchi showed in [Bü62] that the monadic second-order theory of the ordinal $(\omega, <)$ can be analyzed with concepts from automata theory. He proved that a monadic second-order formula can be converted effectively into an equivalent finite automaton (over ω-words) and concluded that the monadic theory of $(\omega, <)$ is decidable. Later papers (e.g., [Bü65]) extended the method also to greater countable ordinals, and it was shown, for example, that the (unrestricted) monadic second-order theory of each countable ordinal is decidable. The automata theoretic method was taken up by many researchers and turned out successful especially in the context of theoretical computer science, regarding areas like generalized formal language theory, program verification, and concurrency theory (see e.g. [Th90] and the articles in [MB96]).

In this track of research it was barely realized, however, that a model theoretic approach based on the Ehrenfeucht-Fraïssé technique, called "composition method", could as well be applied, avoiding the use of automata. This extension of [Fr54] and [Ehr61] was started by Läuchli in [Lä68] and was further developed by Shelah in his celebrated and difficult paper [Sh75]. Shelah gave alternative proofs for Büchi's results and proved many more, also on dense orderings. For example, he showed that the monadic theory of the real number ordering is decidable when the set quantifiers are restricted to countable sets only, but that without this restriction the theory is undecidable. More results were obtained by Gurevich and Shelah in [Gu79], [GS79], [GS83], [GS85], and with Magidor in [GMS83]. Other applications appeared in [Th80], [Gu82], [CFGS82], and [Ze94]. (This list is not claimed to be complete; in particular, there may be further work of Shelah on the topic which is not cited here.)

Although the subject was exposed in Gurevich's concise survey [Gu85], it did not attract much attention among theoretical computer scientists. Preference was (and still is) given to the automata theoretic method: by its connection with a computational model it looks more intuitive, it incorporates "programs" in the form of state-transition systems, and it does not involve frightening logical technicalities as one finds them in [Sh75]. Thus there is a tendency that the merits of the model theoretic approach are overlooked. This is unfortunate, because it excludes some interesting applications. For instance, the model theoretic treatment of dense orderings (which transcends the domain of discrete automata theory) may be of interest in the formal analysis of systems in which conditions on dense time are involved.

We take this as a motivation to review the composition method in the study of monadic theories of orderings (here in particular: ordinals) again, trying to make it more accessible to readers with a background in theoretical computer science. We give an intuitive introduction, including simple examples, in order to enable the reader to follow the results and the proofs of Shelah and Gurevich. We assume that the reader is familiar with the fundamentals of first-order logic, has some idea of first-order Ehrenfeucht games, as presented e.g. in [Ehr61] or in textbooks like [EFT84, EF95], and is acquainted with basic recursion theory, concerning relative recursiveness and the arithmetical hierarchy (see e.g. [Ro67], [Od89]).

After some preparations, we present three results: First, we (re)prove in a new form the "composition theorem" of [Sh75, Gu79] on monadic theories of concatenated orderings. Then we recapitulate Shelah's elegant (and "automata-free") proof of Büchi's theorem on the decidability of the monadic theory of the structure $(\omega, <)$. Finally, we analyze the situation where an ordinal ordering is expanded by unary predicates $P_1, \dots, P_m$; such a structure $(\alpha, <, P_1, \dots, P_m)$ can be considered as an *ordinal word*, or to be more precise, as an α-word over the alphabet $\{0,1\}^m$. We show that a recursive ω^n-word (where the underlying predicates P_i are recursive) can have an undecidable monadic theory (even first-order theory), but that its complexity is bounded: Such a monadic theory belongs to the $2n$-th level of the arithmetical hierarchy (Kleene-Mostowski-hierarchy); and we prove that in general this bound cannot be improved.

2 The monadic theory of labelled orderings

The structures considered in this paper are expansions of nonempty linear orderings $\mathcal{A} = (A, <^A)$ by subsets $P_1^A, \dots P_m^A$. When no confusion arises we cancel the superscript ^{A}, use the abbreviating notation $\overline{P}$ for the set tuple $(P_1^A, \dots, P_m^A)$, and write $(\mathcal{A}, \overline{P})$. Such a structure can be regarded as a labelled ordering with labels in $\{0,1\}^m$: the element $a \in A$ has the label $(b_1, \dots, b_m)$ defined by $b_i = 1$ iff $a \in P_i^A$. Special structures of this type are labelled ordinal orderings (short: ordinal words); in this case A is an ordinal number, considered as the set containing all smaller ordinals as elements, and $<^A$ is the usual ordering of these elements.

The standard first-order and monadic second-order language for structures of this signature is built up as follows, using the relation symbols $<$ and $P_1, \dots, P_m$: We have first-order variables $x, y, \dots$ for elements of structures, monadic second-order variables $X, Y, \dots$ for sets of elements of structures, and the atomic formulas are of the form $x = y$, $x < y$, $x \in P_i$, and $x \in Y$, with the canonical interpretation. First-order formulas are formed by using the connectives $\neg$, $\vee$, $\wedge$, $\rightarrow$ and by applying the quantifiers $\exists$, $\forall$ to first-order variables; if application of these quantifiers to monadic second-order variables is also allowed, one obtains the monadic second-order formulas. If $\varphi(x_1, \dots, x_k, X_1, \dots X_l)$ is a formula with (at most) the indicated free variables and $q_1, \dots, q_k$ are elements and $Q_1, \dots Q_l$ are subsets of A, we write

$$(A, <^A, P_1^A, \dots, P_m^A, q_1, \dots, q_k, Q_1, \dots, Q_l) \models \varphi(x_1, \dots, x_k, X_1, \dots X_l)$$

if φ is satisfied when interpreting x_i by q_i and X_j by Q_j. A sentence is a formula without free variables.

For the analysis of monadic second-order theories it will be convenient to work with a slightly modified (but expressively equivalent) set-up, in which the first-order variables are cancelled. We allow only monadic second-order variables and take as atomic formulas the following: Nonempty$(X \cap Y)$, $X \subseteq Y$, $X < Y$, and $Y_1 \cup \ldots \cup Y_k = \text{All}$ (for distinct Y_i). These are interpreted, respectively, as "$X \cap Y$ is nonempty", "X is a subset of Y", "some element of X is smaller than some element of Y", and "the union of $Y_1, \ldots, Y_k$ is the universe". For better readability, we shall often write Nonempty(X) instead of Nonempty$(X \cap X)$. In the context of a fixed indexing $X_1, X_2, \ldots$ of variables, we allow formulas Nonempty$(X_{i_1} \cap X_{i_2})$ or $X_{i_1} \cup \ldots \cup X_{i_k} = \text{All}$ with distinct i_j only for $i_1 < i_2 < \ldots < i_k$, to spare some redundancies in the set of atomic formulas. By the particular interpretation of $<$ between sets we have, for example, $\{10, 11\} < \{10, 11\}$ but not $\emptyset < \emptyset$ and not $\{10\} < \{10\}$. In fact, we may use the defined predicate "X is a singleton" (written Sing(X)), introduced by

$$\text{Sing}(X) :\leftrightarrow \text{Nonempty}(X) \wedge \neg\, X < X.$$

The use of the unary relation symbols P_i will be avoided, by taking free set variables X_i instead. Thus, we shall use labelled orderings $(\mathcal{A}, \overline{P})$ as interpretations of monadic formulas $\varphi(\overline{X})$. For instance, the formula (from the standard language)

$$\varphi(Z) := \forall x(x \in Z \rightarrow \exists y(x < y \wedge y \in P_1))$$

will now be written as

$$\varphi'(X_1, Z) := \forall X(\text{Sing}(X) \wedge X \subseteq Z \;\rightarrow\; \exists Y(\text{Sing}(Y) \wedge X < Y \wedge Y \subseteq X_1)).$$

The monadic theory of an ordered structure $(\mathcal{A}, \overline{P})$, denoted MTh$(\mathcal{A}, \overline{P})$, is now the set of formulas $\varphi(\overline{X})$ which are satisfied in $(\mathcal{A}, \overline{P})$ (when interpreting X_i by P_i).

In order to classify formulas by an appropriate measure of "complexity", we refer to the prenex normal form: Each monadic formula $\varphi(X_1, \ldots, X_m)$ can be written as

$$\text{Qu}_1\overline{Y_1}\text{Qu}_2\overline{Y_2}\ \ldots\ \text{Qu}_n\overline{Y_n}\ \psi(\overline{X}, \overline{Y_1}, \overline{Y_2}, \ldots, \overline{Y_n}),$$

where ψ is quantifier-free, each Qu$_i$ is either $\exists$ or $\forall$, $\overline{X}$ stands for $X_1, \ldots, X_m$ and $\overline{Y_i}$ for $Y_{i1}, \ldots, Y_{ik_i}$. Referring to the sequence $\overline{k} = (k_1, \ldots, k_n)$ of the lengths of the quantifier blocks, we call such a formula a *$\overline{k}$-formula*. The set of all $\overline{k}$-formulas which are satisfied in a structure is called its $\overline{k}$-theory. We write

$$(\mathcal{A}, \overline{P}) \equiv_{\overline{k}} (\mathcal{B}, \overline{R})$$

if the two labelled orderings $(\mathcal{A}, \overline{P})$, $(\mathcal{B}, \overline{R})$ satisfy the same $\overline{k}$-formulas $\varphi(\overline{X})$, i.e., have the same $\overline{k}$-theory.

3 The monadic Ehrenfeucht game

In the present context, an Ehrenfeucht game serves to verify that two structures are $\equiv_{\overline{k}}$-equivalent. Here we consider labelled linear orderings (until Section 7 not necessarily well-orderings), say $(\mathcal{A}, \overline{P})$ and and $(\mathcal{B}, \overline{R})$ with $\overline{P} = (P_1, \ldots, P_m)$ and $\overline{R} = (R_1, \ldots, R_m)$. The corresponding game $G_{\overline{k}}((\mathcal{A}, \overline{P}), (\mathcal{B}, \overline{R}))$ is played on the two structures by two players I and II, also called Spoiler and Duplicator. If $\overline{k} = (k_1, \ldots, k_n)$ there are n rounds in a play of the game. In round i, Spoiler begins by picking subsets $P_{i1}, \ldots, P_{ik_i}$ in A or subsets $R_{i1}, \ldots, R_{ik_i}$ in B, to which Duplicator responds by picking k_i sets in the other structure (i.e, sets $R_{i1}, \ldots, R_{ik_i}$ in B, respectively $P_{i1}, \ldots, P_{ik_i}$ in A). After the n rounds, Duplicator has won the play if the truth of atomic formulas is preserved when passing from the P_i, P_{ij} to the corresponding sets R_i, R_{ij}. More formally (and writing P_{i0}, R_{i0} instead of P_i, R_i, respectively) we should have:

- $P_{ij} \cap P_{i'j'} \neq \emptyset$ iff $R_{ij} \cap R_{i'j'} \neq \emptyset$,
- $P_{ij} < P_{i'j'}$ iff $R_{ij} < R_{i'j'}$,
- $P_{ij} \subseteq P_{i'j'}$ iff $R_{ij} \subseteq R_{i'j'}$, and
- $P_{i_1 j_1} \cup \ldots \cup P_{i_k j_k} = A$ iff $R_{i_1 j_1} \cup \ldots \cup R_{i_k j_k} = B$.

We indicate that Duplicator has a winning strategy in this game by writing $(\mathcal{A}, \overline{P}) \sim_{\overline{k}} (\mathcal{B}, \overline{R})$.

Let us note two differences to the classical (first-order) Ehrenfeucht game: First, sets of elements are chosen in the moves of a play, rather than individual elements. This reflects the fact that we deal with set quantifiers. Similarly, in the weak second-order game of [Ehr61], Ehrenfeucht allowed the choice of finite sets by the two players in each move. Secondly, the quantifiers are not treated one at a time (i.e., sets are not chosen one at a time) but block-wise, in correspondence with the entries of a sequence $\overline{k} = (k_1, \ldots, k_n)$. This approach will prove useful later on, when the usual induction parameter, the number of (nested) quantifiers, is replaced by the alternation depth of a quantifier sequence.

The basic result on the game, the "Ehrenfeucht-Fraïssé Theorem", states that existence of a winning strategy for Duplicator in the game $G_{\overline{k}}((\mathcal{A}, \overline{P}), (\mathcal{B}, \overline{R}))$ characterizes $\equiv_{\overline{k}}$-equivalence between $(\mathcal{A}, \overline{P})$ and $(\mathcal{B}, \overline{R})$. It is formulated here just for orderings expanded by unary predicates, but is valid for relational structures of any finite signature. The standard proof proceeds in two steps: First one applies Fraïssé's characterization ([Fr54]) of $\equiv_{\overline{k}}$-equivalence by the existence of families of partial isomorphisms with the "back-and-forth property", secondly one shows that this latter condition just means that Duplicator wins the corresponding Ehrenfeucht game. The details are supplied by a straightforward adaptation of the classical case (first-order logic), as treated e.g. in the textbooks [EFT84], [EF95], to the present context of monadic logic.

Theorem 1. (Ehrenfeucht-Fraïssé Theorem)
For any $\overline{k} = (k_1, \ldots, k_n)$ and any two linear orderings $\mathcal{A} = (A, <^A)$ and $\mathcal{B} = (B, <^B)$ expanded by sequences $\overline{P} = (P_1, \ldots, P_m)$, respectively $\overline{R} = (R_1, \ldots, R_m)$ of subsets we have: $(\mathcal{A}, \overline{P}) \equiv_{\overline{k}} (\mathcal{B}, \overline{R})$ iff $(\mathcal{A}, \overline{P}) \sim_{\overline{k}} (\mathcal{B}, \overline{R})$.

4 Hintikka formulas

In this section we show how $\equiv_{\overline{k}}$-equivalence classes can be defined in monadic logic. The defining formulas are sometimes called *Hintikka formulas* (reminding of their first appearance, in [Hi53], in the framework of first-order logic).

As before, we use letter $\mathcal{A}$ for linear orderings $(A, <^A)$ and write expansions by tuples $\overline{P} = (P_1, \ldots, P_m)$ of subsets in the form $(\mathcal{A}, \overline{P})$. For such a structure and any sequence $\overline{k} = (k_1, \ldots, k_n)$, we define a formula $\varphi^{\overline{k}}_{(\mathcal{A},\overline{P})}(X_1, \ldots, X_m)$, which is satisfied precisely in those structures $(\mathcal{B}, \overline{R})$ that are $\equiv_{\overline{k}}$-equivalent to $(\mathcal{A}, \overline{P})$. In other words, the formula $\varphi^{\overline{k}}_{(\mathcal{A},\overline{P})}(\overline{X})$ has to describe which $\overline{k}$-formulas $\psi(\overline{X})$ are true, respectively false, in the structure $(\mathcal{A}, \overline{P})$.

The task is easy if $\overline{k}$ is the empty sequence λ: Here we have to state which atomic formulas (whose free variables are among $X_1, \ldots, X_m$) are true in $(\mathcal{A}, \overline{P})$ and which are false in this structure. Thus we take the formula

$$\varphi^{\lambda}_{(\mathcal{A},\overline{P})}(\overline{X}) := \bigwedge_{\substack{\varphi(\overline{X}) \text{ atomic,} \\ (\mathcal{A},\overline{P}) \models \varphi(\overline{X})}} \varphi(\overline{X}) \;\wedge \bigwedge_{\substack{\varphi(\overline{X}) \text{ atomic,} \\ (\mathcal{A},\overline{P}) \models \neg\varphi(\overline{X})}} \neg\varphi(\overline{X}).$$

Example 1. Consider the structure $\mathcal{A} = (\omega, <)$ expanded by $\overline{P} = (P_1, P_2)$ where $P_1 = \{10\}$ and $P_2 = \{11\}$. The relevant atomic formulas for $\varphi^{\lambda}_{(\mathcal{A},P_1,P_2)}(X_1, X_2)$ are the following: Nonempty(X_1), Nonempty(X_2), Nonempty$(X_1 \cap X_2)$, $X_i \subseteq X_j$ for $i, j \in \{1, 2\}$, $X_i < X_j$ for $i, j \in \{1, 2\}$, $X_1 = \text{All}$, $X_2 = \text{All}$, and $X_1 \cup X_2 = \text{All}$. The conjunction of satisfied formulas is

$$\text{Nonempty}(X_1) \wedge \text{Nonempty}(X_2) \wedge X_1 \subseteq X_1 \wedge X_2 \subseteq X_2 \wedge X_1 < X_2,$$

whereas the other atomic formulas (Nonempty$(X_1 \cap X_2)$, $X_1 \subseteq X_2$, $X_2 \subseteq X_1$, $X_1 < X_1$, $X_2 < X_2$, $X_2 < X_1$, $X_1 = \text{All}$, $X_2 = \text{All}$, and $X_1 \cup X_2 = \text{All}$) are false and assembled in negated form in the second conjunction. In the sequel we denote this formula $\varphi^{\lambda}_{(\mathcal{A},P_1,P_2)}(X_1, X_2)$ just by $\varphi_0(X_1, X_2)$. □

Let us define $\varphi^{\overline{k}}_{(\mathcal{A},\overline{P})}(\overline{X})$ for nonempty sequences $\overline{k}$. If $\overline{k}$ has just a single entry, say $\overline{k} = (k_1)$, then the desired Hintikka formula has to describe which different situations for the truth of atomic formulas can be generated by expanding the structure $(\mathcal{A}, \overline{P})$ by further predicates $\overline{Q} = (Q_1, \ldots Q_{k_1})$. The set of all such k_1-tuples of sets from the powerset of A will be denoted by $\mathcal{P}(A)^{k_1}$. For each choice of $\overline{Q} \in \mathcal{P}(A)^{k_1}$, we shall state the existence of $\overline{Y}$ with $\varphi^{\lambda}_{(\mathcal{A},\overline{P},\overline{Q})}(\overline{X}, \overline{Y})$; moreover, the existence of other tuples has to be excluded, in the sense that for each tuple $\overline{Y}$ one of the above formulas $\varphi^{\lambda}_{(\mathcal{A},\overline{P},\overline{Q})}(\overline{X}, \overline{Y})$ holds.

Example 2. (continued) Let us consider the simple case $\overline{k} = (1)$ and sketch the formula $\varphi^{(1)}_{(\mathcal{A},P_1)}$ where $\mathcal{A} = (\omega, <)$ and $P_1 = \{10\}$ are as above. We shall write down a conjunction of formulas in which, for example, $\exists X_2 \varphi_0(X_1, X_2)$ with $\varphi_0(X_1, X_2)$ from above occurs, representing the choice of $\{11\}$ for X_2. One sees that the *same* formula arises if we take any singleton $\{n\}$ with $n > 11$ instead. But we get different formulas if we consider (for X_2) one of the following eleven sets: $\emptyset$, $\{9\}$, $\{8, 9\}$, $\{9, 10\}$, $\{10\}$, $\{10, 11\}$, $\{11, 12\}$, $\{9, 11\}$, $\{9, 10, 11\}$, $\omega \setminus \{10\}$, and ω. One verifies that the twelve options altogether exhaust all different possibilities to satisfy the atomic formulas in X_1, X_2, given the set $P_1 = \{10\}$ as interpretation of X_1. Similarly to $\exists X_2 \varphi_0(X_1, X_2)$ we can write down corresponding existence claims $\exists X_2 \varphi_i(X_1, X_2)$ for the remaining eleven cases $i = 1, \ldots, 11$. (For instance, the formula $\varphi_3(X_1, X_2)$, which corresponds to the choice of $\{8, 9\}$ for X_2, is the conjunction

$$\mathrm{Nonempty}(X_1) \wedge \mathrm{Nonempty}(X_2) \wedge X_1 \subseteq X_1 \wedge X_2 \subseteq X_2 \wedge X_2 < X_2 \wedge X_2 < X_1,$$

while the other atomic formulas are adjoined in negated form.) Now we may set

$$\varphi^{(1)}_{(\mathcal{A},P_1)} := \bigwedge_{i=0}^{11} \exists X_2 \varphi_i(X_1, X_2) \ \wedge \ \forall X_2 \bigvee_{i=0}^{11} \varphi_i(X_1, X_2).$$

Here the second part expresses that the list of twelve options is complete, i.e. that any choice of a subset occurs in the list. □

The general step, from a sequence $\overline{k} = (k_1, \ldots, k_n)$ to its extension $\overline{k}^\wedge k_{n+1}$, is handled in precisely the same way, replacing λ by $\overline{k}$, and taking into account that k_{n+1} sets are chosen simultaneously. So we obtain the following definition:

$$\varphi^{\overline{k}^\wedge k_{n+1}}_{(\mathcal{A},\overline{P})}(\overline{X}) := \bigwedge_{\overline{Q} \in \mathcal{P}(A)^{k_{n+1}}} \exists \overline{Y} \varphi^{\overline{k}}_{(\mathcal{A},\overline{P},\overline{Q})}(\overline{X}, \overline{Y}) \wedge \forall \overline{Y} \bigvee_{\overline{Q} \in \mathcal{P}(A)^{k_{n+1}}} \varphi^{\overline{k}}_{(\mathcal{A},\overline{P},\overline{Q})}(\overline{X}, \overline{Y}).$$

The disjunction and conjunction in this formula are finite. To verify this, note that there are only finitely many atomic formulas involving variables from a finite set $\{X_1, \ldots, X_r\}$, and that (as verified by induction on the length of $\overline{k}$) the number of logically non-equivalent $\overline{k}$-formulas $\psi(X_1, \ldots, X_r)$ (for a given number of free variables) is finite. Thus the disjunction and the conjunction (over $\overline{Q} \in \mathcal{P}(A)^{k_{n+1}}$) in the definition of $\varphi^{\overline{k}^\wedge k_{n+1}}_{(\mathcal{A},\overline{P})}(\overline{X})$ both range only over finitely many formulas $\varphi^{\overline{k}}_{(\mathcal{A},\overline{P},\overline{Q})}(\overline{X}, \overline{Y})$ and hence specify monadic formulas. By the same reason, for any $\overline{k}$ and any length of the tuple $\overline{X}$, only finitely many different Hintikka formulas $\varphi^{\overline{k}}_{(\mathcal{A},\overline{P})}(\overline{X})$ exist which arise from the infinitely many possible structures $(\mathcal{A}, \overline{P})$. Any such structure will satisfy precisely one Hintikka formula.

Let us formulate some further key properties of Hintikka formulas. First, a Hintikka formula $\varphi^{\overline{k}}_{(\mathcal{A},\overline{P})}(\overline{X})$ fixes truth, respectively falsehood, of any given $\overline{k}$-formula in $(\mathcal{A}, \overline{P})$.

Example 3. (continued) To illustrate this, consider the formula $\varphi^{(1)}_{(\mathcal{A},P_1)}(X_1)$ from the example above. We check that the (1)-formula

$$\psi(X_1) := \exists Y (X_1 \subseteq Y \wedge X_1 < Y)$$

is true in $(\mathcal{A}, P_1)$ $(= (\omega, <, \{10\}))$: From the conjunction member $\exists X_2 \varphi_6(X_1, X_2)$ of $\varphi^{(1)}_{(\mathcal{A},P_1)}(X_1)$ (corresponding to the choice $\{10, 11\}$ for X_2) the formula ψ follows and hence is true in $(\mathcal{A}, P_1)$. On the other hand, $\forall Y(X_1 \subseteq Y \to X_1 < Y)$ is not true in $(\mathcal{A}, P_1)$: This universal statement holds if it is true in all twelve options listed above. By inspection of $\varphi_0, \ldots, \varphi_{11}$ we see that occurrence of the conjunction member $X_1 \subseteq X_2$ does not imply that also $X_1 < X_2$ occurs as a conjunction member; a counterexample is $\varphi_0(X_1, X_2)$ (arising from the choice of $\{10\}$ for X_2). □

So we obtain the following lemma (which is proved by induction on $\overline{k}$):

Lemma 2. (Hintikka formulas)
Let $(\mathcal{A}, \overline{P})$ be the expansion of the ordering $\mathcal{A}$ by unary predicates $P_1, \ldots, P_m$ and let $\overline{k} = (k_1, \ldots, k_n)$.
(a) *If $\psi(\overline{X})$ is a $\overline{k}$-formula, then*

$$(\mathcal{A}, \overline{P}) \models \psi(\overline{X}) \quad \textit{iff} \quad \varphi^{\overline{k}}_{(\mathcal{A},\overline{P})}(\overline{X}) \models \psi(\overline{X})$$

(i.e., iff the second formula follows from the first); moreover, this can be checked effectively.
Conversely, $(\mathcal{A}, \overline{P}) \models \neg\psi(\overline{X})$ iff $\varphi^{\overline{k}}_{(\mathcal{A},\overline{P})}(\overline{X}) \models \neg\psi(\overline{X})$.
(b) *The Hintikka formula $\varphi^{\overline{k}}_{(\mathcal{A},\overline{P})}(\overline{X})$ characterizes the $\equiv_{\overline{k}}$-class of $(\mathcal{A}, \overline{P})$ in the sense that for any labelled ordering $(\mathcal{B}, \overline{R})$ with $\overline{R} = (R_1, \ldots, R_m)$ we have:*

$$(\mathcal{A}, \overline{P}) \equiv_{\overline{k}} (\mathcal{B}, \overline{R}) \quad \textit{iff} \quad (\mathcal{B}, \overline{R}) \models \varphi^{\overline{k}}_{(\mathcal{A},\overline{P})}(\overline{X}).$$

Considering a $\overline{k}$-formula $\psi(\overline{X})$, we may collect those $\overline{k}$-Hintikka formulas from which the formula $\psi(\overline{X})$ follows (as in part (a) of the Lemma). Since each structure $(\mathcal{B}, \overline{R})$ satisfies precisely one Hintikka formula $\varphi^{\overline{k}}_{(\mathcal{A},\overline{P})}(\overline{X})$ of appropriate signature, we obtain:

Lemma 3. (Distributive normal form of $\overline{k}$-formulas)
A $\overline{k}$-formula $\psi(\overline{X})$ is equivalent to the disjunction of those $\overline{k}$-Hintikka-formulas of appropriate signature from which $\psi(\overline{X})$ follows; this disjunction can be computed effectively from $\psi(\overline{X})$.

5 $\overline{k}$-types

Besides the Hintikka-formulas, there is an alternative description of $\equiv_{\overline{k}}$-classes, first used by Läuchli [Lä68] for weak monadic logic and then by Shelah [Sh75] for unrestricted monadic logic. The formalism amounts to a compact representation of Hintikka formulas. Instead of inductively piling up nested disjunctions and conjunctions, one collects the relevant information by iterative formations of subsets of the power set, starting from a set of atomic formulas in certain variables $X_1, \ldots, X_m$. The (hereditarily finite) sets built up in this way to define $\equiv_{\overline{k}}$-classes are called $\overline{k}$-types.

For a sequence $\overline{k} = (k_1, \ldots, k_n)$ and a number m, we define the finite set $\mathcal{T}^{\overline{k}}(m)$ of "all formally possible $\overline{k}$-types with m set parameters". Let Atom_m be the set of atomic formulas in the variables $X_1, \ldots, X_m$ and set

- $\mathcal{T}^{\lambda}(m) := \mathcal{P}(\text{Atom}_m)$
- $\mathcal{T}^{\overline{k}\,\hat{}\,k_{n+1}}(m) := \mathcal{P}(\mathcal{T}^{\overline{k}}(m + k_{n+1}))$

Now we associate with a sequence $\overline{k} = (k_1, \ldots, k_n)$ and a labelled ordering $(\mathcal{A}, P_1, \ldots, P_m)$ an element of $\mathcal{T}^{\overline{k}}(m)$, called the *$\overline{k}$-type* $T^{\overline{k}}(\mathcal{A}, \overline{P})$, again defined inductively over the length of $\overline{k}$:

- $T^{\lambda}(\mathcal{A}, \overline{P}) := \{\varphi(\overline{X}) \mid \varphi(\overline{X}) \text{ atomic}, (\mathcal{A}, \overline{P}) \models \varphi(\overline{X})\}$
- $T^{\overline{k}\,\hat{}\,k_{n+1}}(\mathcal{A}, \overline{P}) :=$

$$\{T^{\overline{k}}(\mathcal{A}, \overline{P}, Q_1, \ldots, Q_{k_n+1}) \mid (Q_1, \ldots, Q_{k_n+1}) \in \mathcal{P}(A)^{k_{n+1}}\}.$$

Example 4. (continued) Consider again the structure $(\omega, <, P_1, P_2)$ with $P_1 = \{10\}$ and $P_2 = \{11\}$. The type $T^{\lambda}(\omega, P_1, P_2)$ contains the following formulas:

$$\text{Nonempty}(X_1), \text{Nonempty}(X_2), X_1 \subseteq X_1, X_2 \subseteq X_2, X_1 < X_2.$$

This information suffices for specifying the Hintikka formula $\varphi^{\lambda}_{(\omega,P_1,P_2)}(X_1, X_2)$. We call this λ-type τ_0 and now form the types $\tau_1, \ldots, \tau_{11}$ for other interpretations of X_2 than $P_2 = \{11\}$, in accordance with the eleven Hintikka formulas $\varphi_1(X_1, X_2), \ldots, \varphi_{11}(X_1, X_2)$ in the example above. For instance, we have

$$\tau_1 = \{X_1 \subseteq X_1, X_2 \subseteq X_2, X_2 \subseteq X_1\},$$

corresponding to the choice of $\emptyset$ for X_2. The (1)-type of the structure (ω, P_1) is the set $T^{(1)}(\omega, P_1) = \{\tau_0, \ldots, \tau_{11}\}$. □

Whereas the $\overline{k}$-type of $(\mathcal{A}, P_1, \ldots, P_m)$ belongs to $\mathcal{T}^{\overline{k}}(m)$, this set $\mathcal{T}^{\overline{k}}(m)$ contains other elements which are not $\overline{k}$-types of structures: There may be inconsistencies, for example when the formula $X_i < X_i$ occurs whereas Nonempty(X_i) is missing. So $\mathcal{T}^{\overline{k}}(m)$ is defined disregarding the question of satisfiability.

It is useful to note the precise correspondence between $\overline{k}$-types and $\overline{k}$-Hintikka formulas. The nesting of set formation in $\overline{k}$-types captures the nesting of conjunctions and disjunctions in Hintikka formulas; here the $\overline{k}$-types are slightly more "economic" than Hintikka formulas, because the possibilities of extension are collected as a set, whereas the description in a Hintikka formula has to refer to these possibilities twice: in the existence claim (conjunction of existential formulas) and in the completeness claim (universal formula over a disjunction).

One may regard the $(k_1, \ldots, k_n)$-type of a structure $(A, <, P_1, \ldots, P_m)$ as a finite tree of height n, where the levels are numbered from 0 (for the leaves) to n (for the root). On the level 0 (of leaves), there are λ-types of structures $(A, <, \overline{P}, \overline{Q})$ where the length of $\overline{Q}$ is $k_1 + \ldots + k_n$. (These types are sets of atomic formulas in the variables $X_1, \ldots, X_l$ where $l = m + k_1 + \ldots + k_n$.) In general, level i contains $(k_1, \ldots, k_i)$-types of structures $(A, <, \overline{P}, \overline{Q})$ where the length of $\overline{Q}$ is $k_{i+1} + \ldots + k_n$, and at the root (level n) we find $T^{\overline{k}}(A, <, \overline{P})$. Such a tree of extension possibilities of a given structure is sometimes called the *Fraïssé tree* of the structure.

In analogy to Lemma 2 (b) above, the $\overline{k}$-type of a structure $(\mathcal{A}, \overline{P})$ determines effectively for any $\overline{k}$-formula $\varphi(\overline{X})$ whether $(\mathcal{A}, \overline{P}) \models \varphi(\overline{X})$. The algorithm to determine the truth value is best explained by an example.

Example 5. Consider the structure $(\omega, <, P_1, P_2)$ where $P_1 = \{10\}$ and P_2 is the set of even numbers. In this structure, the formula

$$\forall X_3(X_1 < X_3 \rightarrow \exists X_4(X_3 < X_4 \wedge X_4 \subseteq X_2))$$

is true; it expresses that for each set S containing a number > 10 there is a set of even numbers containing a number which is greater than some element of S. The prenex normal form of the formula is

$$\forall X_3 \exists X_4(X_1 < X_3 \ \rightarrow \ X_3 < X_4 \wedge X_4 \subseteq X_2),$$

and we verify its truth in $(\omega, <, P_1, P_2)$ by checking the following property of the type $T^{(1,1)}(\omega, P_1, P_2)$: For each element of this $(1,1)$-type (i.e. for each (1)-type occurring in it), there is an element (which is then a λ-type) which satisfies the following: if the formula $X_1 < X_3$ occurs in it, so do $X_3 < X_4$ and $X_4 \subseteq X_2$.

So the quantifiers over sets in a structure correspond to quantifiers over finite sets of types within a given type. It may also happen that a $(k_1, \ldots, k_n)$-type determines an atomic formula ψ to hold even for $n > 1$. This means that in all sets which occur as leaves in the associated Fraïssé tree the atomic formula ψ appears. We say in this case that the $\overline{k}$-type *induces* ψ. Altogether we note that the $\overline{k}$-theory of a structure $(\mathcal{A}, \overline{P})$ is obtained effectively from its $\overline{k}$-type. Thus we obtain the following connection with decidability:

Lemma 4. *The monadic second-order theory* $\mathrm{MTh}(\mathcal{A}, \overline{P})$ *of a labelled ordering* $(\mathcal{A}, \overline{P})$ *is decidable iff the function*

$$f_{(\mathcal{A},\overline{P})} : \overline{k} \ \mapsto \ T^{\overline{k}}(\mathcal{A}, \overline{P})$$

is computable; more generally, $\mathrm{MTh}(\mathcal{A}, \overline{P})$ *and* $f_{(\mathcal{A},\overline{P})}$ *are recursive in each other.*

For a given finite structure $(\mathcal{A}, \overline{P})$, it is clear that the types $T^{\overline{k}}(\mathcal{A}, \overline{P})$ can be generated effectively, so the function $f_{(\mathcal{A},\overline{P})}$ is computable and the theory $\mathrm{MTh}(\mathcal{A}, \overline{P})$ decidable.

6 Composition of $\overline{k}$-types

In this section we consider the concatenation of labelled orderings, and show how to obtain the $\overline{k}$-theory of such a composed labelled ordering from the $\overline{k}$-theories of the component orderings. Here a $\overline{k}$-theory will be represented by the corresponding $\overline{k}$-type. The possibility of such an algorithmic "composition of $\overline{k}$-types" provides the background for decidability results on monadic theories of orderings.

The composition of $\overline{k}$-types involves two aspects: First, it has to be shown that the $\overline{k}$-types of labelled orderings, say $(\mathcal{A}_\iota, \overline{P}_\iota)$ for $\iota = 1, \ldots, r$, determine uniquely the $\overline{k}$-type of the concatenation of these structures in the order given by the indices, written $(\mathcal{A}_1, \overline{P}_1) + \ldots + (\mathcal{A}_r, \overline{P}_r)$. Secondly, an effective procedure is required which produces the $\overline{k}$-type of the composed structure from the $\overline{k}$-types of the components.

The first claim means that $\equiv_{\overline{k}}$ is a congruence with respect to concatenation of labelled orderings. This can be shown elegantly by means of the Ehrenfeucht game: Given labelled orderings $(\mathcal{A}_\iota, \overline{P}_\iota)$ and $(\mathcal{B}_\iota, \overline{R}_\iota)$ with $(\mathcal{A}_\iota, \overline{P}_\iota) \sim_{\overline{k}} (\mathcal{B}_\iota, \overline{R}_\iota)$ for $\iota = 1, \ldots, r$ one has to show that

$$(\mathcal{A}_1, \overline{P}_1) + \ldots + (\mathcal{A}_r, \overline{P}_r) \sim_{\overline{k}} (\mathcal{B}_1, \overline{R}_1) + \ldots + (\mathcal{B}_r, \overline{R}_r).$$

This means that from winning strategies $S_1, \ldots, S_r$ for Duplicator in the games $G_{\overline{k}}((\mathcal{A}_\iota, \overline{P}_\iota), (\mathcal{B}_\iota, \overline{R}_\iota))$ one has to compose a winning strategy for Duplicator in the game $G_{\overline{k}}((\mathcal{A}_1, \overline{P}_1) + \ldots + (\mathcal{A}_r, \overline{P}_r), (\mathcal{B}_1, \overline{R}_1) + \ldots + (\mathcal{B}_r, \overline{R}_r))$. The new strategy is the obvious one: "play S_ι on the pair $((\mathcal{A}_\iota, \overline{P}_\iota), (\mathcal{B}_\iota, \overline{R}_\iota))$, simultaneously for all $\iota \in \{1, \ldots, r\}$". One can apply the same idea if the ordered index set is infinite. The idea has to be refined when the index orderings for the $(\mathcal{A}_\iota, \overline{P}_\iota)$ and $(\mathcal{B}_\iota, \overline{R}_\iota)$ are different.

Shelah's calculus of $\overline{k}$-types also covers this possibility of different index orderings and adds the aspect of computability to the composition process.

Such a composition of theories is familiar from first-order model theory in connection with direct (and reduced) products. The direct product $\mathcal{A} := \prod_{\iota \in I} \mathcal{A}_\iota$ of relational structures $\mathcal{A}_\iota$ (all of a fixed finite signature σ) has the cartesian product $\prod_{\iota \in I} A_\iota$ as universe, and for a relation symbol R from σ one defines $((a_\iota^1)_{\iota \in I}, \ldots, (a_\iota^n)_{\iota \in I}) \in R^{\mathcal{A}}$ iff $(a_\iota^1, \ldots, a_\iota^n) \in R^{\mathcal{A}_\iota}$ for all $\iota \in I$. The Feferman-Vaught Theorem ([FV59], see also [CK73]) shows, in the simplest case, that the first-order theory of a product $\prod_{\iota \in I} \mathcal{A}_\iota$ can be obtained effectively from the first-order theories of the factors $\mathcal{A}_\iota$ and from the theory of the Boolean algebra $(\mathcal{P}(I), \sim, \cup, \cap)$, as follows: Given a sentence ψ (whose truth value in the product is to be determined), one can can compute sentences $\psi_1, \ldots, \psi_r$ (describing "factor properties") and a formula $\beta(x_1, \ldots, x_r)$ of the first-order language of Boolean algebras such that

$$\textstyle\prod_{\iota \in I} \mathcal{A}_\iota \models \psi \text{ iff } (\mathcal{P}(I), \sim, \cup, \cap, P_1, \ldots, P_r) \models \beta(x_1, \ldots, x_r)$$

where P_j is the set of indices $\iota \in I$ with $\mathcal{A}_\iota \models \psi_j$. In this sense, the truth of ψ in the product can be recovered from the distribution of the truth values of certain factor properties over the index set I.

Shelah [Sh75] developed an analogous composition formalism for monadic formulas; here the structures are labelled orderings $(\mathcal{A}_\iota, \overline{P}_\iota)$ with $\mathcal{A}_\iota = (A_\iota, <^{\mathcal{A}_\iota})$, combined by concatenation via an ordered index set I. One speaks of the *ordered sum* of the structures $(\mathcal{A}_\iota, \overline{P}_\iota)$ with respect to the ordering $(I, <^I)$. Formally, this ordered sum, denoted $\sum_{\iota \in I}(\mathcal{A}_\iota, \overline{P}_\iota)$, is defined as follows (assuming that the structures $\mathcal{A}_\iota$ are disjoint and that $\overline{P}_\iota = (P_{\iota 1}, \ldots, P_{\iota m})$):

$$\textstyle\sum_{\iota \in I}(\mathcal{A}_\iota, \overline{P}_\iota) = (\bigcup_{\iota \in I} A_\iota, <, \bigcup_{\iota \in I} \overline{P}_\iota)$$

where

- $a < b$ holds iff $a \in A_\iota$ and $b \in A_\kappa$ with $\iota <^I \kappa$, or $a, b \in A_\iota$ and $a <^{A_\iota} b$ for some $\iota \in I$,
- $\bigcup_{\iota \in I} \overline{P} := (\bigcup_{\iota \in I} P_{\iota 1}, \ldots, \bigcup_{\iota \in I} P_{\iota m})$.

The task of the desired composition algorithm is to produce the $\overline{k}$-type of an ordered sum

$$\textstyle\sum_{\iota \in I}(\mathcal{A}_\iota, P_{\iota 1}, \ldots, P_{\iota m})$$

from a certain $\overline{r}$-type of the index ordering to which information about the $\overline{k}$-types of the components (summand structures) is added. These component $\overline{k}$-types are all from the set $\mathcal{T}^{\overline{k}}(m)$, and we assume a fixed listing of them in the form $\tau_1, \ldots, \tau_s$. Consider the expansion of the index ordering $(I, <^I)$ by the sets $Q_1, \ldots, Q_s$ where

$$Q_j = \{\iota \in I \mid T^{\overline{k}}(\mathcal{A}_\iota, \overline{P}_\iota) = \tau_j\}.$$

We call this structure $(I, <^I, Q_1, \ldots, Q_s)$ the *$\mathcal{T}^{\overline{k}}(m)$-expansion of $(I, <^I)$ with respect to $(\mathcal{A}_\iota, \overline{P}_\iota)_{\iota \in I}$*. Clearly, these Q_j define a partition of I.

It will turn out that, for a suitable $\overline{r}$, the $\overline{r}$-type of such an expansion suffices to determine $T^{\overline{k}}(\sum_{\iota \in I}(\mathcal{A}_\iota, \overline{P}_\iota))$. We present the result here at a more relaxed pace than in [Sh75] (where it is just stated) and [Gu79], [Gu85] (where the proofs are rather condensed and given in the more abstract framework of arbitrary signatures and general Feferman-Vaught-type theorems). Our specific choice of signature leads to a certain simplification.

Theorem 5. (Composition Theorem, [Sh75], [Gu79])
From a sequence $\overline{k} = (k_1, \ldots, k_n)$ and a number m one can compute a sequence $\overline{r} = (r_1, \ldots, r_n)$ such that for any ordered sum $\sum_{\iota \in I}(\mathcal{A}_\iota, P_{\iota 1}, \ldots, P_{\iota m})$ its $\overline{k}$-type is determined by (and can be computed from) the $\overline{r}$-type of the $\mathcal{T}^{\overline{k}}(m)$-expansion of $(I, <^I)$ with respect to $(\mathcal{A}_\iota, \overline{P}_\iota)_{\iota \in I}$.

Before turning to the proof, let us note an essential technical point in this theorem, namely the preservation of the length n in the step from the sequence $\overline{k}$ to the sequence $\overline{r}$. The *sum* of the $\overline{r}$-entries will in general be larger than that of the $\overline{k}$-entries. Speaking in terms of quantifiers this means: In order to determine which formulas are true in an ordered sum, one may have to know "more complicated" facts about the index ordering with respect to the number of individual quantifiers, however with respect to the number of quantifier alternations, no increase of formula complexity is involved. This is the reason why the present classification of formulas, Ehrenfeucht games, and types does not refer to quantifier depth but rather to quantifier alternation depth.

Proof. First we say how to find $\overline{r}$ from $\overline{k}$ and m. We define a corresponding function $\rho : (\overline{k}, m) \mapsto \overline{r}$ inductively over the length of $\overline{k}$: Let $\rho(\lambda, m) = \lambda$ for all m, and let $\rho(\overline{k}^\wedge k_{n+1}, m) = \rho(\overline{k}, m + k_{n+1})^\wedge |\mathcal{T}^{\overline{k}}(m + k_{n+1})|$. For $\overline{r} := \rho(\overline{k}, m)$ we verify the claim of the theorem inductively on the length n of $\overline{k}$.

For $n = 0$ we have to verify that the type $T^\lambda(\sum_{\iota\in I}(\mathcal{A}_\iota, \overline{P}_\iota))$ can be determined effectively from the λ-type of the $\mathcal{T}^\lambda(m)$-expansion of $(I, <^I)$ with respect to $(\mathcal{A}_\iota, \overline{P}_\iota)_{\iota\in I}$. This expansion is the structure $(I, <^I, Q_1, \ldots, Q_s)$ where Q_k assembles those indices $\iota \in I$ for which the ι-th component of the ordered sum has λ-type τ_k (in the listing of types).

Let us check first when a formula Nonempty$(X_i \cap X_j)$ is true in the ordered sum (i.e., belongs to $T^\lambda(\sum_{\iota\in I}(\mathcal{A}_\iota, \overline{P}_\iota))$). This holds iff some index ι exists such that in $(\mathcal{A}_\iota, \overline{P}_\iota)$ the formula Nonempty$(X_i \cap X_j)$ is true. So the type τ_k of this summand should contain Nonempty$(X_i \cap X_j)$, i.e. the corresponding subset Q_k of I, to which ι belongs, is nonempty. Altogether, Nonempty$(X_i \cap X_j)$ is true in the ordered sum iff for some $k \in \{1, \ldots, s\}$ with Nonempty$(X_i \cap X_j) \in \tau_k$, the formula Nonempty$(X_k)$ is in the type $T^\lambda(I, <^I, Q_1, \ldots, Q_s)$.

The other three kinds of atomic formulas are handled similarly: A formula $X_i \subseteq X_j$ is true in the ordered sum iff it is true for each summand iff only those Q_k are nonempty (i.e., we have Nonempty$(X_k) \in \mathcal{T}^\lambda(I, <^I, Q_1, \ldots, Q_s)$) where type τ_k contains $X_i \subseteq X_j$. A formula $X_i < X_j$ is true in the ordered sum iff

- either we have Nonempty$(X_k) \in \mathcal{T}^\lambda(I, <^I, Q_1, \ldots, Q_s)$ for some k such that type τ_k contains $X_i < X_j$ (i.e., $X_i < X_j$ holds already in some component, say of type τ_k),
- or we have $X_k < X_{k'} \in \mathcal{T}^\lambda(I, <^I, Q_1, \ldots, Q_s)$ for some pair (k, k') such that τ_k contains Nonempty(X_i) and $\tau_{k'}$ contains Nonempty(X_j) (this is the case where some component $\mathcal{A}_\iota$ (of type τ_k) exists before a component $\mathcal{A}_{\iota'}$ (of type $\tau_{k'}$), such that there is a X_i-element in $\mathcal{A}_\iota$ and a X_j-element in $\mathcal{A}_{\iota'}$).

Finally, a formula $X_{i_1} \cup \ldots \cup X_{i_l} =$ All is true in the ordered sum iff it holds in all summands iff Nonempty(X_k) occurs in $\mathcal{T}^\lambda(I, <^I, Q_1, \ldots, Q_s)$ only for those k where τ_k contains $X_{i_1} \cup \ldots \cup X_{i_l} =$ All.

Altogether we obtain an effective method to extract $T^\lambda(\sum_{\iota\in I}(\mathcal{A}_\iota, \overline{P}_\iota))$ from $T^\lambda(I, <^I, Q_1, \ldots, Q_s)$.

In the induction step we want to compute the $\overline{k}^\wedge k_{n+1}$-type of a structure $\sum_{\iota\in I}(\mathcal{A}_\iota, P_{\iota 1}, \ldots, P_{\iota m})$ from a certain $\overline{r}^\wedge r_{n+1}$-type of an expansion of the index ordering $(I, <^I)$, namely of the $\mathcal{T}^{\overline{k}^\wedge k_{n+1}}(m)$-expansion of $(I, <^I)$ with respect to $(\mathcal{A}_\iota, \overline{P}_\iota)_{\iota\in I}$. Following the inductive definition of ρ, we fix $\overline{r}$ and r_{k+1}:

$$\overline{r} := \rho(\overline{k}, m + k_{n+1}), \quad r_{n+1} := |\mathcal{T}^{\overline{k}}(m + k_{n+1})|.$$

The idea for the computation of $T^{\overline{k}^\wedge k_{n+1}}(\sum_{\iota\in I}(\mathcal{A}_\iota, P_{\iota 1}, \ldots, P_{\iota m}))$ is as follows: We recall that this type is the set of all types

$$T^{\overline{k}}(\textstyle\sum_{\iota\in I}(\mathcal{A}_\iota, P_{\iota 1}, \ldots, P_{\iota m}, R_{\iota 1}, \ldots, R_{\iota k_{n+1}}))$$

for all possible choices of $(R_{\iota 1}, \ldots, R_{\iota k_{n+1}})_{\iota\in I}$. By induction, we can compute each such type from the $\overline{r}$-type of the $\mathcal{T}^{\overline{k}}(m + k_{n+1})$-expansion of $(I, <^I)$ with respect to $(\mathcal{A}_\iota, P_{\iota 1}, \ldots, P_{\iota m}, R_{\iota 1}, \ldots, R_{\iota k_{n+1}})_{\iota\in I}$. It suffices to know the (finite) collection C of all these $\overline{r}$-types of such expansions of $(I, <^I)$, as induced by all possible choices of $(R_{\iota 1}, \ldots, R_{\iota k_{n+1}})_{\iota\in I}$.

We obtain this collection $\mathcal{C}$ from the object we are given by assumption: This object is

$$T^{\overline{r}^\wedge r_{n+1}}(I, <^I, Q_1, \dots, Q_s)$$

where for $h = 1, \dots, s$

$$Q_h = \{\iota \in I \mid T^{\overline{k}^\wedge k_{n+1}}(\mathcal{A}_\iota, P_{\iota 1}, \dots, P_{\iota m}) = \tau_h\}$$

(referring to the list $\tau_1, \dots, \tau_s$ of the types in $\mathcal{T}^{\overline{k}^\wedge k_{n+1}}(m)$).

The elements of this type $T^{\overline{r}^\wedge r_{n+1}}(I, <^I, Q_1, \dots, Q_s)$ are, by definition, the types

$$T^{\overline{r}}(I, <^I, Q_1, \dots, Q_s, Q'_1, \dots, Q'_t)$$

where $t = r_{n+1} = |\mathcal{T}^{\overline{k}}(m + k_{n+1})|$ and $(Q'_1, \dots, Q'_t) \in \mathcal{P}(I)^t$. The corresponding variables occurring in this type are written as $X_1, \dots, X_s, X'_1, \dots, X'_t$. By our choice of t, we refer to the correspondence between the sets $Q'_1, \dots, Q'_t$ and the types in $\mathcal{T}^{\overline{k}}(m + k_{n+1})$ (which we assume listed, say as $\sigma_1, \dots, \sigma_t$). Now, if $\iota \in Q'_j$ is to indicate that $T^{\overline{k}}(\mathcal{A}_\iota, \overline{P}_\iota, \overline{R}_\iota) = \sigma_j$, then the sets Q'_j have to define a partition of I and to meet a certain compatibility with the Q_h (and hence are no more arbitrary): Suppose $\iota \in Q_h$; then $T^{\overline{k}^\wedge k_{n+1}}(\mathcal{A}_\iota, \overline{P}_\iota) = \tau_h$, and $\iota \in Q'_j$ must mean that the type σ_j of the ι-th component originates from τ_h via some expansion $\overline{R}_\iota$ (so that $\tau_h = T^{\overline{k}^\wedge k_{n+1}}(\mathcal{A}_\iota, \overline{P}_\iota)$ and $\sigma_j = T^{\overline{k}}(\mathcal{A}_\iota, \overline{P}_\iota, \overline{R}_\iota)$). In other words: the $\overline{k}$-type σ_j is an element of the $\overline{k}^\wedge k_{n+1}$-type τ_h. So the compatibility condition says: The sets Q'_j form a partition of I, and if $Q_h \cap Q'_j$ is nonempty, then $\sigma_j \in \tau_h$. It is easy to see that this condition not only is necessary , but also sufficient for satisfiability by an appropriate ordered sum $\sum_{\iota \in I}(\mathcal{A}_\iota, \overline{P}_\iota, \overline{R}_\iota)$.

So from the elements of the given type $T^{\overline{r}^\wedge r_{n+1}}(I, <^I, Q_1, \dots, Q_s)$ we find the desired collection $\mathcal{C}$ by assembling just those $\overline{r}$-types ($\in \mathcal{T}^{\overline{r}}(s + t)$) where

- the formula $X'_1 \cup \dots \cup X'_t = \text{All}$ occurs but no formula $\text{Nonempty}(X'_{j_1} \cap X'_{j_2})$ occurs for distinct j_1, j_2 (the partition property),
- the occurrence of the formula $\text{Nonempty}(X_h \cap X'_j)$ implies that σ_j is an element of τ_h.

(Definability of the partition property on the quantifier-free level is the point where the atomic formulas $X_{i_1} \cup \dots \cup X_{i_l} = \text{All}$ are useful. Clearly they are definable in terms of $\text{Nonempty}(X, Y)$ and $X \subseteq Y$ (even the latter suffices), however at the cost of quantifiers.)

For the case that $\overline{r}$ is not λ, the items above have to be reformulated: Rather than occurrence of atomic formulas in the $\overline{r}$-type under consideration we have to check whether these formulas are *induced* by the $\overline{r}$-type (i.e. they occur in all leaves of the Fraïssé tree associated with the type).

Strictly speaking, our induction step exhibited an algorithm for the computation of a type $T^{\overline{k}^\wedge k_{n+1}}(\sum_{\iota \in I}(\mathcal{A}_\iota, \overline{P}_\iota))$ from $T^{\overline{r}^\wedge r_{n+1}}(I, <^I, Q_1, \dots, Q_s)$, using (by induction) a corresponding algorithm for the computation of types $T^{\overline{k}}(\sum_{\iota \in I}(\mathcal{A}_\iota, \overline{P}_\iota, \overline{R}_\iota))$ from types $T^{\overline{r}}(I, <^I, Q_1, \dots, Q_s, Q'_1, \dots, Q'_t)$. It should be clear from our description, however, that the different algorithms which enter here can be described uniformly and hence within a single global procedure, such that the induction step corresponds to a specific call of this procedure. □

We shall apply the Composition Theorem only in two rather special cases, namely for the two element index ordering $(\{1,2\},<)$ where the associated component types are arbitrary, and for the ordering $(\omega,<)$ where the component types all coincide.

The uniqueness claim of the Composition Theorem allows to write the $\overline{k}$-type of the ordered sum of two labelled orderings with $\overline{k}$-types σ_1, σ_2 (in this order) simply as *sum type* $\sigma_1 + \sigma_2$, independently of the underlying two labelled orderings which are used for concatenation. Now one uses the obvious fact (noted after Lemma 4) that for the finite ordering $(\{1,2\},<)$ and any expansion of it by given predicates $Q_1, \ldots, Q_s$, one can compute the $\overline{r}$-type of $(\{1,2\},<,Q_1,\ldots,Q_s)$. Hence, by the Composition Theorem, we have

Corollary 6. *From two types $\sigma_1, \sigma_2 \in \mathcal{T}^{\overline{k}}(m)$ one can compute the sum type $\sigma_1 + \sigma_2$.*

By an analogous argument, the $\overline{k}$-type of an ordered sum $\sum_{i\in\omega}(\mathcal{A}_i, \overline{P}_i)$ with $T^{\overline{k}}(\mathcal{A}_i, \overline{P}_i) = \sigma$ for all i only depends on σ and hence can be written as *ω-sum type* $\sum_{i\in\omega}\sigma$. Let us verify that this sum type is computable from σ. (The proof shows that the claim holds as well for any index ordering $(I,<)$ instead of $(\omega,<)$.)

Corollary 7. *From a type $\sigma \in \mathcal{T}^{\overline{k}}(m)$ and the $\overline{r}$-type of $(\omega,<)$, where $\overline{r}$ is chosen as in the Composition Theorem, one can effectively compute the ω-sum type $\sum_{i\in\omega}\sigma$.*

Proof. By the Composition Theorem, the computation of $\sum_{i\in\omega}\sigma$ is possible from $T^{\overline{r}}(\omega,<,Q_1,\ldots,Q_s)$, where Q_j is the set of indices carrying a structure of $\overline{k}$-type τ_j in the listing of $\mathcal{T}^{\overline{k}}(m)$. In the present case we have $Q_j = \omega$ for the unique j with $\tau_j = \sigma$, while otherwise $Q_j = \emptyset$. Our task is to compute $\sum_{i\in\omega}\sigma$ from $T^{\overline{r}}(\omega,<)$ alone (without the expansion). For this, we shall verify:

> Let $\overline{r} = (r_1, \ldots r_n)$. For any m and any structure $(\omega,<,\overline{P})$ with $\overline{P} = (P_1,\ldots,P_m)$ we have: If $Q_j \in \{\omega,\emptyset\}$ for $j \in \{1,\ldots,s\}$, then the type $T^{\overline{r}}(\omega,<,Q_1,\ldots,Q_s,\overline{P})$ is computable from $T^{\overline{r}}(\omega,<,\overline{P})$.

Then we obtain the claim of the Corollary when taking the empty tuple for $\overline{P}$.

The proof proceeds by induction on the length of $\overline{r}$, for all m simultaneously. The type $T^{\lambda}(\omega,<,Q_1,\ldots,Q_s,\overline{P})$ is easily derived from $T^{\lambda}(\omega,<,\overline{P})$, using the information which Q_j are empty and which are ω. In the induction step, we have to compute the $\overline{r}^\wedge r_{n+1}$-type of $(\omega,<,Q_1,\ldots,Q_s,\overline{P})$ from the $\overline{r}^\wedge r_{n+1}$-type of $(\omega,<,\overline{P})$, in other words: the set $\{T^{\overline{r}}(\omega,<,\overline{Q},\overline{P},\overline{R}) \mid \overline{R} \in \mathcal{P}(\omega)^{r_{n+1}}\}$ from the set $\{T^{\overline{r}}(\omega,<,\overline{P},\overline{R}) \mid \overline{R} \in \mathcal{P}(\omega)^{r_{n+1}}\}$. We can generate the first set from the second set, element by element, using the inductive assumption for the tuple length $m + r_{n+1}$. □

7 Decidability of the monadic theory of $(\omega, <)$

The previous results have shown how to compute certain types of structures from other given types. For a decidability result, however, an algorithm is desired to compute types without using auxiliary information. In the present section we show that such an algorithm exists for the computation of $T^{\overline{k}}(\omega, <)$ for given $\overline{k}$.

As a preparation, we need a result on types of finite orderings. Let Fin(m) be the class of labelled orderings $(A, <, P_1, \ldots, P_m)$ with finite universe A, and set

$$\mathcal{T}^{\overline{k}}(\text{Fin}(m)) := \{T^{\overline{k}}(\mathcal{A}, \overline{P}) \mid (\mathcal{A}, \overline{P}) \in \text{Fin}(m)\}.$$

Lemma 8. *There is an algorithm to compute $\mathcal{T}^{\overline{k}}(\text{Fin}(m))$ for given $\overline{k}$ and m.*

Proof. We proceed by the increasing size of the structures $(\mathcal{A}, \overline{P})$. Since types are invariant under isomorphism we need just to compute $\{T^{\overline{k}}(\{1, \ldots, r\}, <, \overline{P}) \mid \overline{P} \in \mathcal{P}(\{1, \ldots, r\})^m\}$ for $r = 1$, $r = 2$, etc. For each r and m this can be done effectively, because only finitely many different structures occur. Since $\mathcal{T}^{\overline{k}}(m)$ is finite, one finds effectively the smallest number r_0 such that all types in

$$\{T^{\overline{k}}(\{1, \ldots, r_0 + 1\}, <, \overline{P}) \mid \overline{P} \in \mathcal{P}(\{1, \ldots, r_0 + 1\})^m\}$$

already occur as types of labelled orderings with $\leq r_0$ elements. It follows that for each labelled ordering $(\{1, \ldots, r\}, <, \overline{P})$ with $r > r_0$ there is a shorter one of same $\overline{k}$-type. (Namely, if $r > r_0 + 1$, write the structure as an ordered sum $(\{1, \ldots, r_0 + 1\}, <, \overline{P}) + (\mathcal{A}, \overline{R})$; now the first part may be shortened to a $\overline{k}$-equivalent one of length $\leq r_0$, and the claim follows by Corollary 6.) So the desired set of types is obtained by exhausting the labelled orderings up to cardinality r_0. □

For the proof that the monadic theory of $(\omega, <)$ is decidable, a reduction "from infinity to finiteness" is required. Büchi discovered in [Bü62] that for MTh$(\omega, <)$ this reduction can be built on Ramsey's Theorem A ([Ra29]). Shelah [Sh75] developed a more abstract framework of "additive colorings" and showed also other combinatorial results, for instance over dense orderings. Our presentation below concentrates on $(\omega, <)$ and follows [Th81].

A (finite) *coloring* of ω is a map C from the set of unordered pairs of natural numbers to a finite set $\{c_1, \ldots, c_s\}$ of colors. When writing $C(i, j)$ we assume $i < j$. A coloring is *additive* if from $C(i, j) = C(i', j')$ and $C(j, k) = C(j', k')$ we can infer $C(i, k) = C(i', k')$. In this case we may introduce an addition operation $+$ on the set of colors and write $c + d = e$ if there are i, j, k with $C(i, j) = c$, $C(j, k) = d$, $C(i, k) = e$.

An example of an additive coloring, denoted $C_{\overline{k}, \overline{P}}$, is obtained for any sequence $\overline{k}$ and any structure $(\omega, <, P_1, \ldots, P_m)$; here the color $C_{\overline{k}, \overline{P}}(i, j)$ refers to the restriction of $\overline{P}$ to the segment $[i, j]$, written $\overline{P}|[i, j]$. Formally, define the map $C_{\overline{k}, \overline{P}}$ by

$$C_{\overline{k}, \overline{P}}(i, j) = T^{\overline{k}}([i, j], <, \overline{P}|[i, j]).$$

The additivity of this coloring is obvious from the summation result for $\overline{k}$-types (Corollary 6).

Referring to an additive coloring C, we call an infinite set $\{i_1 < i_2 < \ldots\}$ of natural numbers (c,d)*-homogeneous* if $C(0,i_1) = c$ and $C(i_k,i_l) = d$ for any pair $k < l$. Clearly, we have $d + d = d$ in this case.

Theorem 9. (Ramsey Theorem for additive colorings on ω)
For any finite additive coloring on ω there is, for a suitable pair (c,d) of colors, a (c,d)-homogeneous set.

Proof. We use the following definition, referring to the coloring C: Two numbers i,j *merge at* k $(> i,j)$ if $C(i,k) = C(j,k)$; in this case we write $i \sim_C j(k)$. This means, by additivity, that also $C(i,k') = C(j,k')$ for each $k' > k$ (just note that $C(i,k) + C(k,k') = C(j,k) + C(k,k')$). It follows that the merging relation $i \sim_C j$, which holds when $i \sim_C j(k)$ for some k, is an equivalence relation (of finite index, by the finiteness of the set of colors).

The first step of the proof is to verify the following claim: *There is a* (c,d)*-homogeneous set iff the following condition* $H(c,d)$ *holds:*

$$H(c,d): \quad \exists i(C(0,i) = c \wedge \forall l \exists j,k > l(C(i,j) = d \wedge i \sim_C j(k)))$$

The direction from left to right is easy: Given a (c,d)-homogeneous set $\{i_1, i_2, \ldots\}$, let $i = i_1$ and observe that for all $m > 1$, $C(i_1, i_m) = d$ and $i_1 \sim_C i_m(i_{m+1})$, which establishes $H(c,d)$. For the other direction, let us define a suitable sequence $i_1, i_2, \ldots$ inductively, preserving the following property for increasing m: $i_1 \sim_C i_r(k_r)$ for $r = 1, \ldots, m$ with suitable k_r. Let i_1 be the minimal i as guaranteed by $H(c,d)$. If $i_1, \ldots, i_m$ are defined with the above property, choose $l > k_1, \ldots k_m$ so that the i_r all merge pairwise at l. Applying $H(c,d)$, let i_{m+1} be the smallest $j > l$ with $C(i_1, j) = d$ and such that $i_1 \sim_C j(k)$ for suitable $k (= k_{m+1})$. Then $i_1 \sim_C i_r(k_r)$ for $r = 1, \ldots, m+1$. Clearly $C(i_1, i_2) = C(i_2, i_3) = d$ and by $i_1 \sim_C i_2(i_3)$ also $d + d = d$. Hence $C(i_r, i_{r+1}) = d$ for $r \geq 1$, which means that $\{i_1, i_2, \ldots\}$ is (c,d)-homogeneous.

Now it suffices to guarantee a pair (c,d) of colors such that $H(c,d)$ holds. Let M be an infinite $\sim_C$-equivalence class and i_1 its minimal element. Define $c = C(0, i_1)$ and choose d such that for infinitely many $i \in M$ we have $C(i_1, i) = d$. Then $H(c,d)$ holds, as was to be shown. □

Now all preparations for an "automata-free" proof of Büchi's Theorem are done:

Theorem 10. (Büchi's Theorem [Bü62])
The monadic theory MTh$(\omega, <)$ *is decidable.*

Proof. ([Sh75]) By Lemma 4, it suffices to compute the $\overline{k}$-type of $(\omega, <)$ for each $\overline{k}$. Since $T^{\lambda}(\omega, <) = \emptyset$, we consider only nonempty sequences, i.e. of the form $\overline{k}^\wedge m$ (where now $\overline{k}$ may be empty). We shall present an algorithm to compute, inductively over the length of $\overline{k}$ and simultaneously for all numbers m, the type $T^{\overline{k}^\wedge m}(\omega,<)$, which is the set

$$\{T^{\overline{k}}(\omega,<,\overline{P}) \mid \overline{P} \in \mathcal{P}(\omega)^m\}.$$

For $\overline{k} = \lambda$ it is tedious but easy to compile the corresponding sets of atomic formulas $\varphi(X_1,\ldots,X_m)$; the finitely many possibilities to satisfy these formulas by different set tuples $\overline{P}$ can be generated effectively (see the examples in Sections 4 and 5).

Next we have to compute the set $\{T^{\overline{k}}(\omega,<,\overline{P}) \mid \overline{P} \in \mathcal{P}(\omega)^m\}$ for any *nonempty* sequence $\overline{k} = (k_1,\ldots k_n)$ and any m. We can assume that we already know how to compute the corresponding set for any shorter sequence $(k'_1,\ldots k'_{n-1})$ and arbitrary m'. In order to generate all possible $T^{\overline{k}}(\omega,<,\overline{P})$, we consider for each $\overline{P} \in \mathcal{P}(\omega)^m$ the finite additive coloring $C_{\overline{k},\overline{P}}$ with colors in $\mathcal{T}^{\overline{k}}(m)$. By Theorem 9, some pair (τ,σ) of types in this set exists such there is a (τ,σ)-homogeneous set $\{i_1,i_2,\ldots\}$, i.e. $(\omega,<\overline{P})$ is the ordered sum

$$([0,i_1],<,\overline{P}|[0,i_1]) \;+\; ([i_1+1,i_2],<,\overline{P}|[i_1+1,i_2]) \;+\; \ldots$$

with $T^{\overline{k}}([0,i_1],<,\overline{P}|[0,i_1]) = \tau$ and $T^{\overline{k}}([i_l+1,i_{l+1}],<,\overline{P}|[i_l+1,i_{l+1}]) = \sigma$ for $l \geq 1$. *Hence* $T^{\overline{k}}(\omega,<,\overline{P})$ *can be obtained as a sum type* $\tau + \sum_{i\in\omega}\sigma$ *where* $\tau,\sigma \in \mathcal{T}^{\overline{k}}(\mathrm{Fin}(m))$. Conversely, every such sum type clearly leads to some type $T^{\overline{k}}(\omega,<,\overline{P})$.

By Lemma 8, we can effectively generate all the types in $\mathcal{T}^{\overline{k}}(\mathrm{Fin}(m))$, i.e. all candidates for τ,σ. It remains to compute the sums $\tau + \sum_{i\in\omega}\sigma$. Considering a fixed pair (τ,σ), we apply the Composition Theorem. In order to obtain $\sum_{i\in\omega}\sigma$ it suffices, by Corollary 7, to compute $\mathcal{T}^{\overline{r}}(\omega,<)$, where $\overline{r} = (r_1,\ldots,r_n)$ is chosen as in the Composition Theorem. By definition, this type is

$$\{T^{(r_1,\ldots,r_{n-1})}(\omega,<,\overline{R}) \mid \overline{R} \in \mathcal{P}(\omega)^{r_n}\}.$$

Since $(r_1,\ldots,r_{n-1})$ is shorter than $\overline{k}$, we can compute this set by our assumption (setting $m' = r_n$). Thus $\sum_{i\in\omega}\sigma$ is computed, after which $\tau + \sum_{i\in\omega}\sigma$ is obtained by Corollary 6. □

Note that in the computation of types it was essential to proceed inductively by the length of quantifier alternation types (sequences $\overline{k}$): A type set

$$\{T^{(k_1,\ldots,k_n)}(\omega,<,\overline{P}) | \overline{P} \in \mathcal{P}(\omega)^m\}$$

turned out to be computable from

$$T^{(r_1,\ldots,r_n)}(\omega,<) = \{T^{(r_1,\ldots,r_{n-1})}(\omega,<,\overline{R}) \mid \overline{R} \in \mathcal{P}(\omega)^{r_n}\}$$

for suitable $(r_1,\ldots,r_n)$; this reduction from the sequence length n to $n-1$ supplied the inductive computation process for determining the $\overline{k}$-theories of $(\omega,<)$.

It is instructive to compare this proof of Shelah [Sh75] with the original one of Büchi [Bü83]. Both approaches rely on Ramsey's Theorem. They differ in the choice and application of additive colorings. We explain this in more detail.

Büchi's idea is to write monadic formulas $\varphi(X_1, \ldots, X_m)$ in "automata normal form"

$$\exists Y_1 \ldots \exists Y_k \psi(X_1, \ldots, X_m, Y_1, \ldots, Y_k);$$

here ψ is a *first-order* formula which says that $\overline{Y}$ is a run of a nondeterministic automaton on the input $\overline{X}$ (viewed as an ω-word over $\{0,1\}^m$) which visits infinitely often a final state. (Today we speak of *Büchi automata.*) Since the (first-order) quantifier depth of ψ is just 2, Büchi works with formulas of fixed quantifier alternation depth 3 and captures all the expressive power of monadic formulas by a growing length of the tuples $\overline{Y}$, corresponding to a growing number of states in Büchi automata. Where Shelah works upwards in quantifier alternation depth, Büchi shows that the automata normal form is closed under complement at the cost of more states. Indeed, this amounts to a reduction lemma for quantifier alternation depth: A formula $\forall \overline{Z} \exists \overline{Y} \psi(\overline{X}, \overline{Y}, \overline{Z})$, which is equivalent to $\neg \exists \overline{Z} \neg \exists Y \psi(\overline{X}, \overline{Y}, \overline{Z})$, can be written as $\exists \overline{Y'} \psi(\overline{X}, \overline{Y'})$ using two such complementation steps.

Ramsey's Theorem is applied to establish this complementation (and hence, just as in Shelah's approach, for reducing quantifier alternation depth). In treating the negated automata normal form $\neg \exists Y_1 \ldots Y_k \psi(X_1, \ldots, X_m, \overline{Y})$ Büchi defines an additive coloring $C_{k,\overline{P}}$ on any input structure $(\omega, <, P_1, \ldots, P_m)$ which can serve as interpretation of the formula. Such a structure can be considered as an ω-word over $\{0,1\}^m$. The coloring $C_{k,\overline{P}}$ refers to the state set $\{0,1\}^k$ of the automaton described in ψ. While in Shelah's set-up two segments of $(\omega, <, P_1, \ldots, P_m)$ get the same color when their $\overline{k}$-types coincide, Büchi associates the same color to two segments u, v if the automaton described in ψ cannot distinguish them: For any pair p, q of states, the automaton should be able to pass from p to q via u iff this is possible via v, and a visit of a final state should be possible in such a run via u iff this is possible via v. The (finitely many) classes of this equivalence relation over finite words provide the colors for $C_{k,\overline{P}}$, and the additivity is clear from the way finite automata work. By Ramsey's Theorem, it turns out that the set of ω-words which satisfy $\neg \exists Y_1 \ldots Y_k \psi(X_1, \ldots, X_m, Y_1, \ldots, Y_k)$ can be generated as a union of "periodic sets" $U \cdot V^\omega$ where U, V are equivalence classes obtained from the automaton for ψ.

In the framework of $\overline{k}$-types, Shelah shows a similar fact: The structures $(\omega, <, \overline{P})$ of a given $\overline{k}$-type τ_0 can be obtained by collecting all sets $U_\tau \cdot V_\sigma^\omega$ with $\tau + \sum_{i \in \omega} \sigma = \tau_0$, where U_τ contains the finite segments of type τ and V_σ contains the finite segments of type σ.

At this point, where the periodic representation $U \cdot V^\omega$ of definable sets is reached, the decidability proof proceeds in two different ways: Büchi decides non-emptiness of these periodic sets, using the regularity of U and V (i.e., he decides non-emptiness of Büchi automata). Shelah generates the $\overline{k}$-theories of $(\omega, <)$ for increasing $\overline{k}$ by the composition process, and thus shows decidability in a synthetic rather than an analytic way.

There is a third approach to the decidability of MTh$(\omega, <)$, due to Ladner [La77], which can be regarded as a combination of the methods of Büchi and Shelah. Ladner classifies finite segments of orderings by equivalence with respect to formulas of given quantifier depth (not alternation depth), using the corresponding Ehrenfeucht game, and applies Ramsey's Theorem to the associated finite additive coloring. So the equivalence is a logical one as in Shelah's proof. On the other hand, the resulting periodic representation of monadic second-order definable sets of ω-words is applied more in the spirit of Büchi's approach. As in the nonemptiness test for automata, it is analyzed which length of words u and v (depending on the quantifier depth under consideration) suffices to obtain an ω-word $u \cdot v^{\omega}$ as an element in such such a periodic set. The estimation is done directly, however, without passing to automata and using their number of states.

So there is a close relationship between the different proofs. Büchi went so far to detect "automata" also in Shelah's approach to the monadic theory of $(\omega, <)$ and of greater ordinals (cf. concluding section of [Bü83]). This seems to be a too liberal interpretation of "automata", which overrides the subtle use of the quantifier alternation depth measure in Shelah's proof. Rather it seems an open question to the present author whether reasonable "automata" exist on orderings different from ω and the ordinals. A logical view of "automata" (which also Büchi adopted) is to consider them as rather special formulas (of low quantifier alternation depth). Here it seems open whether for theories of dense orderings, where Shelah and Gurevich ([Sh75], [Gu79], [GS79]) showed decidability results with the calculus of $\overline{k}$-types, such "simple formulas" exist, which are as expressive as the full language and can play a similar role as the automata normal form over $(\omega, <)$ or larger ordinals.

8 Recursive ordinal words and their monadic theory

Ramsey's Theorem says that, given a finite additive coloring of ω, a (c, d)-homogeneous set exists for suitable colors c, d. In this section we consider again the colorings $C_{\overline{k},\overline{P}}$ and extend Ramsey's Theorem by a statement on the "difficulty" to determine a pair (c, d) from $\overline{k}$, assuming that $\overline{P}$ is fixed.

Here we use standard terminology from recursion theory, concerning relative computability and the arithmetical hierarchy (cf. [Ro67]). Given sets $P_1, \ldots, P_m$ of natural numbers, we denote by $\overline{P}'$ the *jump* of the recursion theoretic join of $P_1, \ldots, P_m$. This is a set of natural numbers with two properties (and we skip here the existence proof): First, $\overline{P}'$ is Σ_1 relative to $\overline{P}$, i.e. definable in the form

$$k \in \overline{P}' \Leftrightarrow \exists i R(i, k), \quad \text{with } R \text{ recursive in } \overline{P}.$$

Secondly, $\overline{P}'$ has a completeness property: Any set M definable in the form $M = \{k \mid \exists i R(i, n)\}$ with R recursive in $\overline{P}$ is itself recursive in $\overline{P}'$. Similarly, the *n-th jump* of $\overline{P}$, denoted $\overline{P}^{(n)}$, is a Σ_n-set relative to $\overline{P}$, and any such set M, i.e.

with a definition

$k \in M \Leftrightarrow \exists i_1 \forall i_2, \ldots \exists/\forall i_n R(i_1, i_2, \ldots, i_n, k)\}$, where R is recursive in $\overline{P}$,

is recursive in $\overline{P}^{(n)}$.

The sets $\overline{P}', \overline{P}'', \ldots, \overline{P}^{(n)}, \ldots$ mark the levels of the *arithmetical hierarchy over* $\overline{P}$. The classical arithmetical hierarchy is obtained when $\overline{P}$ consists of recursive sets (or just the empty set).

Our aim is to determine on which level of this hierarchy over $\overline{P}$ the monadic theory $\mathrm{MTh}(\omega, <, \overline{P})$ is located. (Here one identifies a set of formulas with a set of natural numbers via an appropriate coding.)

Theorem 11. *The monadic theory* $\mathrm{MTh}(\omega, <, \overline{P})$ *is recursive in* $\overline{P}''$.

Proof. By Lemma 4, we have to show that the function

$$f_{(\omega,\overline{P})} : \overline{k} \mapsto T^{\overline{k}}(\omega, <, \overline{P})$$

is recursive in $\overline{P}''$. This means that we have to describe an algorithm which computes $f_{(\omega,\overline{P})}$ while during its computation has access to a $\overline{P}''$-oracle, i.e. obtains correct answers to questions about membership of concrete numbers in $\overline{P}''$. Of course, the algorithm can also determine, for any given numbers i, j, the type $T^{\overline{k}}([i,j], <, \overline{P}|[i,j])$ (for this, even the weaker $\overline{P}$-oracle would suffice).

In order to compute $T^{\overline{k}}(\omega, <, \overline{P})$, we can apply the fact, proved by means of Ramsey's Theorem, that this type originates as an ordered sum $\tau + \sum_{i \in \omega} \sigma$. Indeed, by the Composition Theorem we can compute $T^{\overline{k}}(\omega, <, \overline{P})$ once τ and σ are found. Here τ and σ are correctly chosen if there is a (τ, σ)-homogeneous set for the coloring $C_{\overline{k},\overline{P}}$, as defined in the proof of Theorem 10. By our proof of Ramsey's Theorem, τ and σ have this property iff the following condition holds (where we write C for $C_{\overline{k},\overline{P}}$):

$$H(\tau, \sigma) : \ \exists i(C(0,i) = \tau \ \wedge \ \forall l \exists j, k > l(C(i,j) = \sigma \wedge i \sim_C j(k)))$$

Our algorithm proceeds as follows, when supplied with input $\overline{k}$: It checks the condition

$$C(0,i) = \tau \ \wedge \ \forall l \exists j, k > l(C(i,j) = \sigma \wedge i \sim_C j(k))),$$

successively for $i = 1, 2, \ldots$, and for each i works through all (finitely many!) type pairs τ, σ from $T^{\overline{k}}(\mathrm{Fin}(m))$. Each such test (for fixed i, τ, σ) involves two questions to the $\overline{P}''$-oracle. The first asks whether i satisfies $C(0,i) = \tau$ (for which even the weaker $\overline{P}$-oracle would suffice). The second question asks about the second conjunct above, which is again a condition on i. The set M of numbers satisfying this condition is a Π_2-set relative to $\overline{P}$, as seen from the formulation

$$i \in M \ \Leftrightarrow \ \forall l \exists k[k > l \wedge \exists j(i < j < k \wedge C(i,j) = \sigma \wedge C(i,k) = C(j,k))];$$

note that the relation in square brackets (in i, k, l) is recursive in $\overline{P}$. So M is the complement of a Σ_2-set relative to $\overline{P}$, and thus membership of a given number i in it is answered by the $\overline{P}''$-oracle.

By (the proof of) Ramsey's Theorem, $H(\tau, \sigma)$ holds for some pair (τ, σ), and hence a corresponding i exists as required in $H(\tau, \sigma)$. The algorithm will detect such i, τ, σ after finitely many steps. From the pair τ, σ it produces $\tau + \sum_{\iota \in I} \sigma$ and hence the desired type $T^{\overline{k}}(\omega, <, \overline{P})$. □

In a next step we lift this result to higher ordinals ω^n.

Theorem 12. *The monadic theory* $\mathrm{MTh}(\omega^n, <, \overline{P})$ *is recursive in* $\overline{P}^{(2n)}$.

Proof. We proceed by induction on $n \geq 1$. The case $n = 1$ is given by the previous theorem. Consider a labelled ordinal ordering $(\omega^{n+1}, <, \overline{P})$. It suffices to compute the function $\overline{k} \mapsto T^{\overline{k}}(\omega^{n+1}, <, \overline{P})$ by means of an algorithm with a $\overline{P}^{(2n+2)}$-oracle. For this purpose we decompose $(\omega^{n+1}, <, \overline{P})$ as the ω-sum of labelled orderings $(\mathcal{A}_i, \overline{P}_i) = (\omega^n, <, \overline{P}_i)$ where $i \in \omega$. Define a corresponding coloring C on ω by

$$C(i, j) = T^{\overline{k}}((\mathcal{A}_i, \overline{P}_i) + \ldots + (\mathcal{A}_j, \overline{P}_j)).$$

By inductive assumption, for any given i the type $T^{\overline{k}}(\mathcal{A}_i, \overline{P}_i)$ can be computed by means of a $\overline{P}^{(2n)}$-oracle. Applying the effective summation of a finite number types (via Corollary 6), we conclude that the function C is recursive in $\overline{P}^{(2n)}$. Now apply the algorithm of the previous theorem for the new coloring C of ω. This algorithm produces $T^{\overline{k}}(\omega^{n+1}, <, \overline{P})$ upon input $\overline{k}$. During computation it checks for $i = 0, 1, \ldots$ the same conditions as before, now for the new C. Since C is recursive in $\overline{P}^{(2n)}$, a $\overline{P}^{(2n+2)}$-oracle suffices where previously the $\overline{P}''$-oracle was asked. (We use here the fact that a set which is recursive in M'' where M is recursive in $N^{(n)}$, is recursive in $N^{(n+2)}$.) Hence $\mathrm{MTh}(\omega^{n+1}, <, \overline{P})$ is recursive in $\overline{P}^{(2n+2)}$. □

Let us consider the special case of recursive predicates on an ordinal ω^n. (The notion of a recursive predicate on an ordinal ω^n is canonical: Since its elements are representable in the form $\omega^{n-1} \cdot c_{n-1} + \ldots + \omega \cdot c_1 + c_0$ with $c_i \in \omega$, they are in effective 1-1-correspondence to the natural numbers; so the notion of recursiveness is transferred from ω to ω^n.) As mentioned in the first two sections, a labelled ordering $(\omega^n, <, \overline{P})$ is viewed as an ω^n-word; for recursive $\overline{P}$ we speak of a recursive ω^n-word. For this case the previous theorem says the following:

Corollary 13. *The monadic theory of a recursive ω^n-word is recursive in $\emptyset^{(2n)}$, the $2n$-th jump of the empty set.*

The previous results can be sharpened slightly. For this purpose, one transforms the condition $H(c, d)$ used in Ramsey's Theorem from the present $\exists\forall\exists$-form (where just the unbounded quantifiers are counted) to a boolean combination of $\exists\forall$-clauses. This transformation is a combinatorial analogue of McNaughton's Theorem in the theory of ω-automata (see Section 1 of [Th81] for details). Given the new form of $H(c,d)$, the reducibility relation "recursive in", as it appears in the results above, can be strengthened to "truth-table reducible in" (for definitions see [Ro67]). A further restriction in the reducibility notion (namely, to bounded truth-table reducibility) cannot be reached, however, as shown in [Th78].

As a final result, we show that the bound $\emptyset^{(2n)}$ in Corollary 13 cannot be improved.

Theorem 14. *For $n \geq 1$ there is a recursive ω^n-word $(\omega^n, <, P_n)$ such that $\emptyset^{(2n)}$ is recursive in* MTh$(\omega^n, <, P_n)$ *(even in the first-order theory of $(\omega^n, <, P_n)$).*

Proof. It is convenient to work with the complement of $\emptyset^{(2n)}$ rather than $\emptyset^{(2n)}$ itself. These two sets are recursive in each other; so either of them can be applied for the claim. While $\emptyset^{(2n)}$ is Σ_{2n}-complete, the complement of $\emptyset^{(2n)}$ is Π_{2n}-complete. For Π_{2n}-sets we use a specific representation which involves n quantifiers $\exists^\omega$, meaning "there exist infinitely many". (For our purpose, it is convenient to index the n quantified variables from $n-1$ down to 0.)

Lemma 15. ([KSW60], see also section 14.8 of [Ro67])
A set M of natural numbers is a Π_{2n}-set iff it can be defined in the form

$$k \in M \;\Leftrightarrow\; \exists^\omega i_{n-1} \ldots \exists^\omega i_0 R(i_{n-1}, \ldots i_0, k)$$

where R is a recursive relation.

Let M_n be the complement of $\emptyset^{(2n)}$ and R_n the corresponding recursive relation according to the Lemma.

For the definition of the desired (recursive) set P_n in ω^n we first treat the case $n = 1$, using the representation of M_1 in the form

$$k \in M_1 \;\Leftrightarrow\; \exists^\omega i_0 R_1(i_0, k), \quad \text{where } R_1 \text{ is recursive.}$$

We build up the predicate P_1 in the form of an ω-sequence with letters 0 and 1, using an enumeration of the pairs $(i_0, k) \in \omega \times \omega$. Suppose that during the enumeration process the pair (i_0, k) is reached. Now check whether (i_0, k) is in R_1. If yes, then add $01^{k+1}0$ to the P_1-prefix built up so far, otherwise just add 0. Clearly the resulting predicate P_1 is recursive. The occurrence of such blocks of $k+1$ letters 1 (enclosed by two zeroes) is expressible even in first-order logic; we use the auxiliary formula "at y starts a $(k+1)$-block" which says "y and the next k successors belong to P_1, but the subsequent successor as well as the predecessor of y do not". (We work here with the standard first-order language and do not take the trouble of translating the formulas back into the monadic second-order framework as used in the previous sections.) We have

$$k \in M_1 \Leftrightarrow (\omega, <, P_1) \models \forall x \exists y (x < y \wedge \text{at } y \text{ starts a } k+1\text{-block}),$$

which shows that M_1 is recursive in the first-order theory of $(\omega, <, P_1)$.

The construction is now generalized to M_n. Recall the representation of M_n by

$$k \in M_n \Leftrightarrow \exists^\omega i_{n-1} \ldots \exists^\omega i_0 R_n(i_{n-1}, \ldots, i_0, k),$$

where R_n is recursive. On ω^n we define a suitable recursive predicate P_n. The elements of ω^n can be represented as $\omega^{n-1} \cdot i_{n-1} + \ldots + \omega \cdot i_1 + i_0$ where $i_k \in \omega$ (Cantor's normal form); we shall just write $(i_{n-1}, \ldots, i_1, i_0)$ instead. Note that the prefix $(i_{n-1}, \ldots, i_1)$ fixes a unique ω-copy within ω^n, consisting of the elements $(i_{n-1}, \ldots, i_1, j)$ where $j \in \omega$; we call $(i_{n-1}, \ldots, i_1)$ the *address* of this ω-copy. For the construction of P_n, in the form of an ω^n-word over $\{0, 1\}$, we now use an enumeration of the $(n+1)$-tuples over ω; while the enumeration proceeds, each ω-copy in the desired word $(\omega^n, <, P_n)$ is built up simultaneously. Suppose we just deal with the tuple $(i_{n-1}, \ldots, i_1, i_0, k)$ during the enumeration. If it does not belong to R_n, then attach 0 to each ω-copy (more precisely, to the P_n-prefix constructed so far on this ω-copy). If the tuple belongs to R_n, then one exception is made, namely on the unique copy with address $(i_{n-1}, \ldots, i_1)$ the word $01^{k+1}0$ is attached. Again, the construction directly shows that P_n is recursive.

To verify that M_n is recursive in the first-order theory of $(\omega^n, < P_n)$ we have to exhibit, for each k, a sentence φ_k such that

$$k \in M_n \Leftrightarrow (\omega^n, <, P_n) \models \varphi_k.$$

By the construction of P_n it will suffice to express the statement: "$\exists^\omega i_{n-1} \ldots \exists^\omega i_1$ such that on the ω-copy with address $(i_{n-1}, \ldots, i_1)$ there are infinitely many $(k+1)$-blocks".

This is reformulated as follows:

"there are infinitely many ω^{n-1}-copies C_{n-1} such that
on C_{n-1} there are infinitely many ω^{n-2}-copies C_{n-2} such that
...
on C_2 there are infinitely many ω-copies C_1 such that
on C_1 there are infinitely many $(k+1)$-blocks"

An ω^i-copy C_i is easily fixed by the ω^i-limit ordinal (or 0) which is its first element and by the ω^i-limit ordinal which succeeds the copy. It is well-known how to define the ω^i-limit ordinals in first-order logic (inductively on i). For an ω^i-copy C_i enclosed by y_i, z_i, the condition "on C_i there exist infinitely many ω^{i-1}-copies C_{i-1} such that ..." is formalized following the pattern

$$\forall x_{i-1} \in [y_i, z_i) \ \exists y_{i-1} \in [x_{i-1}, z_i) \exists z_{i-1} \in [x_{i-1}, z_i)$$
$$(y_{i-1} < z_{i-1} \ \wedge \ y_{i-1}, z_{i-1} \text{ are successive } \omega^{i-1}\text{-limit ordinals } \wedge \ldots)$$

whence our statement is indeed expressible by a sentence φ_k in the monadic (even first-order) language of the structure $(\omega^n, <, P_n)$. □

The concatenation of all the ordinal words $(\omega^n, <, P_n)$ for $n = 1, 2, \ldots$ will result in a recursive ω^ω-word. Every set $\emptyset^{(2n)}$ (for $n = 1, 2, \ldots$) is recursive in the first-order theory of this ω^ω-word. Hence we obtain:

Corollary 16. *There is recursive ω^ω-word whose monadic (and even first-order) theory is not arithmetical.*

9 Conclusion

We have exposed a model theoretic approach to analyze the monadic theory of labelled orderings, based on the Ehrenfeucht-Fraïssé method and further developed into a powerful calculus by Shelah and Gurevich, and we presented some simple applications concerning the monadic theory of the ordinal ω and of ordinal words. The reader is invited to take this paper as a start to study [Sh75], [Gu79], [GS79] and subsequent work, where more powerful results are shown. Among the open fields for further research we mention just two: the extension of the method to more general structures and the investigation of the computational complexity of procedures which manipulate $\overline{k}$-types.

10 Acknowledgment

I thank Yuri Gurevich for fruitful discussions which gave a good motivation and encouragement to write this paper.

References

[Bü62] J.R. Büchi, On a decision method in restricted second order arithmetic, in: *Logic, Methodology and Philosophy of Science*, Proc. 1960 Intern. Congr. (E. Nagel et al. Eds.), Stanford Univ. Press 1962, pp. 1-11.

[Bü65] J.R. Büchi, Decision methods in the theory of ordinals, *Bull. of the Amer. Math. Soc* **71**, (1965), 767-770.

[Bü83] J.R. Büchi, State-strategies for games in $F_{\sigma\delta} \cap G_{\delta\sigma}$, *J. Symb. Logic* **48** (1983), 1171-1198.

[CFGS82] E. M. Clarke, N. Francez, Y. Gurevich, P. Sistla, Can message buffers be characterized in linear temporal logic?, *Symp. on Principles of Distributed Computing*, ACM 1982, 148-156.

[CK73] C.C. Chang, H.J. Keisler, *Model Theory*, North-Holland, Amsterdam 1973.

[EFT84] H.D. Ebbinghaus, J. Flum, W. Thomas, *Mathematical Logic*, 2nd Edition, Springer-Verlag, New York 1993.

[EF95] H.D. Ebbinghaus, J. Flum, *Finite Model Theory*, Springer, New York 1995.

[Ehr59] A. Ehrenfeucht, Decidability of the theory of the linear ordering relation, *Notices of the A.M.S.* **6** (1959), 268.

[Ehr61] A. Ehrenfeucht, An application of games to the completeness problem for formalized theories, *Fund. Math.* **44**, 241-248.

[FV59] S. Feferman, R.L. Vaught, The first order properties of algebraic systems, *Fund. Math.* **47** (1959), 57-103.

[Fr54] R. Fraïssé, Sur quelques classifications des systèmes de relations, *Publ. Sci. Univ. Alger Sér.* A **1** (1954), 35-182.

[Gu79] Y. Gurevich, Modest theory of short chains I, *J. Symb. Logic* **44** (1979), 481-490.

[Gu82] Y. Gurevich, Crumbly spaces, in: *Sixth Intern. Congr. for Logic, Methodology, and Philosophy of Science (1979)*, North-Holland, Amsterdam 1982, pp. 179-191.

[Gu85] Y. Gurevich, Monadic second-order theories, in: *Model-Theoretic Logics* (J. Barwise, S. Feferman, Eds.), Springer-Verlag, Berlin-Heidelberg-New York 1985, pp. 479-506.

[GMS83] Y. Gurevich, M. Magidor, S. Shelah, The monadic theory of ω_2, *J. Symb. Logic* **48** (1983), 387–398.

[GS79] Y. Gurevich, S. Shelah, Modest theory of short chains II, *J. Symb. Logic* **44** (1979), 491-502.

[GS83] Y. Gurevich, S. Shelah, Rabin's uniformization problem, *J. Symb. Logic* **48** (1983), 1105-1119.

[GS85] Y. Gurevich, S. Shelah, The decision problem for branching time logic, *J. Symb. Logic*, **50** (1985), 668-681.

[Hi53] J. Hintikka, *Distributive normal forms in the calculus of predicates*, Acta Philos. Fennica **6** (1953).

[KSW60] G. Kreisel, J. Shoenfield, H. Wang, Number theoretic concepts and recursive well-orderings, *Archiv für Mathematische Logik und Grundlagenforschung* **5** (1960), 42-64.

[Lä68] H. Läuchli, A decision procedure for the weak second order theory of linear order, in: *Contributions to Mathematical Logic, Proc. Logic Colloquium Hannover 1966*, North-Holland, Amsterdam 1968.

[La77] R. Ladner, Application of model theoretic games to discrete linear orders and finite automata, *Inf. Contr.* **33** (1977), 281-303.

[MB96] F. Moller, G. Birtwistle (Eds.), *Logics for Concurrency*, Springer Lecture Notes in Computer Science, Vol. 1043, Springer-Verlag, Berlin 1996.

[Od89] P. Odifreddi, *Classical Recursion Theory*, Noth-Holland, Amsterdam 1989.

[Ra29] F.P. Ramsey, On a problem of formal logic, *Proc. London Math. Soc.* **30** (1929), 264-286.

[Ro67] H. Rogers, *Theory of Recursive Functions and Effective Computability*, McGraw-Hill, New York 1967.

[Sh75] S. Shelah, The monadic theory of order, *Ann. Math.* **102** (1975), 379-419.

[Th78] W. Thomas, The theory of successor with an extra predicate, *Math. Ann.* **237** (1978), 121-132.

[Th80] W. Thomas, On the bounded monadic theory of well-ordered structures, *J. Symb. Logic* **45** (1980), 334-338.

[Th81] W. Thomas, A combinatorial approach to the theory of ω-automata, *Inform. Contr.* **48** (1979), 261-283.

[Th90] W. Thomas, Automata on infinite objects, in: *Handbook on Theoretical Computer Science*, Vol. A (J. v. Leeuwen, ed.), Elsevier, Amsterdam 1990.

[Ze94] R.S. Zeitman, *The Composition Method*, PhD Dissertation, Wayne State Univ., Michigan, 1994.

Monadic Second Order Logic and Node Relations on Graphs and Trees

Roderick Bloem* and Joost Engelfriet

Department of Computer Science, Leiden University
P.O.Box 9512, 2300 RA Leiden, The Netherlands
e-mail: engelfri@wi.leidenuniv.nl

Abstract. A formula from monadic second-order (MSO) logic can be used to specify a binary relation on the set of nodes of a tree. It is proved that, equivalently, such a relation can be computed by a finite-state tree-walking automaton, provided the automaton can test MSO properties of the nodes of the tree. For graphs, if a binary relation on the nodes of a graph can be computed by a finite-state graph-walking automaton, then it can be specified by an MSO formula, but, in general, not vice versa.

1 Introduction

Computer science can be viewed as the art of turning specifications into implementations, or, on a higher level, as the art of finding algorithms that can do this automatically. In the latter case, the specification language and the machine model have to be formalized, and, ideally, the machine model should "fit" the specification language in the sense that, vice versa, every activity of the machine can be described in the language. In the late fifties of this century one of the most beautiful and basic examples of this type was discovered in [Büc, Elg]: a set of strings can be defined in monadic second order logic (the specification language) if and only if it can be recognized by a finite-state automaton (the machine model). In the sixties this result was generalized in [Don, TW] to sets of node-labelled ordered trees, with an appropriate (in an algebraic sense) generalization of the finite-state automaton to a bottom-up finite-state tree automaton. The trees considered are the usual representations of terms over a finite set Σ of operators. In this paper we extend the result of [Don, TW] to the specification and implementation of binary relations on the nodes of trees.

A monadic second order (MSO) formula with two free (object) variables specifies a binary relation on the set of nodes of each tree over Σ, and we are looking for a simple machine model that computes this relation. Here we say 'computes' rather than 'recognizes', because we are mainly interested in functions on nodes of trees and view relations as a nondeterministic variant of functions. Thus, for a given tree t and a given node u of t, we wish the machine to find the (or a)

* The present address of the first author is: Department of Computer Science, University of Colorado at Boulder, P.O.Box 430, Boulder, CO 80303, email: Roderick.Bloem@colorado.edu

node v such that the pair (u, v) is in the relation specified by the MSO formula on t.

Note that there is an easy and well-known answer to the recognition problem: it is not difficult to see, using the result of [Don, TW], that there is a finite-state tree automaton that, for given t, u, and v, finds out whether u and v satisfy the MSO formula in t, provided the nodes u and v are indicated by special labels. Since the finite-state tree automaton is basically a parallel recognition device, it does not seem to be useful for the purpose of sequential computation. Instead, we propose a variation of the tree-walking automaton introduced in [AU]. A tree-walking automaton A is a finite-state automaton that walks on the tree from node to node, following the edges of the tree. It can read the label of the current node x. It can also test whether x is the root of the tree, and if not, whether it is the first, second, ..., or last argument of its parent node. Depending on the outcome of these tests, A can decide to move up to the parent of x or down to a specific child of x. The automaton A computes the relation that consists of all pairs (u, v) such that when A is dropped on node u in its initial state, it makes a walk on the tree and reaches node v in a final state. It can be shown that this automaton model is not powerful enough. Therefore we strengthen its computation power by allowing it to test any MSO property of the current node (rather than just the ones mentioned above), where an MSO property is specified by an MSO formula with one free (object) variable. Our main result is that the nondeterministic (deterministic) "tree-walking automaton with MSO tests" computes exactly the MSO definable tree-node relations (functions, respectively).

There are two reasons for our interest in functions and relations on the nodes of trees. The first concerns trees with additional pointers. In a compiler the program is represented by its syntax tree, where each node may be viewed as a record with a field containing syntactic information and fields with pointers to the children of the node. However, it can be convenient to have an additional pointer field in the record. For example, records that contain an identifier could have a pointer to the node where that identifier is declared. Clearly, such a new pointer field defines a (partial) function on the nodes of the tree. The general idea of defining data types consisting of trees with additional pointer fields is investigated in [KS], where a pointer field is defined by a tree-walking automaton (without MSO tests), or rather by its corresponding regular expression (called routing expression). By our results, such a pointer field may as well be specified by an MSO formula.

The second reason for our interest in node relations on (graphs and) trees is more theoretical. In the theory of context-free graph grammars (i.e., grammars that generate sets of graphs), a notion of MSO definable graph transduction has been developed (see, e.g., [Eng2, EO2, Cou2, Cou3]) closely related to the well-known notion of interpretation of one logical structure in another (see [ALS] for the history of this notion). The output graph g' of such a transduction is defined in terms of the input graph g by means of MSO formulas, to be interpreted in g, as follows. Roughly speaking, the nodes of g' form a subset of the nodes of g, viz. all nodes of g that satisfy a given MSO formula with one free (object) variable. The edges of g' are defined by a given MSO formula with two free (object) variables: this formula defines a binary relation on the nodes of g, which (restricted to the nodes of g') is taken as the set of edges of g'. In the case that the edges of g' are labeled, there is such an MSO formula for every edge label. It is shown in [EO1, EO2] (for one type of context-free graph grammar) and in [CE] (for another type) that a set of graphs can be generated by a context-free graph grammar if and only if it is the image of a regular tree language (i.e., a tree language recognized by a finite-state tree automaton) under an MSO definable graph transduction. Thus, using our results, the edges of these graphs can be computed by (nondeterministic) tree-walking automata. As a special case of MSO graph transductions, one may consider the case that both the input and output graphs are trees: MSO tree transductions. Since the edges of the output tree can be viewed as pointers, they correspond to functions on the nodes of the input tree, as explained above. Thus, they can be computed by deterministic tree-walking automata. This is used in [Blo] to show that the MSO tree transductions can be computed by two-phase attribute grammars: in the first phase the MSO tests on the nodes of the input tree are evaluated (by attributes of type boolean), and in the second phase the output tree is computed (by attributes of type tree).

2 Preliminaries

In this section we recall some well-known concepts concerning finite (tree) automata and monadic second order logic on graphs.

$\mathbb{N} = \{0, 1, 2, \dots\}$, and for $n \in \mathbb{N}$, $[1, n] = \{i \mid 1 \leq i \leq n\}$. For binary relations R_1 and R_2, their composition is $R_1 \circ R_2 = \{(x, z) \mid \exists y : (x, y) \in R_1 \text{ and } (y, z) \in R_2\}$; note that the order of R_1 and R_2 is nonstandard. The transitive reflexive closure of a binary relation R is denoted R^*.

Graphs and trees

We consider finite, directed graphs with labeled nodes and edges. Let Σ and Γ be alphabets (of node labels and edge labels, respectively). A *graph* over (Σ, Γ) is a triple (V, E, lab), with V a finite set of nodes, $E \subseteq V \times \Gamma \times V$ the set of labeled edges, and $\text{lab} : V \to \Sigma$ the node-labelling function. For a given graph g, its nodes, edges, and node-labelling function are denoted V_g, E_g, and lab_g, respectively. The set of all graphs over (Σ, Γ) is denoted $G_{\Sigma,\Gamma}$.

The trees we consider are the usual graphical representations of terms, which form the free algebra over a set of operators. An *operator alphabet* Σ is an alphabet Σ together with a *rank function* $\text{rk} : \Sigma \to \mathbb{N}$. For all $k \in \mathbb{N}$, $\Sigma_k = \{\sigma \in \Sigma \mid \text{rk}(\sigma) = k\}$ is the set of operators of rank k, i.e., with k arguments. The *rank interval* of the operator alphabet Σ is $\text{rki}(\Sigma) = [1, m]$ where m is the maximal rank of the elements of Σ.

The nodes of a tree over Σ are labeled by operators. To indicate the order of the arguments of an operator, we label the edges by natural numbers. A *tree* over Σ is an acyclic connected graph g over $(\Sigma, \text{rki}(\Sigma))$ such that (1) no node of g has more than one incoming edge, and (2) for every node u of g and every $i \in [1, \mathrm{rk}(\mathrm{lab}_g(u))]$, u has exactly one outgoing edge with label i, and u has only outgoing edges with labels in $[1, \mathrm{rk}(\mathrm{lab}_g(u))]$. The set of all trees over Σ is denoted T_Σ. A subset of T_Σ is also called a *tree language.*

The root of a tree t is denoted $\mathrm{root}(t)$. Note that the direction of the edges of t is from $\mathrm{root}(t)$ to the leaves of t. For nodes u and v of t, if $(u, i, v) \in E_t$, then v is called the i-th child of u, and is denoted by $u \cdot i$. The least common ancestor of nodes u and v is denoted $\mathrm{lca}(u, v)$.

As observed above, trees over Σ are the usual graphical representations of terms over Σ. In fact, if $\sigma \in \Sigma_k$ and $t_1, \ldots, t_k \in T_\Sigma$, then the term $\sigma t_1 \cdots t_k$ can be used to denote the tree t with $\mathrm{lab}_t(\mathrm{root}(t)) = \sigma$ and $\mathrm{root}(t) \cdot i = \mathrm{root}(t_i)$ for every $i \in [1, k]$. We will, however, not use this notation.

Finite automata

We consider (nondeterministic) finite automata on strings and (deterministic) finite tree automata, respectively (see, e.g., [HU] and [GS], respectively).

Let Σ be an (ordinary) alphabet. A *finite automaton* over Σ is a quintuple $A = (Q, \Sigma, \delta, I, F)$, where Q is a finite set of states, Σ is the input alphabet, $\delta \subseteq Q \times \Sigma \times Q$ is the transition relation, $I \subseteq Q$ is the set of initial states, and $F \subseteq Q$ is the set of final states. The elements of δ are called transitions. For every string $w \in \Sigma^*$, A induces a state transition relation $R_A(w) \subseteq Q \times Q$, as follows. For $\sigma \in \Sigma$, $R_A(\sigma) = \{(q, q') \mid (q, \sigma, q') \in \delta\}$. For the empty string ε, $R_A(\varepsilon)$ is the identity on Q. For $\sigma_1, \ldots, \sigma_n \in \Sigma$, $R_A(\sigma_1 \cdots \sigma_n) = R_A(\sigma_1) \circ \cdots \circ R_A(\sigma_n)$. The language recognized by A is $L(A) = \{w \in \Sigma^* \mid R_A(w) \cap (I \times F) \neq \emptyset\}$. $L(A)$ is called a *regular language.*

The tree automata that we consider are the usual total deterministic bottom-up finite-state tree automata. Let Σ be an operator alphabet. A *finite tree automaton* over Σ is a quadruple $A = (Q, \Sigma, \delta, F)$, where Q is a finite set of states, Σ is the input alphabet, $\delta = \{\delta_\sigma\}_{\sigma \in \Sigma}$ where, for $\sigma \in \Sigma_k$, $\delta_\sigma : Q^k \to Q$ is the transition function for σ, and $F \subseteq Q$ is the set of final states. For every tree $t \in T_\Sigma$ and node $u \in V_t$, the *state in which A reaches u*, denoted $\mathrm{state}_{t,A}(u)$, is defined recursively as follows: if $\mathrm{lab}_t(u) = \sigma \in \Sigma_k$, then $\mathrm{state}_{t,A}(u) = \delta_\sigma(\mathrm{state}_{t,A}(u \cdot 1), \ldots, \mathrm{state}_{t,A}(u \cdot k))$. The language recognized by A is $L(A) = \{t \in T_\Sigma \mid \mathrm{state}_{t,A}(\mathrm{root}(t)) \in F\}$. $L(A)$ is called a *regular tree language.*

For a tree $t \in T_\Sigma$ and a node $u \in V_t$, the set of *successful states* of A at u, denoted $\mathrm{succ}_{t,A}(u)$, is defined by top-down recursion as follows: $\mathrm{succ}_{t,A}(\mathrm{root}(t)) = F$, and if $\mathrm{lab}_t(u) = \sigma \in \Sigma_k$ and $1 \leq i \leq k$, then $\mathrm{succ}_{t,A}(u \cdot i)$ is the set of all states $q \in Q$ such that $\delta_\sigma(q_1, \ldots, q_{i-1}, q, q_{i+1}, \ldots, q_k) \in \mathrm{succ}_{t,A}(v)$, where $q_j = \mathrm{state}_{t,A}(u \cdot j)$ for $1 \leq j \leq k$, $j \neq i$. Intuitively, q is in $\mathrm{succ}_{t,A}(u)$ if the automaton, when it reaches u in state q (rather than in $\mathrm{state}_{t,A}(u)$), will reach the root of t in a final state. It is easy to see (by a top-down recursion) that for every node u of t, $t \in L(A)$ iff $\mathrm{state}_{t,A}(u) \in \mathrm{succ}_{t,A}(u)$.

Monadic second order logic

Monadic second order logic is used to describe properties of graphs (see, e.g., [Cou1, Cou4, Eng2, Eng4, EO2]). For alphabets Σ and Γ, we use the language MSOL(Σ, Γ) of monadic second order (MSO) formulas over (Σ, Γ). Formulas in MSOL(Σ, Γ) describe properties of graphs over (Σ, Γ). This logical language has node variables $x, y, \ldots$, and node-set variables $X, Y, \ldots$. For a given graph g over (Σ, Γ), node variables range over the elements of V_g, and node-set variables range over the subsets of V_g.

There are four types of atomic formulas in MSOL(Σ, Γ): $\mathrm{lab}_\sigma(x)$, for every $\sigma \in \Sigma$, denoting that x has label σ; $\mathrm{edg}_\gamma(x, y)$, for every $\gamma \in \Gamma$, denoting that there is an edge labelled γ from x to y; $x = y$, denoting that x equals y, and $x \in X$, denoting that x is an element of X. The formulas are built from the atomic formulas using the connectives $\neg$, $\wedge$, $\vee$, and $\rightarrow$, as usual. Both node variables and node-set variables can be quantified with $\exists$ and $\forall$. We will use $\mathrm{edg}(x, y)$ to denote the disjunction of all $\mathrm{edg}_\gamma(x, y)$, $\gamma \in \Gamma$.

For every n, the set of MSO formulas over (Σ, Γ) with n free node variables and no free node-set variables is denoted $\mathrm{MSOL}_n(\Sigma, \Gamma)$. Since we are predominantly interested in trees, an MSO formula over $(\Sigma, \mathrm{rki}(\Sigma))$, where Σ is an operator alphabet, will also simply be called an MSO formula over Σ. Also, $\mathrm{MSOL}(\Sigma, \mathrm{rki}(\Sigma))$ and $\mathrm{MSOL}_n(\Sigma, \mathrm{rki}(\Sigma))$ will be abbreviated to $\mathrm{MSOL}(\Sigma)$ and $\mathrm{MSOL}_n(\Sigma)$, respectively.

For a closed formula $\phi \in \mathrm{MSOL}_0(\Sigma, \Gamma)$ and a graph $g \in G_{\Sigma,\Gamma}$, we write $g \models \phi$ if g satisfies ϕ. Given a graph g, a valuation ν is a function that assigns to each node variable an element of V_g, and to each node-set variable a subset of V_g. We write $(g, \nu) \models \phi$, if ϕ holds in g, where the free variables of ϕ are assigned values according to the valuation ν. If a formula ϕ has free variables, say, x, X, y and no others, we also write $\phi(x, X, y)$. Moreover, we write $(g, u, U, v) \models \phi(x, X, y)$ for $(g, \nu) \models \phi(x, X, y)$, where $\nu(x) = u$, $\nu(X) = U$, and $\nu(y) = v$.

Let Σ be an operator alphabet. The tree language defined by a closed formula $\phi \in \mathrm{MSOL}_0(\Sigma)$ is $L(\phi) = \{t \in T_\Sigma \mid t \models \phi\}$. $L(\phi)$ is called an MSO *definable tree language*.

The following classical result from [Don, TW] shows the equivalence between monadic second order logic and finite tree automata, as a means of defining tree languages by specification and computation, respectively. It was first shown for the special case of string languages in [Büc, Elg]. See Sections 3 and 11 of [Tho], and [Eng3], for a discussion of such results.

Proposition 1. *A tree language is* MSO *definable if and only if it is regular.*

3 MSO Node Relations

The aim of this paper is to investigate ways of defining binary relations on the nodes of graphs, and in particular on the nodes of trees. One way of doing this is by an MSO formula $\phi(x, y)$ with two free node variables: for a given graph g, two nodes u and v are in the corresponding relation if $(g, u, v) \models \phi(x, y)$.

Another way is by some kind of graph-walking automaton: then u and v are in the corresponding relation if the automaton can walk from u to v, following the edges of g.

For alphabets Σ and Γ, a *graph-node relation* over (Σ, Γ) is a subset of $\{(g, u, v) \mid g \in G_{\Sigma,\Gamma} \text{ and } u, v \in V_g\}$. Thus, a graph-node relation associates with each graph g a binary relation on the nodes of g.

Let $\phi(x, y) \in \text{MSOL}_2(\Sigma, \Gamma)$ be an MSO formula with two free node variables. For each graph $g \in G_{\Sigma,\Gamma}$, $\phi(x, y)$ defines the node relation $R_g(\phi) = \{(u, v) \in V_g \times V_g \mid (g, u, v) \models \phi(x, y)\}$. In this way, $\phi(x, y)$ defines the graph-node relation $R_{\text{GR}}(\phi) = \{(g, u, v) \mid g \in G_{\Sigma,\Gamma}, (u, v) \in R_g(\phi)\}$. $R_{\text{GR}}(\phi)$ is called an MSO *definable graph-node relation.* The set of all MSO definable graph-node relations is denoted MSO-GNR.

Analogous definitions hold for the special case of trees, as follows. For an operator alphabet Σ, a *tree-node relation* over Σ is a subset of $\{(t, u, v) \mid t \in T_\Sigma \text{ and } u, v \in V_t\}$. An MSO formula $\phi(x, y) \in \text{MSOL}_2(\Sigma)$ defines the tree-node relation $R_{\text{TR}}(\phi) = \{(t, u, v) \mid t \in T_\Sigma, (u, v) \in R_t(\phi)\}$. $R_{\text{TR}}(\phi)$ is called an MSO *definable tree-node relation.* The set of MSO definable tree-node relations is denoted MSO-TNR.

We will need the following basic, well-known fact: if a node relation is MSO definable, then so is its transitive reflexive closure (see, e.g., [Cou1]). For a formula $\phi(x, y) \in \text{MSOL}_2(\Sigma, \Gamma)$, we define the formula $\phi^*(x, y) = \forall X((\text{closed}(X) \wedge x \in X) \rightarrow y \in X)$, where $\text{closed}(X) = \forall x, y((x \in X \wedge \phi(x, y)) \rightarrow y \in X)$.

Lemma 2. *Let Σ, Γ be alphabets. For all $g \in G_{\Sigma,\Gamma}$ and $\phi(x, y) \in \text{MSOL}_2(\Sigma, \Gamma)$, $R_g(\phi^*) = R_g(\phi)^*$.*

4 Graph-Walking Automata

A graph-walking automaton is a finite-state automaton that walks on a graph from node to node, following the edges of the graph (along or against the direction of the edges). The automaton can test the label of the current node, and it can test the labels of the edges that are incident with the current node. To enhance the power of this basic model, we will allow the graph-walking automaton to test any MSO definable property of the current node (using MSO formulas with one free node variable). For a given graph g, two nodes u and v are in the node relation computed by the automaton if the automaton can walk from u to v in g, starting in an initial state and ending in a final state.

We start by defining the (infinite) set of instructions of our graph-walking automata; they will be called directives (as in [KS]). Let Σ and Γ be alphabets. The set of *directives* over Σ and Γ is

$$D_{\Sigma,\Gamma} = \{\downarrow_\gamma \mid \gamma \in \Gamma\} \cup \{\uparrow_\gamma \mid \gamma \in \Gamma\} \cup \text{MSOL}_1(\Sigma, \Gamma).$$

A directive is an instruction of how to move from one node to another: $\downarrow_\gamma$ means "move along an edge labelled γ", $\uparrow_\gamma$ means "move against an edge labelled γ", and $\psi(x)$ means "check that ψ holds for the current node". Formally, we define for each $g \in G_{\Sigma,\Gamma}$ and each directive $d \in D_{\Sigma,\Gamma}$ the node relation $R_g(d) \subseteq V_g \times V_g$, as follows:

$$R_g(\downarrow_\gamma) = \{(u,v) \mid (u,\gamma,v) \in E_g\},$$
$$R_g(\uparrow_\gamma) = \{(v,u) \mid (u,\gamma,v) \in E_g\}, \text{ and}$$
$$R_g(\psi(x)) = \{(u,u) \mid (g,u) \models \psi(x)\}.$$

Syntactically, a graph-walking automaton is just an ordinary finite automaton (on strings) with a finite subset of $D_{\Sigma,\Gamma}$ as input alphabet. However, the symbols of $D_{\Sigma,\Gamma}$ are interpreted as instructions on the input graph as explained above.

Let Σ and Γ be alphabets. A ***graph-walking automaton*** (*with* MSO *tests*) over (Σ,Γ) is a finite automaton A over a finite subset Δ of $D_{\Sigma,\Gamma}$.

For a graph-walking automaton $A = (Q, \Delta, \delta, I, F)$ and a graph g, an element (u, q) of $V_g \times Q$ is a ***configuration*** of the automaton. It signifies that A is at node u in state q. A configuration (u, q) is ***initial*** if $q \in I$, and ***final*** if $q \in F$. A pair of nodes (u, v) is in the relation defined by A if A can move from an initial configuration (u,q) to a final configuration (v,q'). This is now formalized, in the obvious way. Recall that, for a directive d, the node relation $R_g(d)$ is defined above, and the state transition relation $R_A(d)$ is defined in Section 2.

Let $g \in G_{\Sigma,\Gamma}$. One step of $A = (Q, \Delta, \delta, I, F)$ on g is defined by the binary relation $\twoheadrightarrow_{g,A}$ on the set of configurations, as follows. For every $u, u' \in V_g$ and $q, q' \in Q$,

$$(u,q) \twoheadrightarrow_{g,A} (u',q') \text{ iff } \exists d \in \Delta : (u,u') \in R_g(d) \text{ and } (q,q') \in R_A(d).$$

To indicate the directive that is executed by A in this step, we also write $(u,q) \overset{d}{\twoheadrightarrow}_{g,A} (u',q')$.

For each graph $g \in G_{\Sigma,\Gamma}$, A computes the node relation $R_g(A) = \{(u,v) \in V_g \times V_g \mid (u,q) \twoheadrightarrow^*_{g,A} (v,q') \text{ for some } q \in I \text{ and } q' \in F\}$. Thus, A computes the graph-node relation $R_{\mathrm{GR}}(A) = \{(g,u,v) \mid g \in G_{\Sigma,\Gamma}, (u,v) \in R_g(A)\}$. $R_{\mathrm{GR}}(A)$ is called a ***regular graph-node relation***. The set of all regular graph-node relations is denoted REG-GNR. Analogous definitions hold for the special case of trees, to be considered in the next section.

Example 1. Let Σ be the operator alphabet $\Sigma_0 \cup \Sigma_2$, with $\Sigma_0 = \{\text{red}, \text{black}\}$ and $\Sigma_2 = \{\sigma\}$. We consider a graph-walking automaton A over $(\Sigma, \text{rki}(\Sigma))$, and we will in particular be interested in the behaviour of A on trees in T_Σ, i.e., binary trees with red and black leaves. For a tree t over Σ, the automaton A connects certain leaves of t, i.e., $R_t(A)$ consists of pairs (u, v) where both u and v are leaves of t. If t has exactly one red leaf, say v_{red}, then all leaves have a pointer to that red leaf, i.e., $R_t(A)$ consists of all (u, v_{red}) where u is a leaf. Otherwise (i.e., if there is no red leaf, or if there is more than one), all leaves are linked in left-to-right circular order, i.e., $R_t(A)$ consists of all (u, v) such that v is the next leaf after u, in that order. Note that $R_t(A)$ is a partial function for every $t \in T_\Sigma$.

Let root(x) be an MSO formula that expresses that x is the root of the tree, e.g., $\text{root}(x) = \neg\exists y(\text{edg}(y,x))$. Let leaf($x$) be an MSO formula that expresses that x is a leaf, e.g., $\text{leaf}(x) = \text{lab}_{\text{red}}(x) \vee \text{lab}_{\text{black}}(x)$. And let orl be an MSO formula that is true iff there is exactly one red leaf, i.e., $\text{orl} = \exists y(\text{lab}_{\text{red}}(y) \wedge \forall z(\text{lab}_{\text{red}}(z) \rightarrow z = y))$. We define $A = (Q, \Delta, \delta, I, F)$ with $Q = \{q_{\text{in}}, r, u, l, d, q_{\text{fin}}\}$, $I = \{q_{\text{in}}\}$, $F = \{q_{\text{fin}}\}$, $\Delta = \{\downarrow_1, \downarrow_2, \uparrow_1, \uparrow_2\} \cup \{\text{leaf}(x) \wedge \text{orl}, \text{leaf}(x) \wedge \neg\text{orl}, \text{lab}_{\text{red}}(x), \text{root}(x), \text{leaf}(x)\}$, and δ consists of the following transitions:

$(q_{\text{in}}, \text{leaf}(x) \wedge \text{orl}, r)$, $(q_{\text{in}}, \text{leaf}(x) \wedge \neg\text{orl}, u)$,
(r, d, r) for all $d \in \{\downarrow_1, \downarrow_2, \uparrow_1, \uparrow_2\}$, $(r, \text{lab}_{\text{red}}(x), q_{\text{fin}})$,
$(u, \uparrow_2, u)$, $(u, \text{root}(x), d)$, $(u, \uparrow_1, l)$, $(l, \downarrow_2, d)$, and
$(d, \downarrow_1, d)$, $(d, \text{leaf}(x), q_{\text{fin}})$.

On a given tree, the automaton A first checks that it is at a leaf, and tests whether or not there is exactly one red leaf. If so, it walks nondeterministically on the tree (in state r) until it happens to find that red leaf. If not, it walks to the "next" leaf along the shortest path: it first moves upwards over 2-labelled edges as far as possible (in state u); then it either is at the root or it moves to the other child of its parent; and finally it moves downwards along 1-labelled edges as far as possible (in state d). It is left to the reader to imagine the behaviour of A on graphs in $G_{\Sigma,\text{rki}(\Sigma)}$ that are not in T_Σ. □

We first show that the graph-node relation computed by a graph-walking automaton depends only on the language (of strings of directives) it recognizes. In other words, for graph-walking automata A_1 and A_2, if $L(A_1) = L(A_2)$ then $R_{\text{GR}}(A_1) = R_{\text{GR}}(A_2)$. We prove this by associating a graph-node relation with every language of strings of directives, in a natural way (as is well known from program scheme theory, cf. [Engl]).

Let Σ and Γ be alphabets. A *walking language* over (Σ, Γ) is a (string) language over a finite subset of $D_{\Sigma,\Gamma}$. A *walking string*, i.e., a string from a walking language, can be viewed as a sequence of instructions to be executed as a walk on a graph; the language itself can be viewed as a nondeterministic choice between such instruction sequences. This is formalized next. Let $g \in G_{\Sigma,\Gamma}$. A walking string $w \in D^*_{\Sigma,\Gamma}$ computes the relation $R_g(w) \subseteq V_g \times V_g$, defined as follows: $R_g(\varepsilon)$ is the identity on V_g, and, for $d_1, \ldots, d_n \in D_{\Sigma,\Gamma}$, $R_g(d_1 \cdots d_n) = R_g(d_1) \circ \cdots \circ R_g(d_n)$. A walking language $W \subseteq D^*_{\Sigma,\Gamma}$ computes the node relation $R_g(W) = \bigcup\{R_g(w) \mid w \in W\}$. Now, a walking language W defines the graph-node relation $R_{\text{GR}}(W) = \{(g, u, v)) \mid g \in G_{\Sigma,\Gamma} \text{ and } (u, v) \in R_g(W)\}$.

Example 2. We define a walking language W that computes the same graph-node relation as the graph-walking automaton A from Example 1. Let Δ be defined as in Example 1. Then W is the regular language $W_1 \cup W_2$ over Δ, where

$$W_1 = (\text{leaf}(x) \wedge \text{orl}) \cdot \{\downarrow_1, \downarrow_2, \uparrow_1, \uparrow_2\}^* \cdot \text{lab}_{\text{red}}(x), \text{ and}$$
$$W_2 = (\text{leaf}(x) \wedge \neg\text{orl}) \cdot \uparrow_2^* \cdot \{\uparrow_1 \cdot \downarrow_2, \text{root}(x)\} \cdot \downarrow_1^* \cdot \text{leaf}(x).$$

Note that $W = L(A)$. According to the next lemma this implies that $R_{\text{GR}}(W) = R_{\text{GR}}(A)$ (which should also intuitively be clear). □

We now show that the graph-node relation computed by a graph-walking automaton depends only on the language it recognizes.

Lemma 3. *For every graph-walking automaton A, $R_{GR}(A) = R_{GR}(L(A))$.*

Proof. Let $A = (Q, \Delta, \delta, I, F)$. We have to show that $R_g(A) = R_g(L(A))$ for every graph g. It is straightforward to show that, for all $u, u' \in V_g$, $q, q' \in Q$, and $w = d_1 \cdots d_n$ with $d_i \in \Delta$, the following equivalence holds: there is a walk $(u_0, q_0) \overset{d_1}{\twoheadrightarrow}_{g,A} (u_1, q_1) \overset{d_2}{\twoheadrightarrow}_{g,A} \cdots \overset{d_n}{\twoheadrightarrow}_{g,A} (u_n, q_n)$ of A on g with $(u_0, q_0) = (u, q)$ and $(u_n, q_n) = (u', q')$ if and only if $(u, u') \in R_g(w)$ and $(q, q') \in R_A(w)$. From this equivalence it follows that $R_g(A) = \{(u, u') \mid \exists q \in I, q' \in F, w \in \Delta^* : (u, u') \in R_g(w)$ and $(q, q') \in R_A(w)\} = \{(u, u') \mid \exists w \in L(A) : (u, u') \in R_g(w)\} = R_g(L(A))$. □

As a corollary we obtain the fact that a graph-node relation can be computed by a graph-walking automaton iff it can be computed by a regular walking language (i.e., a walking language that can be recognized by a finite automaton).

Corollary 4. REG-GNR *is the set of all $R_{GR}(W)$ where W is a regular walking language.*

This allows us to show in a straightforward way (using Kleene's theorem) that all regular graph-node relations are MSO definable.

Theorem 5. REG-GNR $\subseteq$ MSO-GNR.

Proof. Let Δ be a finite subset of $D_{\Sigma,\Gamma}$. By Kleene's theorem, the class of regular (string) languages over Δ is the smallest class of languages that is closed under union ($W_1 \cup W_2$), concatenation ($W_1 \cdot W_2$), and Kleene star (W^*), and contains the empty language and every language $\{d\}$, with $d \in \Delta$. By induction on this characterization we define for every regular walking language W a formula $\phi_W(x, y) \in \mathrm{MSOL}_2(\Sigma, \Gamma)$, such that $R_{GR}(\phi_W) = R_{GR}(W)$. We use 'false' to stand for any formula that never holds (such as $\exists x : \neg(x = x)$). Recall also the definition of the transitive reflexive closure $\phi^*(x, y)$ of a formula $\phi(x, y)$ from Section 3. Let $\gamma \in \Gamma$, $\psi(z) \in \mathrm{MSOL}_1(\Sigma, \Gamma)$, and let W, W_1, W_2 be walking languages over (Σ, Γ). Then we define

$$\begin{aligned}
\phi_\emptyset(x, y) &= \text{false} \\
\phi_{\{\downarrow_\gamma\}}(x, y) &= \mathrm{edg}_\gamma(x, y) \\
\phi_{\{\uparrow_\gamma\}}(x, y) &= \mathrm{edg}_\gamma(y, x) \\
\phi_{\{\psi(z)\}}(x, y) &= \psi(x) \wedge (x = y) \\
\phi_{W_1 \cup W_2}(x, y) &= \phi_{W_1}(x, y) \vee \phi_{W_2}(x, y) \\
\phi_{W_1 \cdot W_2}(x, y) &= \exists z(\phi_{W_1}(x, z) \wedge \phi_{W_2}(z, y)) \\
\phi_{W^*}(x, y) &= \phi_W^*(x, y)
\end{aligned}$$

The correctness of the first five cases should be clear. The remaining two cases follow from the equations $R_{GR}(W_1 \cdot W_2) = R_{GR}(W_1) \circ R_{GR}(W_2)$ and $R_{GR}(W^*) =$

$R_{\mathrm{GR}}(W)^*$ (which can easily be proved), together with Lemma 2. In words, the concatenation of two walking languages computes the composition of the corresponding relations, and the Kleene star of a walking language computes the transitive reflexive closure of the corresponding relation. □

Of course, the inclusion in Theorem 5 is proper. For instance the relation that holds between two nodes u and v when they are in different connected components of the graph, is MSO definable, e.g. by the formula $\neg\phi^*(x,y)$ where $\phi(x,y) = \mathrm{edg}(x,y) \vee \mathrm{edg}(y,x)$. But a graph-walking automaton cannot compute this relation, simply because it cannot walk from u to v. However, even for connected graphs there are graph-node relations that can be expressed by an MSO formula, but cannot be computed by a graph-walking automaton, and, in fact, not by any walking language (whether it is regular or not). We now show this, for an even smaller class of graphs. For alphabets Σ and Γ, let $\mathrm{ACG}_{\Sigma,\Gamma}$ be the set of all acyclic, connected graphs over (Σ, Γ).

Theorem 6. *Let Σ and Γ be alphabets with $\#\Sigma \geq 1$ and $\#\Gamma \geq 2$. There is an* MSO *formula $\phi(x,y) \in \mathrm{MSOL}(\Sigma,\Gamma)$ for which there is no walking language W over (Σ,Γ) such that $R_g(W) = R_g(\phi)$ for every $g \in \mathrm{ACG}_{\Sigma,\Gamma}$.*

Proof. An MSO formula that cannot be simulated by any walking language is $\phi(x,y) = \neg(x = y)$, with $R_g(\phi) = \{(u,v) \mid u,v \in V_g, u \neq v\}$ for every g. We will prove this by introducing a set of graphs that have a circular structure, and are all very much alike, apart from the number of their nodes. The walking language will not be able to tell all of these graphs apart, since it can use only finitely many unary MSO formulas. Then, we find two indistinguishable graphs, such that any walking string that circles the first graph exactly halfway, fully circles the other graph. This leads to a contradiction, since any walking language that computes the relation $u \neq v$ has to have a string circling the first graph halfway, but cannot have a string that circles the second graph fully.

We prove the proposition by contradiction. Suppose there is a walking language W over (Σ,Γ) such that $R_g(W) = \{(u,v) \mid u,v \in V_g, u \neq v\}$ for every $g \in \mathrm{ACG}_{\Sigma,\Gamma}$. Let $W \subseteq \Delta^*$ where Δ is a finite subset of $D_{\Sigma,\Gamma}$. Only finitely many unary MSO formulas are in Δ, say $\psi_1(x),\ldots,\psi_m(x)$. Thus, every node u of a graph g has a *type*, $\mathrm{type}_g(u) = (b_1,\ldots,b_m) \in \{0,1\}^m$, with $b_i = 1 \Leftrightarrow (g,u) \models \psi_i(x)$. There are 2^m different types of nodes.

Let $\sigma \in \Sigma$ and $a,b \in \Gamma$. Consider now for every even n, the graph $g_n = (V_n, E_n, \mathrm{lab}_n)$ over (Σ,Γ) with $V_n = \{u_i \mid 0 \leq i \leq n-1\}$, $\mathrm{lab}_n(u_i) = \sigma$ for every i, and $E_n = \{(u_i, a, u_{i+1}) \mid i \text{ is even}\} \cup \{(u_i, b, u_{i-1}) \mid i \text{ is even}\} \cup \{(u_0, b, u_{n-1})\}$. Thus, g_n is a circle with n nodes, in which the directions and labels of the edges alternate. See Figure 1 for a picture of g_8. Obviously, g_n is in $\mathrm{ACG}_{\Sigma,\Gamma}$. Walking from u_i to u_{i+1} (and from u_{n-1} to u_0) will be referred to as "anti-clockwise" walking.

Clearly, for a fixed graph g_n, all $u_i \in g_n$ with even i have the same type, and all $u_i \in g_n$ with odd i have the same type. This is because, for even l, $f_l(u_i) = u_{i+l \bmod n}$ is an automorphism of g_n. The *type* of g_n is $(\mathrm{type}_{g_n}(u_0), \mathrm{type}_{g_n}(u_1))$. There are 2^{2m} different types of graphs.

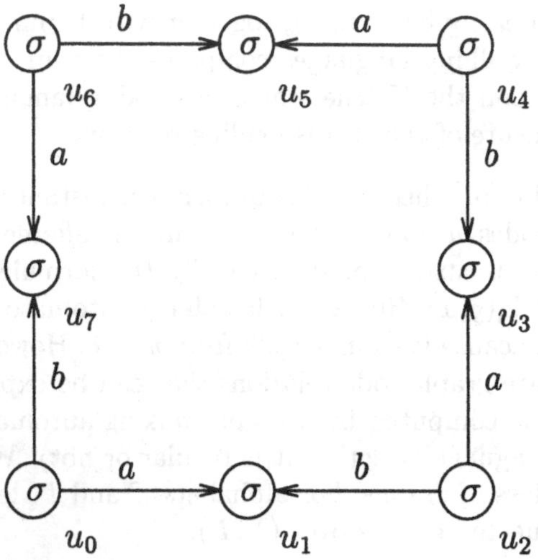

Fig. 1. Graph g_8

For every $w \in \Delta^*$, let $\mathrm{nr}(w) = \#_{\downarrow_a}(w) + \#_{\uparrow_b}(w) - (\#_{\uparrow_a}(w) + \#_{\downarrow_b}(w))$ where, for a directive d, $\#_d(w)$ is the number of occurrences of d in w. Intuitively, $\mathrm{nr}(w)$ is the net number of anti-clockwise steps w takes in a graph of the above form. The following facts can easily be proved: (1) for any graph g_n, and any $w \in \Delta^*$, if $(u_0, u_j) \in R_{g_n}(w)$, then $\mathrm{nr}(w) \bmod n = j$, and (2) for any two graphs g_n and $g_{n'}$ of the same type, and any $w \in \Delta^*$, if $\mathrm{nr}(w) = i$ and $(u_0, u_{i \bmod n'}) \in R_{g_{n'}}(w)$, then $(u_0, u_{i \bmod n}) \in R_{g_n}(w)$. The second statement can be proven by induction on the length of w, and basically depends on the fact that w cannot discern "on" which graph it is, judging by the information of the nodes, since they have the same type and the same incident edges in both graphs.

Consider now the sequence of graphs $g_2, g_4, g_8, g_{16}, \ldots$. Since the number of graph types is finite, there are two graphs g_n, $g_{n'}$ of the same type. Let $n' = n \cdot 2^k$ for some $k \geq 1$. Since $u_0 \neq u_{n'/2}$ in $g_{n'}$, there is a $w \in W$ such that $(u_0, u_{n'/2}) \in R_{g_{n'}}(w)$, by assumption. So, by fact (1), $\mathrm{nr}(w) \bmod n' = n'/2$, i.e., $\mathrm{nr}(w) = n'/2 + l \cdot n'$ for some $l \in \mathrm{N}$. But now, since $(n'/2 + l \cdot n') \bmod n = 0$, fact (2) implies $(u_0, u_0) \in R_{g_n}(w)$, a contradiction. □

It remains an open problem for which classes of graphs regular walking languages and binary MSO formulas have the same strength. In the next section we show that it holds for the class of trees.

5 Tree-Walking Automata

In this section we show our main result: a tree-node relation can be defined by an MSO formula iff it can be computed by a tree-walking automaton.

Let Σ be an operator alphabet. A *tree-walking automaton* (*with* MSO *tests*) over Σ is a graph-walking automaton over $(\Sigma, \text{rki}(\Sigma))$. A tree-walking automaton A computes the tree-node relation $R_{\text{TR}}(A) = \{(t, u, v) \mid t \in T_\Sigma, (u, v) \in R_t(A)\}$. $R_{\text{TR}}(A)$ is called a *regular tree-node relation.* The set of all regular tree-node relations is denoted REG-TNR.

A walking language over $(\Sigma, \text{rki}(\Sigma))$ is also called a walking language over Σ. A walking language W over Σ computes the tree-node relation $R_{\text{TR}}(W) = \{(t, u, v)) \mid t \in T_\Sigma, (u, v) \in R_t(W)\}$. From Corollary 4 it follows that REG-TNR is the set of all $R_{\text{TR}}(W)$ where W is a regular walking language.

For examples of a tree-walking automaton and a walking language over an operator alphabet Σ, see Examples 1 and 2, respectively.

From Theorem 5 it follows that REG-TNR $\subseteq$ MSO-TNR, i.e., every tree-node relation that is computed by a tree-walking automaton, is MSO definable. It remains to show the reverse: every MSO definable tree-node relation can be computed by a tree-walking automaton. In order to implement an MSO formula $\phi(x, y)$ by a tree-walking automaton, we first use the classical result of Proposition 1 to construct a finite tree automaton that recognizes all trees $\text{mark}(t, u, v)$ such that $(t, u, v) \models \phi(x, y)$. Here $\text{mark}(t, u, v)$ is the tree t in which the nodes u and v are marked by special labels. This is a slight variation of a well-known technique, used in the proof of Proposition 1.

Let Σ be an operator alphabet, and $k \geq 1$. Define $B_k = \{0, 1\}^k \setminus \{0\}^k$. Let $\Sigma \cup (\Sigma \times B_k)$ be the operator alphabet such that $\text{rk}(\sigma) = \text{rk}_\Sigma(\sigma)$ for every $\sigma \in \Sigma$, and $\text{rk}(\langle \sigma, b_1, \ldots, b_k \rangle) = \text{rk}_\Sigma(\sigma)$ for all $\langle \sigma, b_1, \ldots, b_k \rangle \in \Sigma \times B_k$. Note that label $\langle \sigma, 0, \ldots, 0 \rangle$ is excluded. We use this alphabet to attach k different marks to the labels of the nodes of a tree. Let t be a tree over Σ, and let $u_1, \ldots, u_k \in V_t$. The *marked tree* $\text{mark}(t, u_1, \ldots, u_k)$ over $\Sigma \cup (\Sigma \times B_k)$ is defined as (V_t, E_t, lab) where, for every $v \in V_t$, $\text{lab}(v) = \text{lab}_t(v)$ if $v \neq u_i$ for all i, and $\text{lab}(v) = \langle \text{lab}_t(v), (v = u_1), \ldots, (v = u_k) \rangle$ otherwise (where $(v = u_i) = 1$ iff v equals u_i).

The following lemma allows us to apply Proposition 1 to formulas with free node variables.

Lemma 7. *Let $k \geq 1$. For every $\phi(x_1, \ldots, x_k) \in \text{MSOL}_k(\Sigma)$ there is a closed* MSO *formula $\psi \in \text{MSOL}_0(\Sigma \cup (\Sigma \times B_k))$ such that*

$$(t, u_1, \ldots, u_k) \models \phi(x_1, \ldots, x_k) \textit{ iff } \text{mark}(t, u_1, \ldots, u_k) \models \psi,$$

and vice versa, i.e., for every formula $\psi \in \text{MSOL}_0(\Sigma \cup (\Sigma \times B_k))$ there exists a formula $\phi(x_1, \ldots, x_k) \in \text{MSOL}_k(\Sigma)$ such that the above holds.

Proof. The proof is straightforward. From an MSO formula $\phi(x_1, \ldots, x_k)$ we construct ψ as follows:

$$\psi = \forall x_1 \ldots x_k ((\text{marked}_1(x_1) \wedge \cdots \wedge \text{marked}_k(x_k)) \rightarrow \phi'(x_1, \ldots, x_k)),$$

where $\text{marked}_i(x) = \exists \langle \sigma, b_1, \ldots, b_k \rangle \in \Sigma \times B_k : b_i = 1 \wedge \text{lab}_{\langle \sigma, b_1, \ldots, b_k \rangle}(x)$, and $\phi'(x_1, \ldots, x_k)$ is obtained from $\phi(x_1, \ldots, x_k)$ by replacing all occurrences of all $\text{lab}_\sigma(y)$ by $(\text{lab}_\sigma(y) \vee \exists (b_1, \ldots, b_k) \in B_k : \text{lab}_{\langle \sigma, b_1, \ldots, b_k \rangle}(y))$.

The other way around is similar. Let us assume that the variables $x_1, \ldots, x_k$ do not occur in ψ. We construct $\phi(x_1, \ldots, x_k)$ by replacing all occurrences of all $\text{lab}_\sigma(y)$ by $(\text{lab}_\sigma(y) \wedge \forall i \in [1,k] : y \neq x_i)$, and all occurrences of all $\text{lab}_{\langle\sigma,b_1,\ldots,b_k\rangle}(y)$ by $(\text{lab}_\sigma(y) \wedge \forall i \in [1,k] : (b_i = 1 \rightarrow y = x_i))$. □

We now prove the main result of this paper. The idea of the proof is similar to the one of Lemma 12 of [EO2] (which is a key lemma in the proof of the main result of [EO1, EO2]: the equivalence between context-free graph grammars and MSO transductions of trees). In that lemma, there is no tree-walking automaton, but an appropriate relabelling of the nodes of the tree (instead of the MSO tests) and a regular language of strings of node labels (instead of the finite control of the automaton).

Theorem 8. MSO-TNR = REG-TNR.

Proof. By Theorem 5 it suffices to show MSO-TNR $\subseteq$ REG-TNR. Let Σ be an operator alphabet and let $\phi(x,y) \in \text{MSOL}_2(\Sigma)$. By Lemma 7 (for $k = 2$) and Proposition 1 (applied to ψ) there is a finite tree automaton M over $\Sigma \cup (\Sigma \times B_2)$ such that, for every tree $t \in T_\Sigma$ and nodes u and v of t, $(t, u, v) \models \phi(x, y)$ if and only if $\text{mark}(t, u, v) \in L(M)$.

Let $M = (Q, \Sigma \cup (\Sigma \times B_2), \delta, F)$. We will prove that there exists a tree-walking automaton A over Σ with $R_{\text{TR}}(A) = R_{\text{TR}}(\phi)$, i.e., for every tree $t \in T_\Sigma$ and nodes u and v of t, $(u, v) \in R_t(A)$ iff $\text{mark}(t, u, v) \in L(M)$. This tree-walking automaton A behaves in a special way: it walks along paths in the tree t in which no edge occurs more than once, that is, for every pair of nodes $(u, v) \in R_t(A)$ the shortest path from u to v is always taken. In fact, A simulates the finite tree automaton M on the path from u to v, using unary MSO formulas to get information on the behaviour of the automaton M on the rest of the tree. More precisely, A simulates the computation of M on the path from u to the least common ancestor $\text{lca}(u, v)$ of u and v, and then simulates the reverse computation of M on the path from $\text{lca}(u, v)$ to v (nondeterministically). Recall that $B_2 = \{(1,0), (0,1), (1,1)\}$ where $(1,0)$ indicates u and $(0,1)$ indicates v (if $u \neq v$) and $(1,1)$ indicates u and v (if $u = v$).

Before defining A we show how the behaviour of M on the parts of the tree outside the shortest path from u to v, can be expressed by unary MSO formulas. Note that the nodes in these parts are labeled by symbols from Σ. Recall from Section 2 that, for a tree t and a node w of t, $\text{state}_{t,M}(w)$ is the state in which M reaches w. We now claim that, for each $q \in Q$, there is an MSO formula $\text{state}_q(x) \in \text{MSOL}_1(\Sigma)$ such that for all $t \in T_\Sigma$ and $w \in V_t$: $(t, w) \models \text{state}_q(x)$ iff $\text{state}_{t,M}(w) = q$. In fact, since it is easy to construct from M a finite tree automaton M_q over $\Sigma \cup (\Sigma \times B_1)$ such that $\text{mark}(t, w) \in L(M_q)$ iff $\text{state}_{t,M}(w) = q$, this follows directly from Proposition 1 (to obtain a closed formula ψ with $L(\psi) = L(M_q)$) and Lemma 7 (for $k = 1$). The tree automaton M_q simulates M, but if it reaches the marked node, i.e., the node with some label $(\sigma, 1)$ (recall that $B_1 = \{1\}$), it verifies whether or not M is in state q, and continues to the root in an accepting or rejecting state, respectively.

The formulas $\text{state}_q(x)$ can be used to compute the state of M at all those nodes that are children of nodes on the shortest path, but do not lie on that path themselves. We also need a unary MSO formula that checks whether a given state q is successful at the node $w = \text{lca}(u, v)$. Recall from Section 2 that $\text{succ}_{t,M}(w)$ is the set of successful states at w, i.e., the set of all states q such that A, when reaching node w in state q, will reach the root in a final state. As above, we now claim that, for each $q \in Q$, there is an MSO formula $\text{succ}_q(x) \in \text{MSOL}_1(\Sigma)$ such that for all $t \in T_\Sigma$ and $w \in V_t$: $(t, w) \models \text{succ}_q(x)$ iff $q \in \text{succ}_{t,M}(w)$. This follows from Proposition 1 and Lemma 7 in the same way as above, because it is easy to construct from M a finite tree automaton M'_q over $\Sigma \cup (\Sigma \times B_1)$ such that $\text{mark}(t, u) \in L(M'_q)$ iff $q \in \text{succ}_{t,M}(w)$: M'_q simulates M, but if it reaches the marked node, it switches to state q.

The reader should keep realizing that M works on $\text{mark}(t, u, v)$ whereas A walks on t. Thus, the formulas $\text{state}_q(x)$ and $\text{succ}_q(x)$ were defined for t, and not for $\text{mark}(t, u, v)$. However, since (as observed above) the parts of t outside the shortest path from u to v are labeled by symbols from Σ, it is straightforward to see that (1) if w is neither an ancestor of u nor of v, then $\text{state}_{\text{mark}(t,u,v),M}(w) = \text{state}_{t,M}(w)$, and (2) if w is an ancestor of both u and v, then $\text{succ}_{\text{mark}(t,u,v),M}(w) = \text{succ}_{t,M}(w)$. This guarantees that A will test the correct information on the behaviour of M.

We now define the tree-walking automaton $A = (Q_A, \Delta_A, \delta_A, I_A, F_A)$ over Σ. To simplify the description of A we allow its transition relation δ_A to contain transitions of the form $(q, d_1 \cdots d_n, q')$ with $q, q' \in Q_A$ and $d_1, \ldots, d_n \in D_{\Sigma,\text{rki}(\Sigma)}$; such a transition stands for the m transitions

$$(q, d_1, q_1), (q_1, d_2, q_2), \ldots, (q_{m-2}, d_{m-1}, q_{m-1}), (q_{m-1}, d_m, q')$$

where $q_1, \ldots, q_{m-1}$ are (unique) new states.

The set of states of A is $Q_A = \{q_{\text{in}}, q_{\text{fin}}\} \cup \{\text{up}_q \mid q \in Q\} \cup \{\text{down}_q \mid q \in Q\}$, with $I_A = \{q_{\text{in}}\}$ and $F_A = \{q_{\text{fin}}\}$. The states up_q are used by A when walking from a node u to $\text{lca}(u, v)$ and simulating M, whereas it uses the states down_q when walking from $\text{lca}(u, v)$ to the node v and simulating M reversely. We define Δ_A to be the set of all directives that occur in δ_A.

It remains to define the transitions in δ_A. Let $k \in \mathbb{N}$, $\sigma \in \Sigma_k$, and $q_1, \ldots, q_k, q \in Q$. For each such choice, δ_A contains the transitions as specified below. Apart from the formulas $\text{succ}_q(x)$, we will use the unary formula

$$\text{test}(x) = \text{lab}_\sigma(x) \wedge \forall m \in [1, k] : \forall y (\text{edg}_m(x, y) \rightarrow \text{state}_{q_m}(y))$$

which expresses that the current node has label σ and $q_1, \ldots, q_k$ are the states in which M reaches its children. Similarly, for $i, j \in [1, k]$ we will use the formulas $\text{test}_i(x)$ and $\text{test}_{i,j}(x)$ which are the same as $\text{test}(x)$ except that the quantification of m is restricted to $[1, k] \setminus \{i\}$ and $[1, k] \setminus \{i, j\}$, respectively.

The transitions in δ_A are the following.

(1) Start moving up:
if $\delta_{\langle\sigma,1,0\rangle}(q_1,\ldots,q_k) = q$, then

$$(q_{\text{in}}, \text{test}(x), \text{up}_q)$$

(2) Move one step up:
if $\delta_\sigma(q_1,\ldots,q_k) = q$, then, for all $i \in [1,k]$,

$$(\text{up}_{q_i}, \uparrow_i \cdot \text{test}_i(x), \text{up}_q)$$

(3) Turn around:
if $\delta_\sigma(q_1,\ldots,q_k) = q$, then, for all $i,j \in [1,k]$ with $i \neq j$,

$$(\text{up}_{q_i}, \uparrow_i \cdot (\text{test}_{i,j}(x) \wedge \text{succ}_q(x)) \cdot \downarrow_j, \text{down}_{q_j})$$

(4) Move one step down:
if $\delta_\sigma(q_1,\ldots,q_k) = q$, then, for all $i \in [1,k]$,

$$(\text{down}_q, \text{test}_i(x) \cdot \downarrow_i, \text{down}_{q_i})$$

(5) Stop moving down:
if $\delta_{\langle\sigma,0,1\rangle}(q_1,\ldots,q_k) = q$, then

$$(\text{down}_q, \text{test}(x), q_{\text{fin}}).$$

Transition types (1-5) deal with the case that u and v are independent nodes, i.e., one is not a descendant of the other. M walks from u up to some node z, turns around, and walks from z down to v. The fact that, when turning around, M moves to a different child of z (in (3): $i \neq j$) guarantees that $z = \text{lca}(u,v)$. In the next transition types we additionally deal with the cases that v is a (proper) descendant of u, that v is a (proper) ancestor of u, and that $v = u$, respectively.

(6) Start moving down:
if $\delta_{\langle\sigma,1,0\rangle}(q_1,\ldots,q_k) = q$, then, for all $i \in [1,k]$,

$$(q_{\text{in}}, (\text{test}_i(x) \wedge \text{succ}_q(x)) \cdot \downarrow_i, \text{down}_{q_i})$$

(7) Stop moving up:
if $\delta_{\langle\sigma,0,1\rangle}(q_1,\ldots,q_k) = q$, then, for all $i \in [1,k]$,

$$(\text{up}_{q_i}, \uparrow_i \cdot (\text{test}_i(x) \wedge \text{succ}_q(x)), q_{\text{fin}})$$

(8) Start and stop:
if $\delta_{\langle\sigma,1,1\rangle}(q_1,\ldots,q_k) = q$, then

$$(q_{\text{in}}, \text{test}(x) \wedge \text{succ}_q(x), q_{\text{fin}}).$$

This ends the description of the tree-walking automaton A. □

We have proved that every MSO specification $\phi(x, y)$ of a binary relation on the nodes of a tree can be implemented by a tree-walking automaton A. However, in general the automaton is nondeterministic, i.e., chooses nondeterministically its way in the tree. This is of course unavoidable if, for some tree t, $R_t(\phi)$ is not a function: then there are distinct (u, v_1) and (u, v_2) in $R_t(\phi)$ and, after starting at node u, A has to walk either to v_1 or to v_2, nondeterministically. We now show that every functional MSO specification can be implemented on a deterministic tree-walking automaton.

A tree-node relation R is *functional* if, for every tree t, the relation $\{(u, v) \mid (t, u, v) \in R\}$ is a partial function. We also say that a tree-walking automaton or a walking language is functional if it computes a functional tree-node relation. The tree-walking automaton of Example 1 and the walking language of Example 2 are functional. The walking languages in [KS], which are described by so-called routing expressions (similar to those in Example 2), are all functional.

A tree-walking automaton $A = (Q, \Delta, \delta, I, F)$ over Σ is *deterministic* if (1) I is a singleton, (2) if $(q, d, q') \in \delta$, then $q \notin F$, and (3) for all distinct transitions $(q, d_1, q_1), (q, d_2, q_2) \in \delta$, d_1 and d_2 are two mutually exclusive formulas in $\text{MSOL}_1(\Sigma)$. Here, two formulas $\phi_1(x), \phi_2(x) \in \text{MSOL}_1(\Sigma)$ are *mutually exclusive* if $(t, u) \models \neg(\phi_1(x) \wedge \phi_2(x))$ for every $t \in T_\Sigma$ and $u \in V_t$. Obviously, every deterministic tree-walking automaton is functional.

Theorem 9. *For every functional tree-walking automaton A over Σ there is a deterministic tree-walking automaton A' with $R_{\text{TR}}(A') = R_{\text{TR}}(A)$.*

Proof. Let $A = (Q, \Delta, \delta, I, F)$. We may assume that I is a singleton (if not, add transitions $(q_{\text{in}}, \text{true}, q)$, $q \in I$, to δ where q_{in} is the new initial state and 'true' is any unary formula that always holds, such as $x = x$).

The automaton A' simulates A, but whenever A can choose between the execution of several different transitions, A' executes just one of these transitions, after having tested that its execution can be continued with a successful walk of A on the tree. To see that this can be tested by unary MSO formulas, let, for every $q \in Q$, $\text{succ}_q(x)$ be the MSO formula such that $(t, u) \models \text{succ}_q(x)$ iff A has a successful walk on t that starts in configuration (u, q), i.e., $(u, q) \twoheadrightarrow^*_{t,A} (v, q')$ for some final configuration (v, q'). Note that $\text{succ}_q(x) = \exists y(\phi_q(x, y))$, where $\phi_q(x, y)$ is the MSO formula corresponding to the tree-walking automaton $A = (Q, \Delta, \delta, \{q\}, F)$ according to Theorem 5. Now, for a transition $(p, d, q) \in \delta$, let $\text{succ}_{d,q}(x)$ be the MSO formula defined as follows: if $d = \downarrow_\gamma$ then $\text{succ}_{d,q}(x) = \exists y(\text{edg}_\gamma(x, y) \wedge \text{succ}_q(y))$, if $d = \uparrow_\gamma$ then $\text{succ}_{d,q}(x) = \exists y(\text{edg}_\gamma(y, x) \wedge \text{succ}_q(y))$, and if $d = \psi(x)$ then $\text{succ}_{d,q}(x) = \psi(x) \wedge \text{succ}_q(x)$. Clearly, the formula $\text{succ}_{d,q}(x)$ expresses that the execution of transition (p, d, q) can be continued with a successful walk.

Let $(p, d_1, q_1), \ldots, (p, d_n, q_n)$ be all transitions in δ that start with a state $p \notin F$. Corresponding to these the automaton A' has the following transitions, where $p_1, \ldots, p_n$ are new states:

$(p, \text{succ}_{d_1,q_1}(x), p_1)$, $(p, \neg\,\text{succ}_{d_1,q_1}(x) \wedge \text{succ}_{d_2,q_2}(x), p_2)$, $\ldots$,
$(p, \neg\,\text{succ}_{d_1,q_1}(x) \wedge \cdots \wedge \neg\,\text{succ}_{d_{n-1},q_{n-1}}(x) \wedge \text{succ}_{d_n,q_n}(x), p_n)$,

and $(p_1, d_1, q_1), \ldots, (p_n, d_n, q_n)$.
The automaton A' has no other transitions; it has the same initial state and final states as A. Obviously, A' is deterministic and computes the same tree-node relation as A. □

Since every deterministic tree-walking automaton is functional, Theorems 8 and 9 together show that the class of tree-node relations that can be computed by deterministic tree-walking automata is exactly the class of functional MSO definable tree-node relations.

We end this paper by mentioning that the MSO tests of the tree-walking automaton are necessary: the usual tree-walking automaton from the litterature (see, e.g., [AU, ERS]) cannot compute all MSO definable tree-node relations: it is shown in [Blo] that the functional tree-node relation of Example 1 cannot be computed by such a tree-walking automaton.

References

[ALS] S. Arnborg, J. Lagergren, D. Seese; Easy problems for tree-decomposable graphs, J. of Algorithms 12 (1991), 308–340

[AU] A. V. Aho, J. D. Ullman; Translations on a context-free grammar, Inf. and Control 19 (1971), 439–475

[Blo] R. Bloem; *Attribute Grammars and Monadic Second Order Logic*, Master's Thesis, Leiden University, June 1996

[Büc] J. Büchi; Weak second-order arithmetic and finite automata, Z. Math. Logik Grundlag. Math. 6 (1960), 66–92

[CE] B. Courcelle, J. Engelfriet; A logical characterization of the sets of hypergraphs defined by hyperedge replacement grammars, Math. Systems Theory 28 (1995), 515–552

[Cou1] B. Courcelle; Graph rewriting: an algebraic and logic approach, in *Handbook of Theoretical Computer Science*, Vol.B (J.van Leeuwen, ed.), Elsevier, 1990, 193–242

[Cou2] B. Courcelle; The monadic second-order logic of graphs V: On closing the gap between definability and recognizability, Theor. Comput. Sci. 80 (1991), 153–202

[Cou3] B. Courcelle; Monadic second-order definable graph transductions: a survey, Theor. Comput. Sci. 126 (1994), 53–75

[Cou4] B. Courcelle; The expression of graph properties and graph transformations in monadic second-order logic, Chapter 5 of the *Handbook of Graph Grammars and Computing by Graph Transformation*, Volume 1: *Foundations* (G.Rozenberg, ed.), World Scientific, 1997

[Don] J. Doner; Tree acceptors and some of their applications, J. of Comp. Syst. Sci. 4 (1970), 406–451

[Elg] C. C. Elgot; Decision problems of finite automata and related arithmetics, Trans. Amer. Math. Soc. 98 (1961), 21–51

[Engl] J. Engelfriet; *Simple Program Schemes and Formal Languages*, Lecture Notes in Computer Science 20, Springer-Verlag, Berlin, 1974

[Eng2] J. Engelfriet, A characterization of context-free NCE graph languages by monadic second-order logic on trees, in *Graph Grammars and their Application to Computer Science* (H.Ehrig, H.-J.Kreowski, G.Rozenberg, eds.), Lecture Notes in Computer Science 532, Springer-Verlag, Berlin, 1991, 311–327

[Eng3] J. Engelfriet; A regular characterization of graph languages definable in monadic second-order logic, Theor. Comput. Sci. 88 (1991), 139–150.

[Eng4] J. Engelfriet; Context-free graph grammars, Chapter 3 of the *Handbook of Formal Languages*, Volume 3: *Beyond Words* (G.Rozenberg, A.Salomaa, eds.), Springer-Verlag, 1997

[EO1] J. Engelfriet, V. van Oostrom; Regular description of context-free graph languages, J. of Comp. Syst. Sci. 53 (1996), 556–574

[EO2] J. Engelfriet, V. van Oostrom; Logical description of context-free graph languages, Tech. Report 96–22, Leiden University, August 1996

[ERS] J. Engelfriet, G. Rozenberg, G. Slutzki; Tree transducers, L systems, and two-way machines, J. of Comp. Syst. Sci. 20 (1980), 150–202

[GS] F. Gécseg, M. Steinby; *Tree automata*, Akadémiai Kiadó, Budapest, 1984

[HU] J . E.Hopcroft, J. D. Ullman; *Introduction to Automata Theory, Languages, and Computation*, Addison-Wesley, Reading, Mass., 1979

[KS] N. Klarlund, M. L. Schwartzbach; Graph Types, Proc. of the 20th Conference on Principles of Programming Languages, 1993, 196–205

[Tho] W. Thomas; Automata on infinite objects, in *Handbook of Theoretical Computer Science*, Vol.B (J.van Leeuwen, ed.), Elsevier, 1990, 133–192

[TW] J. W. Thatcher, J. B. Wright; Generalized finite automata theory with an application to a decision problem of second-order logic, Math. Systems Theory 2 (1968), 57–81

Approximating the Volume of General Pfaffian Bodies

Marek Karpinski[1]* and Angus Macintyre[2]**

[1] Dept. of Computer Science, University of Bonn, 53117 Bonn and International Computer Science Institute, Berkeley.
Email: marek@theory.cs.uni-bonn.de

[2] Mathematical Institute, University of Oxford, Oxford OX1 3LB.
Email: ajm@maths.ox.ac.uk

Abstract. We introduce a new method of approximating volume (and integrals) for a vast number of geometric bodies defined by boolean combinations of Pfaffian conditions. The method depends on the VC Dimension of the underlying classes of bodies. The resulting approximation algorithms are quite different in spirit from previously known methods, and give randomized solutions even for such seemingly intractable problems of statistical physics as computing the volume of sets defined by the systems of exponential (or more generally Pfaffian) inequalities.

1 Introduction

In [KM95] we gave a powerful method for estimating VC dimension of classes defined by Boolean combinations of C^∞ (infinitely differentiable) conditions $f(\overline{v}) > 0$. In this paper we exploit these results to give randomized polynomial time algorithms for (distribution independent) *absolute approximation* of various volumes and integrals defined "in the Pfaffian category". These algorithms are quite different in spirit from the well-known ones for convex bodies cf., e.g., [DF88], [DFK91] and in many concrete situation very efficient.

In addition we consider the issue of formulas for ε-approximation of volume of definable sets (and integrals of definable functions), and we do this not only in *real* analysis but also in p-adic analysis. Due to our current ignorance of good bounds for VC dimension in p-adic algebra or analysis, we cannot yet give efficient randomized algorithms in that setting.

The first of our results were obtained in October-November 1994 just prior to the Dagstuhl meeting on Neural Computing. In the real case they overlap with these of Pascal Koiran [K95], with whom we had at Dagstuhl some technical discussions on approximate definition of volume via VC dimension.

* Research partially supported by the DFG Grant KA 673/4-1, and by ESPRIT BR Grants 7097 and EC-US 030.

** Research supported in part by a Senior Research Fellowship of the SERC.

2 Measure and VC Dimension

2.1 We assume familiarity with the basics about VC dimension [AB92], [L92], [GJ93], with our paper [KM95b], and the basics on approximating the volume of [DF88], [DFK91]. We refer to [H50] for the most basic notions of the measure theory, and to [H76] for the basic notions from differential topology.

For a given class $\mathcal{C}$ of subsets of a universe X, we say that a set $S \subseteq X$ is *shattered* by $\mathcal{C}$ if $\{S \bigcap C : C \in \mathcal{C} = \mathcal{P}(\mathcal{S})$. The VC dimension of $\mathcal{C}$ is the maximal size of any set S that can be *shattered* by $\mathcal{C}$, or ∞ if arbitrary large subsets may be shattered.

Our usual setting will involve a set X and a class $\mathcal{C}$ of subsets of X, of finite VC dimension d (VC-Dim($\mathcal{C}$) $= d$). Model theory provides a wide selection of such X and $\mathcal{C}$. If T is a theory without the *independence* property (cf. [L92]), and $\mathcal{M}$ is any model of T, we get X and $\mathcal{C}$ as follows:
$X = \mathcal{M}^k$, some k;
$\mathcal{C}$ is the class of all subsets of $\mathcal{M}^k$ defined by $\phi(v_1, \ldots, v_k, \beta_1, \ldots, \beta_\ell)$, for a fixed formula ϕ, and $\tilde{\beta}$ ranging over $\mathcal{M}^\ell$.

For this see [L92], where (*astronomical*) bounds on VC-Dim($\mathcal{C}$) are established.

(We note that one can generalize significantly. X could be a set *interpretable* in $\mathcal{M}$. In analytic situations, this would give us access to manifolds, and not just affine spaces).

There are two notable classes of examples:

1. T is o-minimal [L92]. An important example, with enormous expressive power, is the theory of the real exponential field, with primitives for all restricted analytic functions [DMM94];
2. T is the theory of p-adic field with primitives for all restricted analytic functions [DD88]. That this does not have the independence property follows from [D84], [DD88], [L92], and [DHM97].

A conspicuous difference between (1) and (2) is that in (1) one has good bounds for VC-Dim($\mathcal{C}$) in many cases [KM95b], whereas in (2) one has no good bounds so far.

Recently Wilkie [W96] proved the long sought result that the structure of real Pfaffian functions is o-minimal. Earlier we have given very sharp results in the case of sets, defined by quantifier-free Pfaffian conditions [KM95b].

2.2 We have stressed the real and p-adic cases, because $\mathbb{R}$ and $\mathbb{Q}_p$, being locally compact groups, carry Haar measures (cf. [H50]). For convenience, fix, in each case, such a measure μ giving measure 1 to the unit ball.

Lemma 1. *Suppose that $\mathcal{M}$ is either*

i) $\mathbb{R}$ with some o-minimal structure, or
ii) $\mathbb{Q}_p$ with the structure of restricted analytic functions.

Then every definable subset of $\mathcal{M}^k$ is μ-measurable.

Proof. By cell-decomposition [KPS86], [DD88] every definable subset is Borel.

2.3 The basis for the randomized approximation algorithms below is the fundamental [V82], [BEHW90]:

Theorem 2. *Suppose C is a class of subsets of X, of VC dimension d. Let P be a probability distribution on X, and P^n the product distribution on X^n. Suppose P satisfies the condition (#) (to be given below). Then if $n \geq max(\frac{4}{\varepsilon}\log\frac{2}{\delta}, \frac{8d}{\varepsilon}\log\frac{13}{\varepsilon})$*

$$P^n(\{(x_1,\ldots,x_n) : sup_{C\in\mathcal{C}}\left|\frac{1}{n}\sum_i \chi_c(x_i) - P(C)\right| < \varepsilon\}) > 1-\delta$$

(Here χ_c is the characteristic function of C, and (#) ensures that the set displayed above is measurable).

The condition (#) is just that for each n, the functions

$sup_{C\in\mathcal{C}}\left|\frac{1}{n}\sum_i \chi_c(x_i) - P(C)\right|$, and

$sup_{C\in\mathcal{C}}\left|\frac{1}{n}\sum_i \chi_c(x_i) - \frac{1}{n}\sum_i \chi_c(y_i)\right|$

are measurable.

Lemma 3. *(#) holds for o-minimal $\mathcal{M}$, and for the p-adic subanalytic case.*

Proof. [KM97].

A notable variant of Theorem 2, relating to so-called *ε-nets*, is:

Theorem 4. *(Assumptions as in Theorem 2) If $n \geq \frac{8d}{\varepsilon}\log\frac{13}{\varepsilon}$ there is a subset $\{x_1,\ldots,x_n\}$ of X such that for all $C \in \mathcal{C}$ with $P(C) \geq \varepsilon$, $C \cap \{x_1,\ldots,x_n\} \neq \emptyset$.*

3 A General Approximation Formula

3.1 Koiran [K95] works relative to a first-order structure on the $\mathbb{R}$ including $+,-,<$ as primitives. Let L be the language involved.

[K95] considers a measurable $E \subseteq \mathbb{R}^p$, a formula $\varphi(v)$ of L, and $\varepsilon > 0$ in $\mathbb{R}$.

Definition 1. *φ defines an ε-approximate volume for E if*

a) $\mathbb{R} \models \varphi(r) \longrightarrow |r - \mu^p(E)| < \varepsilon$
b) $|r - \mu^p(E)| < \frac{\varepsilon}{4} \longrightarrow \mathbb{R} \models \varphi(r)$

He points out the well-known fact that even for semi-algebraic E one cannot demand
$\mathbb{R} \models \varphi(r)| \longleftrightarrow |r - \mu^p(E)| < \varepsilon \qquad (*)$
if φ is semi-algebraic, because π is undefinable in the semi-algebraic category.

One can give more elaborate examples, for richer structure on $\mathbb{R}$ (using, e.g. the undefinability results in [DMM94]). There is however the intriguing possibility that for some huge o-minimal structure on $\mathbb{R}$ $(*)$ is possible for definable E. This may have something to do with the error function.

[K95], by two methods, proves the results of the following shape:

If E is definable, there is a formula $\varphi(v)$ which defines an ε-approximate volume for E.

We will elaborate these methods, and compare them. The second method [K95], Theorem 4 was known to us, and is related to more general model theoretic considerations. The first is very elementary.

3.2 First Method. It appears that [K95] here does not wish to appeal to o-minimality. He defines $\kappa(E)$ as the maximum number of connected components of $E \cap L$, for any axis parallel line L. It is however worth observing that you cannot have all $\kappa(E)$ finite, for definable E, without Γ being o-minimal. On the other hand, if E is semi-Pfaffian, $\kappa(E)$ is finite and can be usefully estimated [K91].

In both cases one has a uniformity for $\kappa(E)$.

Lemma 5. *Suppose either*

i) *$\mathcal{M}$ is o-minimal, and $\Phi(v_1, \cdots, v_p, w_1, \cdots, w_l)$ is arbitrary, or*

ii) *Φ is defined by a Boolean combination of Pfaffian primitives.*

For $\tilde{\beta} \in \mathcal{M}^l$, let $E_{\tilde{\beta}}$ be defined by $\Phi(\tilde{v}, \tilde{\beta})$. Then there is an integer κ_Φ depending only on Φ such that for any line L parallel to an axis in $\mathbb{R}^p$, and any $\tilde{\beta}$.

$E_{\tilde{\beta}} \cap L$ has $\leq \kappa_\Phi$ connected components.

Proof. For (a), see [KPS86], for (b) [K91].

Now let $I = [0,1]$, and assume $E \subset I^P$, E μ-measurable, and $\kappa(E)$ finite. Let N be an integer, and put on I^P the grid of size $h = \frac{1}{N}$. Let μ_h be the probability measure on I^P uniformly distributed on the N^P vertices of the grid. By a simple Fubini argument [K95] proves the following, (generalizing the case of convex E when $\kappa(E) = 1$.

$| \mu_h(E) - \mu(E) | \leq ph\kappa(E).$

Proof. See [K95].

We now quickly consider algorithmic aspects of this. To calculate $\mu_h(E)$ one must sample N^P points.

Fix $\epsilon > 0$, and let N be chosen so $\frac{pk}{N} < \frac{\epsilon}{4}$. Then $| \mu_h(E) - \mu(E) | < \frac{\epsilon}{4}$. Suppose $| \mu_h(E) - r | < \frac{\epsilon}{2}$. Then $| r - \mu(E) | < \epsilon$

Conversely, suppose $| r - \mu(E) | < \frac{\epsilon}{4}$. Then $| r - \mu_h(E) | < \frac{\epsilon}{4} + \frac{\epsilon}{4} = \frac{\epsilon}{2}$. So $| r - \mu_h(E) | < \frac{\epsilon}{2}$.
ϵ - approximates $\mu(E)$, and is clearly first-order.

Moreover, in terms of conventional logical classifications, when E is defined by $\Phi(\tilde{v}, \tilde{\beta})$ as done earlier, we note that $| r - \mu_h(E) | < \frac{\epsilon}{2}$ is both Σ_m and Π_m in the $\tilde{\beta}$, of Φ is both Σ_m and Π_m in $\tilde{v}, \tilde{\beta}$.

But, the formula is very long. A routine calculation shows that its length is of order $2^{N^P} \cdot N^P \cdot | \Phi |$, and so is exponential in $\frac{\kappa}{\epsilon}$.
We summarize:

Theorem 7. *(Notation as above).*

(i) *If Φ is both Σ_m and Π_m, then given ϵ there is a formula $\varphi(v, w_1, \ldots, w_e)$, also both Σ_m and Π_m, such, for any $\tilde{\beta}$, $\varphi(v, \tilde{\beta})$ defines an ϵ–approximate volume for $\Phi_{\tilde{\beta}}$.*
(ii) *φ has length of order $2^{N^P} \cdot N^P \cdot | \Phi |$, where N is $\frac{4p\kappa}{\epsilon}$.*

Remarks.

a) (i) is clearly of theoretical interest.
b) That the length bound in (ii) is exponental in $\frac{1}{\epsilon}$ makes it useless for the algorithmic purpose of approximating the volume of $E_{\tilde{\beta}}$, at least if one adopts any conventional technique of looking for an r – satisfying $| r - \mu^P(E_{\tilde{\beta}}) | < \epsilon$. Moreover, the size of N makes the method useless even as a randomized algorithm.
c) In the literature one is interested mainly in algorithms polynomial in p, the dimension of the ambient space [DF88, DFK91].
d) In many interesting cases K may be reasonably small. This is certainly so in the semi–algebraic case, if Φ is given as quantifier – free. Suppose $\Phi(\tilde{v}, \tilde{w})$ is a Boolean combination of conditions $f_i(\tilde{v}, \tilde{w}) > 0$,$i \leq s$. Let d_{ij} be the v_j-degree of f_i. Then it is easily seen that

$$max_j 2(\textstyle\sum_{i \leq s} d_{ij}) + 1$$

If Φ is quantifier – free in the language of exponentation, Hardy's method [DF88, DFK91] can be applied to give bounds which are not too satisfactory. In the quantifier – free Pfaffian case, Khovanski's method applies, as in [KM95].

3.3 The Translation Method. We maintain the preceding notation.

The translation method is more subtle, and very congenial to a model theorist who defines generic types in terms of group translation [P83]. The idea is of course to derandomize the sampling algorithm by relating *generic* and *of large measure*.

As before, the procedure is uniform in $\tilde{\beta}$, so we write E for $E_{\tilde{\beta}}$.

We can interpret $(\mathbb{R}/\mathbb{Z})^p$ in the L-Structure, using representations in $[0,1[$. Let $\oplus, \ominus$ be the interpretations of the group operation, and the inverse. On $([0,1[)^p$ these operations are quantifier-free definable, using +, -, <.

Following [K95], we define, for $k \in \mathbb{N}$, and $\nu, \alpha \in I$

$S_{\nu,\alpha} = \{(\chi_1, \dots \chi_k) \in (I^p)^k : \mid \frac{1}{k}\sum_{i=1}^k \chi_E(\chi_i) - \nu \mid \leq \alpha\}$

(There is a hidden uniform dependence on $\tilde{\beta}$. If E is Σ_n und Π_n, so is $S_{\nu,\alpha}$.)

Let Θ be the family of all $\chi \ominus S_{\nu,\alpha}$, $\chi \in (I^p)^k$. This is a family of sets in $(I^p)^k$, indexed by ν, α, β, χ.

Note that the length of (the natural definition of) $S_{\nu,\alpha}$ is of order $k \cdot 2^k \cdot \mid \Phi \mid$, as is that of $\chi \ominus S_{\nu,\alpha}$.

Finally, fix an m, and consider the natural formula C_m expressing that a union of m translates of $S_{\nu,\alpha}$ covers $(I^p)^k$. This formula is Σ_{n+2}, if E is both Σ_n and Π_n. Moreover, the first block of $\exists$'s has length m, and the first block of $\forall$'s has length pk. Inside this there is a disjunction of m formulas $S_{\nu,\alpha} \oplus t_i$, essentially of length that of $S_{\nu,\alpha}$.

Theorem 4 of [K95] is (essentially):

Theorem 8. *Let $d \geq$ VC–dim Θ. Let $m \geq (16 \log 26)d$, and $k \geq \frac{max(12 ln\, m, 48 ln\, 2)}{\epsilon^2}$. Then $C_m(\nu, \frac{\epsilon}{2})$ defines an ϵ–approximate volume for $E_{\tilde{\beta}}$.*

Remarks.

(a) The most obvious issue is what is gained by using C_m, logically more complex than the formula of Theorem 7. The length of C_m is essentially a constant times $m + pk + k \cdot 2^k \cdot \mid \Phi \mid$, so, with minimal choice of m, k, a constant times $d + \frac{p \log d}{\epsilon^2} + \frac{p \log d}{\epsilon^2} \cdot d^{\frac{12}{\epsilon^2}} \cdot \mid \Phi \mid$.

(b) d can in some cases be estimated. [K95] shows that $d \leq c \cdot d' \cdot k \log k$, where C is an absolute constant and d' is the VC dimension of the family $\chi \ominus E_{\tilde{\beta}}$. Note that this estimate is valid *before* we impose the constraints in Theorem 8.

d' can be nicely estimated in some cases where Φ is quantifier–free, and in the semi–algebraic case quite generally, using refined results on quantifier elemination [GJ93]. Let us briefly recall the most notable result of [KM95b]: Suppose $\Phi(\tilde{\nu}, \tilde{w})$ is a Boolean combination of s many formulas $f_i(\tilde{\nu}, \tilde{w}) > 0$, $1 \leq i \leq s$, where each f_i is a polynomial of degree $\leq \Delta$ in $\tilde{\nu}, \tilde{w}$ and q many (q independent of i) $h(\tilde{\nu}, \tilde{w})$ which occur in a Pfaffian chain of length q and degree D. Then the VC dimension of $E_{\tilde{\beta}}$ is bounded by

$$\begin{gathered} 2(ql)(ql-1) + 2l \, log \, \Delta + 2l \, log \, (l\Delta + lD + 1) \\ +2ql \, log \, l + 2ql \, log \, (l\Delta + lD + 1) \\ +l(16 + 2 \, log \, s). \end{gathered} \tag{**}$$

Note that this does not depend on the length of $\tilde{\nu}$!

To deal with the VC dimension d of family $\chi \ominus E_{\tilde{\beta}}$ some minor adjustments are needed, mainly to take account of the interpretation of $\mathbb{R}/\mathbb{Z}$ via [0,1]. We omit constantly the details of the argument, but the only change needed in (**) is to replace s by $s + 4p$.

(c) The main point to notice vis-a-vis $| C_m |$ is that there is a term exponential in $\frac{1}{\epsilon^2}$, but in the Pfaffian case there is no exponential dependence on p. So there is a slight gain over the earlier formula. *But*, if one tries, in the semi-algebraic case, to eliminate quantifiers à la Renegar [R92], the cost is great, because the block of $\forall$'s has length kp, leading to a time cost exponential in p and $\frac{1}{\epsilon^2}$.

(d) Though it seems to be of no practical importance, it is interesting to note that for fixed ϵ, and $| \nu - \mu(E_{\tilde{\beta}}) | < \frac{\epsilon}{4}$ the block of m $\exists$'s can always be instantiated by a tuple *independent of ν and $\tilde{\beta}$*. (This follows easily by the ϵ-net argument.) The (unsolved) problem is to find this tuple.

3.4 p–adic version Delon [D89] proved that the field $\mathbb{Q}_p$ does not have the *independence property*, and from this, using [DD88], [L92], and [DHM97], one easily deduces that $\mathbb{Q}_p$ with the subanalytic structure of van den Dries–Denef [DD88] also does not have the independence property. This of course gives finiteness of VC dimension for families of sets interpretable in $\mathbb{Q}_p$ with subanalytic structure. But, till now, even for the pure field $\mathbb{Q}_p$, one does not have *good* bounds. This is, to some extent, connected with the subtlety of notions like ***(definably) connected component.***

Let V be the unit ball of $\mathbb{Q}_p$ and μ the normalized Haar measure with $\mu(V) = 1$. All p–adic subanalytic sets are μ–measurable, by quantifier-elimination and in [KM97] we establish the condition (#).So the problem of closely approximating the measure of a definable set is natural. But some modifications must be made to compared to [K95], since the values of the measure are not elements of the model. That is, you cannot just substitute real r for ν in $| \nu - \mu(E) | < \epsilon$ and make sense inside $\mathbb{Q}_p$.

We propose the following modification. We follow the notations of [K95], except that we replace I by V. V is now an additive group, so we avoid the details of interpretability. We work in a power $((\mathbb{Q}_p)^t)^k$ as before (replacing p by t for obvious reasons!). The relative frequency $\frac{1}{k} \cdot \sum_{i=1}^{k} \chi_E(\chi_i)$ is one of the rationals $\frac{j}{k}$ $0 \le j \le k$. The least distance between these numbers is $\frac{1}{k}$, and,if $E \subseteq V$, $\mu(E)$ is no more than $\frac{1}{k}$ from some $\frac{j}{k}$. If $\frac{1}{k} < \frac{\epsilon}{4}$, then $\mu(E)$ is within $\frac{\epsilon}{4}$ of some ***unique*** $\frac{j}{k}$.

Now we can just follow notation of [K95] prior to his Theorem 4. Observe that his argument there works even better in the p–adic case. It just does not give a formula of the intended language. All we can do is to define (in the language), for each $j \le k$, the set $\tilde{S}_j$ of $\bar{x}$ in $((\mathbb{Q}_p)^r)^k$ whose relative frequency is $\frac{j}{k}$. Choose the unique j such that $| \frac{j}{k} - \mu(E) | < \frac{\epsilon}{2}$. So, as in [K95] $\mu(S_{\frac{j}{k},\frac{\epsilon}{2}}) > 1 - e^{-\frac{k\epsilon^2}{48}}$, so

$\mu(\tilde{S}_j) > 1 - e^{-\frac{k\epsilon^2}{48}}$, so by [K95] argument (for m, k as there) m translates of $\tilde{S}_j$ cover $(V^t)^k$. Conversely, as in [K95], if m translates of $\tilde{S}_j$ cover $(V^t)^k$, $\mu(\tilde{S}_j) \geq \frac{1}{m} > e^{-\frac{k\epsilon^2}{12}}$. So $\mu(S_{\frac{j}{k},\frac{\epsilon}{2}}) > e^{-\frac{k\epsilon^2}{12}}$ so $\mid \frac{j}{k} - \mu(E) \mid < \epsilon$ as in [K95].

So we have proved, with the same notation as in Theorem 8.

Theorem 9. *If m translates of $\tilde{S}_j$ cover $(V^t)^k$ then $\mid \frac{j}{k} - \mu(E) \mid < \epsilon$, and if $\mid \frac{j}{k} - \mu(E) \mid < \frac{\epsilon}{4}$ then m translates of $\tilde{S}_j$ cover $(V^t)^k$.*

Remarks.

(a) This has for now even less computational significance than Theorem 7, since we don't know how to calculate the VC dimensions.

(b) We don't know any analogue of Lemma 6, though we hope to find one.

4 Integrals

4.1 In the real case, the preceding methods are readily adapted to "express" certain integrals. The relevant integrals are $\int_{\mathcal{A}} f d\mu^P$, for $\mathcal{A}$ a definable subset of I^p, and f a definable function on $\mathcal{A}$ with $\mid f(\bar{x}) \mid \leq 1$, $\bar{x} \in \mathcal{A}$. By splitting $\mathcal{A}$ into the definable sets where f is negative or nonnegative, we reduce to case where $0 \leq f(\bar{x}) \leq 1$ on $\mathcal{A}$. Then $\int_{\mathcal{A}} f d\mu^p = \mu^{p+1}((\bar{x}, y) : 0 \leq y \leq f(\bar{x}), \bar{x} \in \mathcal{A})$, thereby making available all the results of Section 3.

Note that definable functions are measurable, in o–minimal theories!

4.2 The preceding of course covers the case of lengths of definable curves in o–minimal theories.

4.3 It is however much less clear how to deal with various areas of curved surfaces, if these are not surfaces of revolution.

4.4 **p–adic integrals.** There is a considerable body of work on rationality of p–adic Poincare' series, or equivalently rationality in P^{-s} of $\int_{\mathcal{A}} \mid f(\bar{x}) \mid^s d\mu^t$, where $\mathcal{A}$, f are p–adic definable, and s is a real variable. See particularly [D89], [DD88].

For *rational* s one can use the method of 4.1, combined with the ideas of Section 3, to get "expressibility" in the p–adic analytic language. Here there is an interesting problem of taking account of *variation in s*. We discuss this in [KM97].

5 Randomized Algorithms

5.1 For us the main significance of our results [KM95a,b] is that they lead to powerful randomized algorithms in a wide range of geometric problems with Pfaffian data. There is a noticeable difference between this approach and that of, say,

Dyer–Frieze–Kannan [DFK91] who have as data blackboxes for arbitrary convex bodies, and are concerned with *relative* approximation. One of their desiderata is an algorithm running in time polynomial in p, the dimension of the ambient space, and this establishes a point of comparison with our method, where convexity plays no role.

We restrict to sets definied by quantifier–free conditions in the language with primitives for all total Pfaffian functions.

5.2 For us a typical Θ is defined by a Boolean combination of conditions $f_i(\tilde{\nu}, \tilde{w}) > 0$ as in [KM95a,b], where the f_i are polynomials of degree $\leq \Delta$ in $\tilde{\nu}, \tilde{w}$ and q many Pfaffian functions occuring in a chain of length q and degree $\leq$ D.

Our bound for VC-dim($\mathcal{C}_\Theta$) [KM95], following [K91, p. 91] is

$$q(q-1) + 2l \log \Delta + 2(l+q) \log(\Delta + \mathrm{D}) + 2l \log(l+1) + (16 + 2 \log sl),$$

in which k doesn't occur. Let d be the above estimate. Then if a sample of size $m \geq \max(\frac{4}{\epsilon} \log \frac{2}{\delta}, \frac{8d}{\epsilon} \log \frac{13}{\epsilon})$ is chosen, and its relative frequency, for C in $\mathcal{C}_\Theta$, is tested, this is within ϵ of $\mu(C)$ with probability $\geq 1 - \delta$.

So here one has a *randomized* algorithm working with infinite real precision, taking time mp (with the *blackbox convention*) giving an *absolute* error $< \epsilon$ with probability $\geq 1 - \delta$. If we unpack d, and so m, we see that the time of the algorithm, say for $\delta = \frac{1}{4}$, is (up to absolute constants):

i) linear in p;
ii) $\frac{1}{\epsilon} \log(\frac{1}{\epsilon})$ in ϵ;
iii) quadratic in q;
iv) $l \log l$ in l.

We summarize our results as follows.

Theorem 10. *(Notation as above.) There is a randomized algorithm for absolute ϵ-approximation of the volume of semi-Pfaffian sets under arbitrary probability distribution (see Theorem 2), with running time* $O\big(p \cdot q^2 \frac{l}{\epsilon} \log l \log(\frac{1}{\epsilon})\big)$.

5.3 Analogue for Integrals Since integrals
$$\int_A f \, d\mu^p \quad , \text{ where } 0 \leq f \leq 1$$
are special cases of volumes, there is an obvious analogue of Theorem 10 for integration of Pfaffian f over semi-Pfaffian A.

5.4 Doing better It seems likely that one can use the deep results of [V82, Chapter 7] to get randomized algorithms based on computation of means

$$\frac{1}{n} \sum f(x_i) \chi_A(x_i) \quad ,$$

instead of via 5.3. In the o–minimal case one has of course the finiteness of capacity (cf. [V82]). We have not looked closely at the relative merits of this and the idea of 5.3.

5.5 Surface areas, etc. In contrast to the situation for expressibility, where translation invariance was crucial, the randomized algorithm method works For Borel measures on compact Θ-manifolds where Θ is a geometric category [DM96], and in an analogous Pfaffian situation. In the latter case, fast algorithms will exist, via [KM95].

Acknowledgment

It gives us great pleasure to dedicate to Andrzej Ehrenfeucht a paper involving model theory, VC dimension, and randomized approximation algorithms, the subjects illuminated so much by his special imagination.
We thank also Martin Dyer, Mark Jerrum, Pascal Koiran, Eduardo Sontag, and Andy Yao for many stimulating discussions.

References

[AB92] M. Anthony and N. Biggs, Computational Learning Theory: An Introduction, Cambridge University Press, 1992.

[AS93] M. Anthony and J. Shawe-Taylor, A Result of Vapnik with Applications, Discrete Applied Math. **47** (1993), pp. 207–217.

[BEHW90] A. Blumer, A. Ehrenfeucht, D. Haussler and M. Warmuth, Learnability and the Vapnik-Chervonenkis Dimension, J. ACM **36** (1990), pp. 929–965.

[BT90] A. Borodin, P. Tiwari, On the Decidability of Sparse Univariate Polynomial Interpolation, Proc. 22nd ACM STOC (1990), pp. 535–545.

[D89] F. Delon, Définissabilité avec Paramètres Extérieurs dans $\mathbb{Q}_p$ et $\mathbb{R}$, Proc. AMS **106** (1989), pp. 193–198.

[D84] J. Denef, The Rationality of the Poincaré Series associated to the p-adic Points on a Variety, Invent. Math. 77 (1984), pp. 1–23.

[DD88] J. Denef and L.P.D. van den Dries, p-adic and real subanalytic Sets, Annals of Mathematics 128 (1988), 79–138.

[D92] L. van den Dries, Tame Topology and o-Minimal Structures, preprint, University of Illinois, Urbana, 1992; to appear as a book.

[DHM97] L. van den Dries, D. Haskell, H.D. Macpherson, On Dimensional p-adic Subanalytic Sets, to appear.

[DMM94] L. van den Dries, A. Macintyre and D.Marker, The Elementary Theory of Restricted Analytic Fields with Exponentation, Annals of Mathematics **140** (1994), pp 183-205.

[DM96] L. van den Dries, C. Miller, Geometric Categories and o-Minimal Theories, Duke Journal 84 (1996), 497-540.

[DF88] M. E. Dyer and M. Frieze, On the Complexity of Computing the Volume of a Polyhedron, SIAM J. Comput. **17** (1988), pp. 967–974.

[DFK91] M. Dyer, A Frieze and R. Kannan, A Random Polynomial-Time Algorithm for Approximating the Volume of Convex Bodies, J. ACM **83** (1991), pp. 1–17.

[GJ93] P. Goldberg and M. Jerrum, Bounding the Vapnik Chervonenkis Dimension of Concept Classes Parametrized by Real Numbers. Machine Learning, 1995. A preliminary version appeared in Proc. 6th ACM Workshop on Computational Learning Theory, pp. 361–369, 1993.

[H50] P. Halmos, Measure Theory, Chelsey-New York, 1950.

[H12] G. H. Hardy, Properties of Logarithmic-Exponential Functions, Proc. London Math. Soc. 10 (1912), pp. 54–90.

[H92] D. Haussler, Decision Theoretic Generalizations of the PAC Model for Neural Net and other Learning Applications, Information and Computation **100**, (1992), pp. 78–150.

[HKP91] J. Hertz, A. Krogh and R. G. Palmer, Introduction to the Theory of Neural Computation, Addison-Wesley, 1991.

[H76] M. W. Hirsch, Differential Topology, Springer-Verlag, 1976.

[KM95] M. Karpinski and A. Macintyre, Bounding VC Dimension for Neural Networks: Progress and Prospects (Invited Lecture), Proc. EuroCOLT'95, Lecture Notes in Artificial Intelligence Vol.**904**, Springer-Verlag, 1995, pp. 337-341.

[KM95a] M. Karpinski and A. Macintyre, Polynomial Bounds for VC Dimension of Sigmoidal Neural Networks, Proc. 27th ACM STOC (1995), pp.200-208.

[KM95b] M. Karpinski and A. Macintyre, Polynomial Bounds for VC Dimension of Sigmoidal and General Pfaffian Neural Networks, J. Comput. Syst. Sci. **54** (1997), to appear.

[KM97] M. Karpinski, A. Macintyre, Approximating Volumes and Integrals in o-Minimal and p-Minimal Theories, manuscript in preparation.

[KW93] M. Karpinski and T. Werther, VC Dimension and Uniform Learnability of Sparse Polynomials and Rational Functions, SIAM J. Computing **22** (1993), pp 1276–1285.

[K91] A. G. Khovanski, Fewnomials, American Mathematical Society, Providence, R.I., 1991.

[KPS86] J. Knight, A. Pillay and C. Steinhorn, Definable Sets and Ordered Structures II, Trans. American Mathematical Society **295** (1986), pp.593-605.

[K95] P. Koiran, Approximating the Volume of Definable Sets, Proc. 36th IEEE FOCS (1995), pp.258-265.

[KS95] P. Koiran and E. D. Sontag, Neural Networks with Quadratic VC Dimension to appear in Advances in Neural Information Processing Systems (NIPS '95), 1995.

[L92] M. C. Laskowski, Vapnik-Chervonenkis Classes of Definable Sets, J. London Math. Society **45** (1992), pp 377–384.

[LS90] L. Lovasz and M. Simonovits, The Mixing Rate of Markov Chains, an Isoperimetric Inequality, and Computing the Volume, Proc. 31 IEEE FOCS (1990), pp. 346–355.

[M93a] W. Maass, Perspectives of Current Research about the Complexity of Learning on Neural Nets, in: Theoretical Advances in Neural Computation and Learning, V. P. Roychowdhury, K. Y. Siu, A. Orlitsky (Editors), Kluwer Academic Publishers, 1994, pp. 295–336.

[M93b] W. Maass, Bounds for the Computational Power and Learning Complexity of Analog Neural Nets, Proc. 25th ACM STOC (1993), pp. 335–344.

[MSS91] W. Maass, G. Schnitger and E. D. Sontag, On the Computational Power of Sigmoidal versus Boolean Threshold Circuits, Proc. 32nd IEEE FOCS (1991), pp. 767–776.

[MS93] A. J. Macintyre and E. D. Sontag, Finiteness results for Sigmoidal Neural Networks, Proc. 25th ACM STOC (1993), pp.325–334.

[M64] J. Milnor, On the Betti Numbers of Real Varieties, Proc. of the American Mathematical Society **15** (1964), pp 275–280.

[M65] J. Milnor, Topology from the Differentiable Viewpoint, Univ.Press, Virginia, 1965.

[P83] B. Pouzat, Groupes stables, avec types géneriques réguliers, J. S. L. 48 (1983), pp. 339–355.

[R92] J. Renegar, On the Computational Complexity and Geometry of the First-Order Theory of the Reals, Parts I, II, III, J. of Symb. Comput. **3** (1992), pp. 255–352.

[S42] A. Sard, The Measure of the Critical Points of Differentiable Maps, Bull. Amer. Math. Soc. **48** (1942), pp. 883–890.

[S-T94] J. Shawe-Taylor, Sample Sizes for Sigmoidal Neural Networks, Preprint, University of London, 1994, to appear in Proc. ACM COLT, 1995.

[S92] E. D. Sontag, Feedforward Nets for Interpolation and Classification, J. Comp. Syst. Sci. **45** (1992), pp. 20–48.

[V82] V. Vapnik, Estimation of Dependencies Based on Empirical Data, Springer Series in Statistics, 1982.

[W68] H. E. Warren, Lower Bounds for Approximation by Non-linear Manifolds, Trans. of the AMS **133** (1968), pp. 167–178.

[W94] A. J. Wilkie, Model Completeness Results of Restricted Pfaffian Functions and the Exponential Function; to appear in Journal of the AMS, 1994.

[W96] A.J. Wilkie, Handwritten manuscript, Oxford 1996.

Complement-Equivalence Classes on Graphs

Ross McConnell

Department of Computer Science,
Willamette University,
Salem, OR 97301 USA

Abstract. A *complementation operation* on a vertex of a digraph changes all outgoing edges into nonedges, and outgoing nonedges into edges. A similar operation is defined for incoming edges. This defines an equivalence relation where two graphs are equivalent if one can be obtained from the other by a sequence of such operations. We show that given an adjacency-list representation of a digraph G, many fundamental graph algorithms can be carried out on any member of G's equivalence class in $O(size(G))$ time. We use this fact to obtain a simple $O(n + m \log n)$ algorithm for modular decomposition of undirected graphs.

1 Introduction

Ehrenfeucht and Rozenberg have pointed out that graphs are not the most appropriate abstraction on which to examine some problems that had previously been thought of as graph problems, such as decomposition of a graph into modules or total orders [8, 9, 6]. In particular, the modular decomposition of a graph G is the same as the modular decomposition of $\overline{G}$. Thus, from the point of view of modular decomposition, the distinction between edges and non-edges is an artificial one, and G and $\overline{G}$ are best viewed as a single structure. This structure is simply a partition of $E_2(V) = V \times V - \{(v, v)|v \in V\}$ into two sets, where the distinction of these classes as *edges* and *nonedges* is best dropped.

The two-structure point of view played a role in the development of a linear-time algorithms for decomposing two-dimensional partial orders and permutation graphs into two total orders, and recognizing whether a graph is the complement of an interval graph [12]. The key to these bounds was the development of data structures that allow one to ignore the differences between a graph and its complement. This allows computation of many properties of the complement in linear time, instead of the $\Omega(n^2)$ *best-case* time required to construct an adjacency-list representation of the complement explicitly. The problems were solved in linear time by combining properties of G with properties of $\overline{G}$ that were obtained in this way.

In [7], the same authors examine a transformation that operations on the edges out of a vertex or else on the edges into a vertex. They define an equivalence relation on two-structures, where two two-structures are in the same equivalence class iff one of them can be obtained from another by a sequence of these operations. These structures are called *dynamical two-structures*.

A special case of this is the operation of *complementing* a vertex, which is the operation of replacing all outgoing edges with non-edges, and all outgoing non-edges with edges. The inwardly directed edges can be complemented with an analogous operation. Two graphs are *complement-equivalent* if one can be obtained from the other by a sequence of such operations.

This induces a set of equivalence classes on graphs. Let $size(G)$ denote the number of vertices and edges of G. The main result of the present paper is the following:

Claim 1. *Many fundamental algorithms can be run on any member F of G's complement-equivalence class in time that is linear in the size of G.*

What makes this remarkable is that the size of members of a class can differ greatly. An extreme case occurs when $\{A, B\}$ is a partition of V such that $|A| = |B|$, each member of A has a directed edge to all other vertices of the digraph, and no member of B has a directed edge to any other vertex. The size of this graph is $\Omega(n^2)$. However, outwardly complementing the vertices of A yields the empty graph on n vertices, and the size of this graph is $O(n)$. $O(n)$ and $\Omega(n^2)$ graphs can thus appear in the same equivalence class, so given a way of obtaining a small member G in each of a sequence of classes, one might get a sublinear-time algorithms for other members of the classes.

2 Definitions

We let n denote the number of nodes and m the number of edges, and let $size(G)$ denote $n + m$. Let $V(G)$ and $E(G)$ denote, respectively, the number of vertices and edges of G. If G is a digraph, then its *transpose*, G^T, is the digraph that is obtained by reversing the directions of all edges in G. The complement $\overline{G}$ is obtained by changing each edge into a non-edge and each non-edge into an edge. A *non-edge* of G is an edge of $\overline{G}$.

An undirected graph is *complete* if there is an edge between every pair of distinct vertices. It is *empty* if it has no edges. A digraph is a *linear order* if it has no cycles, and between each pair $\{u, v\}$ if distinct nodes, either (u, v) or (v, u) is a directed edge.

Undirected graphs will be treated as a special case of digraphs, where each undirected edge (u, v) is really two directed edges (u, v) and (v, u). Thus, algorithms that we describe for digraphs will be general also to undirected graphs.

If G is a digraph and $X \subseteq V$, then $G|X$ denotes the *subgraph of G induced by X*, namely, the graph whose vertices are X and whose edges are those edges of G that go from one member of X to another. That is, $G|X$ is the graph $(X, (X \times X) \cap E)$. If $\mathcal{F}$ is a partition of $V(G)$ and X is a union of partition classes in $\mathcal{F}$, then $\mathcal{F}|X$ is the set of partition classes that are contained in X. The *quotient* $G/\mathcal{F}$ denotes the graph whose vertices are the members of $\mathcal{F}$, and where (X, Y) is an edge of $G/\mathcal{F}$ iff there is some edge of G from a member of X to a member of Y.

If G is undirected, the set of vertices that are reachable on a single directed edge from a vertex v is denoted $N_G^+(v)$. $\overline{N_G^+}(v)$ denotes the vertices that are not reachable on a single edge from v, that is, $V - (N_G^+(v) \cup \{v\})$. The set of vertices that can reach v on a single directed edge is denoted $N_G^-(v)$, and the ones that cannot are denoted $\overline{N_G^-}(v)$. When G is undirected, we let $N_G(v) = N_G^+(v) = N_G^-(v)$.

3 Modular decomposition

A *module* in an undirected graph $G = (V, E)$ is a set M of vertices such that for each vertex $v \in V - M$, M is either contained in v's neighborhood or disjoint from v's neighborhood. Trivial examples of modules are V, the singleton sets $\{\{v\} | v \in V\}$, and the empty set. A union of connected components of G or of $\overline{G}$ is also an example of a module. However, G may have nontrivial modules even when G and $\overline{G}$ are nonempty.

It is easy to see that if M_1 and M_2 are disjoint modules, then either every member of $M_1 \times M_2$ is an edge or none is. Thus, two disjoint modules are either *adjacent* or *nonadjacent*. It follows that if there is a partition $\mathcal{P}$ of V into disjoint modules, the quotient $G/\mathcal{P}$ uniquely specifies all edges that connect members of different partition classes. We will call such a quotient *modular*. The subgraph induced in G by one of these modules is called a *factor*. Since G can be reconstructed uniquely from the factors and the quotient, this gives a decomposition of G.

If X is a module of G, then $S \subseteq X$ is a set of vertices in the factor $G|X$, and *inverse image* of a S in G is S. If $\mathcal{F}$ is a partition of G where each partition class is a module, then $\mathcal{F}' \subseteq \mathcal{F}$ is a set of vertices of $G/\mathcal{F}$, and the inverse image of $\mathcal{F}'$ in G is $\bigcup \mathcal{F}'$. The following says that modules are preserved under inverse images:

Theorem 2. *If X is a module of G, then $S \subseteq X$ is a module of $G|X$ iff its inverse image in G is a module of G. If $\mathcal{F}$ is a partition of $V(G)$ into a set of modules, then $\mathcal{F}'$ is a module of $G/\mathcal{F}$ iff its inverse image is a module of G.*

The factors and quotient may themselves contain modules, so this decomposition may be applied recursively in each of them. By Theorem 2, this discovers additional modules of G. Among all ways of performing this recursive decomposition, there is a unique canonical one that represents all others, called the *modular decomposition*. This decomposition has been rediscovered in various contexts, and is also known as the *substitution decomposition* [13], or the *prime tree family* [8, 9]. From this tree-like hierarchy of quotients, it is possible to reconstruct G and enumerate all of its modules. We give the following definition of the structure in the special case of undirected graphs:

```
MD(G)
  If G has only one vertex return v
  else
      create a tree node r
      if G has more than one connected component then
          let 𝓕 be the connected components
          let r be labeled a parallel degenerate node
      else if Ḡ has more than one connected component then
          let 𝓕 be the connected components of Ḡ
          let r be labeled a series degenerate node
      else
          Let 𝓕 be the maximal modules of G
          Let r be labeled a prime node
      for each X ∈ 𝓕
          Let r_x = MD(G|x)
          add r_x as a child of r
      return r
```

We may think of r as synonymous with $V(G)$; applying this idea recursively, we see that the nodes of the tree can be thought of as a tree-like hierarchy of subsets of $V(G)$, where each node is just the subset of $V(G)$ that make up its leaf descendants. It is not hard to see that any union of the children of a degenerate node in the returned tree is a module. The following is somewhat less immediate:

Theorem 3. *[9, 13] A subset of $V(G)$ is a module of G iff it is either a node of the decomposition tree, or a union of children of a degenerate node.*

(In the case of digraphs, a third type of *linear* tree is node is also possible, where there is a linear ordering on children of the node, and a set may be a module if it is a union of consecutive children in that ordering [9, 13].)

The *two-edges* of a set V, denoted $E_2(V)$, is the set $V \times V - \{(v,v)|v \in V\}$. A *two-structure* on V is a coloring, or partition, of the the two-edges of V. A digraph $G = (V, E)$ is a coloring of $E_2(V)$ with colors 0 and 1, but unlike a digraph, a two-structure gives no special status to either of these classes.

4 Outward, inward, and two-way equivalence classes on digraphs

If G is a digraph, we *outwardly complement* v if we replace $N_G^+(v)$ with $V - N_G^+(v) - \{v\}$, but $N_G^+(w)$ unchanged for each $w \neq v$. That is, we complement the outgoing edges without changing any of the incoming edges. We *inwardly complement* v if we change $V - N_G^-(v) - \{v\}$ and leave $N_G^-(w)$ the same for all other vertices.

Let us say that two digraphs are ***outward-equivalent*** if one can be obtained from the other by a sequence of outward vertex complementation operations. This is clearly an equivalence relation, so this partitions the set of undirected graphs into equivalence classes, which we may call ***outward equivalence classes.*** Given an undirected graph $G = (V, E)$, where $V = \{v_1, v_2, ..., v_n\}$, we may represent any member of F of G's class in $O(size(G))$ space by the pair $(G, \overline{b})$, where $\overline{b}$ is a boolean vector that has a 1 in position i if v_i must be outward complemented to get F. Similarly, two digraphs are ***inward-equivalent*** if one is obtained from the other by a sequence of inward complementation operations. This defines the ***inward classes*** on digraphs.

Two digraphs are ***two-way*** equivalent if one is obtained from the other by a mixed sequence of outward and inward complementations, and this defines the ***two-way classes*** on digraphs. We may represent any member of G's two-way class in $O(size(G))$ space with the triple $(G, \overline{b}_1, \overline{b}_2)$, where the vectors $\overline{b}_1$ and $\overline{b}_2$ have a 1 in position i if vertex v_i is inward and outward complemented, respectively. For any pair (v_i, v_j), let the ***net complement*** $NC(v_i, v_j)$ be obtained by assigning one point if v_i is inwardly complemented, 1 point if it is outwardly complemented, and 1 point if (v_i, v_j) is not an edge of G. Then (v_i, v_j) is an edge in the represented graph iff $NC(v_i, v_j) \bmod 2 = 0$.

We call these representations ***partially complemented representations***, and call G the ***base graph*** of the representation.

Theorem 4. *For any digraph G, $\overline{G}$ is in G's outward class.*

Proof. Immediate, since $\overline{G}$ is obtained from G by either complementing the list of outgoing edges of each vertex, or by complementing the list of incoming edges of each vertex. □

Theorem 5. *If digraph F is in the equivalence class of graph G, then F^T is in the two-way equivalence class of G^T.*

Proof. Let $(G, \overline{b}_1, \overline{b}_2)$ be a partially-complemented representation of F. We claim that $(G^T, \overline{b}_2, \overline{b}_1)$ represents F^T. Let (i, j) be an ordered pair of elements of V. Then (i, j) is an edge of F iff $NC(i, j)$ is odd. Let $NC'(j, i)$ denote the net complementation of (j, i) in $(G^T, \overline{b}_2, \overline{b}_1)$. Clearly, $NC'(j, i)$ is the same as $NC(i, j)$. □

5 Algorithms on members of an outward equivalence class

To introduce some basic data structures tricks, I will now illustrate how to get an $O(size(G))$ algorithms for a member F of G's *outward* equivalence class. I also give an example of an application, which is obtaining an $O(n + m \log n)$ bound on the modular decomposition algorithm of [5]. The ultimate goal, which is explored in the next section, is to obtain these bounds when F is any member of G's full two-way equivalence class, since the two-way classes are examples of dynamic labeled two-structures.

5.1 Breadth-first search

As a starting exercise, let us examine the running time of breadth-first search on a member F of G's outward class. This solves the single-source shortest paths problem when the length of a path is the number of edges on it.

As in the standard algorithm, we divide V into ***undiscovered*** nodes, ***queued*** nodes, and ***processed nodes.*** Initially all nodes but the start node are undiscovered, and the start node is queued. To process a node, we remove one from the front of the queue, and insert any undiscovered neighbors at the back of the queue. Keep the undiscovered nodes in a linked list. When you process an uncomplemented node, proceed as in the standard case. When you process a complemented node v, mark the members of $N_G^+(v) = N_F^+(v)$. Then traverse the list of undiscovered nodes, splicing out any unmarked nodes and inserting them on the queue. Then remove the marks from $N_G^+(v)$. Marking and unmarking nodes can be charged to the edges of G that are responsible for applying the marks. Touching an unmarked node can happen at most n times, since the node is removed from the list of undiscovered nodes whenever this happens. This gives the $O(size(G))$ bound.

5.2 Depth-first search

Let us examine an algorithm discovered by Elias Dahlhaus, Jens Gustedt and me for depth-first search on F, when $F = (G, \bar{b})$.

One way to obtain a depth-first forest on G is to use a stack of vertices. Insert an initial vertex on the stack. While the stack is not empty, pop a vertex. If it is not already marked visited, mark it visited and push its neighbors on the stack. From the sequence of visits, we can create a ***depth-first tree***, where the parent of each vertex is the one that had most recently pushed it at the time it was marked visited. If not all vertices were marked, the operation can be applied recursively on the remainder of the graph, producing a ***depth-first forest***, which is a forest of ordered, rooted trees.

Let $d[u]$ for each vertex u be a numbering that gives the order in which the vertices were marked visited. This order is sometimes called ***discovery times***, and is a preorder numbering of the forest that also records the order in which trees are generated. With a second $O(n)$ traversal of the forest, it is trivial to get a similar postorder numbering, called the ***finishing times.*** The finishing time of vertex u is denoted $f[u]$.

Note that when a vertex is pushed to the stack, lower instances become irrelevant, since the vertex will be marked visited by the time the lower instances are reached. Let us define a ***non-duplicating push*** to a stack to be like a standard push operation, except that whenever a value is pushed, any lower occurrences of the value in the stack are deleted. This guarantees that at most one occurrence of any value sits on the stack at any time, and, when used in the depth-first algorithm, does not affect the forest produced.

If we are given G, and we want a depth-first forest for F, then when we visit a vertex v, we must push $N_G^+(v)$ if v is not outwardly complmented in F, and $\overline{N_G^+}(v)$ if v is outwardly complemented. All operations except the cost of pushing $\overline{N_G^+}(v)$ may obviously be charged to vertices and edges of G as in the standard algorithm. We show next that pushing $\overline{N_G^+}(v)$ can also be accomplished in $O(N_G^+(v))$ time, amortized, giving the $O(size(G))$ bound for depth-first a forest on F.

Definition 6. A **complement stack** is a data structure that supports the following operations:

- *Initstack*(V): initializes and returns an empty stack S.
- *Push*(X, S): removes any occurrences of members of X from S, then pushes the members of X to the top of S.
- *Cpush*(X, S) performs *Push*($V - X, S$).
- *Pop*(S) returns the top element of S.
- *Time*(S) returns a timestamp that tells when the top element of S was pushed.

Theorem 7. *Complement stacks can be implemented in such a way that:*

- *Initstack requires $O(V)$ time.*
- *Push(X, S) requires $O(|X|)$ time.*
- *Cpush(X, S) requires $O(|X|)$ time, amortized.*
- *Pop and Time require $O(1)$ time each.*

Proof. Let L be the members of V that are not on the stack. A, T, B are doubly-linked lists that partition the members of the stack. A holds those elements on the stack that where last pushed by a call to Push. T holds those elements that were pushed by the most recent call to Cpush. B holds those elements that were last pushed by a call to Cpush, but not by the most recent call to Cpush.

Each element of V keeps track of which of L, A, T, B contains it. If A or B contains it, it carries a **timestamp** that holds the time when the element was last pushed. To keep time, a global variable is incremented after each call to Push or Cpush.

We maintain a credit invariant: *Each member of L, A, B carries a credit.*

- *Initstack*(V) creates a doubly-linked list L of the elements of V and assigns a credit to each of them. It creates empty doubly-linked lists A, T, B.
- *Push*(X, S) traverses each member of X, splices it from the list in $\{L, A, T, B\}$ that it currently resides in, pushes it to the front of A together with a timestamp, and assigns a credit to it. This is clearly $O(|X|)$.
- *Cpush*(X, S) must incur only $O(|X|)$ amortized cost. It marks each member of X and assigns a credit to it on top of any that it already has. It traverses X, removing those members of $X \cap T$ to auxiliary list T'. It then traverses L, A, and B, moving unmarked members to T, and pays for visiting their members by using up a credit at each. This still leaves credits on elements that remain in L, A, or B, since they are members of X and have just received an extra credit from X. It then assigns each member of T' a credit and labels

it with a timestamp and moves it to front of B. If T is now nonempty, T is timestamped.
This requires $O(|X|)$ amortized time, since all operations are $O(|X|)$ except for traversing L, A, B, which is paid for by using a credit sitting on each visited item.

- Pop assigns x to be an element of T if T is nonempty, or else the top element of B if B is nonempty, or else null. If x came from T, then it gets T's timestamp. Let y be the top element of A if it is nonempty, else null. Return the member of $\{x, y\}$ that has the most recent timestamp. This clearly requires $O(1)$ time.

The time bound is observed, since each operation maintains the credit invariant, and Cpush pays for its operations either with its budget of $|X|$ new credits or with other credits it frees up from the structure. □

5.3 Topological sort

A *dag*, or *directed acyclic graph*, is a digraph that has no cycles. A *topological sort* of a digraph is an ordering of its vertices so that every edge goes from an earlier to a later vertex in that ordering. A digraph has a topological sort iff it is a dag, and, ordering the vertices of a dag in descending order of finishing time in any depth-first search gives a topological sort [3]. Thus, we may find a topological sort of any dag in G's outward class in $O(size(G))$ time.

5.4 Longest paths in a dag

We will let the *length* of a path denote the number of edges in it. The length of the longest path in a digraph is finite iff it is a dag; otherwise a cycle exists, and by repeatedly traveling the cycle, one may demonstrate paths of arbitrarily large lengths.

The following well-known approach finds a longest path in a dag F, given its adjacency-list representation in standard form. Label each vertex with the length of the longest path originating at that vertex. Given a topological sort, we may observe that as long as vertices are labeled in reverse topological order, then when it is time to label any vertex v, the members of $N_G^+(v)$ are already labeled. We may then apply the rule that the length of a maximum-length path beginning at v is one plus the maximum of the labels of $N_G^+(v)$. Given an adjacency-list representation of G, we may charge the cost of labeling v to edges out of v, and we get an $O(n + m)$-time algorithm.

We now show that if F is represented as $F = (G, \bar{b})$, we may find a longest path in F in $O(size(G))$ time. To find longest paths in F, there is not time to apply the foregoing algorithm directly, since we do not always have time to examine the labels of vertices in $N_F^+(v)$ when it is time to assign a longest-path label to v. We may nevertheless solve the problem by keeping all labeled vertices bucket-sorted according to their label. A key observation is that if a vertex w has label $k > 0$, then there is a member of $N_F^+(w)$ with label $k-1$. By induction, the buckets from k down to 0 are all nonempty. When it is time to label vertex v, then if v is not outwardly complemented, apply the standard approach of visiting the labels of $N_F^+(v)$. If v is outwardly complemented, then mark its neighbors in G; these are the vertices that are not neighbors in F. Then, starting at the highest nonempty bucket, descend through the buckets sequentially until an unmarked vertex is found. This is a neighbor in F with highest label, so assign v's label to be one plus this label, and add v to the appropriate bucket. In either case, the operation can be charged to v and edges out of v in G. Thus it takes $O(size(G))$.

5.5 Connectivity

Let us define an equivalence relation R on vertices of a digraph G where for vertices u, v, we say uRv iff there is a directed path from u to v and a directed path from v to u. In each equivalence class there is always a path from any vertex to any other, but this is never the case for two vertices in different classes. The equivalence classes are known as ***strongly-connected components.*** If $G = (V, E)$ is a digraph and $\mathcal{P}$ are the strongly connected components of G, then the *component graph* $G/\mathcal{P}$ is a dag. The component graph tells which components have a path to which.

An $O(size(G))$ algorithm for finding strongly-connected components in a member $F = (G, \bar{b})$ of G's outward equivalence class is obtained by modifying a standard algorithm [1]. Define the following function on vertices of G:

$$Low[v] = min(\{d[v]\} \cup \{Low[w] | w \text{ is a child of } v\} \cup \{d[u] : u \in N_G^+(v)\}. \quad (1)$$

This function differs somewhat from a similar function defined in [1], but inspection of their algorithm shows that it computes $Low[v]$ for each vertex v in postorder on the depth-first forest. This guarantees that the children's labels are in place when it is time to label v. As it goes, it occasionally removes from the forest the subtree rooted at the vertex being labeled, using the following rule: remove v's current subtree if $Low[v] \geq d[v]$. Moreover, if $\mathcal{F}$ is the set of strongly-connected components, the vertices of this subtree are a sink in the *component graph* $G/\mathcal{F}$. It then recurses on what remains of the graph, continuing to label in postorder applying the inductive rule for $Low[v]$ on the reduced vertex set. The vertices of each removed tree give the vertices of a strongly-connected component.

If $\mathcal{F}$ is the set of strongly-connected components, then $G/\mathcal{F}$ is a dag [3]. Below, we will find it useful to have a topological sort of this graph available. Because the removed component is always a sink in what remains of the component graph $G/\mathcal{F}$, it follows that the order in which the components are removed gives a topological sort of $G/\mathcal{F}$.

If G is a digraph, and we want the strongly-connected components of a digraph F in G's outward class, then there is time to find the minimum of $d[v]$ and $Low[w]$ for each child of v, since we may charge these lookups to v and to tree edges out of v. Thus, we will incur $O(n)$ time looking up these values and finding their minimum at each vertex v. We must also find $min(\{d[u] : u \in N_G^+(v)\})$. If v is not outwardly complemented, we may do so and charge the cost to edges of G that are incident to v. If v is outwardly complemented, there is not time to visit $N_G^+(v)$. Instead, we assume that we have pre-assembled a list D of vertices of G in ascending order of discovery time. We mark the members of $N_G^+(v)$; this gives all non-neighbors of v in F. We then traverse D in ascending order until we encounter an unmarked vertex. This gives the earliest discovery time of any vertex in $N_F^+(G)$, thereby allowing us to complete the computation of $Low[v]$. We may clearly charge the cost to v and edges out of v in G. Thus, this gives $O(size(G))$ time to find the strongly-connected components of F. Moreover, since the order in which components of F are removed is unaffected by the representation, this algorithm gives a topological sort of the component graph for F.

If F is undirected, then the biconnected components of F can be computed in a similar way. A standard algorithm is given in [1]. The details of adapting it are left to the reader.

5.6 An application to modular decomposition

A number of $O(n+m)$ algorithms are known for modular decomposition [2, 4, 11]. One of them makes use of a variant of the depth-first search algorithm on outward equivalence classes that I have described above [4]. Though this algorithm is arguably simpler than the others, it is still conceptually complex.

In this section, we show that if we are willing to settle for an $O(n + m \log n)$ bound, we can use one that is conceptually simple. Some of the algorithms on outward equivalence that I have described above are critical elements in the time bound.

Let G be an undirected or directed graph, and let v be a vertex. Let $\mathcal{P}(G, v)$ denote the partition of $V(G)$ given by $\{v\}$ and the maximal modules of G that do not contain v. That this is a partition of $V(G)$ is easy to verify using Theorem 3.

Since $\mathcal{P}(G, v)$ is a partition of $V(G)$, we may decompose G into the quotient $G/\mathcal{P}(G, v)$ and $\{G|X : X \in \mathcal{P}(G, v)\}$. Using Theorem 2, it is also easy to verify that inverse images of the non-singleton modules of $G/\mathcal{P}(G, v)$ are precisely the ancestors of v in the modular decomposition of G.

Here is the restriction to undirected graphs of the modular decomposition algorithm of [5]. For the moment, assume the existence of a *Partition* operation for finding $\mathcal{P}(G, v)$.

Algorithm 8. *Compute the modular decomposition of G*

Decomp(G)
    If G has only one node, return a one-node tree
    else
        Select a vertex v of G
        $\mathcal{P} = \mathcal{P}(G, v) = Partition(G, v)$
        For each ancestor $\mathcal{U}$ of $\{v\}$ in $G/\mathcal{P}$ in descending order
        Create a tree node t_u

Make t_u a child of the tree node corresponding to its parent in $G/\mathcal{P}$
For each member X of $\mathcal{P}$ that is a child of U
$t_x = Decomp(G|X)$
Let t_x be a child of t_u
return the root of the resulting tree

This algorithm produces a tree that is correct, except that it allows a parallel degenerate node to be a child of another, and a series degenerate node to be a child of another. To fix this, visit the nodes of the tree in postorder, deleting any such node and letting its parent inherit its children.

To get the time bound, we will precompute the recursion tree for *Decomp*(G) (Algorithm 8). with the following:

Algorithm 9. *RecursivePartition: Precompute the recursion tree for Decomp(G) (Algorithm 8).*

RecursivePartition (G)
Create a single root node r
If G has more than one vertex
Select a vertex v
$\mathcal{P} = Partition(G, v)$
For each class X in $\mathcal{P}$
Let $t = RecursivePartition(G|X)$;
Let t be a child of r
return r.

Once we have done this, we may precompute all quotients referred to as $G/\mathcal{P}$ in recursive calls, by applying an off-line least-common ancestors algorithm on edges of G. We have thus reduced the problem to bounding the cost of calls to *Partition*, and of finding ancestors of v in the quotient $G/\mathcal{P}(G, v)$.

Graph partitioning For finding $\mathcal{P}(G, v)$, we can use a variant of an algorithm for the problem that is due to Spinrad [16, 11]. Spinrad's original formulation was designed to give an $O(n + m \log n)$ bound for a single call to the *Partition* function. We must modify it slightly in order to assign this same bound to the cost of all calls to this function that are generated in *RecursionTree* (Algorithm 9). A different modification for this same problem is given in [10].

We begin with a partition $\mathcal{P}$ of $V(G)$, and start refining $\mathcal{P}$ in a way that guarantees that no module that is contained in one class is ever split up into two partition classes. The procedure halts when no further refinement is possible; at this point, the partition consists of the maximal modules of G that are contained in a single partition class of the initial partition. By setting $\mathcal{P} = \{\{v\}, V(G) - \{v\}\}$ initially, this yields $\mathcal{P}(G, v)$.

The basic driving operation is $Pivot(G, u, X)$, where X is a union of partition classes in $\mathcal{P}$, and $u \in V(G) - X$. For each partition class C contained in X, we refine $\mathcal{P}$ by splitting C into two classes C_a and C_n, which contain the vertices of C that are adjacent and nonadjacent to u, respectively.

Algorithm 10. *Given a partition $\mathcal{P}$ of $V(G)$, find the maximal modules of G that are contained in a single partition class.*

```
Partition(G, P)
  if |P| = 1 return P
  else
      Let M be a maximal member of P
      Select X ∈ P − {M}
      Let Y = ∪(P − {X})
      For each x ∈ X do
          Pivot (G, x, Y)
      For each y ∈ Y do
          Pivot (G, y, X)
      return Partition (G|X, P|X) ∪ Partition (G|Y, P|Y)
```

Correctness: if $|\mathcal{P}| = 1$ then its only member is $V(G)$, which is a module. Otherwise, the pivots in X and Y refine the members of $\mathcal{P}$ in a way that does not split up any modules that are contained in a class. After these pivots, no subsequent pivot on a member of X can further split any member of Y, and *vice versa,* which shows that the operation may be completed with the two recursive calls.

Time bound: We show an $O(size(G) \log n)$ bound. Let $a(X)$ be the sum of cardinalities of the adjacency lists of members of X. We first show that the cost of the pivots in the main call is $O(|X| + a(X))$.

Assume that all classes and adjacency lists are implemented with doubly-linked lists, and each list is labeled with its cardinality. Since G is undirected, we may assume that each edge (u, v) in u's adjacency list has a *twin pointer* to edge (v, u) in v's adjacency list. Mark vertices in X. Remove and collect all edges that go to Y from the adjacency lists in X that go to Y. Using the twin pointers, remove the twins from adjacency lists in Y. This gives all edges connecting X and Y. Using pre-initialized buckets to group (but not sort) these edges by beginning vertex requires $O(a(X))$ time: distribute the edges to buckets, and keep a linked list of nonempty buckets in order to avoid having to touch empty buckets when they are retrieved. This gives a list of vertices in X and Y that have neighbors in Y and X, respectively, and for each of these, a list of those neighbors. The total number of elements in these lists is $O(|X| + a(X))$.

For each neighbor y of x in Y, remove it from its current partition class L and move it to a twin class L', which must be created if L doesn't already have a twin class. Update the size labels of L and L', and delete L if y was the last element of L. This takes $O(1)$ time, and, by repeating it on all edges in x's reduced adjacency list, we get the result of the pivot that uses x. Performing the pivots on all members of X is thus $O(a(X))$. The analysis for elements of Y that have neighbors in X is similar.

Thus, we incur $O(|X| + a(X))$ time, exclusive of recursive calls. If $|\mathcal{P}| = 1$, then no charges are made to X. If $|\mathcal{P}| > 1$, we charge the cost by assigning $O(1)$ to each member of X and each element of an adjacency list in X, and the members of X go to a recursive call on a graph with at most half as many vertices, since $|X| \leq |Z|$. This is true even if the next charge occurs in a separate call to *Partition* that is generated by *RecursionTree* on a subset of X. We conclude that each time a member of X is charged, it is in a recursive call with at most half as many vertices as the last time it was charged, which gives an $O(\log n)$ bound on the number of times it or its adjacency-list elements can be charged in *RecursionTree*. Summing this over all vertices and adjacency-list elements, we get an $O(size(G) \log n)$ bound. The one-time cost of initializing the buckets that are used in the pivot operations is $O(n)$.

Finding the ancestors of a vertex The foregoing shows that we may find the set $\mathcal{P}(G, v)$, To complete the time bound of *Decomp* (Algorithm 8), we must find the ancestors of $\{v\}$ in $G' = G/\mathcal{P}(G, v)$. We described above why we may assume that we have G' available in adjacency-list form when it is time to do this step.

Let $F(G, v)$ be the graph on $V(G)$ where $v \in V(G)$ is isolated, and where for each $x, y \in V - \{v\}$ there is an arc from x to y if y is adjacent to exactly one of x and v in G. It follows that if there is an edge of $F(G, v)$ from x to y, then every module that contains x and v must also contain y. On the other hand, if $X \subset V$ contains v, and there is no edge in F from $X - \{v\}$ to $V - X$, then X is a module: any vertex y in $V - X$ must have the same relationship to all members of X as it does to v.

Lemma 11. *[5] If $x \in X \subseteq V(G)$, then X is a module iff $X - \{v\}$ has no outgoing directed edges in $F(G, v)$.*

From the fact that every module of G that does not contain v is contained in a single class of $\mathcal{P}(G, v)$, it follows that all nontrivial modules of $G' = G/\mathcal{P}(G, v)$ contain $\{v\}$. If A is a degenerate ancestor of v in the modular decomposition of G, then the union X of all of its children except the one that contains v is a maximal module of G that does not contain v, by Theorem 3. So X is the inverse image of a single node of G'. It follows by Theorems 2 and 3 that the inverse images of the modules of G' in G are the ancestors of v in $MD(G)$.

Thus, the nontrivial modules of $G/\mathcal{P}(G, v)$ are nested. The following is now immediate from Lemma 11:

Theorem 12. *[5] Let G be an undirected graph, let $G' = G/\mathcal{P}(G, v)$, $F = F(G', \{v\})$ be $\{v\}$'s forcing graph on G', and let S be the strongly connected components of F.*

1. *The topological sort of F/S is unique;*

2. *$X \subseteq V$ is an ancestor of v in $MD(G)$ iff it is the union of $\{v\}$ and the members of a suffix of this sort.*

The problem for getting an $O(n + m \log n)$ bound is that we cannot guarantee that $size(F) = O(size(G))$; F may be considerably larger. For two years after development of the algorithm of [5], this appeared to be an insurmountable obstacle to bringing the time bound of that algorithm down to $O(n + m \log n)$. The solution is completely trivial in light of the foregoing results about outward equivalence classes, however:

Lemma 13. *Let G be an undirected graph, and let $F = F(G, v)$ be v's forcing graph. Then F^T is in the outward equivalence class of G*

Proof. If (y, x) is an edge of F^T, then y is adjacent to exactly one of x and v. Thus, if v is nonadjacent to y, the neighbors of y in F^T are the same as they are in G, namely $N_G^+(y)$. If v is adjacent to y, the neighbors of y in F^T are $\overline{N_G^+}(y)$, and thus obtained by outwardly complementing y. □

Since we have available an adjacency-list representation of $G' = G/\mathcal{P}(G, v)$, we may compute a partially complemented representation of $F^T = (G', \{v\})$ of G' in $O(size(G'))$ time. We may then find a topological sort of the strong component graph of F^T in $O(size(G'))$ time, by the results of Section 5.5. This gives the topological sort of $\mathcal{F}$ in reverse order. We may charge this cost to edges of G that connect members of $\mathcal{P}(G, v)$, and those edges are not charged in recursive calls. Over all recursive calls of *Decomp* (Algorithm 8), this step thus contributes $O(size(G))$ to the running time.

This shows that the time bound for graph partitioning is the bottleneck for further improvements in the resulting $O(n + m \log n)$ time bound for this algorithm.

6 Algorithms on Two-Way Equivalence Classes

The two-way equivalence classes have interesting mathematical properties not shared by the outward equivalence classes. Unlike the outward classes, they are a special case of dynamic-labeled two-structures. As shown in [7], there is a tree-like representation of the modules of the entire class, similar to the one for a single graph or two-structure, but unrooted.

A *comparability graph* is an undirected graph whose edges can be oriented to give a transitive digraph, hence a partial order. A *permutation graph* is a graph where adjacencies represent pairwise inversions in a permutation, and and a *circular permutation graph* is a cyclic generalization. Rotem and Urutia show the following:

Theorem 14. *[15] Let v be a vertex in G. Let G' be the result of both inwardly and outwardly complementing each member of $N_G(v)$.*

1. *If G is a comparability graph, then so G'.*
2. *G is a circular permutation graph iff it is a comparability graph and G' is a permutation graph.*

In the case of a digraph F in G's outward-equivalence class, we obtained the time bounds by allowing each vertex v to carry a list corresponding either to $N_F^+(v)$ or to $\overline{N_F^+}(v)$. In the case of a member of a two-way class, the N_F^+ depends not only on the outward complementation of v, but also on the inward complementation status of every other member of V. We can nevertheless achieve many of the same time bounds if we keep a separate doubly-linked lists for vertices that are inwardly complemented and inwardly uncomplemented.

6.1 Splitting a set of vertices

Given a vertex v in a graph G, and a representation $F = (G, \overline{b}_1, \overline{b}_2)$ in G's two-way equivalence class, we will want to perform the pivot operation on F and be able to charge it to elements of our data structure.

Let u be a vertex on which we wish to perform a pivot, let $\mathcal{P}$ be a partition of G, and let $\mathcal{P}' \subseteq \mathcal{P}$ be a set of classes we wish to split. Suppose each member $P \in \mathcal{P}'$ is represented with a pair of lists, P_c and P_u, where P_c contains members of P that are inwardly complemented and P_u contains the members that are not inwardly complemented. For each such P, we wish to create a similar representation of $A = P \cap N_F^+(u)$ and $B = P - N_F^+(u)$. If u is outwardly complemented, then $A_c = P_c \cap N_G^+(u)$, $B_c = P_c - N_G^+(u)$, $A_n = P_n - N_G^+(u)$, and $B_n = P_n \cap N_G^+(u)$. If u is not outwardly complemented, the computation is the same, but the roles of A and B are reversed. Suppose there are k directed edges of G from u to P, and we are given these edges as a list. If $k \geq 1$, it takes $O(k)$ time to split each of P_c and P_u according to adjacency in G, and then to label and pair the four resulting lists, as described, to obtain the desired representations of P_a and P_n.

6.2 Breadth-first search

When visiting a vertex u, deciding which nodes must move from the undiscovered list to the queue requires splitting the set U of undiscovered vertices into a set $U_a = U \cap N_F^+(u)$ and $U_n = U - N_F^+(u)$. This takes time proportional to $N_G^+(u)$ by the technique of section 6.1. The cost of visiting u can be charged to edges out of u in G, so breadth-first search on F takes $O(size(G))$ time.

6.3 Depth-first search, topological sort, connectivity

A data structure for depth-first search on digraph F in G's two-way class was developed by Dahlhaus, Gustedt and me in a closely related context [4]. We keep two complement stacks, one for V_u, the set of vertices that are not inwardly complemented, and and one for V_c, the set of vertices that are inwardly complemented. Together the two stacks simulate a single meta-stack with non-duplicating pushes: a pop operation looks at the timestamps of the tops of the two stacks, and pops the most recent, and a *Push* or *Cpush* operation on a vertex v performs the non-duplicating *Cpush* on one of $\{V_u, V_c\}$ and the *Push* on the other. Which one gets the *Push* and which gets the *Cpush* depends on whether v is outwardly complemented, since the goal is to perform a non-duplicating push on all members of $N_F^+(v)$. The depth-first search algorithm then proceeds as before, except that it uses the meta-stack to generate the next vertex to be visited.

Topological sort, strongly-connected components, and biconnected components when F is undirected now follow in $O(size(G))$ time, using the same algorithms as before. However, I have not been able to think of an $O(size(G))$ bound for longest paths in F, when F is only required to be in G's two-way class, rather than its outward class.

6.4 Graph Partitioning

A module in a directed graph is a set X of vertices such that for each $v \in (V(G) - X)$, $X \times \{v\}$ consists of either all edges or all nonedges, and $\{v\} \times X$ consists of either all edges or all nonedges. It is possible for one of these sets to consist of edges, and the other to consist of nonedges.

We wish to have an efficient algorithm to perform graph partitioning on a graph F in G's two-way equivalence class in $O(size(G))$ time. To modify the graph partitioning algorithm, we need only change the *Pivot* step. If u is the pivot vertex, we need to consider both $N_F^+(u)$ and $N_F^-(u) = N_{F^T}^+(u)$. By Theorem 5, we may represent F^T using G^T. It takes $O(size(G))$ time to obtain G^T with a bucket-based grouping of edges by ending vertex number.

We may now use a generalization of the technique of section 6.1, where each class Y is made up of four doubly-linked lists, one for each combination of inward and outward complementation bits. Each of these lists can be split according to membership in $N_G^+(u)$, and the resulting eight lists can then be recombined into two groups of four according to membership in $N_F^+(u)$. Which set of each split pair goes into which group depends on the outward complementation status of u. This operation can then be repeated in a similar way with G^T in order to further split according to membership in $N_F^-(u) = N_{F^T}^+(u)$. When this is complete, Y has been split into four classes (some of them possibly empty), where each is represented with four doubly-linked lists corresponding to each possible combination of inward and outward complementation status.

If there are k edges of G in $\{u\} \times V$ and $V \times \{u\}$, and we are given a list of these edges, then this operation takes $O(k)$ time, using the techniques of Section 5.6. It is then easily verified that the proof of time bound for *RecursionTree* in section 5.6 now yields an $O(size(G))$ time bound for *RecursionTree* on F.

7 Future Work

An open question is what other graph algorithms can be generalized to these representations of two-way equivalence classes. If G is a digraph, it appears likely that one may obtain an $O(size(G))$ time bound for modular decomposition of any digraph F in G's two-way equivalence class, by modifying a linear-time modular decomposition algorithm. An $O(size(G) \log n)$ bound is strongly suggested by the results of Section 6.4. In addition, in [12], an $O(n+m)$ algorithm is given for the *Partition* procedure, but it is quite complicated. Nevertheless, it may be possible to adapt the algorithm using techniques from section 5.6, in order to obtain an $O(size(G))$ bound for graph partitioning on a member F of G's two-way equivalence class. A number of other combinatorial problems are shown in [12] to reduce in linear time to modular decomposition and graph partitioning, but it remains to establish whether the reductions generalize to members of G's equivalence class in $O(size(G))$ time.

References

1. Aho, A.V., Hopcroft, J.E., Ullman, J.D.: *The Design and Analysis of Computer Algorithms*, Addison-Wesley (1974)
2. Cournier, A., Habib, M.: A new linear algorithm for modular decomposition, CAAP '94: 19th international colloquium, lecture notes in computer science, Sophie Tison, ed., Edinburgh, UK (1994) 68-82
3. Cormen, T.H., Leiserson, C.E., Rivest, R.L.: *Algorithms*, MIT Press (1990)
4. Dahlhaus, E., Gustedt, J., McConnell, R.M.: Efficient and practical modular decomposition, Proc. 8th annual ACM-SIAM Symposium on Discrete Algorithms, New orleans (1997) 26-35
5. Ehrenfeucht, A., Gabow, H.N., McConnell, R.M., Sullivan, S.J.: An $O(n^2)$ divide-and-conquer algorithm for the prime tree decomposition of two-structures and modular decomposition of graphs, Journal of Algorithms **16** (1994) 283-294
6. Ehrenfeucht, A., Rozenberg, G: Angular 2-structures, Theoretical Computer Science **92** (1992) 227-248
7. Ehrenfeucht, A., Rozenberg, G.: Dynamic labeled 2-structures, Math. Struct. in Comp. Science **4** (1994) 433-455
8. Ehrenfeucht, A., Rozenberg, G.: Theory of 2-structures, part 1: clans, basic subclasses, and morphisms, Theoretical Computer Science **70** (1990) 277-303
9. Ehrenfeucht, A., Rozenberg, G.: Theory of 2-structures, part 2: representations through labeled tree families, Theoretical Computer Science **70** (1990) 305-342
10. Habib, M., Paul C., Viennot, L.: Lex-BFS, a partition refining technique, manuscript.
11. McConnell, R.M., Spinrad, J.P.: Linear-time modular decomposition and efficient transitive orientation of comparability graphs, Proc. 5th Annual ACM-SIAM Symposium on Discrete Algorithms, Arlington, Virginia (1994) 536-545
12. McConnell, R.M., Spinrad, J.P.: Linear-time transitive orientation, Proc. 8th Annual ACM-SIAM Symposium on Discrete Algorithms", New Orleans (1997) 19-25
13. Moehring, R.H.: Algorithmic aspects of comparability graphs and interval graphs, in Graphs and Orders, I. Rival, ed., D. Reidel (1985) 41-101.

14. Rose, D., Tarjan, R.E., Leuker, G.S.: Algorithmic aspects of vertex elimination on graphs, SIAM J. Comput. **5** (1976) 266-283
15. Rotem, D., Urrutia, J.: Circular permutation graphs, Networks **12** (1982) 429-437
16. Spinrad, J.P.: Vertex partitioning, Manuscript.

On Compact Directed Acyclic Word Graphs

Maxime Crochemore and Renaud Vérin

Institut Gaspard Monge
Université de Marne-La-Vallée,
2, rue de la Butte Verte, F-93160 Noisy-Le-Grand.
http://www-igm.univ-mlv.fr

Abstract. The Directed Acyclic Word Graph (DAWG) is a space-efficient data structure to treat and analyze repetitions in a text, especially in DNA genomic sequences. Here, we consider the Compact Directed Acyclic Word Graph of a word. We give the first direct algorithm to construct it. It runs in time linear in the length of the string on a fixed alphabet. Our implementation requires half the memory space used by DAWGs.

1 Introduction

One of the most surprising facts related to pattern matching and discovered by Ehrenfeucht *et al.* [2] is that the size of the minimal automaton accepting the suffixes of a word is linear. The surprise is due to the maximal number of subwords that may occur in a word: it is quadratic according to the length of the word. This is obviously true if the alphabet is unbounded, but still holds if the alphabet contains at least two letters. In addition to the previous result, Ehrenfeucht *et al.* proved that the automaton can be built in linear time, which is indeed a consequence of the previous fact but does not come readily from it.

In the present article, we consider the compact implementation of the automaton and show that it has a direct construction that runs in linear time. Fast and space-economical methods for this construction are important because the automaton serves as an index on the underlying word, and, as such, is involved in several combinatorial algorithms on words.

Historically, the first linear-size graph to represent the subwords of a word, called the *Directed Acyclic Word Graph* (DAWG), was described in [2] together with a linear-time construction. When terminal states are added to the DAWG, as shown in [8], the structure becomes the minimal automaton accepting the suffixes of the word. Regarded as an automaton accepting the subwords of the word, *i.e.* setting all states as terminal states, the DAWG is not always a minimal automaton. Indeed, this latter automaton can be slightly smaller, but its construction satisfies the same properties ([8, 3, 9]) though the algorithms become a bit more tricky.

Basically, DAWGs provide an implementation of indexes on texts [4]. The index on a text T helps searching it for various patterns. For instance, it leads to an efficient solution to the string-matching problem, searching text T for a word w. The typical running time of a query is $\mathcal{O}(|w|)$ on a fixed alphabet, and is $\mathcal{O}(|w| \log |\Sigma|)$ if the alphabet Σ of the text is unbounded.

Many other efficient solutions to problems on words are applications of DAWGs. They include (see [12]): computing the number of subwords of a word, computing the longest repeated subword of a word, backward DAWG-matching, finding repetitions in words [6], searching for a square [7, 9], computing the longest common subword of a finite set of words and on-line subword matching [10], approximate string-matching [21].

The *suffix tree* is an alternative representation of the subwords of a word that shares with the DAWG essentially the same applications. McCreight [18] introduced the notion and gave an efficient construction after the seminal work of Weiner [22] on a similar structure.

Suffix trees have been more extensively studied than DAWGs, probably because they display positions of the word in a simpler way although the branching from nodes is not uniform as it is from states of DAWGs. Apostolico [1] lists over forty references on suffix trees, and Manber and Myers [17] mention several others (see also [19]). Several variants or implementations of suffix trees have been developed, like ***suffix arrays*** [17], ***PESTry*** [16], ***suffix cactus*** [15], or ***suffix binary search trees*** [14]. Ukkonen [20] designs an on-line construction of suffix trees, and Farach [13] proposes a novel approach leading to a linear-time construction on integer alphabets.

In computational biology, DNA sequences are often only viewed as words over the alphabet $\{a, c, g, t\}$ of nucleotides. In this form, they are objects for linguistic and statistic analysis. For this purpose, suffix automata (or suffix trees) are extremely useful data structures, but the bottleneck to using them is their size. The indexes has to be kept in main memory and their sizes limit their use. The size of available sequences is steadily growing, and therefore saving memory space is wanted both for the construction of the index and for its use.

The *Compact Directed Acyclic Word Graph* (CDAWG) keeps the direct access to information while requiring less memory space. The structure has been introduced by Blumer *et al.* [4, 5]. The implementation is obtained by deleting all states of outdegree one and their corresponding transitions (excepting terminal states).

We present an algorithm that builds directly compact DAWGs. This construction avoids constructing the DAWG first, which makes it suitable for the presently available DNA sequences (about 1.5 million nucleotides long for the longest sequences). Experiments show that our implementation saves half of the memory space required for ordinary DAWGs and suffix trees. At the same time, the reduction of the number of states (2/3 less) and of transitions (about half less) makes the applications run faster. Time and space are saved simultaneously. The memory space used by our implementation of compact DAWGs requires about $6n$ integers for a word of length n. This is to be compared with $7n$ for DAWGs, $8n$ for suffix trees. It is just $2n$ for suffix arrays, but this is paid by a slower access to subwords.

This article is organized as follows. In Section 2 we recall the basic notions on DAWGs. Section 3 introduces the compact DAWG, also called compact suffix automaton, and contains the bounds on its size. We show in Section 3.4 how to build the compact DAWG from the DAWG in linear time with respect to the size of this latter structure. Direct construction algorithm for the compact DAWG is given in Section 4.

2 Definitions

Let Σ be a nonempty alphabet and Σ^* the set of words over Σ, with ε as the empty word. If w is a word in Σ^*, $|w|$ denotes its length, w_i its ith letter, and $w_{i..j}$ its factor (subword) $w_i w_{i+1} \ldots w_j$. If $w = xyz$ with $x, y, z \in \Sigma^*$, then x, y, and z are factors or subwords of w, x is a prefix of w, and z is a suffix of w. $S(x)$ denotes the set of all suffixes of x and $F(x)$ the set of its factors.

For an automaton, the tuple (p, a, q) denotes a transition of label a starting at p and ending at q. A roman letter is used for mono-letter transitions, a greek letter for multi-letter transitions. Moreover, $(p, \alpha]$ denotes a transition from p for which α is a prefix of its label. In this notation the target state is not given.

Here, we recall the definition of the DAWG, and a theorem about its implementation and size both proved in [3] and [9].

Definition 1. The Suffix Automaton of a word x, denoted *DAWG*(x), is the minimal deterministic automaton (not necessarily complete) that accepts $S(x)$, the (finite) set of suffixes of x.

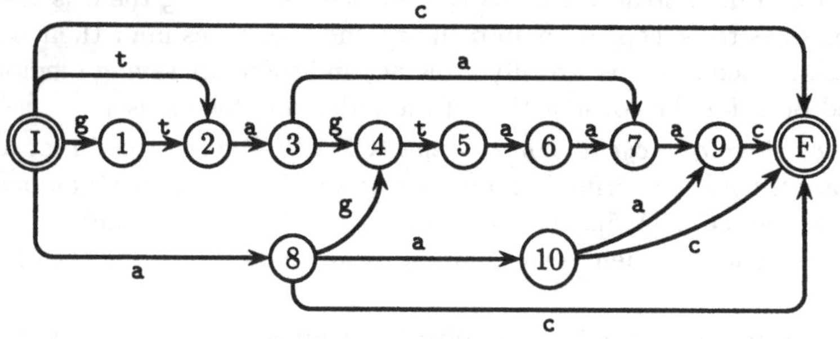

Fig. 1. *DAWG*(gtagtaaac).

For example, Figure 1 shows the DAWG of the word gtagtaaac. States that are double circled are terminal states.

Theorem 2. *The size of the DAWG of a word x is $\mathcal{O}(|x|)$ and the automaton can be computed in time $\mathcal{O}(|x|)$. The maximum number of states of the automaton is $2|x| - 1$, and the maximum number of edges is $3|x| - 4$.*

Recall that the right context (according to $S(x)$) of a factor u of x is $u^{-1}S(x)$. The syntactic congruence associated with $S(x)$ is denoted by $\equiv_{S(x)}$ and is defined, for $x, u, v \in \Sigma^*$, by:

$$u \equiv_{S(x)} v \iff u^{-1}S(x) = v^{-1}S(x).$$

We call *classes of factors* the congruence classes of the relation $\equiv_{S(x)}$. The longest word of a class of factors is called the *representative* of the class. States of *DAWG*(x) are exactly the classes of the relation $\equiv_{S(x)}$. Since this automaton is not required to be complete, the class of words not occurring in x, corresponding to the empty right context, is not a state of *DAWG*(x).

Among the congruence classes we make a selection of classes that are called *strict classes of factors* of $\equiv_{S(x)}$ and that are defined as follows.

Definition 3. Let u be a word of C, a class of factors of $\equiv_{S(x)}$. If at least two letters a and b of Σ exist such that ua and ub are factors of x, then C is called **a strict class of factors** of $\equiv_{S(x)}$.

We also introduce the function $endpos_x$: $F(x) \to \mathrm{N}$, defined, for a word u, by:

$$endpos_x(u) = \min\{|w| \mid w \text{ prefix of } x \text{ and } u \text{ suffix of } w\}$$

and the function $length_x$ defined on states of *DAWG*(x) by:

$$length_x(p) = |u|, \text{ with } u \text{ representative of } p.$$

The word u also corresponds to the concatenated labels of transitions of the longest path from the initial state to p in *DAWG*(x). Transitions that belong to the spanning tree of longest paths from the initial state are called *solid transitions*. Equivalently, for each transition (p, a, q) we have the property:

$$(p, a, q) \text{ is solid} \iff length_x(q) = length_x(p) + 1.$$

The function $length_x$ works as well for multi-letter transitions (transitions labeled by non-empty words), just replacing 1 in the above equivalence by the length of the label of the transition from p to q. This extends the notion of solid transitions to multi-letter transitions:

$$(p, \alpha, q) \text{ is solid} \iff length_x(q) = length_x(p) + |\alpha|.$$

In addition, we define the *suffix link* function on states of *DAWG*(x) by the next statement.

Definition 4. Let p be a state of *DAWG*(x), different from the initial state, and let u be a word of the equivalence class p. The **suffix link** of p, denoted by $s_x(p)$, is the state q which representative v is the longest suffix z of u such that $u \not\equiv_{S(x)} z$.

Note that, consequently to this definition, we have $length_x(q) < length_x(p)$. Then, by iteration, suffix links induce *suffix paths* in *DAWG*(x), which is an important notion used by the construction algorithm. Indeed, as a consequence of the above inequality, the sequence $(p, s_x(p), s_x^2(p), \ldots)$ is finite and ends at the initial state of *DAWG*(x). This sequence is called the *suffix path of* p.

3 Compact Directed Acyclic Word Graphs

3.1 Definition

Compaction of DAWGs is based on the deletion of some states and their outgoing transitions. This is possible by using multi-letter transitions and selecting strict classes of factors defined in the previous section (Definition 3).

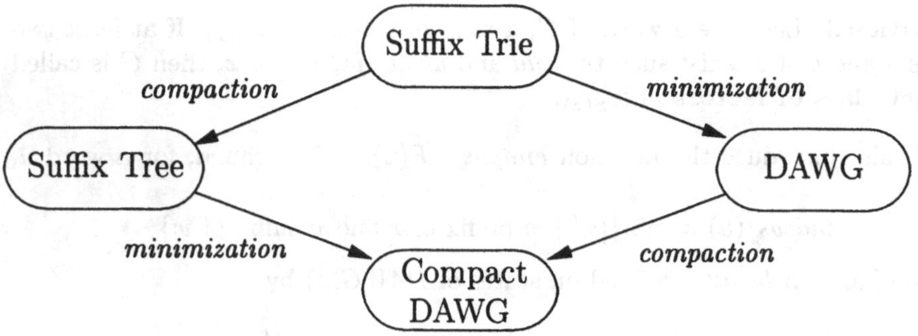

Fig. 2. Consider a word that has an end-marker. Its suffix tree is the compact version of the digital trie of its suffixes. Its DAWG is the minimized (in the sense of automata theory) version of the trie. The compact DAWG can be obtained either by minimizing the suffix tree of the word or by compacting its DAWG.

The definition of CDAWGs parallels the definition of suffix trees obtained from ordinary digital tries of all suffixes of a word. Indeed, disregarding how the end-marker required by suffix trees is managed, the CDAWG may be viewed as well as a compact version of the DAWG or as a minimized (in the sense of automata theory) version of the suffix tree (see Figure 2).

The compact DAWG is defined as follows.

Definition 5. The **Compact Directed Acyclic Word Graph** of a word x, denoted by $CDAWG(x)$, is the compaction of $DAWG(x)$ obtained by keeping only states that are either terminal states or strict classes of factors according to $\equiv_{S(x)}$, and by labeling transitions accordingly.

Consequently to Definition 3, strict classes of factors correspond to states that have an outdegree greater than one. So, we can delete every state having outdegree one exactly, except terminal states. Note that initial and final states are terminal states, so they are not deleted. An example of CDAWG is displayed in Figure 3.

The construction of the DAWG of a word containing repetitions shows that many states have outdegree one only. For example, in Figure 1, the DAWG of the word **gtagtaaac** has 12 states, 7 of which have outdegree one; it has 18

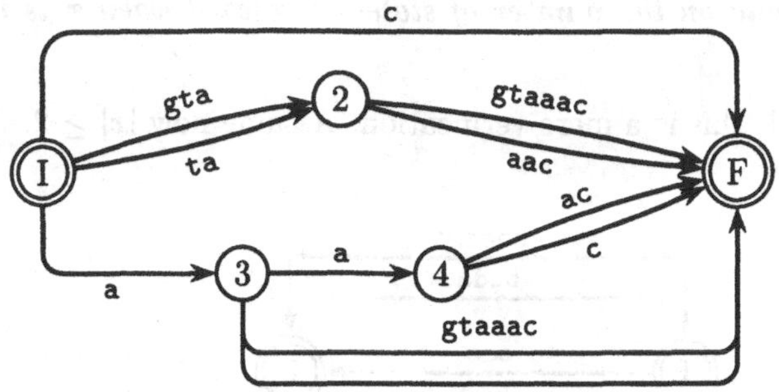

Fig. 3. *CDAWG*(gtagtaaac).

transitions. Figure 3 displays the compacted version, obtained after deletion of the 7 states, using multi-letter transitions. The resulting automaton has only 5 states and 11 edges.

According to experiments made on biological DNA sequences, considering them as words over the alphabet $\Sigma = \{a, c, g, t\}$, we got that more than 60% of states have outdegree one. So, the deletion of these states is worth, it provides an important saving. The average analysis of the number of states and edges in done in [5] in a Bernouilly model of probability.

When a state p is deleted, the deletion of its outgoing edges is realized by concatenating their label to the labels of incoming edges. For example, let r and p be states linked by a transition (r, b, p). The edges (r, b, p) and (p, a, q) are replaced by the edge (r, ba, q) if p is deleted. By recursion, this extends to every multi-letter transition (r, α, p).

In the example of Figure 3, one can note that, inside the word **gtagtaaac**, occurrences of **g** are followed by **ta**, and those of **t** and **gt** by **a**. The word **gta** is the representative of state 2, and there is no state corresponding to subwords **g**, **gt**, nor **t**. State I is directly connected to state 2 by edges (I,**gta**,2) and (I,**ta**,2). States 1 and 2 of Figure 1 no longer exist.

The suffix links defined on states of DAWGs remain valid when we reduce them to CDAWGs due to the next lemma, which proof is straightforward.

Lemma 6. *If p is a state of CDAWG(x), then $s_x(p)$ is a state of CDAWG(x).*

3.2 Size bounds

By Theorem 2 *DAWG*(x) is linear in $|x|$. As we shall see below (Section 3.3), labels of multi-letter transitions are implemented in constant space. So, the size of *CDAWG*(x) is also $\mathcal{O}(|x|)$. Meanwhile, as we delete many states and edges, we review the exact bounds on the number of states and edges of *CDAWG*(x). They are respectively denoted by *States*(x) and *Edges*(x).

Lemma 7. *Given $x \in \Sigma^*$, if $|x| = 0$, then States(x) $= 1$; if $|x| = 1$, then States(x) $= 2$; otherwise $|x| \geq 2$ and $2 \leq$ States(x) $\leq |x| + 1$.*

The upper bound on the number of states is reached when x is in the form $a^{|x|}$, for $a \in \Sigma$.

Proof. For $|x| \leq 1$, this is a mere verification. Assume now $|x| \geq 2$.

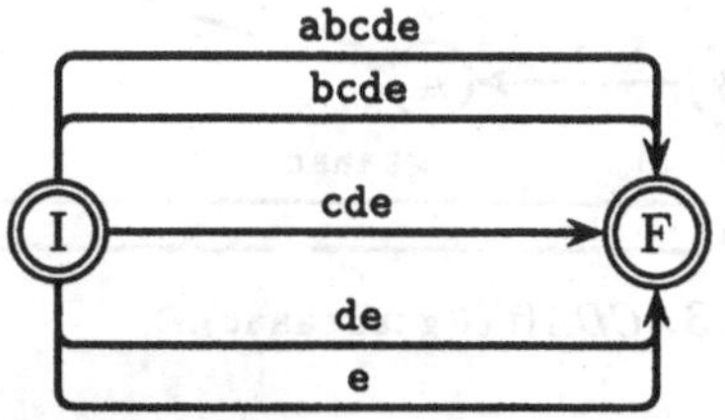

Fig. 4. A CDAWG with the minimum number of states, *CDAWG*(**abcde**).

The lower bound is obvious and obtained when x is composed of pairwise different letters.

Consider the suffix tree of $x\$$, where \$ is a marker. It has exactly $|x|+1$ leaves and at most $|x|$ internal nodes. Its minimization into *CDAWG*(x) compacts all leaves into the final state F, and possibly put together other nodes. Removing the marker does not change the number of states. So, we have *States*(x) $\leq |x|+1$.

The word $a^{|x|}$ satisfies this property since each suffix $a^{|0|}$, $a^{|1|}$, ..., $a^{|x|}$ represents exactly one class. So, we have $|x|+1$ classes and the same number of states.

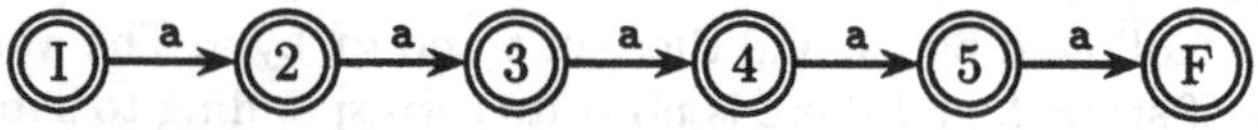

Fig. 5. A CDAWG with the maximum number of states, *CDAWG*(**aaaaa**).

Figures 4 and 5 display CDAWGs whose numbers of states are minimum and maximum, respectively, for words of length 5.

Lemma 8. *Given $x \in \Sigma^*$, if $|x| = 0$, Edges(x) $= 0$; if $|x| = 1$, Edges(x) $= 1$; otherwise $|x| \geq 2$ and Edges(x) $\leq 2|x| - 2$.*

The upper bound on the number of edges is reached when x is in the form $a^{|x|-1}c$, for a and c two different letters of Σ.

Proof. For $|x| \leq 1$, this is a mere verification. Assume now $|x| \geq 2$.

If x is in the form $a^{|x|}$, the number of edges is exactly $|x|$. So, we have to prove the upper bound for a word x containing at least two different letters. Consider the suffix tree of $x\$$. It has exactly $|x|+1$ leaves. It has at most $|x|-1$ internal nodes in this situation (because the root has outdegree 3). The number of edges in the tree is at most $2|x|-1$. After minimization into *CDAWG*(x) and removing the marker, all edges may remain except the edge labeled by \$. This give the upper bound of $2|x| - 2$.

The automaton $CDAWG(a^{|x|-1}c)$, for a and c two different letters of Σ, has $|x|$ states and exactly $2|x| - 2$ edges, distributed as $|x| - 1$ solid edges and $|x| - 1$ non-solid edges.

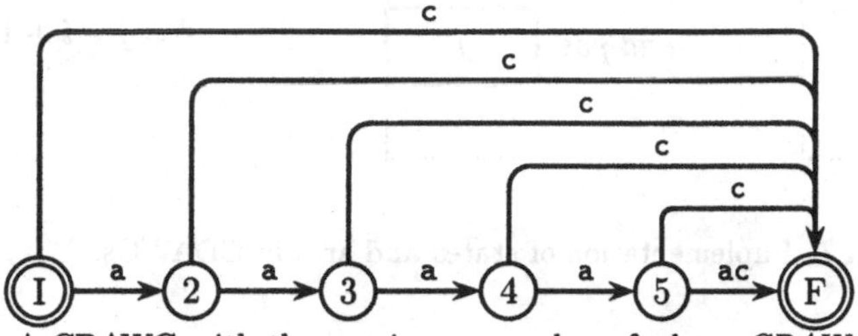

Fig. 6. A CDAWG with the maximum number of edges, *CDAWG*(**aaaaac**).

Figure 6 displays a *CDAWG* having the maximum number of edges for a word of length 6.

3.3 Implementation and experiments

Transition matrices and adjacency lists are two classical implementations of automata. The first one gives a direct access to transitions, but the memory space required is $\mathcal{O}(States(x) \times \text{card}(\Sigma))$. The second implementation stores only the exact number of transitions in memory, but needs $\mathcal{O}(\log\ \text{card}(\Sigma))$ time to access them with standard searching techniques. When the size of the alphabet is great and the transition matrix is sparse, adjacency lists are obviously preferable. Otherwise, like for genomic sequences, transition matrix is a better choice, as shown by the experiments below. So, we only consider here transition matrices to implement CDAWGs.

We now describe the exact implementation of states and edges. We do this on a four-letter alphabet, so characters take 0.25 byte. We use integers encoded with 4 bytes. For each state, to encode the target state of outgoing edges, transitions matrices need a vector of 4 integers. Adjacency lists need, for each edge, 2 integers, one for the target state and another one for the pointer to the next edge.

The basic information required to construct the DAWG is composed of a table to implement the function s_x and one boolean value (0.125 byte) for each edge to know if it is solid or not. For the CDAWG, in order to implement multi-letter transitions, we need one integer for the $endpos_x$ value of each state, and another integer for the label length of each edge. And that is all.

Indeed, we can find the label of a transition by cutting off the length of this transition from the $endpos_x$ value of its target state. Then, we get both the

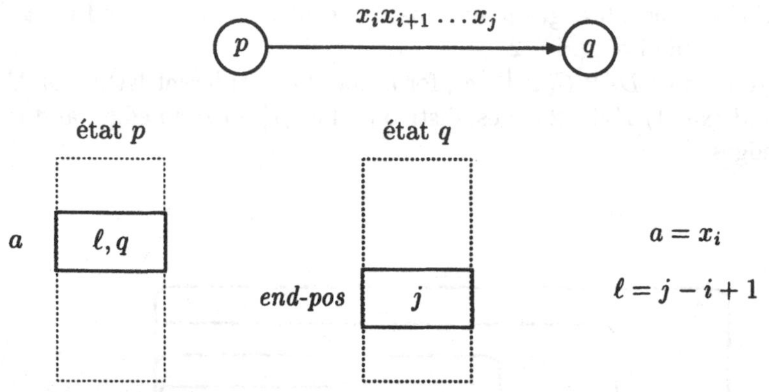

Fig. 7. Implementation of states and arcs in CDAWGs.

position of the label in the source and its length. Figure 7 illustrates this implementation. Keeping the source in memory is negligible considering the global size of the automaton (0.25 byte by character). This is quite a convenient solution also used for suffix trees.

Then, respectively for transitions matrices and adjacency lists, each state requires 20.5 and 17.13 bytes for the DAWG, and 40.5 and 41.21 bytes for the CDAWG. As a reference, suffix trees, as implemented by McCreight [18], need 28.25 and 20.25 bytes per state. Moreover, for CDAWG and suffix trees the source has to be stored in main memory. Theoretical average numbers of states, calculated by Blumer *et al.* ([5]), are $0.54n$ for CDAWG, $1.62n$ for DAWG, and $1,62n$ for suffix trees, when n is the length of x. This gives respective sizes in bytes per character of the source: 45.68 and 32.70 for suffix trees, 33.26 and 27.80 for DAWGs, and 22.40 and 22.78 for CDAWGs.

Considering the complete data structures required for applications, the function $endpos_x$ has to be added for the DAWG and the Suffix Tree. In addition, the occurrence number of each factor has to be stored in each state for all the structures. Therefore, the respective sizes in bytes per character of the source become : 58.66 and 45.68 for suffix trees, 46.24 and 40.78 for DAWGs, and 24.26 and 24.72 for CDAWGs.

Table 1 compares the sizes of implementations of DAWGs and CDAWGs meant for applications to DNA sequences. Sizes for random words of different lengths on a four-letter alphabet are also given. DNA sequences are *Saccharomyces cerevisiae* yeast chromosome II (chro II), a contig of *Escherichia Coli* DNA sequence (coli), and contigs 1 and 115 of *Bacillus Subtilis* DNA sequence (bs). Number of states and edges according to the length of the source and the memory space gain are displayed. Theoretical average ratios are given, computed from [5]. First, we observe there are 2/3 less states in the CDAWG, and near of half edges. Second, the memory space saving is about 50%. Third, the num-

Source x	$\lvert x\rvert$	$\frac{Nb\ states}{\lvert x\rvert}$		$\frac{Nb\ transitions}{\lvert x\rvert}$		$\frac{Nb\ transitions}{Nb\ states}$		memory gain
		dawg	cdawg	dawg	cdawg	dawg	cdawg	
chro II	807188	1,64	0,54	2,54	1,44	1,55	2,66	50,36%
coli	499951	1,64	0,54	2,54	1,44	1,53	2,66	51,95%
bs 1	183313	1,66	0,50	2,50	1,34	1,50	2,66	54,78%
bs 115	49951	1,64	0,54	2,54	1,44	1,55	2,66	50,16%
random	500000	1,62	0,55	2,54	1,47	1,57	2,68	49,53%
random	100000	1,62	0,55	2,55	1,47	1,57	2,68	49,35%
random	50000	1,62	0,54	2,54	1,46	1,56	2,68	49,68%
random	10000	1,62	0,54	2,54	1,46	1,56	2,68	49,47%
theor. aver. ratios		**1,63**	**0,54**	**2,54**	**1,46**	**1,56**	**2,67**	**50,55%**

Table 1. Statistics on the sizes of real DAWGs and CDAWGs.

ber of edges per state is going up to 2.66 when considering CDAWGs. With a four-letter alphabet, this is interesting to note because the implementation by transition matrix requires less space than an implementation by adjacency lists. At the same time, this keeps a direct access to transitions.

3.4 Constructing CDAWGs from DAWGs

The DAWG construction is fully exposed and demonstrated in [3], [9] and [11]. As we show in this section, the CDAWG is easily derived from the DAWG.

Indeed, we just need to apply the definition of the CDAWG. The computation is done by the function *Reduction* below. Observe that, in this function, *state*$(p, a]$ denotes the target state of the transition $(p, a]$. The computation is done during a depth-first traversal of the automaton, and runs in time linear in the number of transitions of $DAWG(x)$. Then, by theorem 2, the computation runs in time linear in the length of the text.

The main drawback of this construction of CDAWGs is that it requires the previous construction of DAWGs. Therefore, the overall construction takes time and memory space proportional to $DAWG(x)$, though $CDAWG(x)$ is significantly smaller. So, it is better to construct the CDAWG directly.

Reduction (state E) returns (ending state, length of redirected edge)
If (E not marked) **Then**
    For all existing edge $(E, a]$ **Do**
        (*state*$(E, a]$, $\lvert$*label*$((E, a])\rvert$) $\leftarrow$ Reduction(ıtastate$(E, a]$);
    mark(E) $\leftarrow$ TRUE;
If (E is of outdegree one) **Then**
    Let $(E, a]$ this edge ;
    Return (*state*$(E, a]$, $1 + \lvert$*label*$((E, a])\rvert$);
Else
    Return (E,1);

4 Direct Construction of CDAWG

In this section, we give the direct construction of CDAWGs. The running time of the algorithm is linear in the size of the input word x on a fixed alphabet. The memory space is proportional to the size of the automaton, and consequently is also linear by Lemmas 7 and 8.

4.1 Algorithm

Since the CDAWG of x is a minimization of its suffix tree, it is rather natural to base the direct construction on McCreight's algorithm [18]. Meanwhile, properties of the DAWG construction are also used, especially the suffix link function (notion that is different from the suffix links of McCreight's algorithm), lengths of longest paths, and positions, as explained in the previous section.

First, we introduce the notions used by the algorithm, some of them are taken from [18]. The algorithm constructs the CDAWG of the word x of length n, noted $x_{0..n-1}$. The automaton is defined by a set of states and transitions, where I and F denotes the initial and the final states respectively. A *partial path* represents a connected sequence of edges between two states of the automaton. A *path* is a partial path that begins at I. The label of a path is the concatenation of the labels of corresponding edges.

The *locus*, or *exact locus*, of a string is the end of the path labeled by the string. The *contracted locus* of a string α is the locus of the longest prefix of α whose locus is defined. -

Preliminary Algorithm Basically, the algorithm that builds *CDAWG*(x) inserts into the current automaton the paths corresponding to all the suffixes of x, from the longest to the shortest suffix. We define suf_i as the suffix $x_{i..n-1}$ of x. We denote by $\mathcal{A}_i$ the automaton constructed after the insertion of all the suf_j for $0 \leq j \leq i$.

Figure 8 displays six steps during the construction of *CDAWG*(aabbabbc). In this figure (and the following), the dashed edges represent suffix links, links that are defined on states and that are used in the next section.

At the beginning of the algorithm the automaton is initialized with the two states I and F only. At step i ($i > 0$), the algorithm inserts a path corresponding to suf_i into $\mathcal{A}_{i-1}$ and produces $\mathcal{A}_i$. The main loop of the algorithm satisfies the following invariant properties:

P1: at the beginning of step i, all suffixes suf_j, $0 \leq j < i$, are paths in $\mathcal{A}_{i-1}$.
P2: at the beginning of step i, the states of $\mathcal{A}_{i-1}$ are in one-to-one correspondence with the longest common prefixes of pairs of suffixes longer than suf_j.

We define $head_i$ as the longest prefix of suf_i which is also a prefix of suf_j for some $j < i$. Equivalently, $head_i$ is the longest prefix of suf_i that is also label of a path in $\mathcal{A}_{i-1}$. We define $tail_i$ as $head_i^{-1} suf_i$.

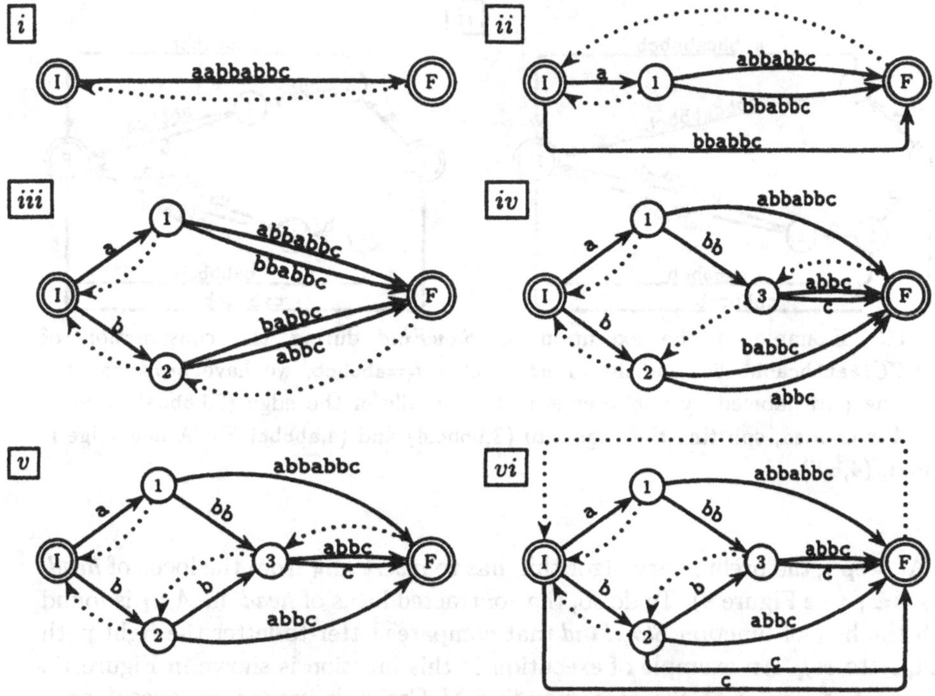

Fig. 8. Six steps during the construction of *CDAWG*(aabbabbc). The pictures display the situation after the insertion of suf_0=aabbabbc (*i*), suf_2=bbabbc (*ii*), suf_3=babbc (*iii*), suf_4=abbc (*iv*), and suf_5=bbc (*v*). *vi* shows the final automaton.

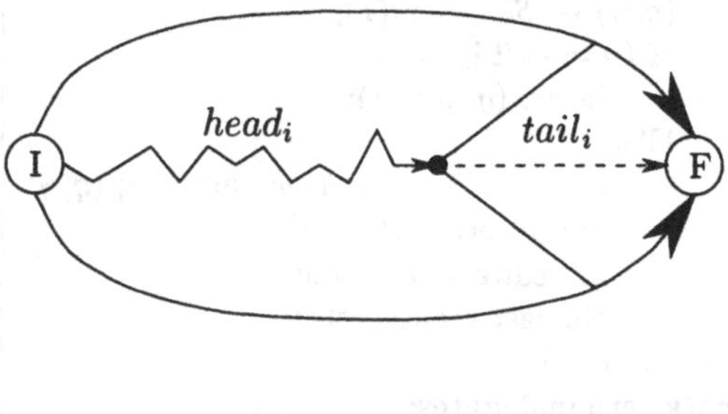

Fig. 9. Scheme of the insertion of suf_i in $\mathcal{A}_{i-1}$: there already is a path labeled by the prefix $head_i$ of suf_i.

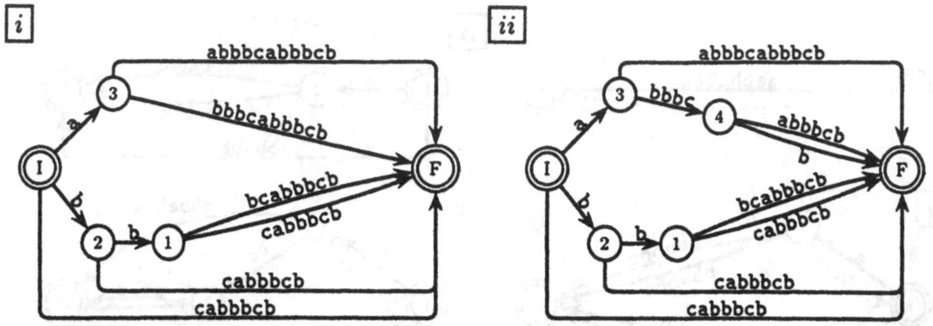

Fig. 10. Example of the execution of *SlowFind* during the construction of *CDAWG*(**aabbbcabbbcb**). For the insertion of suf_6=**abbbcb**, we have $head_6$=**abbbc**. Since the path labeled by **abbbc** ends in the middle of the edge (3,**bbbcabbbcb**,F), state 4 is created, splitting the edge into (3,**bbbc**,4) and (4,**abbbcb**,F). A new edge is created, (4,**b**,F).

At step i, the preliminary algorithm has to insert $tail_i$ from the locus of $head_i$ into $\mathcal{A}_{i-1}$ (see Figure 9). To do so, the contracted locus of $head_i$ in $\mathcal{A}_{i-1}$ is found with the help of function *SlowFind* that compares letter-to-letter the right path of $\mathcal{A}_{i-1}$ to suf_i. An example of execution of this function is shown in Figure 10. This part is similar to the corresponding McCreight's procedure, except on a point discussed below (redirection of edges). If there is a state at the end of the path, it is the locus of $head_i$. Otherwise it is created at the middle of the last encountered edge by splitting it. In any case, an edge labeled by $tail_i$ is created from the locus of $head_i$ to F. The preliminary algorithm is given below.

Preliminary Algorithm

1. **For all** suf_i ($i \in$[0..n-1]) **Do**
2. $(q, \gamma) \leftarrow$ *SlowFind*(I);
3. **If** $(\gamma = \varepsilon)$ **Then**
4. insert (q,$tail_i$,F);
5. **Else**
6. create v locus of $head_i$ splitting $(q, \gamma]$ and insert (v,$tail_i$,F); or redirect $(q, \gamma]$ onto v, the last created state;
7. **End For all**;
8. mark terminal states;

The function *SlowFind* returns a pair (q, γ) such that q is the last encountered state on the path $head_i$, state that is the representative of $head_i\gamma^{-1}$. This keeps accessible the transition that may be split if the state q is not the exact locus of $head_i$, *i.e.* if $\gamma \neq \varepsilon$.

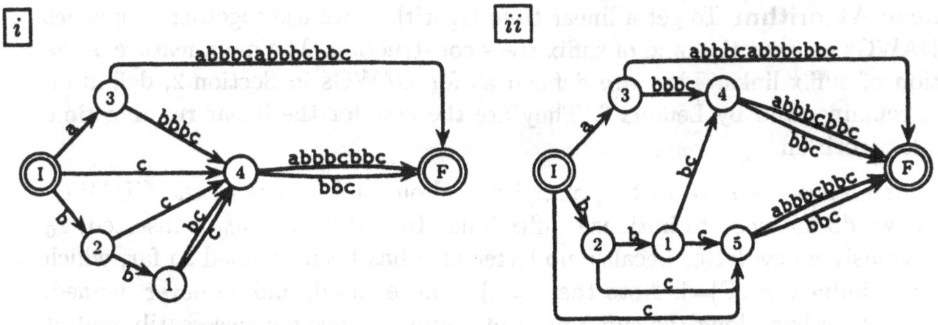

Fig. 11. Example of a duplication in *SlowFind* during the construction of *CDAWG*(**aabbbcabbbcbbc**). The insertion of suf_{11}=**bbc** leads to state 4. As the last edge (1,c,4) is non-solid (*i*), state 4 is cloned into state 5 (*ii*), and the edge (1,c,4) becomes (1,c,5).

If a non-solid edge is encountered during the execution of *SlowFind*, its target state has to be duplicated in a clone and the non-solid edge is redirected to this clone. The redirected transition becomes solid. An example of duplication is given in Figure 11.

In some situation, an edge can be redirected. This happens when a state has just been created at the previous step. The edge is redirected to this state and its label is updated accordingly. Such a situation appears in Figure 8 (case v) for the construction of *CDAWG*(**aabbabbc**) : the insertion of suf_5=**bbc** induces the redirection of the edge (2,babbc,F), which becomes (2,b,3). In the above situation, the suffix link of the last created state is unknown during the insertion of the current suffix. And the redirections go on until the suffix link is found.

Finally, when $tail_i = \varepsilon$ at the end of the construction, terminal states are marked along the suffix path of F.

From the above discussion, a proof of the invariance of properties P1 and P2 can be derived. Thus, at the end of the algorithm all subwords of x and only these words are labels of paths in the automaton (property P1). By property P2, states correspond to strict classes of factors (when the longest common prefix of a pair of suffixes is not equal to any of them) or to terminal states (when the contrary holds). This gives a sketch of the correctness of the algorithm.

The running time of the preliminary algorithm is $\mathcal{O}(|x|^2)$ (with an implementation by transition matrix), like is the sum of lengths of all suffixes of the word x.

Linear Algorithm To get a linear-time algorithm, we use together properties of DAWGs construction and of suffix trees construction. The main feature is the notion of suffix links. They are defined as for DAWGs in Section 2, definition that remains valid by Lemma 6. They are the clue for the linear running time of the algorithm.

Three elements have to be pointed out about suffix links in the CDAWG. First, we do not need to initialize suffix links. Indeed, when suf_0 is inserted, x_0 is obviously a new letter because no letter of x has been scanned so far, which directly induces s_x(F)=I. Note that s_x(I) is never used, and so never defined. Second, traveling along the suffix path of a state p does not necessarily end at state I. Indeed, with multi-letter transitions, if $s_x(p)$=I we have to treat the suffix $a^{-1}\alpha$ ($a \in \Sigma$) where α is the representative of p. And third, suffix links induce the following invariant property satisfied at step i:

P3: at the beginning of step i, the suffix links are defined for each state of $\mathcal{A}_{i-1}$ according to Definition 4, except maybe for the lastly-created state.

The next remark allows redirections without having to search with *SlowFind* for existing states belonging to a same class of factors.

Remark. Let $\alpha\beta$ have locus p and assume that $q = s_x(p)$ is the locus of β. Then, p is the locus of suffixes of $\alpha\beta$ whose lengths are greater than $|\beta|$.

The algorithm has to deal with suffix links each time a state is created. This happens when a state is duplicated, as illustrated by Figure 11, and when a state is created after the execution of *SlowFind*.

During a duplication, suffix links are updated as follows. Let w be the clone of q. In regard to strict classes of factors and Definition 4, the class of w is inserted "between" the ones of q and $s_x(q)$. So, we update suffix links by setting $s_x(w) = s_x(q)$ and then $s_x(q) = w$.

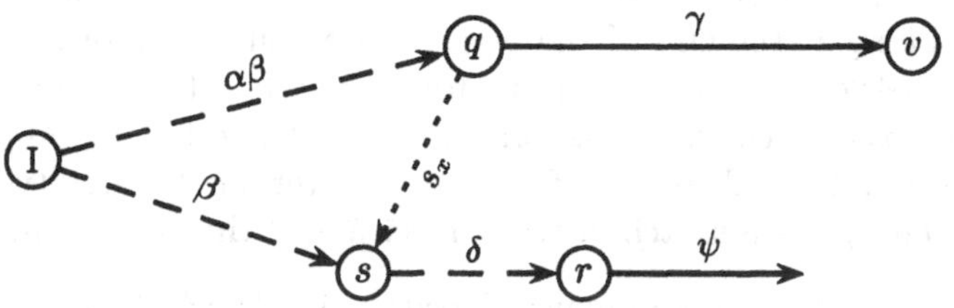

Fig. 12. Searching for $s_x(v)$ using a suffix link.

After the execution of *SlowFind*, if state v is created, we have to compute its suffix link, $s_x(v)$. Let γ be the label of the transition starting at q and ending at v. To compute the suffix link of v, the algorithm goes through the path having label γ from the suffix link of q, $s = s_x(q)$. The operation is repeated if necessary. Figure 12 displays a scheme of this search. The thick dashed edges represent paths in the automaton, and the thin dashed edge represents the suffix link from q to s. The search, as for the duplication, realizes the insertion of a series of suffixes. To travel along the path, we use the function *FastFind*, similar to the one used in McCreight's algorithm [18], that goes through transitions comparing just the first letters of their labels. This function returns the last encountered state and edge.

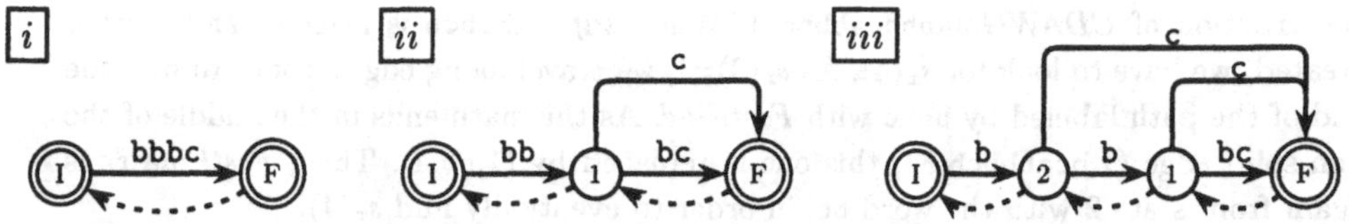

Fig. 13. Example of execution of *FastFind* ending with a solid edge during the construction of *CDAWG*(bbbc). The insertion of suf_1=bbc leads to create state 1. Then *FastFind* works from I with path b. This leads to the middle of the edge (I,bb,1) (*ii*) that is solid. Since we cannot redirect this edge, state 2 is created, splitting (I,bb,1) into (I,b,2) and (2,b,1) (*iii*). The edge (2,c,F) is added, $s_x(1)$ is set to 2, and $s_x(2)$ is set to I.

Let r and $(r, \psi]$ be the state and transition returned by *FastFind*. If r is the exact locus of γ, it is the wanted state, and we set then $s_x(v) = r$. Else, if $(r, \psi]$ is a solid edge, then a new node w is created. The edge $(r, \psi]$ is split, its initial part becomes (r, ψ, w), and the transition (w,$tail_i$,F) is added. Such an example is displayed in Figure 13.

The last situation to consider is when $(r, \psi]$ is non-solid. Then, the edge is replaced by (r, ψ, v). Such an example is displayed in Figure 14.

In the two last cases, since $s_x(v)$ is not found, we run *FastFind* again with $s_x(r)$ and ψ, and this goes on until $s_x(v)$ is eventually found, that is, when $\psi = \varepsilon$.

FastFind is used in the same manner when a state is created by duplication during the execution of *SlowFind*.

The discussion shows how suffix links are updated to insure that property P3 is satisfied. The operations do not influence the correctness of the algorithm, sketched in the last section, but yield the following linear-time algorithm. Its time complexity is discussed in the next section.

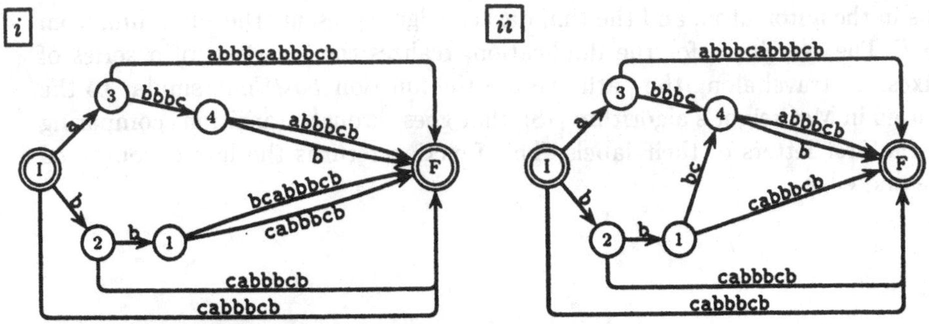

Fig. 14. Example of execution of *FastFind* ending with a non-solid edge during the construction of *CDAWG*(aabbbcabbbcb). When suf_6=abbbcb is inserted and state 4 created, we have to look for $s_x(4)$. As $s_x(3)$=I, we travel along edges from I to find the end of the path labeled by bbbc with *FastFind*. As this path ends in the middle of the non-solid edge (1,bcabbbcb,F), this one is replaced by (1,bc,4). Then, *FastFind* runs again from state 2 with the word bc, in order to eventually find $s_x(4)$.

Linear Algorithm	
1.	$p \leftarrow$ I; $\quad i \leftarrow 0$;
2.	**While** not end of x **Do**
3.	$\quad (q, \gamma) \leftarrow SlowFind(p)$;
4.	\quad **If** $(\gamma = \varepsilon)$ **Then**
5.	$\quad\quad$ insert $(q, tail_i, F)$;
6.	$\quad\quad s_x(F) \leftarrow q$;
7.	$\quad\quad$ **If** $(q \neq$ I$)$ **Then** $p \leftarrow s_x(q)$ **Else** $p \leftarrow$ I;
8.	\quad **Else**
9.	$\quad\quad$ create v locus of $head_i$ splitting $(q, \gamma]$;
10.	$\quad\quad$ insert $(v, tail_i, F)$;
11.	$\quad\quad s_x(F) \leftarrow v$;
12.	$\quad\quad$ find $r = s_x(v)$ with *FastFind*;
13.	$\quad\quad p \leftarrow r$;
14.	\quad update i;
15.	**End While**;
16.	mark terminal states;

4.2 Complexity

Theorem 9. *The algorithm that builds the CDAWG of a word x of Σ^* can be implemented in time $\mathcal{O}(|x|)$ and space $\mathcal{O}(|x| \times \mathrm{card}(\Sigma))$ with a transition matrix, or in time $\mathcal{O}(|x| \times \log \mathrm{card}(\Sigma))$ and space $\mathcal{O}(|x|)$ with adjacency lists.*

Proof. As recalled in section 3.1, the size of *CDAWG*(x) is linear in the length of x, both in term of number of states and number of edges. Tables $endpos_x$,

$length_x$ and s_x take $\mathcal{O}(States(x))$ space. So, an implementation by transition matrix takes $\mathcal{O}(|x| \times \text{card}(\Sigma))$ space. By adjacency lists, it takes $\mathcal{O}(|x|)$ space.

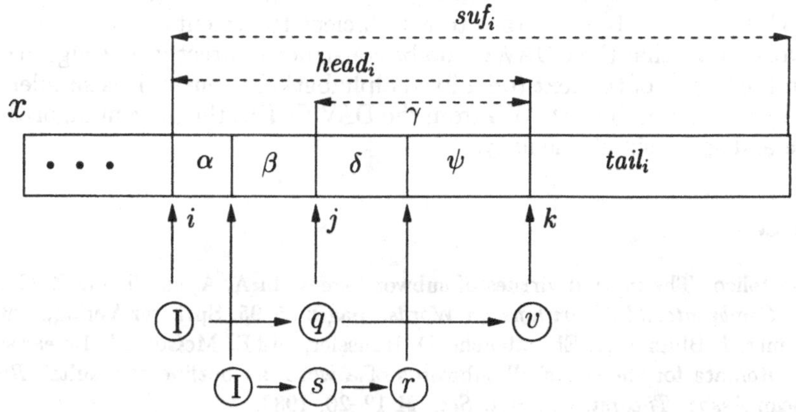

Fig. 15. Positions of labels when suf_i is inserted. States I,q,v represent the scheme of *SlowFind* and states I,s,r represent the scheme of searching for $s_x(q)$, as in Figure 12.

The complexity of the algorithm essentially depends on the number of branchings made on states of the automaton. We prove that this number is linear, which implies the running times of the statement: $\mathcal{O}(|x|)$ with a transition matrix and $\mathcal{O}(|x| \times \log\ \text{card}(\Sigma))$ with adjacency lists.

Branchings during the execution of the algorithm are done during calls to *SlowFind* and *FastFind*. The generic situation is displayed in Figure 15. When *SlowFind* operates, the current letter of x, pointed by k, is compared with a letter of the label of an edge. Doing so, k is strictly incremented, and never after decremented. During calls to *FastFind*, each letter comparison increases strictly the value of j, value that never decreases hereafter. This shows that the number of branchings is linear.

This ends the sketch of the proof.

5 Conclusion

We have considered the Compact Direct Acyclic Word Graph, which is an efficient compact data structure to represent all subwords, or factors, of a word. There are several data structures used to store this set. The present structure provides an interesting space gain compared to the standard DAWG, and also when compared with suffix trees. From the theoretical point of view, the upper bounds are of $|x| + 1$ states and $2|x| - 2$ transitions. This saves $|x|$ states and $|x|$ transitions of the DAWG and at the same time leads to a faster use. From the practical point of view, experiments on genomic DNA sequences and on random strings display a memory space gain of 50% with respect to the DAWG. Moreover, when the size of the alphabet is small, transition matrices do not take more space than adjacency lists, keeping direct access to transitions. Thus, we can construct the data structure of twice larger strings, keeping them in main memory, which is actually important to get efficient treatments.

This work shows that the CDAWG can be constructed directly. The algorithm is linear in the length of the text (on a fixed alphabet). Of course, it is simpler to compute, by reduction, the CDAWG from the DAWG. But the present algorithm saves time and space simultaneously.

References

1. A. Apostolico. The myriad virtues of subword trees. In A. Apostolico & Z. Galil, editor, *Combinatorial Algorithms on Words.*, pages 85–95. Springer-Verlag, 1985.
2. A. Blumer, J. Blumer, A. Ehrenfeucht, D. Haussler, and R. McConnel. Linear size finite automata for the set of all subwords of a word: an outline of results. *Bull. European Assoc. Theoret. Comput. Sci.*, 21:12–20, 1983.
3. A. Blumer, J. Blumer, D. Haussler, A. Ehrenfeucht, M.T. Chen, and J. Seiferas. The smallest automaton recognizing the subwords of a text. *Theoret. Comput. Sci.*, 40:31–55, 1985.
4. A. Blumer, J. Blumer, D. Haussler, and R. McConnell. Complete inverted files for efficient text retrieval and analysis. *Journal of the Association for Computing Machinery*, 34(3):578–595, July 1987.
5. A. Blumer, D. Haussler, and A. Ehrenfeucht. Average sizes of suffix trees and dawgs. *Discrete Applied Mathematics*, 24:37–45, 1989.
6. B. Clift, D. Haussler, R. McDonnell, T.D. Schneider, and G.D. Stormo. Sequence landscapes. *Nucleic Acids Research*, 4(1):141–158, 1986.
7. M. Crochemore. Recherche linéaire d'un carré dans un mot. *C. R. Acad. Sci. Paris Sér. I Math.*, 296:781–784, 1983.
8. M. Crochemore. Optimal factor tranducers. In A. Apostolico and Z. Galil, editors, *Combinatorial Algorithms on Words*, volume 12 of *NATO Advanced Science Institutes, Series F*, pages 31–44. Springer-Verlag, Berlin, 1985.
9. M. Crochemore. Transducers and repetitions. *Theoret. Comput. Sci.*, 45(1):63–86, 1986.
10. M. Crochemore. Longest common factor of two words. In H. Ehrig, R. Kowalski, G. Levi, and U. Montanari, editors, *TAPSOFT*, number 249 in Lecture Notes in Computer Science, pages 26–36. Springer-Verlag, Berlin, 1987.
11. M. Crochemore and C. Hancart. Automata for matching patterns. In G. Rozenberg and A. Salomaa, editors, *Handbook of Formal Languages*. Springer-Verlag, 1997. to appear.
12. M. Crochemore and W. Rytter. *Text Algorithms*, chapter 5-6, pages 73–130. Oxford University Press, New York, 1994.
13. M. Farach. Optimal suffix tree construction with large alphabets. manuscript, October 1996.
14. R. W. Irving. Suffix binary search trees. *Technical report TR-1995-7, Computing Science Department, University of Glasgow*, April 1995.
15. J. Karkkainen. Suffix cactus : a cross between suffix tree and suffix array. *Combinatorial Pattern Matching*, 937:191–204, July 1995.

16. C. Lefevre and J-E. Ikeda. The position end-set tree: A small automaton for word recognition in biological sequences. *CABIOS*, 9(3):343–348, 1993.
17. U. Manber and G. Myers. Suffix arrays: A new method for on-line string searches. *SIAM J. Comput.*, 22(5):935–948, Oct. 1993.
18. E. McCreight. A space-economical suffix tree construction algorithm. *Journal of the ACM*, 23(2):262–272, Apr. 1976.
19. G. A. Stephen. *String searching algorithms*. World Scientific Press, 1994.
20. E. Ukkonen. On-line construction of suffix trees. *Algorithmica*, 14:249–260, 1995.
21. E. Ukkonen and D. Wood. Approximate string matching with suffix automata. *Algorithmica*, 10(5):353–364, 1993.
22. P. Weiner. Linear pattern matching algorithm. In *14th Annual IEEE Symposium on Switching and Automata Theory*, pages 1–11, Washington, DC, 1973.

Metric Entropy and Minimax Risk in Classification

David Haussler[1] and Manfred Opper[2]

[1] Computer Science, UC Santa Cruz, CA 95064, USA
[2] Dept. of Physics, Universität Würzburg, Germany

Abstract. We apply recent results on the minimax risk in density estimation to the related problem of pattern classification. The notion of loss we seek to minimize is an information theoretic measure of how well we can predict the classification of future examples, given the classification of previously seen examples. We give an asymptotic characterization of the minimax risk in terms of the metric entropy properties of the class of distributions that might be generating the examples. We then use these results to characterize the minimax risk in the special case of noisy two-valued classification problems in terms of the Assouad density and the Vapnik-Chervonenkis dimension.

1 Introduction

The most basic problem in pattern recognition is the problem of classifying instances consisting of vectors of measurements into a one of a finite number of types or *classes*. One standard example is the recognition of isolated capital characters, in which the instances are measurements on images of letters and there are 26 classes, one for each letter. Another example is the classification of aircraft according to features extracted from their radar images. Problems of this type are called *classification problems* in statistics. In this paper we derive theoretical bounds on the best performance that can be obtained for statistical methods that perform classification.

Let us denote the entire collection of measurements for a single instance by x. We will refer to x as an *instance* or a *feature vector*. The feature vector is a random quantity that varies from instance to instance, so we will also use the random variable X, or the random variables $X_1, \ldots, X_n$ to refer to a random instance, and n independently selected random instances, respectively. We use the notation $X_i = x$ to denote that the set of measurements of the ith random instance is x. The classes in our preselected family will be numbered $\{1, \ldots, K\}$. For a given instance x, the true class of the instance will be denoted by $y \in \{1, \ldots, K\}$, or by the random variable Y. The pair (x, y), or (X, Y), will be called an *example*. A sequence of n independent random examples will be denoted $(X_1, Y_1), \ldots, (X_n, Y_n)$ or $(x_1, y_1), \ldots, (x_n, y_n)$ and abbreviated by S_n.

One simple view of a method of classification is as a function that takes as input an instance x and outputs a classification $y \in \{1, \ldots, K\}$. However, in practice one cannot be certain of one's classification, so it is better, given an instance

x, to output a probability distribution $\{\hat{P}(Y = y|X = x) : 1 \leq y \leq K\}$ that specifies the estimated probability of each of the possible classes for the instance x. We refer to this distribution as the *predictive distribution*. The predictive distribution tells not only which class is deemed most likely, but how confident the system is in that classification, and what, if any, are good alternative classifications. If, as is nearly always the case, there are different costs associated with making different kinds of misclassifications, then a separate decision making module can use the predicted probabilities produced by the classification system to decide on the optimal action to take. The theory of making optimal decisions from given these probability distributions is quite simple, and is treated fully in standard texts such as that by Duda and Hart [18], so we will not elaborate on it here. Rather we will focus solely on the problem of obtaining accurate predictive distributions, which is the critical part of the problem.

The predictive distribution can be estimated by estimating the joint probability distribution over the random variables X and Y. This joint distribution is usually broken down into a *prior distribution* $\{P(Y = y) : 1 \leq y \leq K\}$ for the values of Y, specifying which of the classes are *a priori* more likely than others, and a *generative model* that gives a conditional probability $P(X = x|Y = y)$ for each x and y, which specifies a distribution over the instances for each class. The estimated predictive probability distribution $\{\hat{P}(Y = y|X = x) : 1 \leq y \leq K\}$ is then obtained by applying Bayes rule.

When a classification method is trained to estimate the joint distribution on X and Y, a set of independent random examples $S_n = (x_1, y_1), \ldots, (x_n, y_n)$ is used. We will refer to this as the *training set*. In this process, one cannot explore all possible joint distributions. Nor would one want to explore all possible distributions, since given only a finite training set, it is impossible, using only a moderate sized training set, to pick out a good distribution from the set of all possible distributions with any kind of statistical reliability in all but trivial cases. Rather, it is up to the designer of the system to use his knowledge to pick a particular class of joint distributions on X and Y from which to choose his statistical model. We will refer to this class as Θ, and let $\theta \in \Theta$ denote a particular model in this class. Formally, each θ is the name or index for a joint probability distribution P_θ on X and Y. We will denote probabilities under the particular distribution indexed by θ by conditioning on θ. For example, the probability that $Y = y$ given that $X = x$, using the joint distribution indexed by θ, will be denoted $P(Y = y|x, \theta)$. Note that we have also abbreviated by conditioning on x only, rather than conditioning on $X = x$. We will also do this henceforth, to shorten our notation.

When assessing the performance of a classification system, there are two key issues to address: How well does the best distribution in Θ approximate the true joint distribution on X and Y, and how close does the method of estimation get to finding the best distribution in Θ. The difference between the best model in Θ and the true joint distribution is called the *approximation error*, and the difference between the model that the estimation method finds and the best model in Θ is called the *estimation error*. There is usually a tradeoff between the approximation and estimation errors: the larger one makes Θ, the more the approximation error can be reduced, but the more the estimation error increases. In order to optimize this tradeoff, it is important that the designer of a classification method have good bounds on the approximation and estimation errors for the class Θ of models he is using. There is a good general theory on approximation error, starting with the fundamental theorems of approximation theory, as given, for example, in the classic book of Lorentz [33] (see also [31, 15]). While in specific cases this error depends strongly on the nature of the true distribution, which is unknown, one can still make statements about the general approximability of functions or distributions in one family by functions or distributions in another. On the other hand, the estimation error can be analyzed somewhat independently from the true distribution, as we will show below. We focus on the estimation error in this paper. We show that rigorous bounds on the estimation error of the best possible classification systems can be established making a minimal set of assumptions.

Because we analyze estimation error, all our performance bounds will be relative to the performance of the best possible model in Θ. For this reason we will refer to Θ as the *comparison class*. If a classification method takes the training examples and produces from them an estimated model $\hat{\theta} \in \Theta$, we will ask how well does $\hat{\theta}$ perform compared to the best model θ^* in the comparison class Θ. Performance will be assessed on further random examples drawn from the same joint distribution used to generate the training examples. However, we will also consider methods that use an estimated model $\hat{\theta}$ that is not a member of the comparison class Θ in order to make their predictions. For example, Bayes methods use a weighted mixture of models in Θ to make predictions, and often this mixture does not correspond to any single model in Θ. In a Bayes method, a prior distribution over the parameter space Θ is specified. Combined with the observed examples, this prior generates a posterior distribution on Θ. The predictive distribution is then obtained by integrating over all the conditional distributions in Θ, weighted according to this posterior distribution (see e.g. [23]). Some of the most successful classification methods are Bayes methods, or computationally efficient approximations to Bayes methods. We will discuss these methods further in the last section of this paper, after we have established the basic theory of estimation error.

This paper is organized as follows: In section 2 we give a minimax definition of the estimation error for a comparison model class Θ. We use relative entropy to measure the difference between the learner's predictive distribution $\hat{P}$ and the best distribution in Θ. Then in the following three sections we develop the theoretical tools needed to determine the rate at which this estimation error converges to zero as a function of the sample size n. The main concepts used are the Hellinger distance, and the metric entropy of Θ with respect to this distance. Then in section 6 we look at the problem of cumulative minimax risk for a series of predictions made on-line by an adaptive classification method. It turns out that tighter estimates of the convergence rate of the estimation error can be made in this case using the general theory. Following this, in section

7, we compare the classification results obtained this way to the results that can be obtained using the Vapnik-Chervonenkis theory [43]. Here we restrict ourselves to a special problem of two-class classification that has been called "noisy concept learning" in the AI and computational learning theory literature [40, 1, 10, 11, 21]. We show how fairly precise, general rates can be obtained for this problem based on a combinatorial parameter known as the Assouad density [2], which is related to the VC dimension [44]. Finally, we review the implications of these results in the closing section, section 8.

The main results given here are derived from results in [28] (see also [35, 27, 22, 34]), where a general theory of minimax estimation error using relative entropy is developed that applies not only to classification problems in the form that we have defined them, but to other important statistical problems, including regression and density estimation.

There is a large statistical literature on minimax rates for estimation error for general statistical problems. However, much of this work has been done using loss functions other than relative entropy, see e.g. the texts [17, 30]. Our line of investigation, based on relative entropy, has its roots in the early work by Ibragimov and Hasminskii, who showed that the cumulative relative entropy risk for Bayes methods for parametric density estimation on the real line is approximately $(d/2)\log n$, where d is the number of parameters and n is the sample size [29]. In this case they were even able to estimate the lower order additive terms in this approximation, which involve the Fisher information and the entropy of the prior. Further related results were given by Efroimovich [20] and Clarke [12]. Clarke and Barron gave a detailed analysis, with applications, of the risk of the Bayes strategy [13], discussing the relation of the cumulative relative entropy loss to the notion of redundancy in information theory, and giving applications to hypothesis testing and portfolio selection theory. These results were extended to the cumulative relative entropy Bayes and minimax risk in [14] (see also [5]). Related lower bounds, which are often quoted, were obtained by Rissanen [37], based on certain asymptotic normality assumptions.

Estimations of the relative entropy risk in nonparametric cases were obtained in [4, 6, 38, 45, 46]. General approaches, for loss functions other than the relative entropy, to minimax risk in nonparametric density estimation were pioneered by Le Cam, who introduced methods using metric entropy and Hellinger distance (see e.g. [32]). This approach is further developed in [7, 8, 25, 41, 9, 6, 45, 42, 28]. The results of sections 2 through 5 show how this theory can be applied to the classification problem.

2 Using relative entropy and minimax risk to define estimation error

Let us summarize the problem we are considering in its abstract setting. The training data is a sequence of examples $S_n = (x_1, y_1), \ldots, (x_n, y_n)$, where each instance x_t is an element of an arbitrary set X, the *instance space* and each outcome y_t is an element of a finite set Y, the *outcome space*. (Here and below we will use X to denote both a random instance, and the instance space from which it is drawn, and similarly for Y. The usage will be clear from the context.) Given this training data and a new instance $x \in X$, a classification method produces an estimated predictive distribution

$$\hat{P}(Y = y|x, S_n)$$

that specifies the estimated probability that the outcome will be y, for each possible outcome $y \in Y$, given that the instance is x, and given the previous training examples. Note that in this notation we explicitly show the dependence of this estimated distribution on the previous training examples, whereas this dependence was implicit in the previous section. To evaluate the performance of the classification method, we have a comparison model class Θ, where each $\theta \in \Theta$ denotes a joint distribution P_θ on X and Y, here viewed as random variables. We are concerned with the estimation error, which we have defined informally as the difference between the performance of the classification method in estimating the outcome Y, and the performance of the best model in Θ. We now formalize this notion.

2.1 General setting

When assessing performance, we need a function that measures how much the predictive distribution $\hat{P}(Y = y|x, S_n)$ differs from the distribution $P(Y = y|x, \theta^*)$ produced by the best model $\theta^* \in \Theta$. While there are several functions that are often used in the literature to measure the difference between two probability distributions, the most natural one to choose here is the *relative entropy* or *Kullback-Leibler divergence*. This measure has a deep and useful information-theoretic interpretation, and it is also arises naturally in related statistical contexts, where loglikelihood ratios play a fundamental role. For two discrete probability distributions $P = (p_1, \ldots, p_K)$ and $Q = (q_1, \ldots, q_K)$, the relative entropy between P and Q is defined by

$$D_{KL}(P||Q) = \sum_{i=1}^{K} p_i \log \frac{p_i}{q_i}.$$

This quantity is nonnegative, and is 0 if and only if $P = Q$. In information theory, $-\log p_i$ is the amount of information contained in the event i under the distribution P, or equivalently, the minimum number of bits (if logarithm base 2 is used) it takes to encode the event i in the optimal (block) code based on the the distribution P is used. The relative entropy $D_{KL}(P||Q)$ is the difference between the average number of bits to encode an event when the true probability distribution is P and the optimal code based on the distribution P is used, and the average number of bits when the true distribution is P, but the optimal code based on the distribution Q is used. This is called *redundancy* in information theory. It is a measure of the regret you have at using the distribution Q to define your code, instead of the optimal (true) distribution P.

We can use the relative entropy to define a regret that is suffered if we use some nonoptimal estimate $\hat{P}(Y = y|x, S_n)$ instead of the best distribution $P(Y = y|x, \theta^*)$. The relative entropy between these two distributions is

$$\sum_y P(Y = y|x, \theta^*) \log \frac{P(Y = y|x, \theta^*)}{\hat{P}(Y = y|x, S_n)},$$

which we can write for short as $D_{KL}(P(\cdot|x, \theta^*)||\hat{P}(\cdot|x, S_n))$. The loglikelihood ratio

$$\log \frac{P(Y = y|x, \theta^*)}{\hat{P}(Y = y|x, S_n)}$$

plays a fundamental role here, and will be referred to as the *loss* for the particular prediction for outcome y. Of course, if the predictive distribution gives higher probability than the distribution P_{θ^*} does to the outcome y that actually occurs, then this loss is negative, and thus may be interpreted as a gain. The relative entropy is the average loss, assuming that the outcome y is generated at random according to the best distribution P_{θ^*}, where θ^* in Θ. This distribution P_{θ^*} is often referred to as the *true distribution*, since it plays that role in this analysis.

The average regret is called the *risk* in statistics. Here, to define the risk, we average over possible training sets S_n and possible instances x. We also assume that these are generated according to the true distribution P_{θ^*}. Thus for all $n \geq 0$ we can define the risk as

$$r_{n+1,\hat{P}}(\theta^*) = \int_{(X\times Y)^n} dP^n_{\theta^*}(S_n) \int_X dP^{(marg)}_{\theta^*}(x) D_{KL}(P(\cdot|x, \theta^*)||\hat{P}(\cdot|x, S_n)).$$

Here $\int_{(X\times Y)^n} dP^n_{\theta^*}(S_n)$ denotes expectation with respect to the random choice of the training set S_n, chosen according to the n-fold product distribution on $X \times Y$ defined by the parameter θ^*, and $\int_X dP^{(marg)}_{\theta^*}$ denotes the expectation with respect to an additional random instance x, chosen according to the marginal distribution on X defined by the parameter θ^*.

From the properties of relative entropy, the risk $r_{n,\hat{P}}(\theta^*)$ is a nonnegative number for every n, and is 0 only when the estimated distribution is the same as the true distribution (with probability 1). Once a comparison class Θ is chosen, the goal in designing a classification method $\hat{P}$ is to make the risk $r_{n,\hat{P}}(\theta^*)$ as small as possible for each n and $\theta^* \in \Theta$. However, any method $\hat{P}$ will work better for some θ^* and worse for others, so there is always some "risk" in choosing a method $\hat{P}$, since one might get a true distribution P_{θ^*} that is unfavorable for that method. To deal with this, when evaluating methods we can look at the *minimax risk*, which is defined as the minimum over all classification methods of the maximum risk over all true distributions in Θ, i.e.

$$r_n^{minimax} = r_n^{minimax}(\Theta) = \inf_{\hat{P}} \sup_{\theta^* \in \Theta} r_{n,\hat{P}}(\theta^*).$$

We define the estimation error for Θ for training samples of size n to be this minimax risk $r_n^{minimax}(\Theta)$. It represents the best possible worst case performance that can be achieved for any classification method using n training examples, when the true distribution is in Θ.

Note that since Y is finite, the minimax risk $r_n^{minimax}$ is bounded by $\log|Y|$, the logarithm of the cardinality of the outcome space Y. To see this, note that we can always set $\hat{P}$ to just predict a uniform distribution on all outcomes in Y, and in this case, no matter what the true conditional distribution on Y given x is, the regret will be at most $\log|Y|$. This is because the relative entropy from any distribution on a finite set to the uniform distribution is at most the logarithm of the cardinality of the set.

2.2 Assumption of a common marginal distribution

It is difficult to obtain an accurate and completely general analysis of the minimax risk for arbitrary Θ. However, since our interest is only in predicting Y given X, and not in predicting X itself, it is reasonable to consider a case where all the joint distributions in Θ share the same marginal distribution on X, and the only difference among the various $\theta \in \Theta$ is in the conditional distribution on Y given X. In this case each joint distribution may be decomposed into a conditional distribution on Y given X, which we denote by $P_\theta(Y = y|x)$ or $P(Y = y|x, \theta)$, and a marginal distribution on X, which we denote $P_\phi(x)$, for a new, fixed parameter ϕ, since this marginal is the same for all θ. The joint distribution on $X \times Y$ will be denoted by $P_{\theta,\phi}(x, y) = P_\theta(Y = y|x)P_\phi(x)$, or, with some abuse of notation, simply by P_θ when the common marginal distribution on X is only implicitly defined. The comparison model class itself, consisting of the set of all joint distributions $\{P_{\theta,\phi}|\theta \in \Theta\}$ will henceforth be represented by the pair (Θ, ϕ) when we wish to make the common marginal distribution on X, P_ϕ, explicit, and otherwise it will be represented simply as Θ, leaving the common marginal distribution implicitly defined. Notation for risk and minimax risk will be similarly extended to include a specific subscript for ϕ when needed. Analysis of this special case focuses attention on the conditional distributions, which are what we are really trying to learn. We restrict our attention to this case for the remainder of this paper.

Now let us consider each labeled example (x, y) as if it were a single random variable $z = (x, y)$, with distribution defined by the parameters θ and ϕ. Consider the problem of estimating the distribution $P_{\theta,\phi}(z) = P_\theta(Y = y|x)P_\phi(x)$ from a random sample $S_n = (x_1, y_1), \ldots, (x_n, y_n) = z_1, \ldots, z_n$, drawn independently according to the unknown distribution $P_{\theta,\phi}(z)$, knowing that $\theta \in \Theta$, and knowing the common marginal distribution ϕ on X. Estimating the distribution of a random variable Z from independent observations $z_1, \ldots, z_n$ is a well studied problem, which we will call the problem of *density estimation*, even if the resulting estimate is in the form of a more general probability distribution, and not a simple density. Here we have defined a special type of density estimation problem. Intuitively, since we already know the marginal distribution on X, this special type of density estimation problem should just boil down to estimating the conditional distribution on Y given X, which is the pattern recognition problem we are studying in this paper. We show this formally below. This reduction allows us to use results derived for the more general density estimation problem when analyzing the pattern recognition problem.

To see how this reduction works, first let us define the risk function for the density estimation problem as in [28]. Denote the estimate for the distribution of Z given a sample S_n by $\hat{P}(z|S_n)$. The risk in density estimation is defined to be the average relative entropy between the true distribution and the predicted distribution, i.e.

$$r^{density}_{n+1,\hat{P}}(\theta^*) = r^{density}_{n+1,\hat{P},\phi}(\theta^*) = \int_{Z^n} dP^n_{\theta^*,\phi}(S_n) D_{KL}(P_{\theta^*,\phi}(\cdot)||\hat{P}(\cdot|S_n)).$$

This is analogous to the definition of the risk $r_{n+1,\hat{P}}(\theta^*)$ defined above for the pattern recognition problem.

Now suppose that $\hat{P}(Y = y|x, S_n)$ is a predictive distribution for the pattern recognition problem. Let us define the corresponding estimate for the density estimation problem by $\hat{P}_\phi(z|S_n) = \hat{P}(Y = y|x, S_n)P_\phi(x)$. We claim that with this choice, the risk of the pattern recognition problem is the same as the risk of the density estimation problem. Indeed

$$\begin{aligned}
r^{density}_{n+1,\hat{P},\phi}(\theta^*) &= \int_{Z^n} dP^n_{\theta^*,\phi}(S_n) \int_Z dP_{\theta^*,\phi}(z) \log \frac{dP_{\theta^*,\phi}(z)}{d\hat{P}_\phi(z|S_n)} \\
&= \int_{Z^n} dP^n_{\theta^*,\phi}(S_n) \int_X \sum_y P_{\theta^*}(Y = y|x) dP_\phi(x) \log \frac{P_{\theta^*}(Y = y|x) dP_\phi(x)}{\hat{P}(Y = y|x, S_n) dP_\phi(x)} \\
&= \int_{Z^n} dP^n_{\theta^*,\phi}(S_n) \int_X dP_\phi(x) \sum_y P_{\theta^*}(Y = y|x) \log \frac{P_{\theta^*}(Y = y|x)}{\hat{P}(Y = y|x, S_n)} \\
&= r_{n+1,\hat{P},\phi}(\theta^*).
\end{aligned}$$

To complete this reduction, we define the minimax risk for the density estimation problem as in [28] by

$$r^{minimax,density}_n(\Theta) = \inf_{\hat{P}} \sup_{\theta^* \in \Theta} r^{density}_{n,\hat{P}}(\theta^*),$$

where the infimum is over all possible estimators of the joint distribution on $Z = X \times Y$. We claim that

$$r^{minimax,density}_n(\Theta) = r^{minimax}_n(\Theta), \tag{1}$$

the minimax risk for the pattern recognition defined above. Indeed, since we have shown above that we can get the same risk for all $\theta^* \in \Theta$ for both the pattern recognition and density estimation problems by the density estimator $\hat{P}_\phi(z|S_n) = \hat{P}(Y = y|x, S_n)P_\phi(x)$, it is clear that $r^{minimax,density}_n(\Theta) \leq r^{minimax}_n(\Theta)$. Now suppose we choose any density estimator $\hat{Q}(z|S_n)$. We may decompose this estimator into $\hat{Q}(z|S_n) = \hat{Q}(Y = y|x, S_n)\hat{Q}(x|S_n)$. Then by the chain rule for relative entropy ([16]), the risk for the density estimation problem can be decomposed into

$$r^{density}_{n+1,\hat{Q},\phi}(\theta^*) = r_{n+1,\hat{Q},\phi}(\theta^*) + \int_{Z^n} dP^n_{\theta^*,\phi}(S_n) \int_X dP_\phi(x) \log \frac{dP_\phi(x)}{d\hat{Q}(x|S_n)}.$$

The last term is never negative, and is zero only when $\hat{Q}(x|S_n) = P_\phi(x)$ for all S_n. In this latter case, $\hat{Q}$ is an estimator of the type used in our reduction. It follows that the minimax risk in density estimation can be obtained by restricting ourselves to such estimators, and hence $r_n^{minimax,density}(\Theta) \geq r_n^{minimax}(\Theta)$. This establishes claim (1).

3 Covering numbers and metric entropy

We now study the asymptotic properties of the minimax risk $r_n^{minimax}$ as the sample size n grows. It is easy to verify that the minimax risk $r_n^{minimax}$ is nonincreasing for all n, and in most cases approaches 0 as n goes to infinity [28]. The rate at which $r_n^{minimax}$ approaches 0 depends primarily on the metric entropy properties of Θ, the topic to which we now turn.

The theory of packing and covering numbers, and the associated metric entropy, was introduced by Kolmogorov and Tikhomirov in [31], and is commonly used in the theory of empirical processes (see e.g. [19, 36, 24, 9, 42]). For the following definitions, let (S, ρ) be any metric space.

Definition 1. (Metric entropy, also called Kolmogorov ϵ-entropy [31]) A partition Π of S is a collection $\{\pi_i\}$ of subsets of S that are pairwise disjoint and whose union is S. The diameter of a set $A \subseteq S$ is given by $\text{diam}(A) = \sup_{x,y\in A} \rho(x, y)$. The diameter of a partition is the supremum of the diameters of the sets in the partition. For $\epsilon > 0$, by $\mathcal{D}_\epsilon(S, \rho)$ we denote the cardinality of the smallest finite partition of S of diameter at most ϵ, or ∞ if no such finite partition exists. The metric entropy of (S, ρ) is defined by

$$\mathcal{K}_\epsilon(S, \rho) = \log \mathcal{D}_\epsilon(S, \rho).$$

We say S is *totally bounded* if $\mathcal{D}_\epsilon(S, \rho) < \infty$ for all $\epsilon > 0$.

Definition 2. (Packing and covering numbers) For $\epsilon > 0$, an *ϵ-cover* of S is a subset $A \subseteq S$ such that for all $x \in S$ there exists a $y \in A$ with $\rho(x, y) \leq \epsilon$. By $\mathcal{N}_\epsilon(S, \rho)$ we denote the cardinality of the smallest finite ϵ-cover of S, or ∞ if no such finite cover exists. For $\epsilon > 0$, an *ϵ-separated subset of S* is a subset $A \subseteq S$ such that for all distinct $x, y \in A$, $\rho(x, y) > \epsilon$. By $\mathcal{M}_\epsilon(S, \rho)$ we denote the cardinality of the largest finite ϵ-separated subset of S, or ∞ if arbitrarily large such sets exist.

The following lemma is easily verified [31].

Lemma 3. *For any $\epsilon > 0$,*

$$\mathcal{M}_{2\epsilon}(S, \rho) \leq \mathcal{D}_{2\epsilon}(S, \rho) \leq \mathcal{N}_\epsilon(S, \rho) \leq \mathcal{M}_\epsilon(S, \rho).$$

It follows that the metric entropy $\mathcal{K}_\epsilon$ (and the condition defining total boundedness) can also be defined using either the packing or covering numbers in place of $\mathcal{D}_\epsilon$, to within a constant factor in ϵ.

Kolmogorov and Tikhomirov also introduced abstract notions of the dimension and order of metric spaces in their seminal paper [31]. These can be used to measure the "massiveness" of both spaces indexed by finite dimensional parameter vectors and infinite dimensional function spaces. In the following, the metric ρ is omitted from the notation, being understood from the context.

Definition 4. The *upper* and *lower metric dimensions* [31] of S are defined by

$$\overline{\mathbf{dim}}(S) = \limsup_{\epsilon \to 0} \frac{\mathcal{K}_\epsilon(S)}{\log \frac{1}{\epsilon}}$$

and

$$\underline{\mathbf{dim}}(S) = \liminf_{\epsilon \to 0} \frac{\mathcal{K}_\epsilon(S)}{\log \frac{1}{\epsilon}},$$

respectively. When $\overline{\mathbf{dim}}(S) = \underline{\mathbf{dim}}(S)$, then this value is denoted $\mathbf{dim}(S)$ and called the *metric dimension* of S. Thus

$$\mathbf{dim}(S) = \lim_{\epsilon \to 0} \frac{\mathcal{K}_\epsilon(S)}{\log \frac{1}{\epsilon}}.$$

For totally bounded S, we say that S is finite dimensional if $\mathbf{dim}(S) < \infty$, else it is infinite dimensional. To measure the massiveness of infinite dimensional spaces, including typical function spaces, further indices were introduced by Kolmogorov and Tikhomirov. The *functional dimension* of S is defined similarly as

$$\mathbf{df}(S) = \lim_{\epsilon \to 0} \frac{\log \mathcal{K}_\epsilon(S)}{\log \log \frac{1}{\epsilon}},$$

with similar upper and lower versions, $\overline{\mathbf{df}}$ and $\underline{\mathbf{df}}$, when this limit does not exist. Finally, the *metric order* of S is defined as

$$\mathbf{mo}(S) = \lim_{\epsilon \to 0} \frac{\log \mathcal{K}_\epsilon(S)}{\log \frac{1}{\epsilon}},$$

with similar upper and lower versions, $\overline{\mathbf{mo}}$ and $\underline{\mathbf{mo}}$.

4 Hellinger distance

We can view the comparison model class Θ as a metric space, and calculate its metric entropy, by specifying a metric on this space. It turns out that the right metric to use is the *Hellinger distance*. If $P = (p_1, \ldots, p_k)$ and $Q = (q_1, \ldots, q_k)$ are two discrete probability distributions, then the Hellinger distance between P and Q is defined as

$$D_{HL}(P, Q) = \left(\sum_{i=1}^{k} (\sqrt{p_i} - \sqrt{q_i})^2 \right)^{1/2}.$$

That is, the Hellinger distance between P and Q is the Euclidean distance between $(\sqrt{p_1}, \ldots, \sqrt{p_k})$ and $(\sqrt{q_1}, \ldots, \sqrt{q_k})$. The Hellinger distance can be generalized to discrete distributions on countably infinite sets, and on continuous sets such as the real line, by using l_2 and L_2 norms, respectively, in place of Euclidean distance. The Hellinger distance is useful because it is a metric, and the squared Hellinger distance approximates the relative entropy distance, which is not a metric. The sense of this approximation is given, e.g., in [28]. This metric has been used to give bounds on the risk of estimation procedures in statistics by many authors, including LeCam [32], Birgé [7, 8], Hasminskii and Ibragimov [25], and van de Geer [41].

Now assume that θ and θ^* are two joint distributions on $X \times Y$ with a common marginal distribution on X. Then, when X is discrete, the Hellinger distance between these two distributions is

$$D_{HL}(\theta, \theta^*) = \left(\sum_{x,y} \left(\sqrt{P(x)P(y|x,\theta)} - \sqrt{P(x)P(y|x,\theta^*)} \right)^2 \right)^{1/2}$$

$$= \left(\sum_x P(x) \sum_y \left(\sqrt{P(y|x,\theta)} - \sqrt{P(y|x,\theta^*)} \right)^2 \right)^{1/2}.$$

This extends naturally to continuous X as well. Using this distance, if all the distributions in Θ are distinct (i.e. differ on a set of positive measure), which we may assume without loss of generality, then (Θ, D_{HL}) is a metric space, else it is a *pseudo* metric space, i.e. a metric space that possibly includes distinct points at distance 0.

5 Rates for minimax risk

We are now in a position to state the main theorem about rates for minimax risk. Let us define the *best exponent in the rate for the minimax risk* by

$$e(\Theta) = \sup\{t : \limsup_{n \to \infty} \frac{r_n^{minimax}(\Theta)}{n^{-t}} \leq 1\}.$$

By calculating this best exponent, we can distinguish the various rates at which the minimax risk approaches 0.

Theorem 5. *Assume Θ is a comparison model class in which all models have a common marginal distribution on X. Then the bounds on $e(\Theta)$ given in the following table are valid.*

size of Θ	bound on exponent
Θ is finite	$e(\Theta) = \infty$
$\mathbf{dim}(\Theta, D_{HL}) = 0$	$e(\Theta) \geq 1$
$\mathbf{dim}(\Theta, D_{HL}) = D$ where $0 < D < \infty$	$e(\Theta) = 1$
$\mathbf{df}(\Theta, D_{HL}) = \beta$ where $1 < \beta < \infty$	$e(\Theta) = 1$
$\mathbf{mo}(\Theta, D_{HL}) = \alpha$ where $0 < \alpha < \infty$	$e(\Theta) = \frac{2}{2+\alpha}$
$\mathbf{mo}(\Theta, D_{HL}) = \infty$	$e(\Theta) = 0$
(Θ, D_{HL}) not totally bounded	$e(\Theta) = 0$

Proof. This follows directly from Theorem 7 in [28], using claim (1). To see that the conditions of Theorem 7 in [28] hold, suppose $|Y| = K$ and let $\hat{P}(Y = y|x) = 1/K$ for all $x \in X$. Then note that in the density estimation problem that corresponds to the pattern recognition problem under consideration, because of the common marginal distribution, for any $\lambda > 0$ and any θ^*,

$$\begin{aligned}\int (dP_{\theta^*,\phi})^{1+\lambda}(d\hat{P}_\phi)^{-\lambda} &= \int_X dP_\phi(x) \sum_y (P_{\theta^*}(Y = y|x))^{1+\lambda}(\hat{P}(Y = y|x))^{-\lambda} \\ &= K^\lambda \int_X dP_\phi(x) \sum_y (P_{\theta^*}(Y = y|x))^{1+\lambda} \\ &\leq K^\lambda \\ &< \infty.\end{aligned}$$

Thus the minimax risk for the regret function $\int (dP_{\theta^*,\phi})^{1+\lambda}(d\hat{P}_\phi)^{-\lambda}$ is finite as required.

Thus for a comparison model class Θ of finite dimension or finite functional dimension, the rate of convergence of the minimax risk (i.e. estimation error) to zero as a function of sample size n is better than $\frac{1}{n^{1-\delta}}$ for all positive δ, but for (larger) model classes of finite metric order α, the best rate is something like $\frac{1}{n^{2/(2+\alpha)}}$, which is much slower for large metric order α. Going to further extremes, convergence is faster than any inverse polynomial for finite model classes Θ (it can be shown to be exponential in n [28]), but model classes of infinite metric order, or that are not even totally bounded, are essentially "unlearnable" with this definition of estimation error as minimax risk: for any learning method there is a choice of true distribution that makes the convergence slower than $\frac{1}{n^\delta}$ for all positive δ. The advantage of this theorem is that is gives a characterization of the best possible rate of convergence entirely in terms of the metric entropy of the model class Θ, without referring to any specific properties of the models themselves. However, it does not give the most precise convergence rates that can be stated for many common cases, especially finite dimensional ones. This is addressed in the following sections.

6 Rates for cumulative minimax risk

More precise bounds on the rate of convergence for the minimax risk can be obtained if we look at the cumulative risk. This is the total minimax average regret (risk) for the first n predictions in a sequential or *on line* prediction setting, in which the examples $(x_1, y_1), \ldots, (x_n, y_n)$ are presented to the learner one at a time, and for each t between 1 and n, after seeing the first $t-1$ examples $S_{t-1} = (x_1, y_1), \ldots, (x_{t-1}, y_{t-1})$ and the tth instance x_t, the learner must produce an estimated predictive probability distribution $\hat{P}(Y_t = y|x_t, S_{t-1})$. The loss in this sequential version of the prediction game is

$$\sum_{t=1}^{n} \log \frac{P(Y_t = y_t|x_t, \theta^*)}{\hat{P}(Y_t = y_t|x_t, S_{t-1})},$$

where P_{θ^*}, with $\theta^* \in \Theta$, is the true distribution. So the cumulative loss is simply the total loss for all individual predictions. When we average this over possible sequences of examples generated independently according to P_{θ^*}, we get the cumulative regret

$$R_{n,\hat{P}}(\theta^*) = \int_{(X\times Y)^n} dP^n_{\theta^*}(S_n) \sum_{t=1}^{n} \log \frac{P(Y_t = y_t|x_t, \theta^*)}{\hat{P}(Y_t = y_t|x_t, S_{t-1})}.$$

It is easily verified, using the linearity of expectation, that

$$R_{n,\hat{P}}(\theta^*) = \sum_{t=1}^{n} r_{t,\hat{P}}(\theta^*).$$

So the cumulative regret for the first n predictions is just the sum of the regrets for each of the predictions for sample sizes t between 1 and n. Finally, just as before, the *cumulative minimax risk* is defined as the minimum over all classification methods of the maximum cumulative regret over all true distributions in Θ, i.e.

$$R_n^{minimax} = R_n^{minimax}(\Theta) = \inf_{\hat{P}} \sup_{\theta^* \in \Theta} R_{n,\hat{P}}(\theta^*).$$

Looking at the cumulative minimax risk provides an alternate way to study the estimation error and its rate of convergence. When comparing the cumulative minimax risk to the minimax risk defined above, called the *instantaneous* minimax risk in [28] to contrast it with the cumulative minimax risk, note that from the definition of the individual minimax risks $r_t^{minimax}$, for $1 \leq t \leq n$, we see that for each separate t we are possibly looking at a different worst case true distribution P_{θ^*} when we compute the minimax risk $r_t^{minimax}$, whereas for the cumulative minimax risk $R_n^{minimax}$, the same true distribution P_{θ^*} must be used for all $1 \leq t \leq n$. Thus the cumulative minimax risk is in some ways a better measure of the sustained difficulty of the learning/prediction problem over a range of sample sizes, while the instantaneous minimax risk for a particular sample size n could in principle reflect the difficulty of the problem due to particular distributions in Θ that are "hard" for that particular sample size n. However, it turns out that this effect cannot be very strong. In particular, it can be shown in general that $R_n^{minimax}$ is nondecreasing in n, and

$$\sum_{t=1}^{n} r_t^{minimax} \geq R_n^{minimax} > n r_n^{minimax}$$

(see [3, 13, 6, 28].) It follows that $R_n^{minimax}$ grows at most linearly in n, since $r_t^{minimax} \leq \log|Y|$ for all t. These inequalities also give fairly tight bounds on $R_n^{minimax}$ in terms of $r_n^{minimax}$ when $r_n^{minimax}$ decreases slowly. For example[3], if $r_n^{minimax} \asymp 1/\sqrt{n}$, then $R_n^{minimax} \asymp \sqrt{n}$. However, if $r_n^{minimax} = D/n$ then the inequalities only tell us that $D \leq R_n^{minimax} \leq D\sum_{t=1}^{n} 1/t \leq D\log(n+1)$. We get a more precise analysis of the cumulative minimax risk $R_n^{minimax}$ by bounding it directly in terms of the metric entropy of Θ, as in the results on the minimax risk in the previous section. (Actually, the results on minimax risk are derived from the results on cumulative minimax risk given here.)

Theorem 6. *Assume Θ is a comparison model class in which all models have a common marginal distribution on X. Then*

1. *If Θ is finite then*
$$R_n^{minimax}(\Theta) \to \log|\Theta| \text{ as } n \to \infty.$$
2. *If* $\mathbf{dim}(\Theta, D_{HL}) = 0$ *then*
$$R_n^{minimax}(\Theta) \in o(\log n).$$
3. *If* $\mathbf{dim}(\Theta, D_{HL}) = D$ *where* $0 < D < \infty$ *then*
$$R_n^{minimax}(\Theta) \sim \frac{D}{2}\log n.$$
4. *If* $\mathbf{df}(\Theta, D_{HL}) = \beta$ *where* $1 < \beta < \infty$ *then*
$$\log R_n^{minimax}(\Theta) \sim \beta \log\log n.$$
5. *If* $\mathbf{mo}(\Theta, D_{HL}) = \alpha$ *where* $0 < \alpha < \infty$ *then*
$$\log R_n^{minimax}(\Theta) \sim \frac{\alpha}{2+\alpha}\log n.$$
6. *If* $\mathbf{mo}(\Theta, D_{HL}) = \infty$ *or* (Θ, D_{HL}) *is not totally bounded, then*
$$\log R_n^{minimax}(\Theta) \sim \log n.$$

Proof. Similar to the proof of Theorem 5, but using Theorem 4 of [28].

Furthermore, analogous results using upper and lower dimensions and orders also hold in the situation when the upper and lower dimensions/orders are different. For example, we can show

[3] For integer or real-valued functions f and g, we say $f \sim g$ if $\lim_{n\to\infty} \frac{f(n)}{g(n)} = 1$, $f \asymp g$ if $\liminf_{n\to\infty} \frac{f(n)}{g(n)} > 0$ and $\limsup_{n\to\infty} \frac{f(n)}{g(n)} < \infty$.

Theorem 7. *Assume Θ is a comparison model class in which all models have a common marginal distribution on X. Then*

$$\limsup_{n\to\infty} \frac{R_n^{minimax}(\Theta)}{\log n} = \frac{\overline{\mathbf{dim}}(\Theta, D_{HL})}{2}$$

and

$$\liminf_{n\to\infty} \frac{R_n^{minimax}(\Theta)}{\log n} = \frac{\underline{\mathbf{dim}}(\Theta, D_{HL})}{2}.$$

Proof. Let $R_n^{minimax} = R_n^{minimax}(\Theta)$ and $\mathcal{K}(\epsilon) = \mathcal{K}_\epsilon(\Theta, D_{HL})$. By Lemma 7 of [28], there is some positive constant c such that for any n and any $\epsilon > 0$,

$$\min\{\mathcal{K}(\epsilon), n\epsilon^2/8\} - \log 2 \leq R_n^{minimax} \leq \mathcal{K}(\epsilon) + c\epsilon^2 n \log n + c.$$

Here we verify the conditions of the lemma again as in the proof of Theorem 5. Now, in the lower bound let $\epsilon = \frac{\log n}{\sqrt{n}}$ and in the upper bound let $\epsilon = \frac{1}{\sqrt{n}\log n}$. Note that if $\overline{\mathbf{dim}}(\Theta, D_{HL}) < \infty$ then $\mathcal{K}(\epsilon)$ is $O(\log(1/\epsilon))$, so $\min\{\mathcal{K}(\epsilon), n\epsilon^2/8\} = \mathcal{K}(\epsilon)$ for large n if $\epsilon = \frac{\log n}{\sqrt{n}}$. It follows that

$$\limsup_{n\to\infty} \frac{\mathcal{K}(\frac{\log n}{\sqrt{n}})}{\log n} \leq \limsup_{n\to\infty} \frac{R_n^{minimax}}{\log n} \leq \limsup_{n\to\infty} \frac{\mathcal{K}(\frac{1}{\sqrt{n}\log n})}{\log n}.$$

Looking at the upper bound, let $m = \sqrt{n}\log n$. Then

$$\limsup_{n\to\infty} \frac{\mathcal{K}(\frac{1}{\sqrt{n}\log n})}{\log n} = \limsup_{m\to\infty} \frac{\mathcal{K}(\frac{1}{m})}{\log \frac{m^2}{f(m)}},$$

where $f(m) \asymp \log^2 m$. Hence

$$\limsup_{m\to\infty} \frac{\mathcal{K}(\frac{1}{m})}{\log \frac{m^2}{f(m)}} = \frac{1}{2} \limsup_{m\to\infty} \frac{\mathcal{K}(\frac{1}{m})}{\log m} = \frac{\overline{\mathbf{dim}}(\Theta, D_{HL})}{2}.$$

Looking at the lower bound, a similar argument shows

$$\limsup_{n\to\infty} \frac{\mathcal{K}(\frac{\log n}{\sqrt{n}})}{\log n} = \frac{\overline{\mathbf{dim}}(\Theta, D_{HL})}{2}.$$

This establishes the first part of the theorem. The second part is established in a similar manner.

7 Vapnik-Chervonenkis entropy and dimension

In this section we examine how the results given here relate to results that can be obtained by another approach that has often been used to analyze the convergence rates of classification methods, namely, the Vapnik-Chervonenkis dimension [44]. For simplicity, in this comparison we restrict ourselves to a special class of classification problems that we call *noisy two-class learning* (also called "noisy concept learning" in the computational learning theory and AI machine learning literature [1]). In noisy two-class learning, the outcome space Y has just two values, which we may designate as $+1$ and -1, instead of an arbitrary finite set of values, as we have been assuming up to this point. Furthermore, not only do the joint distributions in Θ all have the same marginal distribution on X, but the conditional distributions on Y given X all have a special form described as follows:

It is assumed that there is a fixed *noise rate* $0 < \lambda < 1/2$, and for each distribution $\theta \in \Theta$ there is a function $f_\theta : X \to Y$ such that for all instances $x \in X$,

$$P(Y \neq f_\theta(x)|x, \theta) = \lambda.$$

You can view this conditional distribution as being generated by an underlying functional relationship between X and Y, namely, $Y = f_\theta(X)$, composed with an independent noise process that flips the sign of Y independently with probability λ. Thus in this case our examples $(x_1, y_1), \ldots, (x_n, y_n)$ are really a noise corrupted version of an underlying set of random examples $(x_1, f_\theta(x_1)), \ldots, (x_n, f_\theta(x_n))$ of the function f_θ, for some unknown $\theta \in \Theta$. The instances $x_1, \ldots, x_n$ are generated independently at random according to the marginal distribution ϕ on X.

Let us define $\mathcal{F}_\Theta = \{f_\theta : \theta \in \Theta\}$. Vapnik-Chervonenkis theory provides a way of bounding the estimation error of Θ in terms of certain combinatorial properties of the class of functions $\mathcal{F}_\Theta$. The key element of this theory is the *growth function*. For the following definitions, let $\mathcal{F}$ be a family of $\{\pm 1\}$-valued functions on a set X.

Definition 8. For each sequence $x^n = x_1, \ldots, x_n$ in X^n, let $\mathcal{F}_{|x^n} = \{(f(x_1), \ldots, f(x_n)) : f \in \mathcal{F}\}$. The growth function $\Pi_\mathcal{F}(n)$ is defined by

$$\Pi_\mathcal{F}(n) = \max_{x^n \in X^n} |\mathcal{F}_{|x^n}|,$$

where $|S|$ denotes the cardinality of the set S. Thus $\Pi_\mathcal{F}(n)$ is the maximum number of distinct functions that can be obtained by restricting the domain of the functions in $\mathcal{F}$ to n points.

From the growth function we can define the Assouad density of $\mathcal{F}$ [2], and the Vapnik-Chervonenkis (VC) dimension of $\mathcal{F}$. We treat the Assouad density first, relating it to a certain supremum over the metric dimension, and return to the VC dimension later.

Definition 9. The *Assouad density* of $\mathcal{F}$ is defined by

$$dens(\mathcal{F}) = \inf\{d > 0 : \text{ there exists } C > 0 \text{ such that for all } n \geq 1, \Pi_{\mathcal{F}}(n) \leq Cn^d\}.$$

It is easily verified that

$$dens(\mathcal{F}) = \limsup_{n\to\infty} \frac{\log \Pi_{\mathcal{F}}(n)}{\log n} \tag{2}$$

To see this, note that if $r > \limsup_{n\to\infty} \frac{\log \Pi_{\mathcal{F}}(n)}{\log n}$, then there exists $r_0 < r$ and n_0 such that for all $n \geq n_0$, $\frac{\log \Pi_{\mathcal{F}}(n)}{\log n} \leq r_0$, which implies $\Pi_{\mathcal{F}}(n) \leq n^{r_0}$. Hence $r > dens(\mathcal{F})$. On the other hand, if $r < \limsup_{n\to\infty} \frac{\log \Pi_{\mathcal{F}}(n)}{\log n}$ then there exists $r_0 > r$ such that $\Pi_{\mathcal{F}}(n) > n^{r_0}$ infinitely often, and thus $r < dens(\mathcal{F})$. Equation (2) follows.

Now let P_ϕ be the probability distribution on X. For $f, g \in \mathcal{F}_\Theta$, define $D_\phi(f,g) = P_\phi(f(x) \neq g(x))$. Then $(\mathcal{F}_\Theta, D_\phi)$ is a (pseudo) metric space. This metric space is related to the metric space (Θ, D_{HL}) that was central to the results in the previous sections when Θ is the model class for a noisy two-class learning problem. Let the noise rate be λ and let $c_\lambda = 2(\sqrt{\lambda} - \sqrt{1-\lambda})^2$. Then

$$D^2_{HL}(\theta,\theta^*) = \int_X dP_\phi(x) \sum_{y\in Y} \left(\sqrt{P(Y=y|x,\theta)} - \sqrt{P(Y=y|x,\theta^*)}\right)^2 = c_\lambda D_\phi(f_\theta, f_{\theta^*}).$$

Hence the metric entropies of these two spaces are related by

$$\mathcal{K}_\epsilon(\Theta, D_{HL}) = \mathcal{K}_{\epsilon^2/c_\lambda}(\mathcal{F}_\Theta, D_\phi),$$

and thus if Θ is finite dimensional,

$$\mathbf{dim}(\Theta, D_{HL}) = 2\mathbf{dim}(\mathcal{F}_\Theta, D_\phi), \tag{3}$$

and similarly for $\overline{\mathbf{dim}}$ and $\underline{\mathbf{dim}}$. Similar relations can be derived when Θ is infinite dimensional. In this manner, for the noisy two-class learning problem, the results of the previous sections can be restated in terms of the scaling as $\epsilon \to 0$ of the metric entropy of the metric space $(\mathcal{F}_\Theta, D_\phi)$, rather than the metric space (Θ, D_{HL}). A result of Assouad's, given in a monograph by Dudley [19], relates this scaling, in the worst case over distributions P_ϕ, to the Assouad density of $\mathcal{F}_\Theta$. For the following definitions and results, let $\mathcal{F}$ be any class of $\{\pm 1\}$-valued functions on a set X and P_ϕ be any distribution on X.

Definition 10. Let

$s(\mathcal{F}) = \inf\{d > 0 :$ there is a $C > 0$ such that for every P_ϕ and
$0 < \epsilon \leq 1, \mathcal{M}_\epsilon(\mathcal{F}, D_\phi) \leq C\epsilon^{-d}\}$.

Theorem 11. *(Theorem 9.3.1 of [19])*

$$dens(\mathcal{F}) = s(\mathcal{F})$$

Using a similar method, we can also relate $s(\mathcal{F})$ directly to the upper dimension of $(\mathcal{F}, D_\phi)$.

Theorem 12.

$$s(\mathcal{F}) = \limsup_{\epsilon \to 0} \sup_{P_\phi} \frac{\mathcal{K}_\epsilon(\mathcal{F}, D_\phi)}{\log \frac{1}{\epsilon}} = \sup_{P_\phi} \limsup_{\epsilon \to 0} \frac{\mathcal{K}_\epsilon(\mathcal{F}, D_\phi)}{\log \frac{1}{\epsilon}} = \sup_{P_\phi} \overline{\mathbf{dim}}(\mathcal{F}, D_\phi)$$

Proof. The first equality is similar to (2), and the last equality follows directly from the definition of $\overline{\mathbf{dim}}(\mathcal{F}, D_\phi)$. So we need only consider the middle equality. Let $l = \limsup_{\epsilon\to 0} \sup_{P_\phi} \frac{\mathcal{K}_\epsilon(\mathcal{F},D_\phi)}{\log \frac{1}{\epsilon}}$ and $u = \sup_{P_\phi} \limsup_{\epsilon \to 0} \frac{\mathcal{K}_\epsilon(\mathcal{F},D_\phi)}{\log \frac{1}{\epsilon}}$. For any function $f(n,m)$,

$$\limsup_n \sup_m f(n,m) \geq \sup_m \limsup_n f(n,m),$$

so it suffices to show that $l \leq u$. As this inequality is trivial when $l = 0$, we will assume $0 < l \leq \infty$. Let $\{\epsilon_n\}_{n\geq 1}$ be a sequence of positive numbers such that $\epsilon_n \leq 2^{-n}$ and $\{\phi_n\}_{n \geq 1}$ be a sequence of distributions on X such that

$$l = \lim_{n\to\infty} \frac{\log \mathcal{M}_{\epsilon_n}(\mathcal{F}, D_{\phi_n})}{\log \frac{1}{\epsilon_n}}$$

Using Lemma 3 it is clear that such sequences can be found. Suppose $0 < r < t < l$, and $t < \infty$. Let the distribution P_ϕ be defined by

$$dP_\phi(x) = \frac{1}{S} \sum_{n=1}^{\infty} n^{-t/r} dP_{\phi_n}(x),$$

where $S = \sum_{n=1}^{\infty} n^{-t/r} < \infty$. We claim that

$$r \leq \limsup_{\epsilon\to 0} \frac{\mathcal{K}_\epsilon(\mathcal{F}, D_\phi)}{\log \frac{1}{\epsilon}}.$$

Since r can be chosen arbitrarily close to l, or arbitrarily large if $l = \infty$, and the right hand side above is less than or equal to u, this shows that $l \leq u$.

To see that this claim holds, first note that for any set $A \subseteq X$ and any n, $P_\phi(A) \geq \frac{P_{\phi_n}(A)}{Sn^{t/r}}$. Thus any ϵ-separated set of $\mathcal{F}$ under the metric D_{ϕ_n} is an $\frac{\epsilon}{Sn^{t/r}}$-separated set under the metric D_ϕ. For each n let $M(n) = \mathcal{M}_{\epsilon_n}(\mathcal{F}, D_{\phi_n})$. Now set $\gamma_n = \frac{\epsilon_n}{Sn^{t/r}}$. It follows that $\mathcal{M}_{\gamma_n}(\mathcal{F}, D_\phi) \geq M(n)$. Since $\lim_{n\to\infty} \frac{\log M(n)}{\log \frac{1}{\epsilon_n}} = l > t$, there is an n_0 such that for all $n \geq n_0$, $M(n) > \epsilon_n^{-t}$. Hence for large n, $\mathcal{M}_{\gamma_n}(\mathcal{F}, D_\phi) \geq M(n) > \epsilon_n^{-t}$. However, $\gamma_n \to 0$, and

$$\epsilon_n^{-t} = \epsilon_n^{-r} \epsilon_n^{r-t} = \gamma_n^{-r} S^{-r} n^{-t} \epsilon_n^{r-t} > \gamma_n^{-r}$$

for large n, since $\epsilon_n \leq 2^{-n}$ and $r - t < 0$. It follows that $\mathcal{M}_{\gamma_n}(\mathcal{F}, D_\phi) > \gamma_n^{-r}$ for large n, and hence

$$r \leq \limsup_{\epsilon\to 0} \frac{\mathcal{K}_\epsilon(\mathcal{F}, D_\phi)}{\log \frac{1}{\epsilon}}.$$

This establishes the claim.

As a corollary of Theorems 11 and (12), we have

$$dens(\mathcal{F}) = \sup_{P_\phi} \overline{\mathbf{dim}}(\mathcal{F}, D_\phi). \tag{4}$$

It should be noted that this relationship between the growth rate of the maximum size of $\mathcal{F}$ restricted to n points and the metric entropy of $(\mathcal{F}, D_\phi)$ requires that one take the supremum over all distributions P_ϕ. If one does not, then there is no close relationship between these two quantities, even if we use the expected size of $\mathcal{F}$ restricted to n random points. For example, if we let $X = [0,1]$, P_ϕ be the uniform distribution on X and $\mathcal{F}$ be the set of all $\{\pm 1\}$-valued functions that are $+1$ on at most d points, then $\int_{X^n} dP_\phi^n(x^n)|\mathcal{F}_{|x^n}| \asymp n^d$ but $\mathcal{K}_\epsilon(\mathcal{F}, D_\phi) = 0$, since all functions differ only on a set of measure 0, and hence are distance 0 apart under the metric D_ϕ.

The Assouad density is the exponent of the smallest polynomial function that upper bounds the growth function $\Pi_{\mathcal{F}}(n)$. The growth function has a curious combinatorial property: either it is bounded by some polynomial in n, and hence the Assouad density is finite, or it is equal to 2^n for all n (and hence the Assouad density is most decidedly infinite). In fact, if we let $\mathbf{dim}_{VC}(\mathcal{F})$ be the largest n such that $\Pi_{\mathcal{F}}(n) = 2^n$, then if $\mathbf{dim}_{VC}(\mathcal{F}) = d < \infty$, then $\Pi_{\mathcal{F}}(n) \le \sum_{i=0}^{d} \binom{n}{i} \le (en/d)^d$ for all $n \ge d \ge 1$. This result, often cited as Sauer's Lemma [39], was proven independently by Vapnik and Chervonenkis [44] (first in a slightly weaker version). $\mathbf{dim}_{VC}(\mathcal{F})$ is called the *VC dimension* of $\mathcal{F}$. It follows that

$$dens(\mathcal{F}) \le \mathbf{dim}_{VC}(\mathcal{F}). \tag{5}$$

This inequality is often tight, but not always tight. Indeed, for any finite $\mathcal{F}$, $dens(\mathcal{F}) = 0$, yet there are finite $\mathcal{F}$ with arbitrarily large VC dimension. However, for all $\mathcal{F}$, $dens(\mathcal{F})$ is finite if and only if $\mathbf{dim}_{VC}(\mathcal{F})$ is finite.

Finally, we can put the above results together with the results of the previous section to obtain the following characterization of the estimation error for noisy two-class learning problems, defined as the cumulative minimax risk $R_n^{minimax}(\Theta, \phi)$.

Theorem 13. *If Θ is any set of conditional distributions for the noisy two-class learning problem with any noise rate $0 < \lambda < 1/2$, then if $\mathbf{dim}_{VC}(\mathcal{F}_\Theta)$ is finite we have*

$$\sup_{P_\phi} \limsup_{n\to\infty} \frac{R_n^{minimax}(\Theta, \phi)}{\log n} = dens(\mathcal{F}_\Theta) \le \mathbf{dim}_{VC}(\mathcal{F}_\Theta)$$

and if $\mathbf{dim}_{VC}(\mathcal{F}_\Theta)$ is infinite then $\sup_{P_\phi} R_n^{minimax}(\Theta, \phi)$ grows linearly in n.

Proof. If $\mathbf{dim}_{VC}(\mathcal{F}_\Theta)$ is finite then using Theorem 7, the $\overline{\mathbf{dim}}$ version of Equation (3), Equation (4), and Equation (5) in that order, we have

$$\begin{aligned} \sup_{P_\phi} \limsup_{n\to\infty} \frac{R_n^{minimax}(\Theta, \phi)}{\log n} &= \sup_{P_\phi} \overline{\mathbf{dim}}(\mathcal{F}_\Theta, D_\phi) \\ &= dens(\mathcal{F}_\Theta) \\ &\le \mathbf{dim}_{VC}(\mathcal{F}_\Theta). \end{aligned}$$

If $\mathbf{dim}_{VC}(\mathcal{F}_\Theta)$ is infinite then for any n we can choose a distribution P_ϕ that is uniform on a large finite set $X_0 \subseteq X$ that is *shattered* in the sense that $|\mathcal{F}_{|X_0}| = 2^{|X_0|}$. Suppose a function f is chosen uniformly at random from $\mathcal{F}_{|X_0}$. If $|X_0|$ is large enough then the first n instances $x_1, \ldots, x_n$ will be distinct with probability near 1, and all labelings of these points with ± 1 values $y_1, \ldots, y_n$ will be equally likely to occur. Under such conditions, the average instantaneous regret in predicting y_t given $(x_1, y_1), \ldots, (x_{t-1}, y_{t-1})$ and x_t is a positive constant for all $1 \leq t \leq n$ for any noise rate $0 < \lambda < 1/2$ and any prediction method, so $R_n^{minimax}(\Theta, \phi)$ grows linearly in n. Since $R_n^{minimax}(\Theta, \phi)$ cannot grow faster than linear in n for any ϕ, as was remarked in Section 6, it follows that $\sup_{P_\phi} R_n^{minimax}(\Theta, \phi)$ grows linearly in n.

This theorem shows that $\sup_{P_\phi} R_n^{minimax}(\Theta, \phi)$ either grows logarithmically or slower, or it grows linearly. There is no rate in between. Results of this type are also available from the standard Vapnik-Chervonenkis theory [44, 11]. However, what is novel here is that in the case of logarithmic growth, the best possible constant in front of the logarithm is identified here to be the Assouad density. It is difficult to identify such constants with the standard Vapnik-Chervonenkis theory, which relies on uniform convergence of empirical estimates, and therefore gives only indirect bounds on the minimax risk.

Some tighter upper bounds are known for Theorem 13. In particular, in [26] it was shown that

Theorem 14. *If Θ is any set of conditional distributions for the noisy two-class learning problem with any noise rate $0 < \lambda < 1/2$, and $\mathcal{F} = \mathcal{F}_\Theta$, then for all marginal distributions P_ϕ on X*

$$R_n^{minimax}(\Theta, \phi) \leq \int_{X^n} dP_\phi^n(x^n) \log |\mathcal{F}_{|x^n}| \leq \log \Pi_{\mathcal{F}}(n).$$

It is an open problem to obtain tighter lower bounds.

8 Conclusions

We have looked at the performance of the best possible classification method in terms of the minimax relative entropy risk, obtained by comparing the predictive distribution produced by the particular classification method to the best possible distribution in a comparison model class Θ. We are able to characterize the best performance that can be achieved in terms of the metric entropy of the model class Θ.

One important question that remains is: what classification method gives this best possible performance? Recall that a Bayes method is one that employs a prior distribution over the model class Θ, and computes its predictive distribution by averaging over all conditional distributions $P(Y|x, \theta)$, weighted according to the posterior probability of θ given the training examples. It turns out that by careful choice of the prior, we can find Bayes methods that get asymptotically close to the best minimax performance. The priors to use can be found by examining the proof of the lower bound given in Lemma 7 of [28], upon which the lower bound in the result given in Theorem 6 above is based. These place a uniform prior distribution on a finite subset of Θ, chosen to be a maximal ϵ-separated subset with respect to the Hellinger distance for some suitable ϵ. The best value of ϵ to use decreases as the sample size n grows. This idea is quite intuitive: one picks a representative set of models in Θ, uses a uniform, "noninformative", prior on this set, lets the training data focus attention on the best model by computing a posterior distribution over this representative set of models, which will place much higher weight on those models that perform well on the training data, then finally uses an average of these well-performing models to form the predictive distribution for future outcomes. To get more and more accuracy as the number of training examples grows, one chooses larger and larger representative model sets, obtained by using a finer "mesh", i.e. a smaller separation ϵ between representative models. This leads to a kind of sieve method, as discussed in the introduction.

In some cases, in the limit as the sample size n goes to infinity, and hence the separation ϵ goes to zero, the uniform distribution over the maximal ϵ-separated set of models approaches something like a Jeffreys' prior over the model class Θ, which is known already to be asymptotically minimax (called "asymptotically least favorable" for technical reasons) for relative entropy risk in smooth parametric cases [14]. It is an interesting open problem to determine to what extent this holds for more general Θ, and what characterizes asymptotically minimax "generalized Jeffreys' priors".

As a practical classification method, the Bayes method using a uniform prior on an ϵ-separated set has two drawbacks

1. The size of the ϵ-separated set may grow too large too quickly as $\epsilon \to 0$. This happens for higher finite dimensional Θ, and for infinite dimensional Θ. This is kind of a "curse of dimensionality."
2. To compute the ϵ-separated set, it is required that one know the common marginal distribution P_ϕ on the the instance space X that it shared by the models in Θ. Often this distribution is not known, and one wants to do classification using a class of conditional distributions on the outcome Y given X, leaving the marginal distribution on X unspecified.

The first problem is a deep one. In cases where the asymptotically minimax prior is known and the posterior for this prior can be efficiently computed then this prior can be used in place of the priors on individual ϵ-separated sets. In applications of Bayes methods where such computations are not tractable it is common to employ Markov Chain Monte Carlo methods. However, it is difficult to give precise theoretical bounds on the performance of such methods.

The second problem can be handled by either trying to estimate the marginal distribution on the instance space, or by developing a method that works even for the worst case marginal distribution. In section 7 we have outlined what can be achieved with the latter approach in the special case of noisy two-class classification problems, relating our theory to the Vapnik-Chervonenkis theory.

It remains an important open problem to extend this analysis to arbitrary comparison model classes with a common marginal distribution. A related problem is to extend the whole of the theory used in this paper to handle the case where the models do not share a common marginal distribution.

9 Acknowledgements

David Haussler would like to thank Vincent Mirelli for suggesting some of the problems investigated here, and for valuable comments on an earlier draft of this paper.

References

1. D. Angluin and P. Laird. Learning from noisy examples. *Machine Learning*, 2(4):343–370, 1988.
2. P. Assouad. Densité et dimension. *Annales de l'Institut Fourier*, 33(3):233–282, 1983.
3. A. Barron. In T. M. Cover and B. Gopinath, editors, *Open Problems in Communication and Computation*, chapter 3.20. Are Bayes rules consistent in information?, pages 85–91. 1987.
4. A. Barron. The exponential convergence of posterior probabilities with implications for Bayes estimators of density functions. Technical Report 7, Dept. of Statistics, U. Ill. Urbana-Champaign, 1987.
5. A. Barron, B. Clarke, and D. Haussler. Information bounds for the risk of Bayesian predictions and the redundancy of universal codes. In *Proc. International Symposium on Information Theory.*
6. A. Barron and Y. Yang. Information theoretic lower bounds on convergence rates of nonparametric estimators, 1995. unpublished manuscript.
7. L. Birgé. Approximation dans les espaces métriques et théorie de l'estimation. *Zeitschrift fuer Wahrscheinlichkeitstheorie und verwandte gebiete*, 65:181–237, 1983.
8. L. Birgé. On estimating a density using Hellinger distance and some other strange facts. *Probability theory and related fields*, 71:271–291, 1986.
9. L. Birgé and P. Massart. Rates of convergence for minimum contrast estimators. *Probability Theory and Related Fields*, 97:113–150, 1993.
10. A. Blumer, A. Ehrenfeucht, D. Haussler, and M. K. Warmuth. Occam's razor. *Information Processing Letters*, 24:377–380, 1987.
11. A. Blumer, A. Ehrenfeucht, D. Haussler, and M. K. Warmuth. Learnability and the Vapnik-Chervonenkis dimension. *Journal of the Association for Computing Machinery*, 36(4):929–965, 1989.
12. B. Clarke. *Asymptotic cumulative risk and Bayes risk under entropy loss with applications.* PhD thesis, Dept. of Statistics, University of Ill., 1989.
13. B. Clarke and A. Barron. Information-theoretic asymptotics of Bayes methods. *IEEE Transactions on Information Theory*, 36(3):453–471, 1990.
14. B. Clarke and A. Barron. Jefferys' prior is asymptotically least favorable under entropy risk. *J. Statistical Planning and Inference*, 41:37–60, 1994.
15. G. F. Clements. Entropy of several sets of real-valued functions. *Pacific J. Math.*, 13:1085–1095, 1963.

16. T. Cover and J. Thomas. *Elements of Information Theory.* Wiley, 1991.
17. L. Devroye and L. Györfi. *Nonparametric density estimation, the L_1 view.* Wiley, 1985.
18. R. O. Duda and P. E. Hart. *Pattern Classification and Scene Analysis.* Wiley, 1973.
19. R. M. Dudley. A course on empirical processes. *Lecture Notes in Mathematics*, 1097:2–142, 1984.
20. S. Y. Efroimovich. Information contained in a sequence of observations. *Problems in Information Transmission*, 15:178–189, 1980.
21. A. Ehrenfeucht, D. Haussler, M. Kearns, and L. Valiant. A general lower bound on the number of examples needed for learning. *Information and Computation*, 82:247–261, 1989.
22. M. Feder, Y. Freund, and Y. Mansour. Optimal universal learning and prediction of probabilistic concepts. In *Proc. of IEEE Information Theory Conference*, page 233. IEEE, 1995.
23. A. Gelman. *Bayesian Data Analysis.* Chapman and Hall, NY, 1995.
24. E. Giné and J. Zinn. Some limit theorems for empirical processes. *Annals of Probability*, 12:929–989, 1984.
25. R. Hasminskii and I. Ibragimov. On density estimation in the view of Kolmogorov's ideas in approximation theory. *Annals of statistics*, 18:999–1010, 1990.
26. D. Haussler and A. Barron. How well do Bayes methods work for on-line prediction of $\{+1, -1\}$ values? In *Proceedings of the Third NEC Symposium on Computation and Cognition.* SIAM, 1992.
27. D. Haussler and M. Opper. General bounds on the mutual information between a parameter and n conditionally independent observations. In *Proceedings of the Seventh Annual ACM Workshop on Computational Learning Theory*, 1995.
28. D. Haussler and M. Opper. Mutual information, metric entropy, and risk in estimation of probability distributions. Technical Report UCSC-CRL-96-27, Univ. of Calif. Computer Research Lab, Santa Cruz, CA, 1996.
29. I. Ibragimov and R. Hasminskii. On the information in a sample about a parameter. In *Second Int. Symp. on Information Theory*, pages 295–309, 1972.
30. A. J. Izenman. Recent developments in nonparametric density estimation. *JASA*, 86(413):205–224, 1991.
31. A. N. Kolmogorov and V. M. Tihomirov. ϵ-entropy and ϵ-capacity of sets in functional spaces. *Amer. Math. Soc. Translations (Ser. 2)*, 17:277–364, 1961.
32. L. LeCam. *Asymptotic methods in statistical decision theory.* Springer, 1986.
33. G. Lorentz. *Approxiamtion of Functions.* Holt, Rinehart, Winston, 1966.
34. R. Meir and N. Merhav. On the stochastic complexity of learning realizable and unrealizable rules. Unpublished manuscript, 1994.
35. M. Opper and D. Haussler. Bounds for predictive errors in the statistical mechanics of in supervised learning. *Physical Review Letters*, 75(20):3772–3775, 1995.
36. D. Pollard. *Empirical Processes: Theory and Applications*, volume 2 of *NSF-CBMS Regional Conference Series in Probability and Statistics.* Institute of Math. Stat. and Am. Stat. Assoc., 1990.
37. J. Rissanen. Stochastic complexity and modeling. *The Annals of Statistics*, 14(3):1080–1100, 1986.
38. J. Rissanen, T. Speed, and B. Yu. Density estimation by stochastic complexity. *IEEE Trans. Info. Th.*, 38:315–323, 1992.
39. N. Sauer. On the density of families of sets. *Journal of Combinatorial Theory (Series A)*, 13:145–147, 1972.

40. L. G. Valiant. A theory of the learnable. *Communications of the ACM*, 27(11):1134–42, 1984.
41. S. van deGeer. Hellinger-consistency of certain nonparametric maximum likelihood estimators. *Annals of Statistics*, 21:14–44, 1993.
42. A. van der Vaart and J. Wellner. *Weak Convergence and Empirical Processes.* Springer, NY, 1996.
43. V. N. Vapnik. *Estimation of Dependences Based on Empirical Data.* Springer-Verlag, 1982.
44. V. N. Vapnik and A. Y. Chervonenkis. On the uniform convergence of relative frequencies of events to their probabilities. *Theory of Probability and its Applications*, 16(2):264–80, 1971.
45. W. Wong and X. Shen. Probability inequalities for likelihood ratios and convergence rates for sieve MLE's. *Annals of Statistics*, 23(2):339–362, 1995.
46. B. Yu. Lower bounds on expected redundancy for nonparametric classes. *IEEE Trans. Info. Th.*, 42(1), 1996.

Of Periods, Quasiperiods, Repetitions and Covers

Alberto Apostolico[1,2]* and Dany Breslauer[2]

[1] Department of Computer Science, Purdue University, West Lafayette, IN 47907, USA
[2] Dipartimento di Elettronica e Informatica, Università di Padova, Padova, Italy.
[3] Zeta Information Systems, New York, USA.

Abstract. *Quasiperiodic* strings were defined by Apostolico and Ehrenfeucht [3], as *strings which are entirly covered by occurrences of another (shorter) string.* This paper surveys a handful of results on the structure and detection of quasiperiodic strings and on related string covers, attempting to simplify and present in a uniform manner the algorithms being surveyed.

1 Introduction

Periodicities and other regularities in strings arise in various disciplines such as combinatorics, automata and formal language theory, data compression, stochastic process theory, symbolic dynamics, system theory and molecular biology. In the Summer of 1990, A. Ehrenfeucht suggested that some repetitive structures defying the classical characterizations of periods and repetitions could be captured by resort to a germane notion of "quasiperiod". In their paper "Efficient Detection of Quasiperiodicities in Strings" [3] Apostolico and Ehrenfeucht defined *quasiperiodic* strings as *strings which are entirely covered by occurrences of another (shorter) string.* They also gave an $O(n \log^2 n)$ time algorithm to find all maximal quasiperiodic substrings within a given string. Apostolico, Farach and Iliopoulos [4] gave an $O(n)$ time algorithm that finds the quasiperiod of a given string, namely the *shortest string* that covers the string in question. This algorithm was subsequently simplified and improved by Breslauer [9] who gave an $O(n)$ time on-line algorithm, and parallelized by Breslauer [10] and Iliopoulos and Park [19], the latter giving an optimal-speedup $O(\log \log n)$ time parallel CRCW-PRAM algorithm. Moore and Smyth [24] gave an $O(n)$ time algorithm that finds *all strings* that cover a given string. These developments eventually led to the study by Iliopoulos, Moore and Park [18] and by Ben-Amram et al. [5] of covers which are not necessarily aligned with the ends of the string being

* Partially supported by NSF Grant CCR-92-01078, by NATO Grant CRG 900293, by British Engineering and Physical Sciences Research Council grant GR/L19362, by the National Research Council of Italy, and by the ESPRIT III Basic Research Programme of the EC under contract No. 9072 (Project GEPPCOM).

covered, but are rather allowed to overflow on either side. The sequential algorithm for this problem takes $O(n \log n)$ time [18] and the parallel counterpart [5] achieves an optimal speedup taking $O(\log n)$ time, but using superlinear space.

This paper surveys the above mentioned articles, attempting to put the different results in a unified framework, and to simplify the algorithms. The main emphasis is put on some of the sequential algorithms while the parallel counterparts are sketched in lesser detail.

2 Preliminaries

We start by recalling the basic definitions and properties of strings that will be used throughout the paper. Lothaire's book [21] provides an excellent overview of additional periodicity properties of strings.

2.1 Periods and Repetitions

Given a string $w = w_1 w_2 \cdots w_n$, we denote its length by $|w| = n$. The individual symbols w_i are assumed to be taken from some underlying alphabet Σ. We write $w_{[i \ldots j]}$ to specify the substring $w_i w_{i+1} \cdots w_j$, for $i \leq j$, and denote by ϵ the empty string.

The string w is said to have ***a period of length*** π, if $w_i = w_{i+\pi}$, for all feasible values of i. Clearly, by the definition above, $\pi = 0$ is a period length of w and any $\pi \geq |w|$ is also a period length. In additions, $\pi < 0$ is a period length if and only if $-\pi$ is a period length as well. We shall restrict our attention, therefore, only to period lengths π of w, such that $0 \leq \pi \leq |w|$. The integers 0 and $|w|$ are always period lengths of w, and are called the ***trivial period lengths***; any integer in between may or may not be a period length depending on the structure of w.

A non-empty string u, $|u| \leq |w|$, will be called ***a period*** of w if w is a substring of u^k, for some integer $k \geq 1$. Clearly, if u is a period of w, then its length $|u|$ is a period length of w, since $|u|$ is a period length of u^k. Moreover, if $u = xy$, then any ***rotation*** yx of u is also a period of w since $(yx)^{k+1} = y(xy)^k x = yu^k x$ contains w as a substring. Note that this terminology is slightly different from the standard definition of a period, in that the latter requires that u is also a prefix of w. In this paper, a period u of w that is also a prefix of w is called a ***left aligned period***. Clearly, given any period length $\pi > 0$ of w, the prefix $w_{[1 \ldots \pi]}$ is a left aligned period of w.

A period u is in fact a ***regular cover*** of w, where occurrences of u appear in w spaced exactly $|u|$ positions apart (other occurrences are also allowed) and the occurrences on the sides can overflow. Given any period u of w, consider the rotation $\hat{u}$ of u such that $\hat{u}$ is also a prefix of w (in other words, $\hat{u}$ is the rotation of u that is a left aligned period of w). If $w = \hat{u}^k$ for some integer k, namely if the regular cover of w by u is also right aligned, then w is said to have an ***aligned regular cover*** u. If w has no proper aligned regular covers (w itself is always a cover) then w is said to be ***primitive***.

The following Theorem due to Fine and Wilf [16] is of fundamental importance in the study of periodic strings:

Theorem 1. *If a string w has two period of lengths p and q, and $|w| \geq p+q-\gcd(p,q)$, then w has also a period of length $\gcd(p,q)$.*

The shortest non-zero period length of w will be called *the* period length of w and denoted $\pi(w)$. A string w such that $|w| \geq 2\pi(w)$ is said to be *periodic*. By the theorem above, in a periodic string w, all periods lengths that are smaller than $|w|/2$, must be multiples of *the period* length $\pi(w)$.

2.2 General Covers

One may generalize the notion of a period u that covers w with regular occurrences that are $|u|$ positions apart in w, to covers where the occurrences of u in w are not required to be uniformly spaced, and are allowed, in addition, to overflow on either side. For example, the string w = '*aabaabab*' may be covered by occurrences of u = '*aba*', but the positions of these occurrences in w are not regular and in fact *aba* is not a period of w. This type of covers were called *general covers* in [18] where a covering string such as our u above is also termed *a seed* of w.

2.3 Aligned Covers

Some notable families of covers result by considering covering strings u for w that are not necessarily regularly spaced but are aligned on both sides of w and are not allowed to overflow. Such strings u are said to be aligned covers of w. Given the similarity between non-regular covers and regular covers (periods), aligned covers u of w were named *quasiperiods* of w by Apostolico and Ehrenfeucht [3]. In addition, strings that do not have any non-trivial (shorter) aligned covers were called *superprimitive* and strings that have shorter aligned covers were termed *quasiperiodic*. Observe that any periodic string is also quasiperiodic, but not every quasiperiodic string is periodic. Most of the treatment of the present paper is confined to aligned covers, leaving general covers, that are handled differently, to future extensions.

2.4 Borders

We say that a non-empty string z is a *border* of a string w if w begins and ends with an occurrence of z. Namely, $w = zu$ and $w = vz$ for some possibly empty strings u and v. Clearly, a string is always a border of itself. This border is called the *trivial* border.

We describe next few facts about periods, borders, and aligned covers.

Fact 2. *A string w has a period of length π, such that $\pi < |w|$, if and only if it has a non-trivial border of length $|w| - \pi$.*

Proof. Immediate from the definitions of a border and a period.

Fact 3. *If a string z aligned-covers a string w then z is a border of w.*

Proof. Since the first symbol of w must be covered by z, the string w must start with an occurrence of z. Since the last symbol of w must also be covered by z, the string w must also end with an occurrence of z. That is, z is a border of w.

Note that by this last Fact any cover of a string w can be represented by a single integer that is the length of the border of w.

Fact 4. *If a string z covers a string w, then z covers also any possible border v of w such that $|v| \geq |z|$.*

Proof. Given any prefix of w, it is covered by z except possibly at most the last $|z| - 1$ symbols of the prefix. Similarly, given any suffix of w, it is covered by z except possibly at most the first $|z| - 1$ symbols of the suffix. Since v is a border of w, it is both a prefix and a suffix, and it must be covered by z.

Fact 5. *Every string has a unique quasiperiod.*

Proof. Assume that a string w is covered by two strings u and v, and let w.l.o.g. $|u| \leq |v|$. By Fact 3 v is a border of w. By Fact 4 u covers w. Since $u \neq v$, then v is quasiperiodic.

Fact 6. *If a string w has a border z, such that $2|z| \geq |w|$, then z covers w.*

Proof. z covers the first half of w since it is a prefix of w and the last half of w since it is also a suffix. Therefore, all symbols of w are covered by z.

3 Finding Aligned Covers

The essentials of all the algorithms that find quasiperiods are extremely similar and were described in the paper by Apostolico, Iliopolous and Farach [4]. We outline the ideas first and then present one algorithm in details.

1. **Candidates.** By Fact 3 only borders of w are candidates to be aligned covers of w.
2. **Elimination.** Given two borders u and v of w, such that $|u| < |v|$, one of them can be eliminated as being the quasiperiod of w by Fact 4, since if u covers v then v cannot be the quasiperiod of w and if u does not cover v, then surely u cannot cover w.

3.1 Computing Borders and Periods

In order to utilize the basic ideas mentioned above, the algorithms for finding covers of a string need first to find the borders of that string. There are well established algorithms for the more or less explicit computation of borders, and they will only be mentioned here without further details.

The classical Knuth-Morris-Pratt [20] string searching algorithm computes in its pattern processing step the so called *failure function* of the pattern string that is essentially a table of border lengths of every prefix of the pattern. As seen above, this is computationally equivalent to computing the period lengths of all prefixes of the pattern. In fact, we can interpret the Knuth, Morris and Pratt [20] pattern preprocessing algorithm as an on-line algorithm that finds the period length of each prefix of a string $w_{[1...n]}$ while the string is being read in one symbol at a time. The algorithm takes $O(n)$ time and uses linear auxiliary space. The number of symbol comparisons performed by the algorithm is less than $2n$.

The parallel string cover algorithm uses the parallel CRCW-PRAM algorithm of Breslauer and Galil [2, 12] that finds all periods of a string of length n in $O(\log\log n)$ time using $n/\log\log n$ processors.

3.2 Sequential Algorithm for Quasiperiods

The first algorithm for finding the quasiperiod of a string is due to Apostolico, Farach and Iliopoulos [4]. The algorithm used the ideas above in a recursive paradigm resulting in $O(n)$ time. Breslauer [9] devised an algorithm that works *on-line*, i.e., it finds the quasiperiod of all pattern prefixes as these prefixes are are produced consecutively, one symbol at a time. This algorithm, which is outlined next, also requires fewer symbol comparisons.

The idea in the algorithm is to maintain on-line, as soon that the input prefix $w_{[1...k]}$ is given, the quasiperiod of this prefix. When a longer prefix $w_{[1...k]}$ is given, its quasiperiod is computed by observing that is must either be $w_{[1...k]}$ itself (whence $w_{[1...k]}$ is superprimitive) or it must be the quasiperiod of the longest non-trivial border of $w_{[1...k]}$ (in which case $w_{[1...k]}$ is quasiperiodic).

The algorithm scans the input string $w_{[1...n]}$ one symbol at a time. It maintains two arrays: *Quasi*$[i]$ and *Reach*$[i]$. The *Quasi*$[i]$ array stores the quasiperiod of any prefix $w_{[1...i]}$, for $1 \leq i \leq k$. The *Reach*$[i]$ array is used only for superprimitive prefixes of w and it stores, for every such prefix $w_{[1...i]}$, the longest prefix of $w_{[1...k]}$ that is covered by $w_{[1...i]}$. Note that the prefix $w_{[1...i]}$ is superprimitive if and only if *Quasi*$[i] = i$. When the algorithm proceeds to the next symbol $w_{[k]}$, it has compute the quasiperiod of $w_{[1...k]}$ and store it in *Quasi*$[k]$.

As soon that the next input symbol $w_{[k]}$ is reached, the algorithm calls the Knuth-Morris-Pratt algorithm to find the period length π_k of the prefix $w_{[1...k]}$. The only comparisons of input symbols are performed in these calls to the Knuth-Morris-Pratt algorithm. Once the period of the prefix $w_{[1...k]}$ is given, the algorithm can proceed to compute the values of *Quasi*$[k]$ and *Reach*$[$*Quasi*$[k]]$ based only on the values of π_k, *Quasi*$[1 \dots k-1]$ and *Reach*$[1 \dots k-1]$. There are two cases:

1. If $\pi_k = k$, then by Fact 2 the prefix $w_{[1...k]}$ has no non-trivial border. By Fact 3 any string that covers $w_{[1...k]}$ must also be a border. Thus, the prefix $w_{[1...k]}$ is covered only by itself and therefore it is superprimitive. In this case we define *Quasi*$[k] = k$ and *Reach*$[k] = k$.

– The $Quasi[i]$ array stores the quasiperiod of any prefix $w_{[1...i]}$.
– The $Reach[i]$ array stores only for superprimitive prefixes $w_{[1...i]}$ the longest
– prefix of $w_{[1...k]}$, that is covered $w_{[1...i]}$.

```
    k = 1
    while k ≤ n do
        Compute the period π_k of w[1...k] using the Knuth-Morris-Pratt
            pattern preprocessing algorithm.
        – If the prefix w[1...k] has a non-trivial border check if the quasiperiod
        –    of that border covers the whole prefix.
        if π_k < k and Reach[Quasi[k − π_k]] ≥ k − Quasi[k − π_k] then
            – If the quasiperiod of the border w[1...k−π_k] covers the whole
            –    prefix w[1...k], then it is also the quasiperiod of w[1...k].
            Quasi[k] = Quasi[k − π_k]
            Reach[Quasi[k]] = k
        else
            – If the prefix w[1...k] does not have any non-trivial border
            –    or if the quasiperiod of the border does not cover the whole
            –    prefix w[1...k], then it is superprimitive.
            Quasi[k] = k
            Reach[k] = k
        end
        k = k + 1
    end
```

Fig. 1. The quasiperiodicity algorithm.

2. If $\pi_k < k$, then by Fact 2 the prefix $w_{[1...k]}$ has a border of length $k - \pi_k$. Since π_k is the shortest period of $w_{[1...k]}$, the border of length $k - \pi_k$ is the longest non-trivial border of the prefix $w_{[1...k]}$.
If the prefix $w_{[1...k]}$ is quasiperiodic, then by Fact 4 the border $w_{[1...k-\pi_k]}$ must also be covered by the same quasiperiod. The algorithm checks if the quasiperiod of the border $w_{[1...k-\pi_k]}$ can cover the whole prefix $w_{[1...k]}$. If a cover of the prefix $w_{[1...k]}$ by the quasiperiod of $w_{[1...k-\pi_k]}$ is given, then a cover of a shorter prefix of $w_{[1...k]}$ is obtained by removing the last occurrence of that quasiperiod. This means that the quasiperiod of $w_{[1...k-\pi_k]}$ covers a prefix of $w_{[1...k]}$ that is long enough that with one more occurrence of that quasiperiod, the whole prefix $w_{[1...k]}$ is covered.
Thus, all the algorithm has to do is to check if the quasiperiod of the border $w_{[1...k-\pi_k]}$ can cover a prefix of $w_{[1...k]}$ that is long enough. That is, if $Reach[Quasi[k - \pi_k]] \geq k - Quasi[\pi_k]$, then that quasiperiod covers $w_{[1...k]}$ and we define $Quasi[k] = Quasi[k - \pi_k]$ and update $Reach[Quasi[k]] = k$. Otherwise, the prefix $w_{[1...k]}$ is superprimitive and we define $Quasi[k] = k$ and $Reach[k] = k$.

Theorem 7. *The algorithm that is described above and in Figure 1 takes $O(n)$ time and uses linear auxiliary space. The number of comparisons of input symbols is at most $2n$.*

3.3 Parallel Algorithm for Quasiperiods

The parallel algorithms described in this paper are for the concurrent-read concurrent-write parallel random access machine model. We use the weakest version of this model called the *common CRCW-PRAM*. In this model many processors have access to a shared memory. Concurrent read and write operations are allowed at all memory locations. If several processors attempt to write simultaneously to the same memory location, it is assumed that they write the same value.

One of the major issues in the design of PRAM algorithms is the assignment of processors to their tasks. The problem is easier when the input is rigidly allocated, e.g., on an array, as is the case with strings. In this case, we can effectively resort to a powerful general principle which ignores the issue of processor assignment.

Theorem 8. *[Brent [8]] Any synchronous parallel algorithm of time t that consists of a total of x elementary operations can be implemented on p processors in $\lceil x/p \rceil + t$ time.*

The optimal speedup parallel quasiperiodicity algorithm we describe next is a variation on the algorithm given by Iliopolous and Park [19] improving on a similar non-optimal algorithm by Breslauer [10], utilizing a newly discovered parallel string searching algorithm. It uses two other parallel algorithms that are out of the scope of this paper.

1. The parallel string searching algorithm of Cole et al. [14] that finds all occurrences of a pattern string of length m in a text string of length n in constant time using n processors after a pattern preprocessing step that requires $O(\log\log m)$ time using $m/\log\log m$ processors.
   By a lower bound of Breslauer and Galil [11], this algorithm is the fastest possible optimal parallel string matching algorithm on a general alphabet, where input symbols are accessed only by pairwise comparisons. Breslauer [10] shows that the same lower bound also applies to finding the quasiperiod of a string.
   This algorithm is used in conjunction with the following algorithm to test if a given string z covers another string u.
2. The algorithm of Fich, Ragde and Wigderson [15] to compute the minimum of n integers between 1 and n in constant time using n processors.

The parallel algorithm starts by computing all the borders of the input string $w_{[1,\dots,n]}$ in $O(\log\log n)$ time and $n/\log\log n$ processors using the algorithm by Breslauer and Galil [12]. The borders are partitioned according to their length into at most $\log n$ groups $[2^i \cdots 2^{i+1} - 1]$. By Fact 6 and the observations above about elimination of quasiperiodicity candidates, only the shortest border in each group is a candidate to be the quasiperiod of w. This provides an easy elimination of all but at most $\log n$ candidates, which can be carried out in constant time by n processors using the integer minima algorithm that is mentioned above, in each interval separately, but in parallel.

The main step that underlies the rest of the algorithm is a procedure that tests, given two strings u and v, if u covers v. This is carried out by using the constant time string searching algorithm to find all occurrences of u in v using $|v|$ processors, provided that u has been already preprocessed, and then applying the integer minima (maxima) algorithm mentioned above in each interval of consecutive $|u|$ positions of v to check that the occurrences of u are not spaced too far apart in v. Clearly, all the $O(\log n)$ candidates for the quasiperiod can be preprocessed for the string searching algorithm simultaneously in $O(\log\log n)$ and using n processors. This preprocessing can be later used for searching the same preprocessed pattern on multiple occasions.

The algorithm first eliminates all but at most one candidate of those with length smaller than or equal to $n/\log n$. This is done by picking the longest candidate shorter than $n/\log n$ and testing simultaneously in constant time using n-processors if each of the $O(\log n)$ shorter candidates covers that longest candidate. The shortest candidate to cover the longest candidate with length at most $n/\log n$ is the only remaining candidate among those with lengths at most $n/\log n$.

After this elimination step, we are left with at most $1 + \log\log n$ quasiperiod candidates. Specifically, there will be at most one candidate shorter than $n/\log n$ and at most $\log\log n$ longer candidates. The algorithm proceeds by picking in each step the shortest two remaining candidates and eliminating one of them using the cover test above. This takes constant time for each step and $O(\log\log n)$ time in total. The number of operations used sums up to be $O(n)$ since it is bounded by the sum of lengths of all candidates.

The algorithm therefore is composed of several steps taking $O(\log\log n)$ time, using $n/\log\log n$ processors or steps taking $O(\log\log n)$ time and making $O(n)$ operations. The steps can be combined and slowed down to get the following result.

Theorem 9. *The quasiperiod of a string can be found in $O(\log\log n)$ time using $n/\log\log n$ processors and linear space.*

4 Finding All Covers

Moore and Smyth [24] gave an $O(n)$ time algorithm that finds *all covers* of a given string. At the heart of the algorithm is a procedure that finds the longest proper aligned cover of a string in $O(n)$ time, by looking, essentially, for the longest cover of the string which is aligned on its left but may overflows on its right.

We do not give the details of that algorithm in this short survey. We remark, however, that one can also easily find all covers of a string in $O(n \log n)$ time sequentially, and in $O(\log \log n)$ time with optimal speedup in parallel, using the ideas presented in the sequential and parallel algorithms above.

5 Finding Maximal Quasiperiodic Substrings

Assume that we are given a string w of length $n = |w|$. Consider a segment $u = w_{[i \dots i+h-1]}$ of length $h = |u|$, starting at position i, and its quasiperiod z. Apostolico and Ehrenfeucht named the triple (i, z, h) *quasiperiodic span*, and say that the quasiperiodicity is *maximal* if the following two conditions are satisfied:

1. The quasiperiodic span can not be extended on either side by more occurrences of z. Namely, there is no other quasiperiodic span (i', z, h') such that $i' \leq i$ and $i' + h' \geq i + h$.
2. The quasiperiodic span can not be extended on the right by one more symbol when the covering string is also extended by the same symbol. Namely, if $a = w_{i+|u|}$, then za does not cover ua, or in other words, ua does not have the same quasiperiod as za.

Apostolico and Ehrenfeucht were interested in finding all maximal quasiperiodic substrings of a given string. Their algorithm is based on suffix trees and properties of their structures, which are given next. For a survey of other applications of suffix trees see [1, 17].

Suffix trees Suffix trees are a compressed form of ***digital search trees*** that are very useful in many algorithms on strings. The usual definition of a suffix tree is the following:

Let w be a string having as its last symbol a special marker '#' that does not appear anywhere else in w. The *suffix tree* of w is a rooted tree with $|w|$ leaves and $|w| - 1$ internal nodes such that:

1. Each edge is labeled with a non-empty substring of w.
2. No two sibling edges are labeled with substrings that start with the same symbol.
3. Each leaf is labeled with a distinct position of w.
4. The concatenation of the labels of the edges on the path from the root to a leaf labeled with position i yields precisely the suffix $w_{[i \dots |w|]}$.

An example of a suffix tree is given in Figure 2. Note that a suffix tree has no internal nodes with one child. Substrings of w can be represented by their starting and ending positions and therefore a suffix tree can be stored in $O(|w|)$ space.

We shall identify the name of a suffix tree node with the concatenation of the edge labels from the root to that node. If two suffixes $w_{[i \dots k]}$ and $w_{[j \dots k]}$ have the same prefix, namely if $w_{[i \dots i+l-1]} = w_{[j \dots j+l-1]}$ and $w[i + l] \neq w[j + l]$, then the

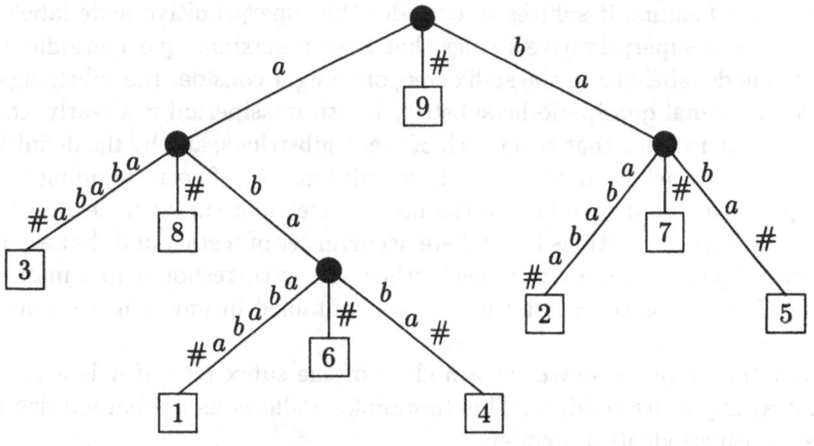

Fig. 2. The suffix tree of w = '*abaababa#*'. The edges are labeled with substrings of w and the leaves with the position in w at which the corresponding suffix starts.

paths from the root to the leaves labeled with positions i and j share an initial segment. The concatenation of the labels of the edges on this initial segment is equal to $w[i \dots i+l-1]$. Similarly, for any substring u of w, there must either be a unique node in w such that the concatenation of labels on the path from the root to the node is equal to uv with v being of minimal length and possibly empty. If there is such a node labeled u, it is called the proper *locus* of u and if it is labeled uv with $v \neq \epsilon$, then it is called the *extended locus* of u.

The special alphabet symbol '#' that was assumed to be the last symbol of the string $w[1..k]$ is normally appended at the end of a given string to guarantee that the suffix tree has distinct leaves that correspond to each suffix. There exist several efficient algorithms to construct suffix trees and important related structures such as inverted files and subword automata [6, 7, 13, 23, 22, 25, 26].

5.1 Suffix Trees and Maximal Quasiperiodicities

We next give some relations, that were given by Apostolico and Ehrenfeucht, between suffix trees and maximal quasiperiodicities.

Lemma 10. *Let $(i, z, |u|)$ be a maximal quasiperiodicity in w. Then, z must have a proper locus in the suffix tree of w.*

Proof. Assume on the contrary that z only has an extended locus zv and let a be the first symbol of v. By properties of the suffix tree, all occurrences of z in w must also be occurrences of za and therefore the substring $w_{[i \dots i+|u|]}$ must be covered by za, what contradicts the maximality of $(i, z, |u|)$.

By the last Lemma, it suffices to consider the superprimitive node labels in the suffix tree as superprimitive strings that have a maximal quasiperiodicity.

Given a node labeled z in the suffix tree, one might consider the substrings of w that are maximal quasiperiodic substrings with quasiperiod z. Clearly, there must be occurrences of z that cover each of these substrings, and by the definition of the suffix tree, each occurrence of z in w will have a leaf corresponding to its starting position as a descendant of the node z. Define a *run* at node z to be a maximal sequence of positions in w where occurrences of u start and that are not more than $|z|$ positions apart from each other. A run corresponds to a maximal substring of w that is covered by u and not contained in any longer substring that is covered by z.

We say that a run *coalesces* at a node z of the suffix tree, if it is a run at z but not at any of its children. This terminology allows us to characterize the maximal quasiperiodicities precisely.

Theorem 11. *(i, z, h) is a maximal quasiperiodicity in w if and only if there is a node in the suffix tree of w labeled z and a run $\{i_1 < i_2 < \cdots < i_k\}$, $k \geq 2$, that coalesces at z, such that:*

1. *$i = i_1$ and $i_k = i + h - |z|$. Namely, $u = w_{[i_1 \ldots i_1+h-1]}$ can not be extended on either side by more occurrences of z.*
2. *There is no ancestor z' of z where the position i_1 is in the same run with the position $i_1 + |u| - |z'|$. Namely, z is not covered by a shorter string z' and therefore, is superprimitive.*

The algorithm of Apostolico and Ehrenfeucht is built around the Theorem above and it finds the maximal quasiperiodicities by maintaining the runs for each node of the suffix tree while climbing bottom-up from the leaves. Observe that as the algorithm progresses computing the runs at a node given the runs of their children, runs of children split since the length of the parent node label is shorter than the children, and other runs merge. Apostolico and Ehrenfeucht gave an algorithm that takes $O(n \log^2 n)$ time.

6 Open Questions

Some remaining open questions are the following:

1. *All Aligned Covers.* Finding all aligned covers of all prefixes of a string in $O(n)$ time; namely, the longest proper aligned cover of each quasiperiodic prefix. An optimal parallel algorithm for finding all aligned covers of a string. These are probably the easier problems in this list.
2. *Maximal Quasiperiodic Substrings.* Improving on the $O(n \log^2 n)$ algorithm that was outlined above. Designing and efficient parallel version of this algorithm.
3. *General Covers.* Improving on the $O(n \log n)$ sequential algorithm and on the exisiting parallel algorithm.

References

1. A. Apostolico. The Myriad Virtues of Subword Trees. In A. Apostolico and Z. Galil, editors, *Combinatorial Algorithms on Words*, volume 12 of *NATO ASI Series F*, pages 85–96. Springer-Verlag, Berlin, Germany, 1985.
2. A. Apostolico, D. Breslauer, and Z. Galil. Optimal Parallel Algorithms for Periods, Palindromes and Squares. In *Proc. 19th International Colloquium on Automata, Languages, and Programming*, number 623 in Lecture Notes in Computer Science, pages 296–307. Springer-Verlag, Berlin, Germany, 1992.
3. A. Apostolico and A. Ehrenfeucht. Efficient Detection of Quasiperiodicities in Strings. *Theoret. Comput. Sci.*, 119:247–265, 1993.
4. A. Apostolico, M. Farach, and C.S. Iliopoulos. Optimal Superprimitivity Testing for Strings. *Inform. Process. Lett.*, 39:17–20, 1991.
5. A. Ben-Amram, O. Berkman, C. Iliopolous, and K. Park. Computing the Covers of a String in Linear Tme. In *Proc. 5th ACM-SIAM Symp. on Discrete Algorithms*, pages 501–510, 1994.
6. A. Blumer, J. Blumer, A. Ehrenfeucht, D. Haussler, M.T. Chen and J. Seiferas. The Smallest Automaton Recognizing the Subwords of a Text, *Theoretical Computer Science* , 40:31–55, 1985.
7. A. Blumer, A., J. Blumer, A. Ehrenfeucht, D. Haussler, and R. McConnell. Complete Inverted Files for Efficient Text Retrieval and Analysis. *Journal of the ACM*, 34(3): 578-595 (1987).
8. R.P. Brent. Evaluation of General Arithmetic Axpressions. *J. Assoc. Comput. Mach.*, 21:201–206, 1974.
9. D. Breslauer. An On-Line String Superprimitivity Test. *Inform. Process. Lett.*, 44(6):345–347, 1992.
10. D. Breslauer. Testing String Superprimitivity in Parallel. *Inform. Process. Lett.*, 49(5):235–241, 1994.
11. D. Breslauer and Z. Galil. A Lower Bound for Parallel String Matching. *SIAM J. Comput.*, 21(5):856–862, 1992.
12. D. Breslauer and Z. Galil. Finding all Periods and Initial Palindromes of a String in Parallel. *Algorithmica*, 1995.
13. M.T. Chen and J. Seiferas. Efficient and Elegant Subword-tree Construction. In A. Apostolico and Z. Galil, editors, *Combinatorial Algorithms on Words*, volume 12 of *NATO ASI Series F*, pages 97–107. Springer-Verlag, Berlin, Germany, 1985.
14. R. Cole, M. Crochemore, Z. Galil, L. Gąsieniec, R. Hariharan, S. Muthukrishnan, K. Park, and W. Rytter. Optimally Fast Parallel Algorithms for Preprocessing and Pattern Matching in One and Two Dimensions. In *Proc. 34th IEEE Symp. on Foundations of Computer Science*, pages 248–258, 1993.
15. F.E. Fich, R.L. Ragde, and A. Wigderson. Relations Between Concurrent-write Models of Parallel Computation. *SIAM J. Comput.*, 17(3):606–627, 1988.
16. N.J. Fine and H.S. Wilf. Uniqueness Theorems for Periodic Functions. *Proc. Amer. Math. Soc.*, 16:109–114, 1965.
17. R. Grossi and G.F. Italiano. Suffix Trees and their Applications in String Algorithms. Manuscript, 1995.
18. C.S. Iliopoulos, D.W.G. Moore, and K. Park. Covering a String. In *Proc. 4th Symp. on Combinatorial Pattern Matching*, number 684 in Lecture Notes in Computer Science, pages 54–62, Berlin, Germany, 1993. Springer-Verlag.

19. C.S. Iliopoulos and K. Park. An Optimal $O(\log \log n)$-time Algorithm for Parallel Superprimitivity Testing. *J. Korea Information Science Society*, 21(8):1400–1404, 1994.
20. D.E. Knuth, J.H. Morris, and V.R. Pratt. Fast Pattern Matching in Strings. *SIAM J. Comput.*, 6:322–350, 1977.
21. M. Lothaire. *Combinatorics on Words.* Addison-Wesley, Reading, MA, U.S.A., 1983.
22. U. Manber and E. Myers. Suffix Arrays: a New Method for On-line String Searches. *Proceedings of the 1st Symposium on Discrete Algorithms*, 319-327, 1990.
23. E.M. McCreight. A Space Economical Suffix Tree Construction Algorithm. *J. Assoc. Comput. Mach.*, 23:262–272, 1976.
24. D. Moore and W.F. Smyth. Computing the Covers of a String in Linear Time. In *Proc. 5th ACM-SIAM Symp. on Discrete Algorithms*, pages 511–515, 1994.
25. E. Ukkonen. Constructin Suffix Trees On-line in Linear Time. *Proceedings of Information Processing 92* , Vol. 1, 484–492, 1992.
26. P. Weiner. Linear Pattern Matching Algorithms. In *Proc. 14th Symposium on Switching and Automata Theory*, pages 1–11, 1973.

Combinatorics of Standard Sturmian Words

Aldo de Luca

Dipartimento di Matematica Università di Roma "La Sapienza"
Piazzale Aldo Moro 2, 00185, Roma, Italy

Abstract. We overview some recent developments of the theory of Sturmian words showing that the 'kernel' of the theory is the combinatorics of the set PER of all finite words w on the alphabet $\mathcal{A} = \{a, b\}$ having two periods p and q which are coprimes and such that $|w| = p + q - 2$. The elements of PER have many surprising structural properties. In particular, the relation $Stand = \mathcal{A} \cup PER\{ab, ba\}$ holds, where $Stand$ is the set of all finite standard Sturmian words. Moreover, PER can be generated by two different procedures. The first uses the operator of left palindrome closure, whereas the second uses some elementary standard morphisms. We prove the existence of a basic correspondence, that we call standard, between these two methods.
Key words. Sturmian words, standard words, standard morphisms.

1 Introduction

Sturmian words have been extensively studied by several authors for at least two centuries. They have many applications in various different fields like Algebra and Theory of numbers, Physics (Symbolic Dynamics, Crystallography) and Computer Science (Computer Graphics and Pattern matching); the study of the structure and combinatorics of these words has become a subject of the greatest interest, with a large literature on it (see, for instance,[2], [5], [4],[12] , [9],[15], [16] and references therein). The term "Sturmian" was first used by G. A. Hedlund [12]. Sturmian words can be defined in several different but equivalent ways. Some definitions are 'geometrical' and others of 'combinatorial' nature. A 'geometrical' definition is the following: A Sturmian word can be defined by considering the sequence of the cuts (*cutting sequence*) in a squared-lattice made by a semi-line having a slope which is an irrational number. A horizontal cut is denoted by the letter b, a vertical cut by a and a cut with a corner by ab or ba. Sturmian words represented by a semi-line starting from the origin are usually called *standard* or *characteristic*.

A combinatorial definition of a Sturmian word can be given in terms of its *subword complexity*. More precisely *Sturmian words are infinite words which are not ultimately periodic and have minimal subword complexity.* We recall that if w is an infinite word on the alphabet $\mathcal{A}$, then the subword complexity of w is the map $f_w : N \to N$, defined as: for each $n \geq 0$

$$f_w(n) = card(F(w) \cap \mathcal{A}^n),$$

where $F(w)$ is the set of all factors, or subwords, of w of length n. In other words for each n, $f_w(n)$ gives the number of factors of w of length n.

An infinite word w is ultimately periodic if it can be expressed as:

$$w = uv^\omega = uvv \dots v \dots .$$

One easily verifies that a word w is ultimately periodic if and only if there exists an integer c such that

$$f_w(n) < c$$

for all $n \geq 0$. Less evident is the following

Fact. An infinite word is ultimately periodic if and only if there exists an integer n for which

$$f_w(n) \leq n.$$

Hence, if an infinite word w is not ultimately periodic then for all n, $f_w(n) \geq n+1$. Sturmian words are infinite words w whose subword complexity f_w is such that

$$f_w(n) = n + 1$$

for all $n \geq 0$, so that they have the minimal possible value for subword complexity without being ultimately periodic. Moreover, since $f_w(1) = 2$ one has that these words are in a two letter alphabet. It is worth noting that between ultimately periodic and Sturmian words there are no other words. A Sturmian word s even though it is not ultimately periodic, is *uniformly recurrent*, i.e. any one of its factors will occur in the word infinitely often with a bounded distance between two consecutive occurrences. More precisely it holds the following property:

There exists a map $k : N \to N$, called *recurrence function* of s such that for any $n \geq 0$ any factor of s of length n will occur in any factor of s of length $k(n)$.

The most famous Sturmian word is the Fibonacci word

$$f = abaababaabaababaababaabaababaabaab \dots .$$

which is the limit, according to a suitable topology (cf.[?]), of the sequence of words $\{f_n\}_{n \geq 0}$, inductively defined as:

$$f_0 = b, \; f_1 = a, \; f_{n+1} = f_n f_{n-1},$$

for all $n > 0$. The words f_n of this sequence are called the *finite Fibonacci words*. The name Fibonacci is due to the fact that for each n, $|f_n|$ is equal to the $(n+1)$ th term of the Fibonacci numerical sequence:

$$1, 1, 2, 3, 5, 8, 13 \dots .$$

Standard Sturmian words can be defined in the following way which is a natural generalization of the definition of the Fibonacci word. Let $c_0, c_1, \dots, c_n, \dots$ be any sequence of natural integers such that $c_0 \geq 0$ and $c_i > 0$ $(i = 1, ..., n)$. We define, inductively, the sequence of words $\{s_n\}_{n\geq 0}$, where

$$s_0 = b,\ s_1 = a,\ \text{and}\ s_{n+1} = s_n^{c_{n-1}} s_{n-1},\ \text{for}\ n \geq 1.$$

The sequence $\{s_n\}_{n\geq 0}$ converges to a limit s which is an infinite standard Sturmian word. Any standard Sturmian word is obtained in this way. The sequence $\{s_n\}_{n\geq 0}$ is called the *approximating sequence* of s and $(c_0, c_1, c_2, ...)$ the *directive sequence* of s. The Fibonacci word f is the standard Sturmian word whose directive sequence is $(1, 1, ..., 1, ...)$. The set of all infinite standard Sturmian words will be denoted by **Stand**. The set of all the words s_n, $n \geq 0$ of any standard sequence $\{s_n\}_{n\geq 0}$ constitutes a language *Stand* which has remarkable and surprising properties. Any word of *Stand* is called *finite standard word*, or *generalized Fibonacci word*.

2 Finite standard words

Infinite standard Sturmian words are of great interest since it has been proved [14] that for any Sturmian word w there exists an infinite standard Sturmian word $s \in$ **Stand** such that $F(w) = F(s)$. If one is interested in the study of the language *St* of the factors of all Sturmian words, one can limit oneself to consider only infinite standard Sturmian words. Indeed, one has:

$$St = \cup_{s\in \mathbf{Stand}} F(s).$$

Since any element of **Stand** is the limit of a sequence of finite standard words, one has that any factor of a word of **Stand** is a factor of a word of *Stand*. Hence,

$$St = F(Stand).$$

The following interesting and useful combinatorial characterization of the language *St* was given by S. Dulucq and D. Gouyou-Beauchamps [10]:

Theorem 1. *The language St is the set of all the words $w \in \mathcal{A}^*$, $\mathcal{A} = \{a, b\}$, such that for any pair (u, v) of factors of w having the same length one has:*

$$||u|_a - |v|_a| \leq 1.$$

In other words the set *St* of all finite Sturmian words coincides with the set of all finite words w over the alphabet $\{a, b\}$ such that for any pair u and v of factors of w having the same length, the modulus of the difference of the number of occurrences of a letter (a or b) in u and in v is at most 1.

The set *Stand* has several characterizations that we shall use in the following, based on quite different concepts:

(a). Palindrome words. Let *PAL* be the set of all palindromes on $\mathcal{A}$. The set Σ is the subset of $\mathcal{A}^*$ defined as:

$$\Sigma = \mathcal{A} \cup (PAL^2 \cap PAL\{ab, ba\}).$$

Thus a word w belongs to Σ if and only if w is a single letter or satisfies the equation:

$$w = AB = Cxy,$$

where $A, B, C \in PAL$ and $\{x, y\} = \{a, b\}$. A remarkable result obtained by Pedersen *et al.* [17] is that there exists, and it is ***unique***, a word w such that

$$w = AB = Cxy,$$

if and only if $gcd(|A| + 2, |B| - 2) = 1$.

It was proved in [9] that

$$Stand = \Sigma.$$

(b). Periodicities of the words. Let $w \in \mathcal{A}^*$ and $\Pi(w)$ be the set of its periods (cf.[13]). We define the set PER of all words w having two periods $p, q \in \Pi(w)$ which are coprimes and such that $|w| = p + q - 2$. Thus a word w belongs to PER if it is a power of a single letter or is a word of maximal length for which the theorem of Fine and Wilf (cf.[13]) does not apply. In the sequel we assume that $\epsilon \in PER$. This is, formally, coherent with the above definition if one takes $p = q = 1$. In [9] de Luca and Mignosi proved that:

$$Stand = \mathcal{A} \cup PER\{ab, ba\}.$$

(c). Special elements. A word $w \in St$ is a ***right (left) special element*** of St if $wa, wb \in St$ $(aw, bw \in St)$. The word w is ***bispecial*** if it is right and left special. It is called ***strictly bispecial*** if

$$awa, awb, bwa, bwb \in St.$$

Let us denote by SBS the set of all strictly bispecial elements of St. It has been proved in [9] that

$$PER = SBS.$$

If S_R (S_L) denotes the set of all right (left) special elements of St one has that (cf.[6]):

$$SBS = S_R \cap PAL = S_L \cap PAL.$$

(d). Lyndon words. We recall that a *Lyndon word* is a primitive word which is minimal, with respect to the lexicographic order, in its conjugation class. This is also equivalent to say that a word is Lyndon if and only if it is lexicographically less than any proper right factor. It has been proved in [1]

$$aPERb \cup \mathcal{A} = St \cap Lynd.$$

(e). Slopes of the words. Let $\mathcal{A} = \{a, b\}$. We consider the map $\rho : \mathcal{A}^* \to \mathbb{Q} \cup \{\infty\}$, defined as:

$$\rho(\epsilon) = 1, \; \rho(w) = |w|_b/|w|_a, \; for\ w \neq \epsilon.$$

We assume that $1/0 = \infty$. For any $w \in \mathcal{A}^*$ we call $\rho(w)$ the 'slope' of w. A *word $w \in \mathcal{A}^*$ is called a Christoffel word if and only if every prefix w' of w has a slope which is maximal with respect to the slope of any other word u such that $|u| = |w'|$ and $\rho(u) \leq \rho(w)$.* We denote by CP the set of all Christoffel words which are primitive. One can prove that for any value of the slope given by an irreducible fraction, there exists a unique Christoffel primitive word having that slope.

It was shown in [1] that

$$aPERb \cup A = CP.$$

Hence, $St \cap Lynd = CP$. A direct proof of this latter equality is due to Borel and Laubie [3].

We can summarize the above results in the following basic theorem.

Theorem 2.
(1) $Stand = \Sigma$,
(2) $\Sigma = \mathcal{A} \cup PER\{ab, ba\}$,
(3) $PER = SBS$,
(4) $aPERb \cup A = St \cap Lynd$,
(5) $aPERb \cup A = CP$.

This theorem has several applications. In particular, one can determine the subword complexity of *Stand* and derive in a simple and purely combinatorial way, the subword complexity formula for *St* (cf.[9]).

Theorem 2 shows that the 'kernel' of the set of standard Sturmian words is the set PER. One has

$$St = F(PER).$$

We recall the following important structure result on the set PER whose proof is in [6]. We denote for a word w by $alph(w)$ the set of the letters occurring in w.

Proposition 3. *Let w be such that $Card(alph(w)) > 1$. Then $w \in PER$ if and only if w can be uniquely represented as:*

$$w = PxyQ = QyxP,$$

with x, y fixed letters in $\{a, b\}$, $x \neq y$ and $P, Q \in PAL$. Moreover, $gcd(p, q) = 1$, where $p = |P| + 2$ and $q = |Q| + 2$.

It appears from the above results that one can describe the elements of PER by two quite different concepts: the periods or the number of occurrences of the letters a and b. This will be a basic duality underlying all the theory even though, as we shall see in the following these two descriptions are strictly linked.

3 Generation of standard words

Finite standard words can be generated in some different ways. The first method is due to Rauzy [20]. We begin by introducing the set of *standard pairs*.

Let $\mathcal{A} = \{a, b\}$. We consider the smallest subset $\mathcal{R}$ of $\mathcal{A}^* \times \mathcal{A}^*$ which contains the pair (a, b) and is closed under the property:

$$(u, v) \in \mathcal{R} \Rightarrow (u, uv), (vu, v) \in \mathcal{R}.$$

We call $\mathcal{R}$ also the set of *standard pairs*. Let us denote by $Trace(\mathcal{R})$ the set:

$$Trace(\mathcal{R}) = \{u \in \mathcal{A}^* \mid \exists v \in \mathcal{A}^* \textit{ such that } (u, v) \in \mathcal{R} \textit{ or } (v, u) \in \mathcal{R}\}.$$

One can prove that [20]:

$$Stand = Trace(\mathcal{R}).$$

We shall give now two different, and in some respects dual, methods which will allows us to construct directly the set of all finite standard words. By Theorem 2 it is sufficient to construct the set PER. Moreover, one has that PER is closed under the automorphism E of $\mathcal{A}^*$ defined as:

$$E : \begin{matrix} a \to b \\ b \to a \end{matrix},$$

thus it is sufficient to construct the set PER_a of all the elements of PER beginning with the letter a in order to be able to construct the set $PER = \{\epsilon\} \cup PER_a \cup E(PER_a)$.

Let us introduce in $\mathcal{A}^*$ the map $(-) : \mathcal{A}^* \to PAL$ which associates with any word $w \in \mathcal{A}^*$ the palindrome word $w^{(-)}$ defined as the shortest palindrome word having the suffix w. We call $w^{(-)}$ the *palindrome left-closure of* w. If P is the greatest palindrome prefix of $w = Pu$, then one has

$$w^{(-)} = u^{\sim} P u,$$

where $u^{\sim}$ is the reversed of u. If X is a subset of $\mathcal{A}^*$ we denote by $X^{(-)}$ the set

$$X^{(-)} = \{w^{(-)} \in \mathcal{A}^* \mid w \in X\}.$$

It holds the following [6]:

Lemma 4. *Let $w \in PER$ and $x \in \{a, b\}$. Then $(xw)^{(-)} \in PER$. Moreover, if $w = PxyQ$, with $P, Q \in PAL$ and $\{x, y\} = \{a, b\}$, then one has*

$$(xw)^{(-)} = QyxPxyQ, \; (yw)^{(-)} = PxyQyxP.$$

Let us now define the map

$$\psi : \mathcal{A}^* \rightarrow PER,$$

as: $\psi(\epsilon) = \epsilon$ and for all $w \in \mathcal{A}^*$, $x \in \mathcal{A}$,

$$\psi(wx) = (x\psi(w))^{(-)}.$$

The word w is called the *generating word* of $\psi(w)$. One has that for all $w, u \in \mathcal{A}^*$

$$\psi(wu) \in \mathcal{A}^*\psi(w).$$

It has been proved in [6] that the map $\psi : \mathcal{A}^* \rightarrow PER$ is a bijection and that the restriction of ψ to $a\mathcal{A}^*$ is a bijection of $a\mathcal{A}^*$ onto PER_a. Hence one has:

$$PER_a = \psi(a\mathcal{A}^*) = \cup_{n\geq 0}\psi(a\mathcal{A}^n).$$

Setting for each $n \geq 0$

$$X_n = \psi(a\mathcal{A}^n),$$

it follows

$$PER_a = \bigcup_{n\geq 0} X_n,$$

where for any $n > 0$ one has:

$$X_n = (\mathcal{A}X_{n-1})^{(-)}.$$

A different, and in some respects dual, way of generating the elements of PER is obtained by introducing a further bijection $\phi : \mathcal{A}^* \rightarrow PER$ constructed by suitable iterations of some endomorphisms of $\mathcal{A}^*$ as follows.

Let us denote by F the Fibonacci morphism $F : \mathcal{A}^* \rightarrow \mathcal{A}^*$ defined as:

$$F : \begin{array}{l} a \rightarrow ab \\ b \rightarrow a \end{array} .$$

The name Fibonacci is due to the fact that for any n, $F^{(n)}(a) = f_{n+1}$, where f_{n+1} is the term of order $n+1$ of the sequence of finite Fibonacci words, $F^1 = F$ and for any n, $F^n = F \circ F^{n-1}$. Hence, one has $F^\omega = f$.

An endomorphism g of $\mathcal{A}^*$ is called a *standard Sturmian morphism*, or simply, *standard* if it preserves finite standard words, i.e.

$$w \in Stand \Rightarrow g(w) \in Stand.$$

It has been proved in [7] that an endomorphism g is standard if and only if

$$g \in \{E, F\}^*;$$

thus g is the identity or it is obtained by composing the morphisms E and F an arbitrary number of times. Moreover, a standard morphism preserves also infinite standard words. The following characterization of the set *Stand* was also proven:

Proposition 5. *Stand is the smallest subset of $\mathcal{A}^*$ containing $\mathcal{A}$ and closed under the Fibonacci morphism F and the automorphism E.*

In the following we shall denote by λ and μ the standard morphisms:

$$\lambda = E \circ F, \quad \text{and} \quad \mu = F.$$

By means of these two morphisms we define the map $\phi : \mathcal{A}^* \to PER$ as follows: $\phi(\epsilon) = \epsilon$ and for all $w \in a\mathcal{A}^*$

$$\phi(wa) = \lambda(\phi(w))a$$

$$\phi(wb) = \mu(\phi(w))a.$$

Moreover, one defines

$$\phi(E(w)) = E(\phi(w)),$$

so that ϕ is defined also on the words of $b\mathcal{A}^*$.

It has been proved in [7] that ϕ is a bijection and that the restriction of ϕ to $a\mathcal{A}^*$ is a bijection of $a\mathcal{A}^*$ in PER_a. Hence,

$$PER_a = \phi(a\mathcal{A}^*) = \bigcup_{n \geq 0} \phi(a\mathcal{A}^n).$$

If we set for any $n \geq 0$, $Y_n = \phi(a\mathcal{A}^n)$, one has for $n > 0$:

$$Y_n = (\lambda(Y_{n-1})) \cup \mu(Y_{n-1}))a.$$

The following theorem holds in [7]:

Theorem 6. *For each $n \geq 0$*

$$X_n = Y_n.$$

This implies that for each $n \geq 0$

$$X_{n+1} = (\mathcal{A}X_n)^{(-)} = (\lambda(X_n) \cup \mu(X_n))a.$$

This result shows that the set X_{n+1} can be generated by the set X_n by two conceptually very different procedures. The first by the operation of left-palindrome closure $(-)$ applied to the set $\mathcal{A}X_n$ and the second by using the morphisms λ and μ applied to X_n followed by a final multiplication with the letter a. The result is very surprising since left-palindrome closure uses, to be performed, a *global* information on the word, whereas the computation of a morphism uses only *local* information.

Since for $n > 0$ one has $X_n = \psi(a\mathcal{A}^{n-1}) = Y_n = \phi(a\mathcal{A}^{n-1})$ there exists a permutation T_n, or simply T,

$$T : a\mathcal{A}^{n-1} \to a\mathcal{A}^{n-1},$$

such that for any $x \in a\mathcal{A}^{n-1}$

$$\phi(x) = \psi(Tx).$$

The determination of the permutation T, that we call the ***standard correspondence***, will be done in the next sections. In the following table we report for $n = 4$ the permutation $T_4 : a\mathcal{A}^3 \to a\mathcal{A}^3$:

word	ϕ	ψ
aaaa	aaaa	aaaa
aaab	abababa	aaabaaa
aaba	aabaabaa	aabaaabaa
aabb	abaabaaba	aabaabaa
abaa	aaabaaa	abaabaaba
abab	ababaababa	abaababaaba
abba	aabaaabaa	ababaababa
abbb	abaababaaba	abababa

4 Arithmetization

In this section we introduce some suitable maps which are bijections of the set PER and the set $\mathcal{I}$ of all irreducible fractions p/q with $p, q > 0$. These representations of PER are of great interest and have remarkable applications. Moreover, they are related to each other and in a very natural way with some basic representations of $\mathcal{I}$ in the complete binary tree.

Rate. The map $\eta : PER \to \mathcal{I}$ is defined as follows: for any $w \in PER$

$$\eta(w) = \frac{|w|_b + 1}{|w|_a + 1}.$$

For any $w \in PER$, $\eta(w)$ that we call the *rate* of w, gives the slope of the standard word wab or of wba or of the Christoffel primitive word awb.

Ratio of the periods. One can introduce two natural maps which give the ratio of the periods of a word of PER.

Ratio ζ. The map $\zeta : PER \to \mathcal{I}$ is defined as follows. For all $n \geq 0$ we set

$$\zeta(a^n) = \frac{1}{n+1}, \quad \zeta(b^n) = n + 1.$$

If $w \in PER$ and $card(alph(w)) > 1$ then from Proposition 3, w can be uniquely factorized as:

$$w = QbaP = PabQ$$

with $P, Q \in PAL$. Then we set:

$$\zeta(w) = \frac{|Q| + 2}{|P| + 2} = \frac{q}{p},$$

where $p = |P| + 2$ and $q = |Q| + 2$.

Ratio θ. The map $\theta : PER \rightarrow \mathcal{I}$ is defined as: for all $n \geq 0$,

$$\theta(a^n) = \frac{1}{n+1}, \quad \theta(b^n) = n+1.$$

If $w \in PER$ and $card(alph(w)) > 1$ then from Proposition 3 w can be uniquely factorized as:

$$w = QyxP = PxyQ$$

with $\{x, y\} = \{a, b\}$, $P, Q \in PAL$ and $|P| < |Q|$. If $w \in PER_a$ then we set:

$$\theta(w) = \frac{|P|+2}{|Q|+2} = \frac{p}{q}.$$

If $w \in PER_b$ then $\theta(w) = q/p$. The meaning of $\theta(w)$ when $w \in PER_a$, is that the ratio of the *minimal* period of w and the period q such that $gcd(p, q) = 1$ and $|w| = p + q - 2$.

One can prove that the maps η, ζ and θ are *bijections*. In order to see this and the relationships existing among them we recall some basic representations of the binary words by irreducible fractions.

First of all we introduce a suitable representation of irreducible fractions by continued fractions. Let $(\alpha_0, \alpha_1, \ldots, \alpha_n)$ be a sequence of integers such that $\alpha_i > 0$, $(i = 1, \ldots, n-1)$ and $\alpha_0, \alpha_n \geq 0$. We shall denote by

$$\langle \alpha_0, \ldots, \alpha_n \rangle$$

the continued fraction $[\alpha_0, \ldots, \alpha_{n-1}, \alpha_n + 1]$ and by FC the set of all such continued fractions with n *even* integer. It is trivial to verify that $\mathcal{I}$ is faithfully represented in FC, so that in the following we shall identify $\mathcal{I}$ with FC. In FC one can naturally introduce the product operation $\oplus$ defined as follows (cf.[6],[8]): Let $\langle a_0, a_1, \ldots, a_n \rangle$, $\langle b_0, b_1, \ldots, b_m \rangle \in FC$ then

$$\langle a_0, .., a_n \rangle \oplus \langle b_0, .., b_m \rangle = \begin{cases} \langle a_0, .., a_{n-1}, a_n + b_0, b_1, .., b_m \rangle & \text{if } nm = 0 \\ & \text{or } a_n + b_0 > 0 \\ \langle a_0, .., a_{n-2}, a_{n-1} + b_1, b_2, .., b_m \rangle & \text{otherwise.} \end{cases}$$

One can easily verify that the operation $\oplus$ is associative and that the continued fraction $\langle 0 \rangle$ is the identity element, so that FC is a monoid with respect to the operation $\oplus$.

Let us now represent the words of $\mathcal{A}^*$ by continued fractions of the set FC. Any word $w \in \mathcal{A}^*$ can be uniquely represented as:

$$w = b^{\alpha_0} a^{\alpha_1} b^{\alpha_2} \ldots a^{\alpha_{n-1}} b^{\alpha_n},$$

where n is an even integer, $\alpha_i > 0$, $(i = 1, \ldots, n-1)$ and $\alpha_0, \alpha_n \geq 0$. We call the sequence of integers $(\alpha_0, \alpha_1, \ldots, \alpha_n)$ the *integral representation* of the word

w. A list $(\alpha_0, \alpha_1, \ldots, \alpha_n)$, with n odd and $\alpha_i > 0$, $(i = 1, \ldots, n)$ and $\alpha_0 \geq 0$ determines the word w whose integral representation is $(\alpha_0, \alpha_1, \ldots, \alpha_n, 0)$. In this case the integral representation $(\alpha_0, \alpha_1, \ldots, \alpha_n)$ is said *incomplete*. For any $w \in \mathcal{A}^*$ we define as the *order* of w the integer $ord(w)$ equal to the number of non zero elements in the integral representation of w.

Let us now introduce the maps:

$$Sb, Ra : \mathcal{A}^* \to \mathcal{I}$$

defined as: if $w \in \mathcal{A}^*$ has the integral representation $(\alpha_0, \alpha_1, \ldots, \alpha_n)$, then

$$Sb(w) = \langle \alpha_0, ..., \alpha_n \rangle, \quad Ra(w) = \langle \alpha_n, ..., \alpha_0 \rangle.$$

For any $w \in \mathcal{A}^*$, $Sb(w)$ and $Ra(w)$ are called, respectively, the *Stern-Brocot number* and the *Raney number* of w. It follows from the definition that

$$Ra(w) = Sb(w^{\sim}).$$

Moreover, one has $Sb(E(w)) = 1/Sb(w)$ and $Ra(E(w)) = 1/Ra(w)$. It was proven in [8] the following:

Proposition 7. *The maps Sb and Ra are, respectively, an isomorphism and an anti-isomorphism of* $\mathcal{A}^*$ *and* FC *with respect to the operation* $\oplus$.

A further bijection of $\mathcal{A}^*$ and FC is given by the following map $Fa : \mathcal{A}^* \to \mathcal{I}$ defined as: for any $w \in \mathcal{A}^*$

$$Fa(w) = \begin{cases} Ra(w) & \text{if } ord(w) \text{ is odd} \\ 1/Ra(w) & \text{otherwise.} \end{cases}$$

For any $w \in \mathcal{A}^*$, $Fa(w)$ has been called the *Farey number* of w. It holds the following [1]

Theorem 8.

$$SB = \psi \circ \eta, \quad Ra = \psi \circ \zeta, \quad Fa = \psi \circ \theta.$$

The theorem trivially implies that the maps η, ζ and θ are bijections.

Example. Let $w = a^2ba^3b$ whose integral representation is $(0, 2, 1, 3, 1)$. One has:

$$u = \psi(w) = aabaaabaaabaaabaabaaabaaabaaabaa.$$

Thus $u = QbaP = PabQ$ with

$$P = aabaaabaaabaa \text{ and } Q = aabaaabaaabaaabaa.$$

Moreover,

$$Sb(w) = \langle 0, 2, 1, 3, 1 \rangle = 9/25, \quad Ra(w) = \langle 1, 3, 1, 2, 0 \rangle = 19/15, \quad Fa(w) = 15/19.$$

Hence:

$$Sb(w) = \eta(u) = (|u|_b + 1)/(|u|_a + 1) = 9/25, \quad Ra(w) = \zeta(u) = q/p = 19/15$$

and

$$Fa(w) = \theta(u) = p/q = 15/19.$$

5 Trees and Matrices

Let us now consider the complete binary tree. Each path from the root to a particular node can be represented by a word $w \in \{a, b\}^*$. More precisely, the sequence of letters of $w = b^{\alpha_0} a^{\alpha_1} \ldots . a^{\alpha_{n-1}} b^{\alpha_n}$, read from left to right, gives the sequence of right and left moves in order to reach the node represented by w starting from the root. Since for every node there exists a unique path going from the root to the node, one has that the nodes are faithfully represented by the words $w \in \{a, b\}^*$; in the following we shall identify the nodes of the tree with the binary words of $\{a, b\}^*$.

Let us now label each node of the tree with an irreducible fraction p/q, p and q positive integers, in the following way. The root has the label $1/1$. If a node has the label p/q, then the 'left son' has the label $p/(p+q)$ and the 'right son' has the label $(p+q)/q$. We call this labeled binary tree the *Raney tree*. The label of each node w, as we shall see in more details in the following, is just the Raney number $Ra(w)$ of w.

Let us now consider the *Stern-Brocot tree* (cf.[11]) which is a complete binary tree labeled by irreducible fractions according to the following rule. The label p/q in a node is given by $(p' + p'')/(q' + q'')$, where p'/q' is the nearest ancestor above and to the left and p''/q'' is the nearest ancestor above and to the right (In order to construct the tree one needs also to add to the binary tree two more nodes labeled by $1/0$ and $0/1$). The label of a node w is the *Stern-Brocot number* $Sb(w)$ of w.

We introduce now the set of all matrices:

$$M = \begin{pmatrix} a & b \\ c & d \end{pmatrix},$$

where a, b, c and d are non-negative integers and such that $det(M) = ad - bc = 1$. This set is usually denoted by $\mathcal{D}_1$, or simply $\mathcal{D}$. As is well known $\mathcal{D}$ is a monoid freely generated by the two matrices:

$$\begin{pmatrix} 1 & 0 \\ 1 & 1 \end{pmatrix}, \begin{pmatrix} 1 & 1 \\ 0 & 1 \end{pmatrix}.$$

Let $\mathcal{A} = \{a, b\}$ be a two letter alphabet. The map $\alpha : \mathcal{A} \to \mathcal{D}$ defined as:

$$\alpha(a) = \begin{pmatrix} 1 & 0 \\ 1 & 1 \end{pmatrix}, \ \alpha(b) = \begin{pmatrix} 1 & 1 \\ 0 & 1 \end{pmatrix},$$

can be extended to an isomorphism of $\mathcal{A}^*$ onto $\mathcal{D}$ (the empty word ϵ is represented by the identity matrix). This representation of $\mathcal{A}^*$ is of great interest and has many applications. We recall that it was used to give a very simple positive solution of the famous Ehrenfeucht conjecture on the existence for any language of a finite test set for morphisms (cf. [18]).

We can then identify, when no confusion arises, each word $w \in \{a, b\}^*$ with the corresponding matrix $\alpha(w)$; we say also that w is the *generating word* of the matrix $\alpha(w)$. We denote also by w' the word $E(w^{\sim})$. The operation $(')$ is an antiautomorphism of $\mathcal{A}^*$. It holds the following [19]

Proposition 9. *Let $w \in \mathcal{A}^*$ and let*

$$\alpha(w) = \begin{pmatrix} a & b \\ c & d \end{pmatrix}.$$

Then:

$$\alpha(w^\sim) = \begin{pmatrix} d & b \\ c & a \end{pmatrix}, \quad \alpha(E(w)) = \begin{pmatrix} d & c \\ b & a \end{pmatrix}, \quad \alpha(w') = \begin{pmatrix} a & c \\ b & d \end{pmatrix}.$$

It holds the following proposition whose proof is trivial:

Proposition 10. *Let p and q be positive integers which are coprimes. Then there exists a unique matrix $M \in \mathcal{D}$ such that*

$$\begin{pmatrix} p \\ q \end{pmatrix} = M \begin{pmatrix} 1 \\ 1 \end{pmatrix}.$$

To each vector $u = \begin{pmatrix} p \\ q \end{pmatrix}$, $p, q > 0$ we associate the number

$$f(u) = p/q.$$

If $M = \begin{pmatrix} a & b \\ c & d \end{pmatrix} \in \mathcal{D}$, then we define

$$f(M) = (a+b)/(c+d) = f(M \begin{pmatrix} 1 \\ 1 \end{pmatrix}).$$

From the general theory (cf.[19], [11]) one has:

Proposition 11. *If $w = b^{\alpha_0} a^{\alpha_1} \ldots a^{\alpha_{n-1}} b^{\alpha_n}$, then $f(\alpha(w))$ has the development in continued fractions given by:*

$$f(\alpha(w)) = \langle \alpha_0, \alpha_1, \ldots, \alpha_n \rangle = Sb(w).$$

From this it follows that

$$f(\alpha(w^\sim)) = \langle \alpha_n, \alpha_{n-1}, \ldots, \alpha_0 \rangle = Sb(w^\sim) = Ra(w).$$

Let $\delta : PER \to \mathcal{D}$ be the map defined as: for all $n \geq 0$

$$\delta(a^n) = \begin{pmatrix} 1 & 0 \\ n & 1 \end{pmatrix}, \quad \delta(b^n) = \begin{pmatrix} 1 & n \\ 0 & 1 \end{pmatrix},$$

and if $w = PbaQ = QbaP$, with $P, Q \in PAL$, then

$$\delta(PbaQ) = \begin{pmatrix} q_b & p_b \\ q_a & p_a \end{pmatrix},$$

having set for $x \in \mathcal{A}$,

$$p_x = |P|_x + 1, \quad q_x = |Q|_x + 1.$$

One can easily prove by Lemma 4 and a simple inductive argument that for any $w = PbaQ = QbaP$, $P, Q \in PAL$

$$q_b p_a - q_a p_b = 1,$$

so that $\delta(PbaQ) \in \mathcal{D}$. Note that the rate η of $PbaQ$ is $(p_b + q_b)/(p_a + q_a)$ and the ratio of periods ζ is given by $(p_a + p_b)/(q_a + q_b)$.

The following theorem whose proof is a straightforward consequence of Proposition 10, holds:

Theorem 12. *The map $\alpha : \mathcal{A}^* \to \mathcal{D}$ can be decomposed as:*

$$\alpha = \psi \circ \delta.$$

A consequence of the preceding theorem is that the map δ is a bijection of PER and $\mathcal{D}$. Moreover, the theorem gives a new insight into the structure of 2×2 unimodular matrices with entries in N; indeed, the elements of these matrices can be always regarded as the number of occurrences of the letter a and b in the standard words Qab and Pab.

6 The standard correspondence

In this section we shall determine the standard correspondence

$$T : a\mathcal{A}^{n-1} \to a\mathcal{A}^{n-1},$$

such that for any $x \in a\mathcal{A}^{n-1}$,

$$\phi(x) = \psi(Tx).$$

Let us introduce the monoid of all 2×2 matrices $\mathcal{M}$ that one can generate from the two matrices:

$$\begin{pmatrix} 1 & 0 \\ 1 & 1 \end{pmatrix}, \begin{pmatrix} 0 & 1 \\ 1 & 1 \end{pmatrix}.$$

Any matrix in $\mathcal{M}$ will have a determinant equal to ± 1.

Let $\mathcal{A} = \{a, b\}$ be a two letter alphabet. The map $\beta : \mathcal{A} \to \mathcal{M}$ defined as:

$$\beta(a) = \begin{pmatrix} 1 & 0 \\ 1 & 1 \end{pmatrix}, \ \beta(b) = \begin{pmatrix} 0 & 1 \\ 1 & 1 \end{pmatrix},$$

can be extended to a morphism $\beta : \mathcal{A}^* \to \mathcal{M}$ of $\mathcal{A}^*$ onto $\mathcal{M}$.

Lemma 13. *For all $w \in a\mathcal{A}^*$ one has:*

$$\eta(\phi(w)) = f(\beta(w^{\sim})).$$

Proof. Let $\sigma : \mathcal{A}^* \longrightarrow N^{[2,1]}$ be the map defined as:

$$\sigma(w) = \begin{pmatrix} |w|_b + 1 \\ |w|_a + 1 \end{pmatrix}.$$

We shall prove, by induction on the length $|w|$ of w that for all $w \in a\mathcal{A}^*$

$$\sigma(\phi(w)) = \beta(w^{\sim}) \begin{pmatrix} 1 \\ 1 \end{pmatrix}.$$

From this the lemma trivially follows since $\eta(\phi(w)) = f(\sigma(\phi(w)))$. We recall that $\phi(wa) = \lambda(\phi(w))a$ and $\phi(wb) = \mu(\phi(w))a$. Moreover, from the definitions of the morphisms λ and μ one has:

$$|\lambda(\phi(w))|_b = |\phi(w)|_b, \quad |\lambda(\phi(w))|_a = |\phi(w)|_a + |\phi(w)|_b,$$

$$|\mu(\phi(w))|_b = |\phi(w)|_a, \quad |\mu(\phi(w))|_a = |\phi(w)|_a + |\phi(w)|_b.$$

The base of the induction, i.e. the case $w = a$ is trivially true. We suppose then $|w| > 1$ and the result true up to $|w| - 1$. One has:

$$\sigma(\phi(wa)) = \begin{pmatrix} |\lambda(\phi(w))|_b + 1 \\ |\lambda(\phi(w))|_a + 2 \end{pmatrix} = \begin{pmatrix} |\phi(w)|_b + 1 \\ |\phi(w)|_a + |\phi(w)|_b + 2 \end{pmatrix} =$$

$$\begin{pmatrix} 1 & 0 \\ 1 & 1 \end{pmatrix} \begin{pmatrix} |\phi(w)|_b + 1 \\ |\phi(w)|_a + 1 \end{pmatrix} = \beta(a)\sigma(\phi(w)).$$

In a similar way one has:

$$\sigma(\phi(wb)) = \begin{pmatrix} 0 & 1 \\ 1 & 1 \end{pmatrix} \begin{pmatrix} |\phi(w)|_b + 1 \\ |\phi(w)|_a + 1 \end{pmatrix} = \beta(b)\sigma(\phi(w)).$$

By using the inductive hypothesis and the fact that $\beta((wx)^{\sim}) = \beta(x)\beta(w^{\sim})$ with $x \in \mathcal{A}$, the result follows.

Let us now introduce some suitable substitution operations on the lists. If $(a_1, \ldots, a_n)$ and $(b_1, \ldots, b_m)$ are two lists of objects then for any $1 \le i \le m$ by $(a_1, \ldots, a_{i-1}, (b_1, \ldots, b_m), a_{i+1}, \ldots, a_n)$ we denote the list which is obtained by inserting the list $(b_1, \ldots, b_m)$ at the place i of the first list, i.e.

$$(a_1, \ldots, a_{i-1}, (b_1, \ldots, b_m), a_{i+1}, \ldots, a_n) = (a_1, \ldots, a_{i-1}, b_1, \ldots, b_m, a_{i+1}, \ldots, a_n).$$

Of course one can do several insertions of different lists in a given list. For instance, $(1, 2, (3, 4), 5, (6, 7, 8), 9)$ will be the list $(1, 2, 3, 4, 5, 6, 7, 8, 9)$. Moreover, for any integer $h > 0$, $(a)^h$ will denote the list:

$$(a)^h = (\underbrace{a, \ldots, a}_{h-times}).$$

If $h = 0$ then $(a)^0$ will denote the empty list $()$.

Let w be a word of $a\mathcal{A}^*$ having the integral representation $(0, \alpha_1, \ldots, \alpha_n)$ (n even integer) we define as Tw the word whose integral representation, possibly incomplete, is for $n = 2$

$$(0, (1)^{\alpha_2}, \alpha_1), \tag{1}$$

and for $n > 2$,

$$(0, (1)^{\alpha_n}, \alpha_{n-1} + 1, (1)^{\alpha_{n-2}-1}, \alpha_{n-3} + 1, \ldots, (1)^{\alpha_2 - 1}, \alpha_1). \tag{2}$$

Let us observe that in the above list the number of terms is

$$d = \sum_{k=1}^{n/2} \alpha_{2k} + 2,$$

so that if $d-1$ is odd, then in order to obtain the complete integral representation of Tw one has to complete the list with a last element 0.

From the definition

$$|Tw| = \alpha_n + (\alpha_{n-1} + 1) + (\alpha_{n-2} - 1) + \ldots + (\alpha_{n-3} + 1) + (\alpha_2 - 1) + \alpha_1$$

$$= \sum_{i=1}^{n} \alpha_i = |w|.$$

If we set $m = |w|$ one has that T is actually a map $T : a\mathcal{A}^{m-1} \to a\mathcal{A}^{m-1}$. The following theorem shows that T is a bijection.

Theorem 14. *For any $w \in a\mathcal{A}^*$,*

$$\phi(w) = \psi(Tw).$$

Proof. First of all we observe that if $p/q \in \mathcal{I}$ has the development in continued fractions $\langle 0, h_1, \ldots, h_n \rangle$ then:

$$f(\begin{pmatrix} 1 & 0 \\ 1 & 1 \end{pmatrix} \begin{pmatrix} p \\ q \end{pmatrix}) = \frac{p}{p+q} = \langle 0, 1 + h_1, h_2, \ldots, h_n \rangle$$

and

$$f(\begin{pmatrix} 0 & 1 \\ 1 & 1 \end{pmatrix} \begin{pmatrix} p \\ q \end{pmatrix}) = \frac{q}{p+q} = \langle 0, 1, h_1, \ldots, h_n \rangle.$$

From this it easily follows that for all $m \geq 0$ one has:

$$f(\beta(a^m) \begin{pmatrix} p \\ q \end{pmatrix}) = \langle 0, h_1 + m, h_2, \ldots, h_n \rangle \tag{3}$$

and

$$f(\beta(b^m) \begin{pmatrix} p \\ q \end{pmatrix}) = \langle 0, (1)^m, h_1, \ldots, h_n \rangle. \tag{4}$$

We show that for any $w \in aA^*$ the rate of $\psi(Tw)$ is equal to the rate of $\phi(w)$, i.e.

$$\eta(\psi(Tw)) = \eta(\phi(w)).$$

Since η is a bijection one will obtain $\psi(Tw) = \phi(w)$. Let $(0, \alpha_1, \dots, \alpha_n)$, n even, be the integral representation of a word of aA^*. The proof is obtained by induction on the even integer n, i.e. we suppose the result true up to n and we prove it for $n+2$. Let $(0, \alpha_1, \dots, \alpha_n, \alpha_{n+1}, \alpha_{n+2})$ be the integral representation of a word $w \in aA^*$. We can factorize w as $w = uv$ with

$$u = a^{\alpha_1} \dots b^{\alpha_n}, \quad v = a^{\alpha_{n+1}} b^{\alpha_{n+2}}.$$

One has $\beta(w^\sim) = \beta(v^\sim)\beta(u^\sim)$ so that:

$$\sigma(\phi(w)) = \beta(w^\sim)\begin{pmatrix}1\\1\end{pmatrix} = \beta(v^\sim)\beta(u^\sim)\begin{pmatrix}1\\1\end{pmatrix} = \beta(v^\sim)\sigma(\phi(u)).$$

By induction

$$f(\sigma(\phi(u))) = \eta(\phi(u)) = \eta(\psi(Tu)),$$

where the word Tu has the integral representation given by Eq.(1) if $n = 2$ or by Eq.(2) if $n > 2$. Hence, $\eta(\psi(Tu)) = p/q$ has a development in continued fractions given for $n = 2$ by

$$\langle 0, (1)^{\alpha_2}, \alpha_1 \rangle$$

and for $n > 2$ by

$$\langle 0, (1)^{\alpha_n}, \alpha_{n-1}+1, (1)^{\alpha_{n-2}-1}, \alpha_{n-3}+1, \dots, (1)^{\alpha_2-1}, \alpha_1 \rangle.$$

We shall consider from now on only the case $n > 2$; the case $n = 2$ is similarly dealt with. From Eq.s(3) and (4) one has:

$$f(\beta(a^{\alpha_{n+1}})\begin{pmatrix}p\\q\end{pmatrix}) = \langle 0, \alpha_{n+1}+1, (1)^{\alpha_n-1}, \alpha_{n-1}+1, \dots, (1)^{\alpha_2-1}, \alpha_1 \rangle = \frac{r}{s}$$

and

$$f(\beta(b^{\alpha_{n+2}})\begin{pmatrix}r\\s\end{pmatrix}) = \langle 0, (1)^{\alpha_{n+2}}, \alpha_{n+1}+1, (1)^{\alpha_n-1}, \alpha_{n-1}+1, \dots, (1)^{\alpha_2-1}, \alpha_1 \rangle.$$

Thus $f(\sigma(\phi(w))) = \eta(\phi(w)) = \eta(\psi(Tw))$ and this proves the induction step. As regards the base of the induction let w have the integral representation $(0, \alpha_1, \alpha_2)$. Let us first suppose that $\alpha_2 = 0$ so that $w = a^{\alpha_1}$. In this case Tw has the integral representation $(0, (1)^0, \alpha_1) = (0, \alpha_1, 0)$ and $\phi(w) = a^{\alpha_1} = \psi(Tw)$. The remaining part of the proof is obtained by a simple induction on the value of the integer α_2.

Example. Let $w = ab^3ab$ whose integral representation is $(0, 1, 2, 1, 1)$. One has that Tw has the integral representation $(0, (1), 2, (1)^2, 1) = (0, 1, 2, 1, 1, 1)$, i.e. $Tw = ab^2aba$. One easily verifies that

$$\phi(w) = \psi(Tw) = ababaababaababaabababaababa.$$

7 Concluding remarks

As we have shown in the previous sections, in the theory of Sturmian words a central role is played by the set PER of all finite words in the binary alphabet which are either powers of a single letter or are words of maximal length having two periods which are coprimes. The elements w of PER have surprising structural properties. They are palindromes in a two letter alphabet. Moreover, a binary word w in which occur more than one letter, belongs to PER if and only if it can be uniquely factorized as $w = PxyQ = QyxP$, with x, y distinct letters and P, Q elements of PER.

The elements w of PER can be faithfully represented by the *ratio* of the periods or by the *rate*; this latter quantity is the *slope* of the word awb, so that it depends on the occurrences of the letters a and b in w. These two descriptions are suitably related and give rise to a *basic duality* in the theory emphasized by the existence of two different procedures for the construction of the elements of PER. The first is based on the palindrome factorization and so on the *periods* of the word; the construction uses the operator of left-palindrome closure. The second is based on the *occurrences* of the letters since it is performed by some suitable morphisms. These procedures are conceptually very different as the latter uses *local* information on the word on which it operates, whereas the former needs a *global* information. Moreover, a basic correspondence (*standard correspondence*) allowing us to pass from the first to the second procedure was determined.

There is an analogy between this duality and the duality existing between the *time (or space) domain* and the *frequency domain* in the Fourier analysis which are related by the *Fourier transform*. In this setting the basic components are harmonic, i.e. they describe essentially uniform circular motions (or harmonic oscillators) each of which is determined by an *amplitude* and a *frequency*.

As we have discussed in the introduction, Sturmian words are not ultimately periodic but are the 'nearest' possible to these words. Any standard Sturmian word is the limit of a sequence of elements of PER each of which has *two periods*. We believe that these words, in analogy to Fourier's analysis, can be usefully utilized in the future as basic components for the construction of more realistic and effective modelling of some classes of complex systems.

References

1. J. Berstel and A. de Luca, Sturmian words, Lyndon words and trees, *Preprint L.I.T.P. 95/24, University of Paris 7, June 95, Theoretical Computer Science*, to appear
2. J. Berstel and P. Séébold, Morphismes de Sturm, *Bull. Belg. Math. Soc.* **1** (1994) 175-189.
3. J.-P. Borel and F. Laubie, Quelques mots sur la droite projective reelle, *Journal de Theorie des Nombres de Bordeaux* **5**(1993), 23-51.
4. T.C. Brown, Descriptions of the characteristic sequence of an irrational, *Canad. Math. Bull.* **36** (1993) 15-21

5. D. Crisp, W. Moran, A. Pollington and P. Shiue, Substitution invariant cutting sequences, *J. théorie des nombres de Bordeaux* **5**(1993) 123-138.
6. A. de Luca, Sturmian words: Structure, Combinatorics, and their Arithmetics, *Theoretical Computer Science, special issue on Formal Languages*, to appear.
7. A. de Luca, On standard Sturmian morphisms, *Preprint 95/18 Dipartimento di Matematica Università di Roma "La Sapienza", Theoretical Computer Science*, to appear; *see also* Lecture Notes in Computer Science, *Proc.s ICALP '96*, vol. 1099, pp.403-415 (Springer Verlag, New York, 1993).
8. A. de Luca, A conjecture on continued fractions, *Preprint 8/96 Dipartimento di Matematica Università di Roma "La Sapienza", Theoretical Computer Science*, to appear.
9. A. de Luca and F. Mignosi, Some Combinatorial properties of Sturmian words, *Theoretical Computer Science* **136**(1994) 361-385.
10. S. Dulucq and D. Gouyou-Beauchamps, Sur les facteurs des suites de Sturm, *Theoretical Computer Science* **71** (1990) 381-400.
11. R.L. Graham, D.E. Knuth and O. Patashnik, *Concrete Mathematics, 2 edition* (Addison-Wesley, Reading, MA, 1994).
12. G.A. Hedlund, Sturmian minimal sets, *Amer. J. Math.* **66**(1944) 605-620.
13. M. Lothaire, *Combinatorics on words*, (Addison-Wesley, Reading, MA, 1983).
14. F. Mignosi, Infinite words with linear subword complexity, *Theoretical Computer Science* **65**(1989) 221-242.
15. F. Mignosi and P. Séébold. Morphismes sturmiens et règles de Rauzy, *J. théorie des nombres de Bordeaux* **5**(1993) 221-233.
16. M. Morse and G.A. Hedlund, Symbolic dynamics II: Sturmian trajectories, *Amer. J. Math.* **62**(1940), 1-42.
17. A. Pedersen, Solution of Problem E 3156, *The American Mathematical Montly* **95** (1988) 954-955.
18. D. Perrin, On the solution of Ehrenfeucht's conjecture, *EATCS Bull.* **27** (1985) 68-70.
19. G. N. Raney, On Continued Fractions and Finite Automata, *Math. Ann.* **206**(1973) 265-283.
20. G. Rauzy, Mots infinis en arithmétique, in M. Nivat and D. Perrin, eds., *Automata in Infinite words*, Lecture Notes in Computer Science, vol.192 (Springer, Berlin, 1984) pp.164-171.

Compactness of Systems of Equations on Completely Regular Semigroups*

Tero Harju[1], Juhani Karhumäki[1] and Mario Petrich[2]**

[1] Department of Mathematics, University of Turku, 20014 Turku, Finland
[2] Departamento de Matemática Pura, Faculdade de Ciências, Universidade do Porto, 4050 Porto, Portugal

Abstract. A semigroup S is said to have the compactness property, or CP for short, if each system of equations over a finite set of variables has an equivalent finite subsystem, that is, having exactly the same solutions in S. We prove that a completely 0-simple semigroup S satisfies CP if and only if the group G in a Rees matrix representation $S = \mathcal{M}^0(I, G, \Lambda; P)$ satisfies this property. Further, a variety of completely regular semigroups in which the finitely generated members have finite matrices is shown to satisfy CP if and only if the corresponding variety of groups satisfies CP. It is then shown that the varieties of bands of groups satisfy the condition on finite matrices.

1 Introduction and Summary

A topological space is compact if every open cover of it has a finite subcover. A most natural analogue of this statement in the context of semigroup equations is the following. A semigroup S has the *compactness property* if any system of semigroup equations $\mathcal{E}$ over a finite set of variables is equivalent to a finite subsystem $\mathcal{F}$ of $\mathcal{E}$, that is, $\mathcal{E}$ and $\mathcal{F}$ have the same set of solutions in S. There is an alternative, nonequivalent, concept which also claims to be a rightful analogue of compactness in topology for the semigroup setting, namely: a semigroup S is *equationally compact*, see [6], if every system of equations $\mathcal{E}$ has a solution in S whenever all of its finite subsystems have a solution in S. This latter notion is actually an analogue of the contrapositive of the definition of compactness in topology: a collection of open sets none of whose finite subcollections covers a set, does not cover the set either.

In order to avoid confusion, we retain the above definition of the compactness property even though in view of the above arguments, it represents a closer analogue of compactness in topology than the concept of equational compactness.

In this article we study the compactness property for completely regular and completely (0-)simple semigroups. This is motivated not only by the rich and well-developed theory of these semigroups, but partly also by an immediate corollary to results of [8] and Andersen's theorem, see [4], stating that if a

* Partly supported by Academy of Finland, grant 14047.
** This study was made while the author was visiting University of Turku.

simple semigroup satisfies CP, then it is completely simple (and thus completely regular).

We conclude this introduction with a short synopsis of our results. In Sect. 2 we give a formal definition of the compactness property, and present some earlier results needed in later sections. In Sect. 3 we state the basic definitions in completely regular semigroups.

Section 4 is devoted to the general study of the compactness property in the varieties of completely regular semigroups. We generalize a result from [8] for these varieties: a variety $\mathcal{V}$ of completely regular semigroups satisfies CP if and only if the finitely generated members in $\mathcal{V}$ satisfy the ascending chain condition for congruences.

In Sect. 5 it is shown that if a semigroup S satisfies CP, so do its Rees matrix semigroups $\mathcal{M}^0(I, S, \Lambda; P)$. The converse of this result holds in the case when S is a group, and in particular it follows that a completely simple semigroup satisfies CP just in case its subgroups do. This line of thought is continued in Sect. 6, but now for the varieties of completely regular semigroups. It is shown that if $\mathcal{V}$ is a variety of completely regular semigroups, where the finitely generated members have finite matrices, then $\mathcal{V}$ satisfies CP if and only if the variety $\mathcal{V} \cap \mathcal{G}$ satisfies CP. As an example the variety $\mathcal{BG}$ of bands of groups satisfies the condition on finite matrices.

2 Systems of Equations

A *semigroup equation* $u = v$ is a pair of words over a finite alphabet $X = \{x_1, \ldots, x_n\}$, say, the elements of which are called *variables*. A *solution* of an equation $u = v$ *in a semigroup* S is an n-tuple $a_1, a_2, \ldots, a_n$ of elements of S such that $u(a_1, \ldots, a_n) = v(a_1, \ldots, a_n)$. Equivalently, a solution of $u = v$ is a morphism $\alpha: X^+ \to S$ from the free semigroup X^+ generated by X into S such that $u\alpha = v\alpha$. Hence an equation generalizes the notion of an *identity*; indeed, an identity $u = v$ is satisfied by S if for all morphisms $\alpha: X^+ \to S$, we have $u\alpha = v\alpha$, that is, the solution set of the equation $u = v$ is $S^n = S \times \ldots \times S$.

Let S be a semigroup, X a finite set of variables and $\mathcal{E}$ a system of equations $\{u_i = v_i\}_{i \in I}$ over X. We say that $\mathcal{E}$ is *equivalent (in S) to its subsystem* $\mathcal{F} \subseteq \mathcal{E}$, if $\mathcal{E}$ and $\mathcal{F}$ have the same set of solutions, *i.e.*, if $\alpha: X^+ \to S$ is a solution of all $u = v \in \mathcal{F}$, then α is a solution of all $u = v \in \mathcal{E}$.

A semigroup S is said to satisfy the *compactness property*, or *CP* for short, if for all finite sets of variables X every system $\mathcal{E}$ over X is equivalent in S to one of its finite subsystems $\mathcal{F} \subseteq \mathcal{E}$.

The set X of variables is always supposed to be finite. Indeed, for infinite sets of variables the compactness property, as defined above, reduces to a triviality, since any finite (sub)system of equations contains only finitely many variables. From the finiteness of X it also follows that each system of equations over X contains denumerably many equations, and therefore each (infinite) system $\mathcal{E}$ can be enumerated, $\{u_i = v_i\}_{i \geq 1}$.

A class $\mathcal{C}$ of semigroups is said to satisfy the *compactness property* if every semigroup $S \in \mathcal{C}$ satisfies it. Further, the class $\mathcal{C}$ satisfies the compactness property *uniformly* if for each system $\mathcal{E}$ of equations there exists a finite subsystem $\mathcal{F}$ such that $\mathcal{F}$ is equivalent to $\mathcal{E}$ in all $S \in \mathcal{C}$, that is, if the semigroups $S \in \mathcal{C}$ have a common equivalent finite subsystem for each $\mathcal{E}$.

As an example, all finite semigroups satisfy CP, but as shown in [8], the class of finite semigroups does not satisfy CP uniformly. The variety of all semigroups does not satisfy CP, but all free semigroups do. Indeed, this is the most striking example, known as the *Ehrenfeucht's compactness theorem*, which was proved independently by Albert and Lawrence [2] and Guba [7]. In [7], see also [5], it was also shown that free groups satisfy CP.

As examples of individual semigroups that do not satisfy CP, we mention the bicyclic semigroup and the free inverse semigroups. Both of these examples follow from a result in [8] stating that if a finitely generated semigroup S satisfies CP, then every chain of idempotents of S is finite.

As it is easy to see, the subsemigroups and finite direct products of semigroups satisfying CP satisfy CP. On the other hand, CP is not closed under taking quotients since the free semigroups do satisfy CP, but *e.g.* the bicyclic semigroup does not. Indeed, as the following example shows, CP is not preserved even for taking Rees quotients.

Example 1. We give an example in which a finitely generated semigroup S and its ideal I both satisfy CP but the Rees quotient S/I does not. For this let A^+ be the free semigroup generated by $A = \{a, b\}$ and define

$$L = \{ab^k ab^j a \mid 1 \le j \le k\} \quad \text{and} \quad I = A^* L A^* \ .$$

Clearly, I is an ideal of A^+. The free semigroup A^+ satisfies CP and I satisfies CP since it is a subsemigroup of A^+. Let $X = \{x, y, z\}$ be a set of variables and

$$\mathcal{E} = \{xy^k xz^j x = xy^k xz^j xx \mid k, j \ge 1\} \ .$$

Let $\mathcal{F} \subseteq \mathcal{E}$ be any finite subsystem. Set

$$m = \max\{j \mid xy^k xz^j x = xy^k xz^j xx \in \mathcal{F}\} \ ,$$

and define $\alpha \colon X^+ \to A^+/I$ by $x\alpha = a$, $y\alpha = b^m$, $z\alpha = b$. Now, by the definition of L, we have $(xy^k xz^j x)\alpha = (xy^k xz^j xx)\alpha$ in S/I for each $xy^k xz^j x = xy^k xz^j xx$ from $\mathcal{F}$, but $(xyxz^{m+1}x)\alpha \ne (xyxz^{m+1}xx)\alpha$, and therefore $\mathcal{F}$ is not equivalent to $\mathcal{E}$. This proves the claim. □

Other examples of semigroups in which CP is known to fail are the nonhopfian semigroups and groups such as the Baumslag-Solitar group, see [8].

We conclude this section with a few technical comments on the compactness property. For this let $u \in X^+$ be a word. The *content* $c(u)$ of u is the set of variables that occur in u, that is, $x \in c(u)$ if and only if $u = u_1 x u_2$ for some words $u_1, u_2 \in X^*$. We denote by $h(u)$ and $t(u)$ the *head* and the *tail* of u, that is, $h(u), t(u) \in X$ and $u \in h(u)X^* \cap X^*t(u)$. Also denote by

$$f(u) = \{xy \mid x, y \in X,\ xy \text{ occurs in } u\}$$

the set of all *2-factors* of u. Define a relation $\sim$ on X^+ by

$$u \sim u' \iff h(u) = h(u'),\ f(u) = f(u'),\ t(u) = t(u') \ .$$

Lemma 1. *Let X be a finite set of variables. The relation $\sim$ is a congruence on X^+ and the quotient $X^+/\sim$ is a finite semigroup.*

Proof. The congruence property of $\sim$ is obvious. For the finiteness of $X^+/\sim$ it is enough to observe that there are only finitely many triples (x, F, y), where $x, y \in X$ and $F \subseteq X^2$, and each such triple (x, F, y) defines uniquely an equivalence class of the relation $\sim$. □

We say that a family $U = \{u_i\}_{i \geq 1}$ of words over X is *locally uniform* if $u_i \sim u_j$ for all i and j. Clearly, if U is locally uniform, then also $c(u_i) = c(u_j)$ for all i, j. A system $\{u_i = v_i\}_{i \geq 1}$ of equations over X is *locally uniform* if both families $\{u_i\}_{i \geq 1}$ and $\{v_i\}_{i \geq 1}$ are locally uniform.

Lemma 2. *A semigroup S satisfies CP if and only if each locally uniform system of equations has a finite equivalent subsystem.*

Proof. The direct part of the lemma is trivial. For the converse statement consider any system $\mathcal{E} = \{u_i = v_i\}_{i \geq 1}$ of equations over a finite set X. By Lemma 1, the sets $U = \{u_i\}_{i \geq 1}$ and $V = \{v_i\}_{i \geq 1}$ are partitioned into finitely many congruence classes $U = U_1 \cup \ldots \cup U_r$ and $V = V_1 \cup \ldots \cup V_s$ by the relation $\sim$, and therefore $\mathcal{E}$ has a finite partition into $\mathcal{E} = \cup_{i,j} \mathcal{E}_{ij}$, where $\mathcal{E}_{i,j} = \mathcal{E} \cap U_i \times V_j$. By the construction each $\mathcal{E}_{ij}$ is locally uniform, and by assumption $\mathcal{E}_{ij}$ has a finite equivalent subsystem, say $\mathcal{F}_{ij}$. It is immediate that $\mathcal{F} = \cup_{i,j} \mathcal{F}_{ij}$ is equivalent to $\mathcal{E}$ in S, and since this union is finite, the claim follows. □

For a semigroup S let S^0 be the semigroup where a new zero element is adjoined.

Lemma 3. *A semigroup S satisfies CP if and only if S^0 satisfies CP.*

Proof. Suppose that S satisfies CP. Let $\mathcal{E} = \{u_i = v_i\}_{i \geq 1}$ be a locally uniform system of equations over X as in Lemma 2, and let $\mathcal{F} \subseteq \mathcal{E}$ be a finite equivalent subsystem in S. Let $\alpha: X^+ \to S^0$ be a morphism. Since $c(u_i) = c(u_j)$ for all i and j, $u_i\alpha = 0$ for some i, then $u_i\alpha = 0$ for all i. The case for the right hand sides v_i is similar. Therefore if $u_1\alpha = v_1\alpha$, then either $u_i\alpha, v_i\alpha \in S$ for all i or $u_i\alpha = 0 = v_i\alpha$ for all i. Thus $\mathcal{F} \cup \{u_1 = v_1\}$ is equivalent to $\mathcal{E}$ in S^0.

The converse part is immediate since S is a subsemigroup of S^0. □

The corresponding result for S^1, where an identity element 1 is adjoined to S (if it is not a monoid), was shown in [8]: A semigroup S satisfies CP if and only if the monoid S^1 satisfies CP.

3 Some Terminology on Semigroups

For the results in theory of semigroups stated in this section, including completely (0-)simple and completely regular semigroups, we refer to [4] and [10].

A semigroup S is *completely regular*, *c.r.* for short, if S is a union of its subgroups. In particular, each element $a \in S$ belongs to a maximal subgroup G of S, where there exists a unique element $a^{-1} \in G$ such that $aa^{-1} = a^{-1}a = e$ with e being the identity element of G. The element e is clearly an idempotent of S, and the set $E(S)$ of idempotents of S consists of the identity elements of its maximal subgroups.

A semigroup S is *completely simple* (respectively, *completely 0-simple*) if it is simple (respectively, 0-simple) and it contains a *primitive idempotent*, that is, an idempotent which is minimal (respectively, minimal among the nonzero idempotents) in the partially ordered set $E(S)$ of idempotents.

It can be easily shown that any $\mathcal{H}$-class H of a completely simple semigroup is a group. Further, each completely simple semigroup is a regular semigroup. Indeed, we have a stronger result which states that a semigroup S is completely simple if and only if it is simple and completely regular.

Let T be a semigroup to which we adjoin a zero element to obtain $T^0 = T \cup \{0\}$. Let I and Λ be two nonempty index sets and $P = (p_{\lambda i})\colon \Lambda \times I \to T^0$ a $\Lambda \times I$-matrix with entries in T^0 such that each row and column of P has at least one nonzero entry. Denote by $\mathcal{M}^0(I, T, \Lambda; P)$ the *Rees matrix semigroup (with zero) over* T defined on $(I \times T \times \Lambda) \cup \{0\}$, where 0 is the zero element and the operation is given by

$$(i, a, \lambda)(j, b, \mu) = \begin{cases} (i, ap_{\lambda j}b, \mu) & \text{if } p_{\lambda j} \neq 0 \ , \\ 0 & \text{if } p_{\lambda j} = 0 \ . \end{cases}$$

Similarly, we define the *Rees matrix semigroup (without zero) over* T as the semigroup $\mathcal{M}(I, T, \Lambda; P)$ on $I \times T \times \Lambda$, where $P\colon \Lambda \times I \to T$.

It is well known that a semigroup is completely (0−)simple if and only if it is isomorphic to a Rees matrix semigroup (with zero) over a group.

It can be shown, see the proof of Theorem IV.2.7 in [10], that each Rees matrix semigroup without zero is isomorphic to a semigroup $S = \mathcal{M}(I, G, \Lambda; P)$, where there exists an element $j \in I \cap \Lambda$ such that $p_{\lambda j} = 1 = p_{ji}$ for all $\lambda \in \Lambda$ and $i \in I$. Here 1 denotes the identity element of the group G. Such a matrix P is said to be *normalized* at j. We shall assume throughout this paper that the matrices are normalized at some j in each Rees matrix semigroup.

A semigroup B is a *band* if all its elements are idempotent, that is, $B = E(B)$. If μ is a congruence on a semigroup S such that $B = S/\mu$ is a band, then the congruence classes $S_{a\mu} = a\mu$ are subsemigroups of S. In this case S is called a *band of semigroups* S_α for $\alpha = a\mu$, $a \in S$. In particular, if each S_α is a group, then S is a *band of groups*. By definition each band of groups is a c.r. semigroup, and for each band of groups S, $\mathcal{H}$ is a congruence on S such that $S/\mathcal{H}$ is a band. Here, as in all c.r. semigroups, the $\mathcal{H}$-classes H_a are the maximal subgroups of S.

A semigroup Y is a *semilattice* if it is a commutative band. If ρ is a congruence on a semigroup S such that the quotient $Y = S/\rho$ is a semilattice, then we write $S = (Y; S_\alpha)$ and say that S is a *semilattice* Y *of semigroups* S_α for $\alpha \in Y$, where S_α is the ρ-class corresponding to α. Note that each semigroup S has a least congruence η for which the quotient S/η is a semilattice.

By an early result of Clifford c.r. semigroups can be characterized as those semigroups that are semilattices $(Y; S_\alpha)$ of completely simple semigroups.

4 Compactness in Varieties of Completely Regular Semigroups

The family of c.r. semigroups is not a variety of semigroups because a subsemigroup of a c.r. semigroup need not be completely regular. A subsemigroup T of a c.r. semigroup S is a completely regular subsemigroup if for all $a \in T$ also $a^{-1} \in T$. However, as it is easily seen, epimorphic images and direct products of c.r. semigroups are completely regular.

A c.r. semigroup S can be regarded as a *unary semigroup* $(S, \cdot, {}^{-1})$ with an additional unary operation $a \mapsto a^{-1}$; c.r. semigroups form a variety of these algebras. We denote by $\mathcal{CR}$ the variety of c.r. semigroups (as unary semigroups). The variety of groups $\mathcal{G}$ is a subvariety of $\mathcal{CR}$, and as it is well known, if $\mathcal{V}$ is a variety of groups, then the c.r. semigroups with subgroups in $\mathcal{V}$ form a variety of c.r. semigroups.

Recall that a semigroup S satisfies the *ascending chain condition* (a.c.c.) for congruences if each properly ascending chain $\rho_1 \subset \rho_2 \subset \dots$ of congruences ρ_i on S is finite.

It was shown in [8] that if a variety $\mathcal{V}$ of semigroups satisfies CP, then $\mathcal{V}$ satisfies CP uniformly, and, moreover, that a variety $\mathcal{V}$ of *monoids* satisfies CP if and only if each finitely generated member of $\mathcal{V}$ satisfies the a.c.c. for congruences.

We argue that the equivalence of CP and the a.c.c. condition is valid also for the varieties of c.r. semigroups (which need not be varieties of semigroups).

Theorem 4. *The following conditions are equivalent for a variety $\mathcal{V}$ of c.r. semigroups.*

(1) *$\mathcal{V}$ satisfies CP.*

(2) *The finitely generated semigroups in $\mathcal{V}$ satisfy the a.c.c. for congruences.*

(3) *The finitely generated free c.r. semigroups of $\mathcal{V}$ satisfy the a.c.c. for congruences.*

Proof. For the equivalence of (1) and (2) we rely on the proof of the corresponding result for monoids in [8] except for one modification. The identity elements of the monoids are needed in the original proof only to ensure the existence of a natural embedding $\beta_j : S_j \to \Pi_{i\geq 1} S_i$ such that, with the projections $\pi_k : \Pi_{i \geq 1} S_i \to S_k$, the mapping $\beta_j \pi_j : S_j \to S_j$ is the identity morphism and $\beta_j \pi_k : S_j \to S_k$ is the constant morphism (images equal to the identity element of S_k) for all $k \neq j$. The latter condition can be substituted in c.r. semigroups by setting $\pi_k \beta_j(s) = e_k$ for each $k \neq j$, where $e_k \in E(S_k)$ is a fixed idempotent.

Further, the proof in [8] does not make use of the property that varieties are closed under taking subsemigroups. Since morphic images and direct products of c.r. semigroups are completely regular, it follows that the proof in [8] applies as such to c.r. semigroups with the above modification.

The nontrivial part of the equivalence of (2) and (3) follows from the observation that if a semigroup S satisfies the a.c.c. for congruences, then so do all its morphic images. □

Note that if a c.r. semigroup S satisfies CP, then so do all its subsemigroups and not only those that are completely regular.

Groups form a subvariety of c.r. semigroups, and thus as a corollary to Theorem 4 we have the following result of Albert and Lawrence [1].

Corollary 5. *A variety $\mathcal{V}$ of groups satisfies CP if and only if the finitely generated groups in $\mathcal{V}$ satisfy the a.c.c. for normal subgroups.*

Since *e.g.* the Baumslag-Solitar group, which has a group presentation $G = \langle a, b \mid b^2a = ab^3 \rangle$, does not satisfy CP (see [8]), we have by Theorem 4 that no variety of c.r. semigroups containing all groups satisfies CP.

5 Compactness in Rees Matrix Semigroups

As the next result states, the completely simple semigroups are natural objects to study in the context of the compactness property.

Lemma 6. *A (0-)simple semigroup S satisfying CP is completely (0-)simple.*

Proof. If S is a (0-)simple semigroup but not completely (0-)simple, then S contains a bicyclic semigroup as its subsemigroup, see [4, Theorem 2.54]. As mentioned above the bicyclic semigroup does not satisfy CP and the claim follows. □

Lemma 7. *Let X be a finite set of variables and $S = \mathcal{M}^0(I, T, \Lambda; P)$ be a Rees matrix semigroup over T. For a locally uniform family $\{u_i\}_{i\geq 1}$ of words over X and a morphism $\alpha: X^+ \to S$, $u_1\alpha = 0$ if and only if $u_i\alpha = 0$ for all $i \geq 0$.*

Proof. Assume $u_1\alpha = 0$ in the notation of the lemma. It is immediate that there exists a 2-factor $xy \in f(u_1)$ such that $x\alpha \cdot y\alpha = 0$. Since $\{u_i\}_{i\geq 1}$ is locally uniform, $xy \in f(u_i)$ and thus $u_i\alpha = 0$ for all i. □

Theorem 8. *Let $S = \mathcal{M}^0(I, T, \Lambda; P)$ be a Rees matrix semigroup over T. If T satisfies CP, so does S. The converse holds if T is a monoid and P has an invertible element.*

Proof. Assume that the semigroup T satisfies CP. Let X be a finite set of variables and $\mathcal{E} = \{u_i = v_i\}_{i\geq 1}$ a locally uniform system of equations over X as allowed by Lemma 2. Further, let $\alpha: X^+ \to S$ be any morphism.

If $u_i\alpha = 0$ for some i, then $u_i\alpha = 0$ for all i (since $f(u_i) = f(u_j)$ for all i, j), and hence α is a solution of $\mathcal{E}$ in S if and only if $u_i\alpha = 0 = v_i\alpha$ for all i. By Lemma 7 it follows that $u_i\alpha = v_i\alpha$ for all i if $v_1\alpha = 0$, and thus also if $u_1\alpha = v_1\alpha$.

The case when $v_i\alpha = 0$ for some i is clearly symmetric to the one above.

Suppose then that $u_i\alpha, v_i\alpha \neq 0$ for all i. Write

$$x\alpha = (i_x, a_x, \lambda_x) \in I \times T \times \Lambda$$

for each $x \in X$ and let

$$Y = X \cup \{z_{xy} \mid x, y \in X\}$$

be an extended set of variables. Define a morphism $\alpha' : Y^+ \to T^0$ by

$$\begin{cases} x\alpha' = a_x \in T, \\ z_{xy}\alpha' = p_{\lambda_x i_y} \in T^0 \ . \end{cases}$$

For each word $u = x_1 x_2 \ldots x_n$ with $x_i \in X$ set

$$u' = x_1 z_{x_1 x_2} x_2 z_{x_2 x_3} \ldots z_{x_{n-1} x_n} x_n \in Y^+ \ .$$

If $u \in X^+$ with $u\alpha \neq 0$, then

$$u\alpha = (i_{x_1}, u'\alpha', \lambda_{x_n}) \tag{1}$$

since $p_{\lambda_k i_{k+1}} \neq 0$ for all k.

Consider the system $\mathcal{E}' = \{u_i' = v_i'\}_{i \geq 1}$ of equations over Y. By assumption $\mathcal{E}'$ has a finite equivalent subsystem $\mathcal{F}' = \{u_i' = v_i'\}_{i \in F}$ in T. Now for a solution $\alpha : X^+ \to S$ of $\mathcal{F} = \{u_i = v_i\}_{i \in F}$ in S, we have, by (1), that α' is a solution of $\mathcal{F}'$ and hence of $\mathcal{E}'$. Again by (1) and the fact that $\mathcal{E}$ is locally uniform (and so $h(u_i) = h(u_j)$, $t(u_i) = t(u_j)$ and $h(v_i) = h(v_j)$, $t(v_i) = t(v_j)$), α is a solution of $\mathcal{E}$. We conclude that $\mathcal{F} \cup \{u_1 = v_1\}$ is a finite equivalent subsystem of $\mathcal{E}$ in S.

For the latter claim we observe that if T has an identity element and the matrix P has an invertible element $p_{\lambda i}$, then the mapping $t \mapsto (i, tp_{\lambda i}^{-1}, \lambda)$ is an embedding of T into S. Indeed, it is plain that this mapping is injective, and that for any two elements $t_1, t_2 \in T$ we have $t_1 t_2 \mapsto (i, t_1 p_{\lambda i}^{-1} p_{\lambda i} t_2 p_{\lambda i}^{-1}, \lambda) = (i, t_1 t_2 p_{\lambda i}^{-1}, \lambda)$. □

The (nonzero) maximal subgroups of a completely (0−)simple semigroup are isomorphic with one another and therefore we have the following

Corollary 9. *Let S be completely (0−)simple. Then S satisfies CP if and only if any of its maximal subgroups does.*

Corollary 10. *A variety $\mathcal{V}$ of completely simple semigroups satisfies CP if and only if the variety $\mathcal{V} \cap \mathcal{G}$ does.*

Using a construction of free completely simple semigroups from [3] we obtain

Corollary 11. *Every free completely simple semigroup satisfies CP.*

6 Compactness in Completely Regular Semigroups

For a variety $\mathcal{V}$ of c.r. semigroups the intersection $\mathcal{V} \cap \mathcal{G}$ consists of subgroups of semigroups in $\mathcal{V}$, and therefore if $\mathcal{V}$ satisfies CP, then so does $\mathcal{V} \cap \mathcal{G}$. Our aim in this section is to consider the converse of this statement. Using Theorem 4 in Sect. 4 we shall find a nontrivial special case when this holds.

We shall first relate the a.c.c. for congruences on a completely simple semigroup to the a.c.c. for its maximal subgroups.

Let $S = \mathcal{M}(I, G, \Lambda; P)$ be a Rees matrix semigroup and let ρ be a congruence on S. We associate with ρ a triple $(r_\rho, N_\rho, \pi_\rho)$ as follows. Define the relations r_ρ on I and π_ρ on Λ by

$$i\, r_\rho\, j \iff \forall \lambda \in \Lambda : \ (i, p_{\lambda i}^{-1}, \lambda)\, \rho\, (j, p_{\lambda j}^{-1}, \lambda) \ ,$$
$$\lambda\, \pi_\rho\, \mu \iff \forall i \in I : \ (i, p_{\lambda i}^{-1}, \lambda)\, \rho\, (i, p_{\mu i}^{-1}, \mu) \ .$$

Further, let
$$N_\rho = \{g \mid (1, g, 1)\rho(1, 1_G, 1)\} \ ,$$
where P is normalized at 1 and 1_G is the identity of G. One can now show that r_ρ is an equivalence relation on I, π_ρ is an equivalence relation on Λ, and N_ρ is a normal subgroup of G.

For the next two lemmas we refer to [9, Sect. IV.4].

Lemma 12. *Let ρ and σ be congruences on $S = \mathcal{M}(I, G, \Lambda; P)$. Then $\rho \subseteq \sigma$ if and only if $r_\rho \subseteq r_\sigma$, $N_\rho \subseteq N_\sigma$ and $\pi_\rho \subseteq \pi_\sigma$. In particular, $\rho = \sigma$ if and only if $(r_\rho, N_\rho, \pi_\rho) = (r_\sigma, N_\sigma, \pi_\sigma)$.*

A triple (r, N, π) is said to be *admissible* if r is an equivalence relation on I, N a normal subgroup of G, and π is an equivalence relation on Λ such that

$$\forall i, j \in I, \ \lambda, \mu \in \Lambda \text{ and either } i\, r\, j \text{ or } \lambda\, \pi\, \mu, \text{ then } p_{\lambda i} p_{\mu i}^{-1} p_{\mu j} p_{\lambda j}^{-1} \in N \ .$$

Lemma 13. *To each admissible triple (r, N, π) there corresponds a unique congruence ρ on S such that $(r_\rho, N_\rho, \pi_\rho) = (r, N, \pi)$.*

For the relation between the congruences on a completely simple semigroup and the normal subgroups of its maximal subgroups, we have

Lemma 14. *A Rees matrix semigroup $S = \mathcal{M}(I, G, \Lambda; P)$ satisfies the a.c.c. for congruences if and only if I and Λ are finite and G satisfies the a.c.c. for normal subgroups.*

Proof. If I is infinite, let $r_1 \subset r_2 \subset \ldots$ be a strictly ascending chain of equivalence relations on I, and let $\omega = \Lambda \times \Lambda$. First, for each $i \geq 1$, (r_i, G, ω) is an admissible triple for S, and hence the corresponding congruences ρ_i form a strictly ascending chain by Lemmas 12 and 13. The case for Λ is dual.

Next assume that $N_1 \subset N_2 \subset \ldots$ is a strictly ascending chain of normal subgroups of G. Let ε denote the identity relation on I and Λ. Clearly, $(\varepsilon, N_i, \varepsilon)$

is an admissible triple for S, and again the corresponding congruences ρ_i on S form a strictly ascending chain.

In the other direction, assume $\rho_1 \subset \rho_2 \subset \dots$ is an ascending chain of congruences on S, and let the corresponding admissible triples be (r_i, N_i, π_i). By Lemma 12, we have

$$r_1 \subseteq r_2 \subseteq \dots, \quad N_1 \subseteq N_2 \subseteq \dots, \quad \pi_1 \subseteq \pi_2 \subseteq \dots,$$

and since I and Λ are finite and G satisfies the a.c.c. for normal subgroups, these chains are finite, which implies that for $i \geq n_0$, $\rho_i = \rho_{i+k}$ for all $k \geq 0$. The assertion follows. □

For a finitely generated c.r. semigroup $S = (Y; S_\alpha)$ the semilattice Y is finite. However, the completely simple subsemigroups S_α need not be finitely generated. An example of such a situation is the 2-generator free c.r. semigroup as constructed by Clifford [3].

We say that a c.r. semigroup $S = (Y; S_\alpha)$, where $S_\alpha = \mathcal{M}(I_\alpha, G_\alpha, \Lambda_\alpha; P_\alpha)$, has *finite matrices* if for each α the index sets I_α and Λ_α are finite. Note that S has finite matrices if and only if every left and every right zero subsemigroup of S is finite.

Lemma 15. *If $S = (Y; S_\alpha)$ is a finitely generated c.r. semigroup with finite matrices, then the Rees matrix semigroups S_α are finitely generated.*

Proof. Let $S_\alpha = \mathcal{M}(I_\alpha, G_\alpha, \Lambda_\alpha; P_\alpha)$ for $\alpha \in Y$, where the matrices P_α are normalized at an element which is denoted simply by α. We use (a part of) the standard representation of c.r. semigroup of [11]: there are three sets of functions

$$\begin{aligned} \langle\ ,\ \rangle &: S_\alpha \times I_\alpha \to I_\beta\ , \\ [\ ,\ ] &: \Lambda_\alpha \times S_\alpha \to \Lambda_\beta\ , \\ \tau &: S_\alpha \to G_\beta \text{ (denoted } a \mapsto a_\beta) \end{aligned}$$

defined for all $\alpha \geq \beta$ such that for all $a \in S_\alpha$ and $b \in S_\beta$,

$$ab = (\langle a, \langle b, \alpha\beta\rangle\rangle, a_{\alpha\beta}\, p_{[\alpha\beta,a]\langle b,\alpha\beta\rangle}\, b_{\alpha\beta}, [[\alpha\beta, a], b])\ , \tag{2}$$

where the function $a \mapsto a_\beta$ satisfies

$$a = (\langle a, i\rangle, a_\alpha, [\lambda, a]) \tag{3}$$

for all $a \in S_\alpha$, and

$$(ab)_\gamma = a_\gamma\, p_{[\gamma,a]\langle b,\gamma\rangle}\, b_\gamma \tag{4}$$

for all $a \in S_\alpha$, $b \in S_\beta$ and $\gamma \leq \alpha\beta$.

Assume that S is generated by a finite set A of elements and let $\gamma \in Y$. Let $c \in S_\gamma$ be such that $c = (i, g, \lambda)$. Now c is a product $c = a^{(1)}a^{(2)} \dots a^{(n)}$ of some

$a^{(j)} \in A$, where $a^{(j)} \in S_{\alpha_j}$ with $\alpha_j \geq \gamma$. From (3) it follows that $g = c_\gamma$, and thus, by (4),

$$g = (a^{(1)}a^{(2)} \dots a^{(n)})_\gamma = a_\gamma^{(1)} p_1 a_\gamma^{(2)} p_2 \dots p_{n-1} a_\gamma^{(n)}$$

for some entries p_j of P_γ. We conclude that G_γ is generated by the finite set

$$A_\gamma = \{a_\gamma \mid a \in A\} \cup \{p \mid p \in P_\gamma\} \ .$$

Since the matrix P_γ is finite, the group G_γ is finitely generated. Moreover the index sets I_γ and Λ_γ are finite, and therefore the finite set $I_\gamma \times A_\gamma \times \Lambda_\gamma$ generates S_γ. □

Recall that for a congruence η on a regular semigroup S

$$\ker \eta = \{a \mid a \eta e \ \text{ for some } e \in E(S)\}$$

is its *kernel.*

Lemma 16. *Let $S = (Y; S_\alpha)$ be a finitely generated c.r. semigroup. If each of the Rees matrix semigroups S_α satisfies the a.c.c. for congruences, then so does S.*

Proof. To prove the claim we use (a part of) the characterization of the congruences on a c.r. semigroup from [12]. This result states that each congruence ρ on S can be represented uniquely by a congruence aggregate $(\xi; \eta_\alpha)$, in notation $\rho = \rho_{(\xi;\eta_\alpha)}$, where ξ is a congruence on Y and η_α is a congruence on S_α such that for all $a \in S_\alpha$ and $b \in S_\beta$,

$$a \rho_{(\xi;\eta_\alpha)} b \iff \alpha \xi \beta \ , ab \, \eta_{\alpha\beta} \, ba, \ \ ab^{-1} \in \ker \eta_{\alpha\beta} \ .$$

Further, η_α is the restriction of ρ on S_α and

$$\alpha \xi \beta \iff a \rho c, \ b \rho d \ \text{ for some } a \in S_\alpha, \ b \in S_\beta, \ c, d \in S_{\alpha\beta} \ .$$

One obtains from these, see again [12], that for all congruences $\rho_1 = \rho_{(\xi^{(1)};\eta_\alpha^{(1)})}$ and $\rho_2 = \rho_{(\xi^{(2)};\eta_\alpha^{(2)})}$,

$$\rho_1 \subseteq \rho_2 \iff \xi^{(1)} \subseteq \xi^{(2)} \ \text{ and } \ \eta_\alpha^{(1)} \subseteq \eta_\alpha^{(2)} \ \text{for all } \alpha \in Y \ . \tag{5}$$

Let then $\rho_i = \rho_{(\xi^{(i)};\eta_\alpha^{(i)})}$, for $i \geq 1$, be congruences on S such that $\rho_i \subseteq \rho_{i+1}$. By (5) we have $\xi^{(i)} \subseteq \xi^{(i+1)}$ and $\eta_\alpha^{(i)} \subseteq \eta_\alpha^{(i+1)}$ for all i. Since S is finitely generated, Y is a finite semilattice, and therefore there exists an integer n_0 such that $\xi^{(i)} = \xi^{(i+k)}$ for all $i \geq n_0$ and $k \geq 0$. On the other hand, if each S_α satisfies the a.c.c. for congruences, then there exists an integer n_1 such that $\eta_\alpha^{(i)} = \eta_\alpha^{(i+k)}$ for all $i \geq n_1$ and $k \geq 0$ since Y is finite. This shows, by (5), that $\rho_i = \rho_{i+k}$ for all $i \geq \max(n_0, n_1)$ and $k \geq 0$, and hence S satisfies the a.c.c. for congruences. □

In conclusion we have the following result.

Theorem 17. *Let $S = (Y; S_\alpha)$ be a finitely generated c.r. semigroup with finite matrices. If each of its finitely generated subgroups satisfies the a.c.c. for normal subgroups, then S satisfies the a.c.c. for congruences.*

Proof. Suppose the finitely generated subgroups of S satisfy a.c.c. on normal subgroups. Then by Lemma 15 each S_α is finitely generated, and by Lemma 14 each S_α satisfies the a.c.c. for congruences. Finally, by Lemma 16, S itself satisfies the a.c.c. for congruences. □

Corollary 18. *Let $\mathcal{V}$ be a variety of c.r. semigroups whose finitely generated members have finite matrices. Then $\mathcal{V}$ satisfies CP if and only if $\mathcal{V} \cap \mathcal{G}$ does.*

Proof. The direct part of the claim is clear since $\mathcal{V} \cap \mathcal{G} \subseteq \mathcal{V}$.

In the reverse direction, suppose that the groups in $\mathcal{V}$ satisfy CP, and thus, since $\mathcal{V} \cap \mathcal{G}$ is a variety of groups, the finitely generated groups in $\mathcal{V}$ satisfy the a.c.c. for normal subgroups by Corollary 5. By Theorem 17 each finitely generated $S \in \mathcal{V}$ satisfies the a.c.c. for congruences, and thus the claim follows from Theorem 4. □

We can now show that the condition of Theorem 17 is satisfied by finitely generated bands of groups.

Lemma 19. *Let $S = (Y; S_\alpha)$ be a finitely generated c.r. semigroup that is a band of groups. Then S has finite matrices.*

Proof. Let $S_\alpha = \mathcal{M}(I_\alpha, G_\alpha, \Lambda_\alpha; P_\alpha)$ for all $\alpha \in Y$. Now $S/\mathcal{H}$ is a band, and since S is finitely generated, so is $S/\mathcal{H}$. Every finitely generated band is finite, see *e.g.* [9, Theorem IV.4.9], and thus $S/\mathcal{H}$ is a finite band. Therefore S is a finite union of groups and so $E(S)$ is finite. The claim follows since the idempotents of S are exactly the identity elements of its maximal subgroups. □

Bands of groups form a subvariety $\mathcal{BG}$ of the variety $\mathcal{CR}$ of c.r. semigroups and thus we have the following generalization of Corollary 5.

Corollary 20. *A variety $\mathcal{V}$ of bands of groups satisfies CP if and only if the variety $\mathcal{V} \cap \mathcal{G}$ does.*

References

1. M.H. Albert and J. Lawrence, The descending chain condition on solution sets for systems of equations in groups, *Proc. Edinburgh Math. Soc.* **29** (1985), 69 – 73.
2. M.H. Albert and J. Lawrence, A proof of Ehrenfeucht's Conjecture, *Theoret. Comput. Sci.* **41** (1985), 121 – 123.
3. A.H. Clifford, The free completely regular semigroup on a set, *J. Algebra* **59** (1979), 434 – 451.
4. A.H. Clifford and G.B. Preston, "The Algebraic Theory of Semigroups", Vol I, Amer. Math. Soc., Survey No. **7**, Providence, 1961.

5. A. de Luca and A. Restivo, On a generalization of a conjecture of Ehrenfeucht, *Bull. EATCS* **30** (1986), 84 – 90.
6. G. Grätzer, "Universal Algebra", Springer, New York, 1979.
7. V.S. Guba, The equivalence of infinite systems of equations in free groups and semigroups with finite subsystems, *Mat. Zametki* **40** (1986), 321 – 324 (Russian).
8. T. Harju, J. Karhumäki and W. Plandowski, Compactness of systems of equations in semigroups, *Internat. J. Algebra and Comput.*, to appear.
9. J.M. Howie, "An Introduction to Semigroup Theory", Academic Press, London, 1976.
10. M. Petrich, "Introduction to Semigroups", Merrill, Columbus, 1973.
11. M. Petrich, A structure theorem for completely regular semigroups, *Proc. Amer. Math. Soc.* **99** (1987), 617 – 622.
12. M. Petrich, Congruences on completely regular semigroups, *Canad. J. Math.* **41** (1989), 439 – 461.

Decision Problems Concerning Algebraic Series with Noncommuting Variables

Juha Honkala
Department of Mathematics
University of Turku
SF-20500 Turku, Finland
jhonkala@sara.cc.utu.fi

Abstract. Equivalence and rationality problems are shown to be decidable for algebraic series with noncommuting variables having bounded supports. As a tool, Parikh simplifying mappings are defined and studied.

1 Introduction

Formal power series play an important role in many diverse areas of theoretical computer science and mathematics, see Salomaa and Soittola [14], Kuich and Salomaa [10], Berstel and Reutenauer [1] and Kuich [9]. The classes of power series studied most often in connection with automata, grammars and languages are the rational and algebraic series.

In language theory formal power series often provide a powerful tool for obtaining deep decidability results, see Salomaa and Soittola [14] and Kuich and Salomaa [10]. A brilliant example is the solution of the equivalence problem for finite deterministic multitape automata given by Harju and Karhumäki [5].

In this paper we consider decision problems concerning algebraic series with noncommuting variables. In their monograph [10] Kuich and Salomaa develop many decidability results for algebraic series by using methods from commutative algebra and algebraic geometry. In most cases, however, it is necessary to assume that the variables commute. Our purpose is to generalize some of the main results of Kuich and Salomaa to algebraic series with noncommuting variables having bounded supports. Here we use the notion of boundedness in the sense customary in the theory of context-free languages. As a tool we define and study Parikh simplifying mappings. We say that a mapping f defined on a language L Parikh simplifies L if no two words in the image $f(L)$ of L have the same Parikh vector. Parikh simplifying mappings provide a method of simplification which makes it possible to lift decision methods for series with commuting variables to series with noncommuting variables.

Standard terminology and notation concerning formal languages and power series will be used in this paper. Whenever necessary, the reader may consult Salomaa [13], Salomaa and Soittola [14], Kuich and Salomaa [10] and Berstel and Reutenauer [1].

2 Parikh simplifying mappings

Suppose that Σ and Δ are finite alphabets. Denote by Δ^* and $\mathcal{P}(\Delta^*)$ the free monoid generated by Δ and the set of all subsets of Δ^*, respectively. A mapping $f : \Sigma^* \to \mathcal{P}(\Delta^*)$ *Parikh simplifies* a language $L \subseteq \Sigma^*$ if the following two conditions are satisfied:
(i) the sets $f(u)$, $u \in L$, are nonempty and pairwise disjoint,
(ii) no two words in $f(L) = \cup_{u\in L} f(u)$ are commutatively equivalent.
Such a mapping f is called *Parikh simplifying* on L.

If $\Delta^\oplus$ is the free commutative monoid generated by Δ and $c : \Delta^* \to \Delta^\oplus$ is the canonical morphism, then condition (ii) requires that $c(u) \neq c(v)$ whenever $u, v \in f(L)$ and $u \neq v$.

Example 1. Consider the language $L = (a^3)^*(ba^2)^*(ab^2)^*$ over the binary alphabet $\Sigma = \{a, b\}$. Note that L does contain words which are commutatively equivalent, for example, the words $a^3(ab^2)$ and $(ba^2)^2$. However, L can easily be Parikh simplified. Indeed, let $\Delta = \{a_1, a_2, a_3\}$ and define the morphism $h : \Delta^* \to \Sigma^*$ by $h(a_1) = a^3$, $h(a_2) = ba^2$, $h(a_3) = ab^2$. Then the mapping $h^{-1} : \Sigma^* \to \mathcal{P}(\Delta^*)$ Parikh simplifies L.

Example 2. Let Σ be an alphabet and $a \in \Sigma$. Define the mapping $f : \Sigma^* \to a^*$ by $f(w) = a^i$ where $w \in \Sigma^*$ is the i'th word in some fixed total order of Σ^*. Then f Parikh simplifies any language $L \subseteq \Sigma^*$. It is clear that Parikh simplifiers of this kind are of little help in attacking various decision problems.

Motivated by Example 2 we say that a mapping $f : \Sigma^* \to \mathcal{P}(\Delta^*)$ satisfies the *length condition* if there is a positive integer C such that for every nonempty word $w \in \Sigma^*$ the length of the shortest word in $f(w)$ is at most $C|w|$ if $f(w)$ is nonempty. Note that, e.g., all rational transductions satisfy the length condition.

Now, a language $L \subseteq \Sigma^*$ is called *Parikh simplifiable* if there exists a mapping f which Parikh simplifies L and satisfies the length condition. Next we show that bounded languages are Parikh simplifiable and characterize Parikh simplifiable context-free languages. Recall that a language $L \subseteq \Sigma^*$ is *bounded* if there exist a positive integer m and words $w_1, \ldots, w_m \in \Sigma^*$ such that

$$L \subseteq w_1^* w_2^* \ldots w_m^*.$$

We need the following result due to Raz [12].

Theorem 1 (Raz). *A context-free language L is bounded if and only if L is sparse.*

By definition, a language L is sparse if there is a polynomial $P(n)$ such that for each $n \geq 0$, L contains at most $P(n)$ words of length n.

Theorem 2. *A bounded language $L \subseteq \Sigma^*$ is Parikh simplifiable. A context-free language $L \subseteq \Sigma^*$ is Parikh simplifiable if and only if L is bounded.*

Proof. Suppose first that

$$L \subseteq w_1^* w_2^* \dots w_m^*$$

where $w_1, w_2, \dots, w_m \in \Sigma^*$. Choose a new alphabet $\Delta = \{a_1, a_2, \dots, a_m\}$ with m letters and define the morphism $h: \Delta^* \to \Sigma^*$ by $h(a_i) = w_i$ for $1 \leq i \leq m$. Consider the mapping $f : \Sigma^* \to \mathcal{P}(\Delta^*)$ defined by

$$f(w) = h^{-1}(w) \cap a_1^* a_2^* \dots a_m^*.$$

Clearly, f satisfies condition (i). Condition (ii) holds because $f(L) \subseteq a_1^* a_2^* \dots a_m^*$. Hence f Parikh simplifies L. Also, f satisfies the length condition.

Suppose then that L is a context-free language and the mapping $f : \Sigma^* \to \mathcal{P}(\Delta^*)$ Parikh simplifies L and satisfies the length condition. First, note that the number of words of length at most n in $\Delta^{\oplus}$ is bounded from above by a polynomial $P(n)$. Next, denote

$$L_n = \{w \in L \,||w| \leq n\}$$

for $n \geq 0$ and define the partial mapping $\overline{f} : \Sigma^* \to \Delta^*$ by letting $\overline{f}(w)$ to be a word of minimal length in $f(w)$ if $f(w)$ is nonempty. Because f satisfies the length condition, there is a positive integer C such that

$$\overline{f}(L_n) \subseteq \{u \in \Delta^* \,||u| \leq Cn\}$$

for $n \geq 1$. Therefore

$$\mathrm{card}(L_n) = \mathrm{card}(\overline{f}(L_n)) \leq P(Cn)$$

for $n \geq 1$, implying that L is sparse. Hence, by Theorem 1, L is bounded. □

Corollary 3. *It is decidable whether or not a given context-free language L is Parikh simplifiable.*

Proof. The claim follows by Theorem 2 because it is decidable whether or not a given context-free language is bounded (see Ginsburg [3]). □

If a context-free language $L \subseteq \Sigma^*$ is Parikh simplifiable, one can effectively find words $w_1, w_2, \dots, w_m \in \Sigma^*$ such that

$$L \subseteq w_1^* w_2^* \dots w_m^*$$

and, hence, effectively obtain a Parikh simplifying mapping on L.

The following result will be needed in the sequel. By definition, a mapping $f : \Sigma^* \to \mathcal{P}(\Delta^*)$ is *single-valued* if $f(u)$ has at most one element for every $u \in \Sigma^*$.

Theorem 4. *Suppose L is a bounded set. Then there exists a single-valued Parikh simplifying mapping f on L.*

Proof. Suppose

$$L \subseteq w_1^* w_2^* \dots w_m^* \tag{1}$$

where $w_1, w_2, \dots, w_m \in \Sigma^*$. Consider the morphism $h : \{a_1, a_2, \dots, a_m\}^* \to \Sigma^*$ defined by $h(a_i) = w_i$, $1 \leq i \leq m$. Then h maps $a_1^* a_2^* \dots a_m^*$ onto $w_1^* w_2^* \dots w_m^*$. By the Cross-Section Theorem due to Eilenberg [2], there exists a rational language $R \subseteq a_1^* a_2^* \dots a_m^*$ such that h maps R bijectively onto $w_1^* w_2^* \dots w_m^*$. Furthermore, the construction of R is effective, given $w_1, w_2, \dots, w_m \in \Sigma^*$. Now, define the mapping $f : \Sigma^* \to \mathcal{P}(\{a_1, \dots, a_m\}^*)$ by $f(w) = h^{-1}(w) \cap R$. Then f is a single-valued Parikh simplifying mapping on L. □

Note that the construction of f in Theorem 4 is effective if words $w_1, w_2, \dots, w_m$ are given such that (1) holds. In particular, f is effectively obtainable if L is a Parikh simplifying context-free language.

3 The equivalence problem

In this section we use Parikh simplifying mappings to solve the equivalence problem for algebraic series with bounded supports.

Suppose A is a commutative semiring and $X = \{x_1, \dots, x_n\}$ is a finite alphabet. The set of formal power series (resp. polynomials) with noncommuting variables in X and coefficients in A is denoted by $A \ll X^* \gg$ (resp. $A < X^* >$). The set of A-algebraic (resp. A-rational) series with noncommuting variables in X is denoted by $A^{\text{alg}} \ll X^* \gg$ (resp. $A^{\text{rat}} \ll X^* \gg$). As a tool in the proofs we consider also A-algebraic and A-rational series with commuting variables in X. The corresponding sets are denoted by $A^{\text{alg}} \ll X^{\oplus} \gg$ and $A^{\text{rat}} \ll X^{\oplus} \gg$, respectively. Furthermore, c is the canonical morphism $c : A \ll X^* \gg \to A \ll X^{\oplus} \gg$. Hence

$$A^{\text{alg}} \ll X^{\oplus} \gg = \{c(r) | r \in A^{\text{alg}} \ll X^* \gg\}$$

and

$$A^{\text{rat}} \ll X^{\oplus} \gg = \{c(r) | r \in A^{\text{rat}} \ll X^* \gg\}.$$

By definition, a series $r \in A \ll X^* \gg$ is *s-bounded* if supp(r) is a bounded set, i.e., there exist a positive integer $m \geq 1$ and words $w_1, w_2, \dots, w_m \in X^+$ such that

$$\text{supp}(r) \subseteq w_1^* w_2^* \dots w_m^*. \tag{2}$$

Note that r being s-bounded has nothing to do with the coefficients of r.

Example 3. Consider the series

$$r_1 = \sum_{n \geq 0} (n+1)^2 a^n b^n - \sum_{n \geq 0} n^2 (ab)^n - \sum_{n \geq 0} (a^2 b^2)^n - \sum_{n \geq 0} (ba^2 b)^n ba$$

and

$$r_2 = 2 \sum_{n \geq 0} n (ba)^n.$$

Because

$$\operatorname{supp}(r_1) \subseteq a^*b^*(ab)^*(a^2b^2)^*(ba^2b)^*(ba)^*$$

and

$$\operatorname{supp}(r_2) \subseteq (ba)^*,$$

both series are s-bounded. Clearly, $r_1 \neq r_2$. However, $c(r_1) = c(r_2)$. Therefore, the equivalence problems for series with noncommuting variables and commuting variables are two different problems.

Lemma 5. *Suppose A is a positive semiring. If $r \in A^{alg} \ll X^* \gg$, it is decidable whether or not r is s-bounded. If r is s-bounded, an integer $m \geq 1$ and words $w_1, w_2, \ldots, w_m \in X^+$ can be effectively found such that (2) holds.*

Proof. Because A is a positive semiring, the support of r is a context-free language (see Salomaa and Soittola [14]). Hence the claim follows because it is decidable whether or not a given context-free language is bounded. □

We next recall some earlier results.

Theorem 6 (Kuich, Salomaa). *For every quasiregular $r \in \mathbf{Q}^{alg} \ll X^{\oplus} \gg$, a unique (up to a factor ± 1) irreducible primitive polynomial*

$$P(x_1, \ldots, x_n, y) \in \mathbf{Z} < (X \cup y)^{\oplus} >$$

can be effectively constructed such that

$$P(x_1, \ldots, x_n, r) = 0.$$

To prove Theorem 6 Kuich and Salomaa use deep methods from commutative algebra and algebraic geometry. However, to apply their algorithm for the construction of the polynomial P only basic knowledge of linear algebra is needed. Honkala [7] generalized Theorem 6 to arbitrary computable fields and as a consequence obtained the next result.

Theorem 7. *Suppose E is a computable field. It is decidable whether or not two given series in $E^{alg} \ll X^{\oplus} \gg$ are equal.*

Intuitively, a field is *computable* if for any $m \geq 1$, polynomials $Q_1, Q_2 \in F[z_1, \ldots, z_m]$ and $c_1, \ldots, c_m \in E$ we can check whether or not

$$Q_1(c_1, \ldots, c_m) = Q_2(c_1, \ldots, c_m).$$

Here F is the prime subfield of E.

It is assumed in what follows that when an s-bounded series r is given, also words $w_1, w_2, \ldots, w_m$ are given such that (2) holds. By Lemma 5 this additional assumption is not needed if $r \in A^{\mathrm{alg}} \ll X^* \gg$ where A is a positive semiring.

Theorem 8. *Suppose E is a computable field. It is decidable whether or not two given s-bounded series $r, s \in E^{alg} \ll X^* \gg$ are equal.*

Proof. Suppose

$$\mathrm{supp}(r) \subseteq w_1^* w_2^* \dots w_m^*$$

and

$$\mathrm{supp}(s) \subseteq v_1^* v_2^* \dots v_p^*,$$

where $w_i, v_j \in X^+$, $1 \le i \le m$, $1 \le j \le p$. Then

$$\mathrm{supp}(r) \cup \mathrm{supp}(s) \subseteq w_1^* w_2^* \dots w_m^* v_1^* v_2^* \dots v_p^*. \tag{3}$$

Let $\Delta = \{a_1, \dots, a_m, b_1, \dots, b_p\}$ be a new alphabet with $m+p$ letters and define the morphism $h : \Delta^* \to X^*$ by $h(a_i) = w_i$ for $1 \le i \le m$ and $h(b_j) = v_j$ for $1 \le j \le p$. Denote by t the characteristic series of $a_1^* a_2^* \dots a_m^* b_1^* b_2^* \dots b_p^*$. Furthermore, denote

$$r_1 = c(h^{-1}(r) \odot t)$$

and

$$s_1 = c(h^{-1}(s) \odot t).$$

By the closure properties of algebraic series, r_1 and s_1 belong to $E^{\mathrm{alg}} \ll \Delta^{\oplus} \gg$. We claim that $r = s$ if and only if $r_1 = s_1$. One direction being obvious suppose that $r_1 = s_1$. Then, for nonnegative integers $i_1, \dots, i_m, j_1, \dots, j_p$ we have

$$(r, w_1^{i_1} \dots w_m^{i_m} v_1^{j_1} \dots v_p^{j_p}) = (h^{-1}(r), a_1^{i_1} \dots a_m^{i_m} b_1^{j_1} \dots b_p^{j_p}) =$$

$$(h^{-1}(r) \odot t, a_1^{i_1} \dots a_m^{i_m} b_1^{j_1} \dots b_p^{j_p}) = (c(h^{-1}(r) \odot t), a_1^{i_1} \dots a_m^{i_m} b_1^{j_1} \dots b_p^{j_p}) =$$

$$(c(h^{-1}(s) \odot t), a_1^{i_1} \dots a_m^{i_m} b_1^{j_1} \dots b_p^{j_p}) = (h^{-1}(s) \odot t, a_1^{i_1} \dots a_m^{i_m} b_1^{j_1} \dots b_p^{j_p}) =$$

$$(h^{-1}(s), a_1^{i_1} \dots a_m^{i_m} b_1^{j_1} \dots b_p^{j_p}) = (s, w_1^{i_1} \dots w_m^{i_m} v_1^{j_1} \dots v_p^{j_p}).$$

(Note that in both sides of the fourth equation, $a_1^{i_1} \dots a_m^{i_m} b_1^{j_1} \dots b_p^{j_p}$ is regarded as an element of $\Delta^{\oplus}$.) Because (3) holds, this implies that $r = s$. Consequently, to decide whether or not $r = s$ it is enough to decide whether or not $r_1 = s_1$. This is possible by Theorem 7. □

For **N**-algebraic series Theorem 8 is due to Raz [11]. For the equivalence problem of **N**-algebraic series with noncommuting variables see also Kuich [8].

4 The rationality problem

In this section we use single-valued Parikh simplifying mappings to solve the rationality problem for s-bounded algebraic series.

Theorem 9. *Suppose $A = \mathbf{Z}$ or $A = \mathbf{Q}$. It is decidable whether or not an s-bounded series $r \in \mathbf{Q}^{alg} \ll X^* \gg$ is A-rational.*

The proof of Theorem 9 is in three steps. First, we establish a normal form for s-bounded rational series. Then we use Parikh simplifying mappings to reduce the problem to series with commuting variables. Finally, we use a corollary of Theorem 6 due to Kuich and Salomaa to decide whether or not the series has the normal form

Lemma 10. *Suppose A is a semiring and $X = \{x_1, \ldots, x_n\}$ is an alphabet. If $r \in A^{rat} \ll X^* \gg$ and*

$$supp(r) \subseteq x_1^* x_2^* \ldots x_n^*,$$

there exist a positive integer t and A-rational series r_{ij} for $1 \leq i \leq t$, $1 \leq j \leq n$, such that

$$r = \sum_{i=1}^{t} r_{i1} r_{i2} \ldots r_{in}$$

and

$$r_{ij} \in A^{rat} \ll x_j^* \gg$$

for $1 \leq i \leq t$, $1 \leq j \leq n$.

Proof. By the Representation Theorem of Schützenberger (see Salomaa and Soittola [14]), there exist a positive integer k, a monoid morphism $\mu : X^* \to A^{k \times k}$ and vectors $F \in A^{1 \times k}$ and $G \in A^{k \times 1}$ such that

$$(r, w) = F\mu(w)G$$

for any $w \in X^*$. For $1 \leq \alpha \leq n$ and $1 \leq k_1, k_2 \leq k$ define the series $r(\alpha, k_1, k_2) \in A^{\mathrm{rat}} \ll x_\alpha^* \gg$ by

$$r(\alpha, k_1, k_2) = \sum_{i \geq 0} \mu(x_\alpha^i)_{k_1, k_2} x_\alpha^i.$$

Then

$$r = \sum_{i_\alpha \geq 0} (r, x_1^{i_1} \ldots x_n^{i_n}) x_1^{i_1} \ldots x_n^{i_n} = \sum_{i_\alpha \geq 0} F\mu(x_1^{i_1} \ldots x_n^{i_n}) G x_1^{i_1} \ldots x_n^{i_n}$$

$$= \sum_{i_\alpha \geq 0, 1 \leq k_\beta \leq k} F_{k_1} \mu(x_1^{i_1})_{k_1, k_2} \mu(x_2^{i_2})_{k_2, k_3} \ldots \mu(x_n^{i_n})_{k_n, k_{n+1}} G_{k_{n+1}} x_1^{i_1} x_2^{i_2} \ldots x_n^{i_n}$$

$$= \sum_{1 \leq k_\beta \leq k} F_{k_1} r(1, k_1, k_2) r(2, k_2, k_3) \ldots r(n, k_n, k_{n+1}) G_{k_{n+1}}.$$

This implies the claim. □

If $A = \mathbf{B}$, where $\mathbf{B}$ is the Boolean semiring, Lemma 10 implies a characterization of bounded regular sets due to Ginsburg and Spanier [4].

Now, suppose $r \in \mathbf{Q}^{\mathrm{alg}} \ll X^+ \gg$ is s-bounded. Let $w_1, w_2, \ldots, w_m \in \Sigma^+$ be nonempty words such that

$$\mathrm{supp}(r) \subseteq w_1^* w_2^* \ldots w_m^*$$

and let $\Delta = \{a_1, a_2, \ldots, a_m\}$ be a new alphabet. Define the morphism $h : \Delta^* \to X^*$ by $h(a_i) = w_i$, $1 \leq i \leq m$. By Theorem 4 there exists a regular language $R \subseteq a_1^* a_2^* \ldots a_m^*$ such that the mapping $f : X^* \to \mathcal{P}(\Delta^*)$ defined by $f(w) = h^{-1}(w) \cap R$ is single-valued and Parikh simplifies $\mathrm{supp}(r)$.

Consider the series

$$s = h^{-1}(r) \odot \mathrm{char}(R).$$

By the closure properties of algebraic series, s belongs to $\mathbf{Q}^{\mathrm{alg}} \ll \Delta^* \gg$. Because $R \subseteq a_1^* a_2^* \dots a_m^*$ we have

$$\operatorname{supp}(s) \subseteq a_1^* a_2^* \dots a_m^*.$$

If $w \in X^*$, then

$$(h(s), w) = \sum_{h(u)=w} (h^{-1}(r) \odot \operatorname{char}(R), u)$$

$$= \sum_{h(u)=w, u \in R} (h^{-1}(r), u) = \sum_{h(u)=w, u \in R} (r, h(u)) = (r, w).$$

Hence, $h(s) = r$. Consequently, r is A-rational if and only if s is A-rational. Thus, it suffices to prove Theorem 9 with the additional assumption that

$$\operatorname{supp}(r) \subseteq x_1^* x_2^* \dots x_n^*.$$

The following lemma completes the second step in the proof of Theorem 9.

Lemma 11. *Suppose $r \in A \ll X^* \gg$ and supp$(r) \subseteq x_1^* x_2^* \dots x_n^*$. Then r is A-rational if and only if there exist a positive integer t and series s_{ij} for $1 \leq i \leq t$, $1 \leq j \leq n$ such that*

$$c(r) = \sum_{i=1}^{t} s_{i1} s_{i2} \dots s_{in} \tag{4}$$

and

$$s_{ij} \in A^{rat} \ll x_j^* \gg \tag{5}$$

for $1 \leq i \leq t$, $1 \leq j \leq n$.

Proof. If r is A-rational, the claim follows by Lemma 10. Suppose then that there exist series s_{ij}, $1 \leq i \leq t$, $1 \leq j \leq n$ such that (4) and (5) hold. Let $r_{ij} \in A \ll X^* \gg$ be series such that $s_{ij} = c(r_{ij})$ and $r_{ij} \in A^{\mathrm{rat}} \ll x_j^* \gg$ for $1 \leq i \leq t$, $1 \leq j \leq n$. We claim that

$$r = \sum_{i=1}^{t} r_{i1} r_{i2} \dots r_{in}. \tag{6}$$

Indeed, the supports of both sides of (6) are included in $x_1^* x_2^* \dots x_n^*$. Furthermore,

$$c(\sum_{i=1}^{t} r_{i1} r_{i2} \dots r_{in}) = \sum_{i=1}^{t} c(r_{i1}) c(r_{i2}) \dots c(r_{in}) = \sum_{i=1}^{t} s_{i1} s_{i2} \dots s_{in} = c(r).$$

This implies the claim. Consequently, r is A-rational. □

For the proof of Theorem 9 it remains to decide whether or not $c(r)$ has the normal form of Lemma 11 if $r \in \mathbf{Q}^{\mathrm{alg}} \ll X^* \gg$ and $\operatorname{supp}(r) \subseteq x_1^* x_2^* \dots x_n^*$. First, decide whether or not $c(r)$ is A-rational. This is possible by Theorem 16.13 (resp. Theorem 16.14) in Kuich and Salomaa [10] if $A = \mathbf{Q}$ (resp. $A = \mathbf{Z}$). Furthermore, if $c(r)$ is A-rational, we can effectively find polynomials $P, Q \in A < X^{\oplus} >$ with no nontrivial common factors such that

$$c(r) = \frac{P}{1-Q}$$

and $(Q, \varepsilon) = 0$. We claim that $c(r)$ can be written in the normal form of Lemma 11 if and only if there exist polynomials $Q_1, Q_2, \ldots, Q_n$ such that

$$1 - Q = (1 - Q_1)(1 - Q_2) \cdots (1 - Q_n) \tag{7}$$

and

$$Q_i \in A < x_i^+ > \tag{8}$$

for $1 \leq i \leq n$. Suppose first that such a factorization exists. Then $\frac{P}{1-Q}$ can be expressed as a finite sum of terms of the form

$$\frac{P_1}{1-Q_1} \frac{P_2}{1-Q_2} \cdots \frac{P_n}{1-Q_n}$$

where $P_i \in A < x_i^* >$ for $1 \leq i \leq n$. Therefore $c(r)$ has the normal form of Lemma 11. Suppose then that $c(r)$ has the normal form. It follows that

$$c(r) = \frac{P'}{(1-Q_1')(1-Q_2')\ldots(1-Q_n')}$$

where $P' \in A < X^{\oplus} >$ and $Q_j' \in A < x_j^+ >$, $1 \leq j \leq n$. Therefore $1 - Q$ divides $(1-Q_1')(1-Q_2')\ldots(1-Q_n')$ and hence is of the form given in (7) and (8). This concludes the proof of the claim.

Now, denote by $1 - \overline{Q}_i$ the polynomial obtained from $1 - Q$ by substituting $x_\alpha = 0$ for $1 \leq \alpha \leq n$, $\alpha \neq i$. If (7) holds, we have

$$1 - Q = (1 - \overline{Q}_1)(1 - \overline{Q}_2)\ldots(1 - \overline{Q}_n). \tag{9}$$

Conversely, if (9) holds, $1 - Q$ can be expressed as in (7) and (8). Hence, it is decidable whether or not a given polynomial $1 - Q$ can be expressed as in (7) and (8). Consequently, it is decidable whether or not $c(r)$ has the normal form of Lemma 11.

This concludes the proof of Theorem 9.

References

[1] Berstel, J. and Reutenauer, C.: *Rational Series and Their Languages* (Springer, Berlin, 1988).

[2] Eilenberg, S.: *Automata, Languages and Machines*, Vol. A (Academic Press, New York, 1974).

[3] Ginsburg, S.: *The Mathematical Theory of Context-Free Languages* (McGraw-Hill, New York, 1966).

[4] Ginsburg, S. and Spanier, E. H.: Bounded regular sets, *Proc. Amer. Math. Soc.* 17 (1966) 1043-1049.

[5] Harju, T. and Karhumäki, J.: The equivalence problem of multitape finite automata, *Theoret. Comput. Sci.* 78 (2) (1991) 347-355.

[6] Honkala, J.: On images of algebraic series, *J. Univ. Comput. Sci.* 2 (4) (1996) 217-223.

[7] Honkala, J.: On the equivalence problem of algebraic series with commuting variables, *EATCS Bulletin* 59 (1996) 157-162.

[8] Kuich, W.: On the multiplicity equivalence problem for context-free grammars, in J. Karhumäki, H. Maurer and G. Rozenberg (eds.): *Results and Trends in Theoretical Computer Science* (Springer, Berlin, 1994) pp. 232-250.

[9] W. Kuich, Semirings and formal power series: their relevance to formal languages and automata, in G. Rozenberg and A. Salomaa (eds.): *Handbook of Formal Languages* (Springer, Berlin, 1997), to appear.

[10] Kuich, W. and Salomaa, A.: *Semirings, Automata, Languages* (Springer, Berlin, 1986).

[11] Raz, D.: Deciding multiplicity equivalence for certain context-free languages, in G. Rozenberg and A. Salomaa (eds.): *Developments in Language Theory* (World Scientific, Singapore, 1994) pp. 18-29.

[12] Raz, D.: Length considerations in context-free languages, *Theoret. Comput. Sci.*, to appear.

[13] Salomaa, A.: *Formal Languages* (Academic Press, New York, 1973).

[14] Salomaa, A. and Soittola, M.: *Automata-Theoretic Aspects of Formal Power Series* (Springer, Berlin, 1978).

Associative Shuffle of Infinite Words*

Alexandru Mateescu[1,2] and George Daniel Mateescu[3]

[1] Turku Centre for Computer Science,
Lemminkäisenkatu 14 A, 20520 Turku, Finland
[2] Department of Mathematics, University of Bucharest, Romania,
email: mateescu@sara.utu.fi
[3] Department of Mathematics, University of Bucharest,
Academiei, 14, sector 1, 70109 Bucharest, Romania,
email: dmateescu@roimar.imar.ro

Abstract. Some shuffle-like operations on infinite (ω-words) are investigated. The operations are introduced using a uniform method based on the notion of an *ω-trajectory*. We prove an interconnection between associative closure of shuffle on ω-trajectories and periodicity of ω-words. This provides a characterization of the ultimately periodic ω-words. Finally, a remarkable property of the Fibonacci ω-word is exhibited.

1 Preliminaries

The operation of parallel composition of words and languages is a fundamental operation in parallel computation and in the theory of concurrency. Usually, this operation is modelled by the shuffle operation or restrictions of this operation, such as literal shuffle, insertion, left-merge, or the infiltration product, [3].

This paper continues our investigations on some shuffle-like operations on ω-words and ω-languages, see [5] and [6]. The shuffle-like operations considered below are defined using syntactic constraints on the ω-shuffle operation. A first approach of this topic was considered in [8].

The constraints on the general shuffle are based on the notion of an *ω-trajectory* and describe the general strategy to switch from one ω-word to another ω-word. Roughly speaking, an ω-trajectory is a line in plane, starting in the origin and continuing parallel with the axis Ox or Oy. The line can change its direction only in points with nonnegative integer coordinates. An ω-trajectory defines how to move from an ω-word to another ω-word when carrying out the shuffle operation. Each set T of ω-trajectories defines in a natural way a shuffle operation over T. Given a set T of ω-trajectories the operation of shuffle over T is not necessarily an associative operation. However, for each set T there exists a smallest set of trajectories $\overline{T}$ such that $\overline{T}$ contains T and, moreover, shuffle over $\overline{T}$ is an associative operations. Such a set $\overline{T}$ is referred to as the associative closure of T. We show that the associative closure of some very simple finite sets leads to the set of periodic or to the set of ultimately periodic ω-words.

* This work has been partially supported by the Project 137358 of the Academy of Finland. All correspondence to Alexandru Mateescu.

The set of nonnegative integers is denoted by ω. If A is a set, then the set of all subsets of A is denoted by $\mathcal{P}(A)$.

Let Σ be an alphabet, i.e., a finite nonempty set of elements called *letters*. The free monoid generated by Σ is denoted by Σ^*. Elements in Σ^* are referred to as *words*. The empty word is denoted by λ.

If $w \in \Sigma^*$, then $|w|$ is the length of w. Note that $|\lambda| = 0$. If $a \in \Sigma$ and $w \in \Sigma^*$, then $|w|_a$ denotes the number of occurrences of the symbol a in the word w. The *mirror* of a word $w = a_1a_2 \ldots a_n$, where a_i are letters, $1 \leq i \leq n$, is $mi(w) = a_n \ldots a_2a_1$ and $mi(\lambda) = \lambda$. A word w is a *palindrome* iff $mi(w) = w$.

Let Σ be an alphabet. An *ω-word* over Σ is a function $f : \omega \longrightarrow \Sigma$. Usually, the ω-word defined by f is denoted as the infinite sequence $f(0)f(1)f(2)\ldots$. An ω-word w is *ultimately periodic* iff $w = \alpha vvvvv\ldots$, where α is a (finite) word, possibly empty, and v is a nonempty word. In this case w is denoted as αv^ω. An ω-word w is referred to as *periodic* iff $w = vvv\ldots$ for some nonempty word $v \in \Sigma^*$. In this case w is denoted as v^ω. The set of all ω-words over Σ is denoted by Σ^ω. An *ω-language* is a subset L of Σ^ω, i.e., $L \subseteq \Sigma^\omega$. The reader is referred to [9], [11] for general results on ω-words and to [10] for notions and results in formal languages.

The *shuffle* operation, denoted by $\sqcup\!\sqcup$, is defined recursively by:

$$(au \sqcup\!\sqcup bv) = a(u \sqcup\!\sqcup bv) \cup b(au \sqcup\!\sqcup v),$$

and

$$(u \sqcup\!\sqcup \lambda) = (\lambda \sqcup\!\sqcup u) = \{u\},$$

where $u ,v \in \Sigma^*$ and $a ,b \in \Sigma$.

The above operation is extended in a natural way to languages: the *shuffle* of two languages L_1 and L_2 is:

$$L_1 \sqcup\!\sqcup L_2 = \bigcup_{u \in L_1, v \in L_2} u \sqcup\!\sqcup v.$$

The *literal shuffle*, denoted by $\sqcup\!\sqcup_l$, is defined as:

$$a_1a_2 \ldots a_n \sqcup\!\sqcup_l b_1b_2 \ldots b_m = \begin{cases} a_1b_1a_2b_2 \ldots a_nb_nb_{n+1} \ldots b_m, & \text{if } n \leq m, \\ a_1b_1a_2b_2 \ldots a_mb_ma_{m+1} \ldots a_n, & \text{if } m < n, \end{cases}$$

where $a_i, b_j \in \Sigma$.

$$(u \sqcup\!\sqcup_l \lambda) = (\lambda \sqcup\!\sqcup_l u) = \{u\},$$

where $u \in \Sigma^*$.

2 Shuffle on ω-trajectories

In this section we introduce the notions of the ω-trajectory and shuffle on ω-trajectories. The shuffle of two ω-words has a natural geometrical interpretation related to lattice points in the plane (points with nonnegative integer coordinates) and with a certain "walk" in the plane defined by each ω-trajectory.

Let $V = \{r, u\}$ be the set of *versors* in the plane: r stands for the *right* direction, whereas, u stands for the *up* direction.

Definition 1. An *ω-trajectory* is an element t, $t \in V^\omega$. A set T, $T \subseteq V^\omega$, is called a set of *ω-trajectories.* □

Let Σ be an alphabet and let t be an ω-trajectory, $t = t_0t_1t_2\ldots$, where $t_i \in V, i \geq 0$. Let α, β be two ω-words over Σ, $\alpha = a_0a_1a_2\ldots, \beta = b_0b_1b_2\ldots$, where $a_i, b_j \in \Sigma, i, j \geq 0$.

Definition 2. The *shuffle of α with β on the ω-trajectory t*, denoted $\alpha ⧢_t \beta$, is defined as follows:

$\alpha ⧢_t \beta = c_0c_1c_2\ldots$, where, if $|t_0t_1t_2\ldots t_i|_r = k_1$ and $|t_0t_1t_2\ldots t_i|_u = k_2$, then

$$c_i = \begin{cases} a_{k_1-1}, & \text{if } t_i = r, \\ b_{k_2-1}, & \text{if } t_i = u. \end{cases}$$

□

Remark. The shuffle on (finite) trajectories of (finite) words is investigated in [6]. In this case a trajectory is an element t, $t \in V^*$.

Let Σ be an alphabet and let t be a trajectory, $t = t_0t_1\ldots t_n$, where $t_i \in V, 1 \leq i \leq n$. Let α, β be two words over Σ, $\alpha = a_0a_1\ldots a_p, \beta = b_0b_1\ldots b_q$, where $a_i, b_j \in \Sigma, 0 \leq i \leq p$ and $0 \leq j \leq q$.

The shuffle of α with β on the trajectory t, denoted $\alpha ⧢_t \beta$, is defined as follows:

if $|\alpha| \neq |t|_r$ or $|\beta| \neq |t|_u$, then $\alpha ⧢_t \beta = \emptyset$, else

$\alpha ⧢_t \beta = c_0c_1c_2\ldots c_{p+q+2}$, where, if $|t_0t_1\ldots t_i|_r = k_1$ and $|t_0t_1\ldots t_i|_u = k_2$, then

$$c_i = \begin{cases} a_{k_1-1}, & \text{if } t_i = r, \\ b_{k_2-1}, & \text{if } t_i = u. \end{cases}$$

Observe that there is an important distinction between the finite case, i.e., the shuffle on trajectories, and the infinite case, i.e., the shuffle on ω-trajectories: sometimes the result of shuffling of two words α and β on a trajectory t can be empty whereas the shuffle of two ω-words over an ω-trajectory is always nonempty and consists of only one ω-word.

If T is a set of ω-trajectories, the *shuffle of α with β on the set T of ω-trajectories*, denoted $\alpha ⧢_T \beta$, is:

$$\alpha ⧢_T \beta = \bigcup_{t \in T} \alpha ⧢_t \beta.$$

The above operation is extended to ω-languages over Σ, if $L_1, L_2 \subseteq \Sigma^\omega$, then:

$$L_1 ⧢_T L_2 = \bigcup_{\alpha \in L_1, \beta \in L_2} \alpha ⧢_T \beta.$$

Notation. If T is V^ω then $⧢_T$ is denoted by $⧢_\omega$.

Example 1. Let α and β be the ω-words $\alpha = a_0 a_1 a_2 a_3 \ldots$, $\beta = b_0 b_1 b_2 b_3 \ldots$ and assume that $t = r^2 u^2 r^3 u r^2 u r u \ldots$. The shuffle of α with β on the trajectory t is:

$$\alpha ⧢_t \beta = \{a_0 a_1 b_0 b_1 a_2 a_3 a_4 b_2 a_5 a_6 b_3 a_7 b_4 \ldots\}.$$

The result has the following geometrical interpretation (see Figure 1): the trajectory t defines a line starting in the origin and continuing one unit right or up, depending of the definition of t. In our case, first there are two units right, then two units up, then three units right, etc. Assign α on the Ox axis and β on the Oy axis of the plane. The result can be read following the line defined by the trajectory t, that is, if being in a lattice point of the trajectory, (the corner of a unit square) and if the trajectory is going right, then one should pick up the corresponding letter from α, otherwise, if the trajectory is going up, then one should add to the result the corresponding letter from β. Hence, the trajectory t defines a line in the plane, on which one has "to walk" starting from the origin O. In each lattice point one has to follow one of the versors r or u, according to the definition of t.

Assume now that $t' = u r^5 u^4 r^3 \ldots$ is another trajectory. Observe that:

$$\alpha ⧢_{t'} \beta = \{b_0 a_0 a_1 a_2 a_3 a_4 b_1 b_2 b_3 b_4 a_5 a_6 a_7 \ldots\}.$$

Consider the set of trajectories, $T = \{t, t'\}$. The shuffle of α with β on the set T of trajectories is:

$$\alpha ⧢_T \beta = \{a_0 a_1 b_0 b_1 a_2 a_3 a_4 b_2 a_5 a_6 b_3 a_7 b_4 \ldots, b_0 a_0 a_1 a_2 a_3 a_4 b_1 b_2 b_3 b_4 a_5 a_6 a_7 \ldots\}.$$

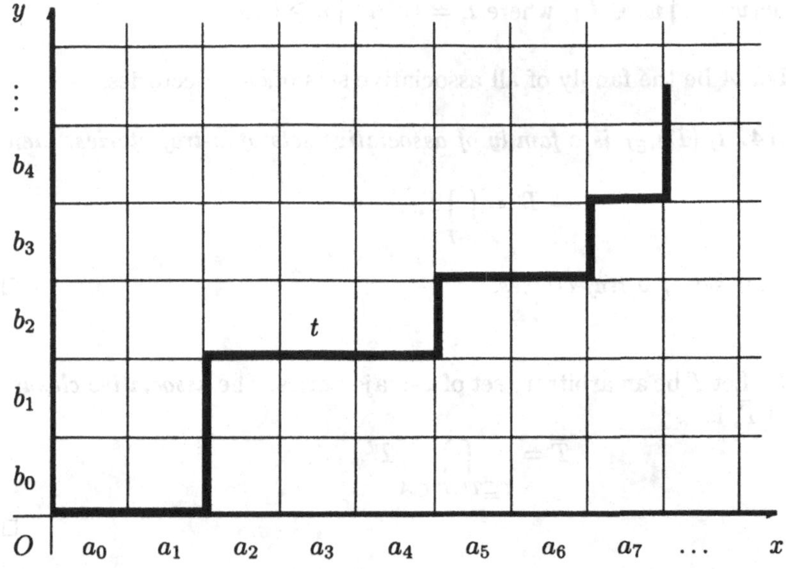

Figure 1

3 Associativity

The main results in this paper deal with associativity. After a few general remarks, we restrict the attention to the set V_+^ω of ω-trajectories t such that both r and u occur infinitely often in t. (It will become apparent below why this restriction is important.) It turns out that associativity can be viewed as stability under four particular operations, referred to as $\diamond$-operations.

Definition 3. A set T of ω-trajectories is ***associative*** iff the operation $⊔⊔_T$ is associative, i.e.,

$$(\alpha ⊔⊔_T \beta) ⊔⊔_T \gamma = \alpha ⊔⊔_T (\beta ⊔⊔_T \gamma),$$

for all $\alpha, \beta, \gamma \in \Sigma^\omega$. □

The following sets of ω-trajectories are associative:

1. $T = \{r, u\}^\omega$.
2. $T = \{t \in V^\omega \mid |t|_r < \infty\}$.
3. $T = \{\alpha_0\beta_0\alpha_1\beta_1 \ldots \mid \alpha_i \in r^*, \beta_i \in u^*$ and, moreover, α_i and β_i are of even length, $i \geq 0\}$.

Nonassociative sets of ω-trajectories are for instance:

1. $T = \{(ru)^\omega\}$.
2. $T = \{t \in V^\omega \mid t$ is a Sturmian ω-word $\}$.

3. $T = \{w_0 w_1 w_2 \dots \mid w_i \in L\}$, where $L = \{r^n u^n \mid n \geq 0\}$.

Notation. Let $\mathcal{A}$ be the family of all associative sets of ω-trajectories.

Proposition 4. *If $(T_i)_{i \in I}$ is a family of associative sets of ω-trajectories, then*

$$T' = \bigcap_{i \in I} T_i ,$$

is an associative set of ω-trajectories. □

Definition 5. Let T be an arbitrary set of ω-trajectories. The *associative closure* of T, denoted $\overline{T}$, is

$$\overline{T} = \bigcap_{T \subseteq T', T' \in \mathcal{A}} T'.$$

□

Observe that for all $T, T \subseteq \{r, u\}^*$, $\overline{T}$ is an associative set of ω-trajectories and $\overline{T}$ is the smallest associative set of ω-trajectories that contains T.

Remark. The function $\overline{-}, - : \mathcal{P}(V^\omega) \longrightarrow \mathcal{P}(V^\omega)$ defined as above is a closure operator. □

Notation. Let V_+^ω be the set of all ω-trajectories $t \in V^\omega$ such that t contains infinitely many occurrences both of r and of u.

Now we give another characterization of an associative set of ω-trajectories from V_+^ω. This is useful in finding an alternative definition of the associative closure of a set of ω-trajectories and also to prove some other properties related to associativity.

However, this characterization is valid only for sets of ω-trajectories from V_+^ω and not for the general case, i.e., not for sets of ω-trajectories from V^ω.

Definition 6. Let W be the alphabet $W = \{x, y, z\}$ and consider the following four morphisms, ρ_i, $1 \leq i \leq 4$, where $\rho_i : W \longrightarrow V_+^\omega$, $1 \leq i \leq 4$, and

$$\rho_1(x) = \lambda , \quad \rho_1(y) = r , \quad \rho_1(z) = u,$$

$$\rho_2(x) = r , \quad \rho_2(y) = u , \quad \rho_2(z) = u,$$

$$\rho_3(x) = r , \quad \rho_3(y) = u , \quad \rho_3(z) = \lambda,$$

$$\rho_4(x) = r , \quad \rho_4(y) = r , \quad \rho_4(z) = u.$$

□

Next, we consider four operations on the set of ω-trajectories, V_+^ω.

Definition 7. Let $\diamond_i$, $1 \leq i \leq 4$ be the following operations on V_+^ω.

$$\diamond_i : V_+^\omega \times V_+^\omega \longrightarrow V_+^\omega, \qquad 1 \leq i \leq 4,$$

Let t, t' be in V_+^ω. By definition:

$$\diamond_1(t, t') = \rho_1((x^\omega \sqcup\!\sqcup_t y^\omega) \sqcup\!\sqcup_{t'} z^\omega),$$

$$\diamond_2(t, t') = \rho_2((x^\omega \sqcup\!\sqcup_t y^\omega) \sqcup\!\sqcup_{t'} z^\omega),$$

$$\diamond_3(t', t) = \rho_3(x^\omega \sqcup\!\sqcup_{t'} (y^\omega \sqcup\!\sqcup_t z^\omega)),$$

$$\diamond_4(t', t) = \rho_4(x^\omega \sqcup\!\sqcup_{t'} (y^\omega \sqcup\!\sqcup_t z^\omega)).$$

□

Definition 8. A set $T \subseteq V_+^\omega$ is *stable* under $\diamond$-operations iff for all $t_1, t_2 \in T$, it follows that $\diamond_i(t_1, t_2) \in T$, $1 \leq i \leq 4$.

□

The following theorem is proved in [5]:

Theorem 9. *Let T be a set of ω-trajectories, $T \subseteq V_+^\omega$. The following assertions are equivalent:*

(i) T is an associative set of ω-trajectories.
(ii) T is stable under $\diamond$-operations.

Remark. We restricted our attention only to the set V_+^ω and not to the general case V^ω. However, if T contains a trajectory t that is not in V_+^ω, then $\diamond_1(t,t)$is not necessarily in V^ω. For instance $t = rur^\omega$, then $\diamond_1(t,t) = ur \notin V^\omega$. Thus the operation $\diamond_1$ is not well defined. A similar phenomenon happens with the operation $\diamond_3$. □

Comment. Observe that $\mathcal{D} = (\mathcal{P}(V_+^\omega), (\diamond_i)_{1 \leq i \leq 4})$ is a universal algebra. If T is a set of ω-trajectories, then denote by $\widetilde{T}$ the subalgebra generated by T with respect to the algebra $\mathcal{D}$.

Proposition 10. *Let $T \subseteq V_+^\omega$ be a set of ω-trajectories.*

(i) $\widetilde{T}$ is an associative set of ω-trajectories and, morover,

(ii) $\widetilde{T} = \overline{T}$, i.e., the associative closure of T is exactly the subalgebra generated by T in $\mathcal{D}$.

□

Proposition 11. *Let $T \subseteq V_+^\omega$ be a set of ω-trajectories.*

(i) If each $t \in T$ is a periodic ω-word, then the associative closure of T, $\overline{T}$, has the same property, i.e., each ω-trajectory in $\overline{T}$ is periodic.

(ii) if additionally, each $t \in T$ has a palindrome as its period, then the associative closure of T, $\overline{T}$, has the same property.

(iii) If T is a set of ultimately periodic ω-trajectories, then the associative closure of T, $\overline{T}$, has the same property, i.e., each ω-trajectory in $\overline{T}$ is ultimately periodic.

□

The above proposition yields:

Corollary 12. *The following sets of ω-trajectories are associative:*

(i) the set of all periodic ω-trajectories from V_+^ω.

(ii) the set of all periodic ω-trajectories from V_+^ω that have as their period a palindrome.

(iii) the set of all ultimately periodic ω-trajectories from V_+^ω.

□

In the sequel we show some interrelations between associativity and commutativity of shuffle operation.

Definition 13. Let *sym* be the following mapping, $sym : V \longrightarrow V$, $sym(r) = u$ and $sym(u) = r$. Also consider the mapping $\varphi : \{x, y, z\} \longrightarrow \{x, y, z\}$, $\varphi(x) = z$, $\varphi(y) = y$ and $\varphi(z) = x$. *sym* and φ are extended to ω-words over V and, respectively over $\{x, y, z\}$.

□

One can easily verify the following properties.

Proposition 14. *The following equalities are true:*

1. $\varphi(\alpha \sqcup\!\sqcup_t \beta) = \varphi(\alpha) \sqcup\!\sqcup_t \varphi(\beta)$, *for all* $\alpha, \beta \in \{x, y, z\}^\omega$ *and* $t \in V_+^\omega$.
2. $\rho_1(\varphi(\alpha)) = sym(\rho_3(\alpha))$, $\alpha \in \{x, y, z\}^\omega$.
3. $\rho_2(\varphi(\alpha)) = sym(\rho_4(\alpha))$, $\alpha \in \{x, y, z\}^\omega$.
4. $\rho_3(\varphi(\alpha)) = sym(\rho_1(\alpha))$, $\alpha \in \{x, y, z\}^\omega$.
5. $\rho_4(\varphi(\alpha)) = sym(\rho_2(\alpha))$, $\alpha \in \{x, y, z\}^\omega$.
6. $\diamond_1(sym(t_1), sym(t_2)) = sym(\diamond_3(t_2, t_1))$, $t_1, t_2 \in V_+^\omega$.
7. $\diamond_2(sym(t_1), sym(t_2)) = sym(\diamond_4(t_2, t_1))$, $t_1, t_2 \in V_+^\omega$.
8. $\diamond_3(sym(t_1), sym(t_2)) = sym(\diamond_1(t_2, t_1))$, $t_1, t_2 \in V_+^\omega$.
9. $\diamond_4(sym(t_1), sym(t_2)) = sym(\diamond_2(t_2, t_1))$, $t_1, t_2 \in V_+^\omega$.

□

Consequently, we obtain the following:

Corollary 15. *Let $T \subseteq V_+^\omega$. Then:*

1. $sym(\overline{T}) = \overline{sym(T)}$.
2. T is associative if and only if $sym(T)$ is associative.
3. If T is commutative, then $\overline{T}$ is commutative. The converse is not true.

□

Next theorem provides a characterization of those ω-trajectories that are periodic. The reader is referred to [5] for the proof of this theorem. As such it is also a direct contribution to the study of ω-words, exhibiting an interconnection between periodicity and associativity.

Theorem 16. *Let t be an ω-trajectory such that $t \neq r^\omega$ and $t \neq u^\omega$. The following assertions are equivalent:*
(i) t is a periodic ω-word.
(ii) t is in the associative closure of $(ru)^\omega$.

□

Corollary 17. *An ω-word $t \in V_+^*$ is a periodic ω-word if and only if t can be obatined from the ω-word $(ru)^\omega$ by finitely many applications of operations $\diamond_i$, $1 \leq i \leq 4$.* □

A similar result is next Theorem 16. It states a characterization of ultimately periodic ω-words in terms of the associative closure of a certain ω-trajectory. This theorem is an extension of a result from [5] where the set of all ultimately periodic is obtained as the associative closure of two trajectories instead of the associative closure of only one trajectory.

Definition 18. Let T be a set of ω-trajectories. A finite sequence $t_1, t_2, \ldots, t_n$ of ω-trajectories is referred to as a formal description of t_n with respect to T iff for each $1 \leq k \leq n$, either $t_k \in T$, or there exists $1 \leq i \leq 4$, and there are $1 \leq p, q < k$, such that $t_k = \diamond_i(t_p, t_q)$. □

Notation. A formal description $t_1, t_2, \ldots, t_n$ of t_n with respect to T is denoted by $< t_1, t_2, \ldots, t_n >_T$.

Remark. Let T be a set of ω-trajectories. One can easily verify that

$$\overline{T} = \{t_n \mid \text{there exists } < t_1, t_2, \ldots, t_n >_T\}.$$

Moreover, if $< t_1, t_2, \ldots, t_n >_T$ and $T \subseteq T'$, then $< t_1, t_2, \ldots, t_n >_{T'}$.

Theorem 19. *Let t be an ω-trajectory such that $t \in V_+^\omega$. The following assertions are equivalent:*
(i) t is a ultimately periodic ω-word.
(ii) t is in the associative closure of the ω-trajectory $r(ru)^\omega$.

Proof. (*ii*) $\Longrightarrow$ (*i*) It follows from Proposition 10, (*ii*) and Proposition 11, (*iii*).

(*i*) $\Longrightarrow$ (*ii*) We start the proof by proving some claims. Consider the notations $T_0 = \{(ru)^\omega\}$, $T_1 = \{r(ru)^\omega\}$, $T_2 = \{(ru)^\omega, r(ru)^\omega\}$ and
$T_3 = \{(ru)^\omega, r(ru)^\omega, u(ru)^\omega\}$.

Claim A. *The sets T_1 and T_2 have the same associative closure, i.e., $\overline{T_1} = \overline{T_2}$.*

Proof of Claim A. One can easily verify the following two equalities:

$$\diamond_1(r(ru)^\omega, r(ru)^\omega) = (uru)^\omega$$

and

$$\diamond_3((uru)^\omega, r(ru)^\omega) = (ur)^\omega$$

Using Corollary 15,1, and Theorem 16 we obtain that $\overline{(ur)^\omega} = \overline{(ru)^\omega}$. Hence $\overline{T_2} \subseteq \overline{T_1}$. The converse inclusion is obviously true. Thus we obtain Claim A.

Claim B. *The sets T_2 and T_3 have the same associative closure, i.e., $\overline{T_2} = \overline{T_3}$.*

Proof of Claim B. Note that $u(ru)^\omega = (ur)^\omega$ and hence $u(ru)^\omega$ is a periodic ω-trajectory. By Theorem 16, $u(ru)^\omega$ is in the associative closure of $(ru)^\omega$. Thus we obtain Claim B.

Claim C. *If $< t_1, t_2, \ldots, t_n >_{T_0}$, then*

$$< t_1, t_2, \ldots, t_n, rt_1, ut_1, rt_2, ut_2, \ldots, rt_n, ut_n >_{T_2} .$$

Proof of Claim C. We show by induction on k that

$$< t_1, t_2, \ldots, t_n, rt_1, ut_1, rt_2, ut_2, \ldots, rt_k, ut_k >_{T_2} .$$

Assume that $k = 1$. Observe that from Definition 18 it follows that $t_1 = (ru)^\omega$. By definition of T_2 and Claim B we conclude that $< t_1, t_2, \ldots, t_n, rt_1, ut_1 >_{T_2}$. Assume by induction that $< t_1, t_2, \ldots, t_n, rt_1, ut_1, rt_2, ut_2, \ldots, rt_{k-1}, ut_{k-1} >_{T_2}$. If $t_k = (ru)^\omega$, then the conclusion follows as in the case $k = 1$. Now, assume that $t_k = \diamond_i(t_p, t_q)$ for some $1 \leq i \leq 4$, $1 \leq p, q < k$ and we show that the trajectory rt_k has a formal description with respect to T_2. Consider all possible situations:

If $i = 1$, that is $t_k = \diamond_1(t_p, t_q)$, then note that

$$rt_k = \rho_1((x^\omega \sqcup\!\sqcup_{ut_p} y^\omega) \sqcup\!\sqcup_{rt_q} z^\omega) = \diamond_1(ut_p, rt_q).$$

Assume that $i = 2$. Hence $t_k = \diamond_2(t_p, t_q)$, and note that

$$rt_k = \rho_2((x^\omega \sqcup\!\sqcup_{rt_p} y^\omega) \sqcup\!\sqcup_{rt_q} z^\omega) = \diamond_2(rt_p, rt_q).$$

Consider now that $i = 3$. Thus $t_k = \diamond_3(t_p, t_q)$, and observe that

$$rt_k = \rho_3((x^\omega \sqcup\!\sqcup_{rt_p} y^\omega) \sqcup\!\sqcup_{t_q} z^\omega) = \diamond_3(rt_p, t_q).$$

If $i = 4$ and therefore $t_k = \diamond_4(t_p, t_q)$, then note that

$$rt_k = \rho_4((x^\omega \sqcup\!\sqcup_{rt_p} y^\omega) \sqcup\!\sqcup_{t_q} z^\omega) = \diamond_4(rt_p, t_q).$$

Finally, we prove how the trajectory ut_k can be obtained. Again, we consider all possible situations.

Assume that $i = 1$ and $t_k = \diamond_1(t_p, t_q)$. Observe that

$$ut_k = \rho_1((x^\omega \sqcup\!\sqcup_{t_p} y^\omega) \sqcup\!\sqcup_{ut_q} z^\omega) = \diamond_1(t_p, ut_q).$$

If $i = 2$, and thus $t_k = \diamond_2(t_p, t_q)$, then note that

$$ut_k = \rho_2((x^\omega \sqcup\!\sqcup_{t_p} y^\omega) \sqcup\!\sqcup_{ut_q} z^\omega) = \diamond_2(t_p, ut_q).$$

Consider that $i = 3$. Hence $t_k = \diamond_3(t_p, t_q)$ and notice that

$$ut_k = \rho_3((x^\omega \sqcup\!\sqcup_{ut_p} y^\omega) \sqcup\!\sqcup_{rt_q} z^\omega) = \diamond_3(ut_p, rt_q).$$

If $i = 4$, that is $t_k = \diamond_4(t_p, t_q)$, then note that

$$ut_k = \rho_4((x^\omega \sqcup\!\sqcup_{ut_p} y^\omega) \sqcup\!\sqcup_{ut_q} z^\omega) = \diamond_4(ut_p, ut_q).$$

Next two claims assert that certain trajectories are in the associative closure of the set $T_2 = \{(ru)^\omega, r(ru)^\omega\}$.

Claim D. If t is a periodic trajectory, $t \in V_+^\omega$, then the trajectories rt and ut are in $\overline{T_2}$.

Proof of Claim D. Let t be a periodic trajectory, $t \in V_+^\omega$. Observe that from Theorem 16 and from Remark 3 there exists a formal description of t with respect to $T_0 = \{(ru)^\omega\}$, say $< t_1, t_2, \ldots, t_n, t >_{T_0}$. From Claim B it follows that $< t_1, t_2, \ldots, t_n, t, rt_1, ut_1, \ldots, rt_n, ut_n, rt, ut >_{T_2}$. From Remark 3 we conclude Claim D.

Claim E. The trajectories $r^2(ru)^\omega$ and $ur(ru)^\omega$ are in $\overline{T_2}$.

Proof of Claim E. Observe that: $r^2(ru)^\omega = \rho_4(x^\omega \sqcup\!\sqcup_{t_1} (y^\omega \sqcup\!\sqcup_{t_2} z^\omega)$, where $t_1 = (ru^3)^\omega$ and $t_2 = r(ru^2)^\omega$. Moreover, $ur(ru)^\omega = \rho_1((x^\omega \sqcup\!\sqcup_{t_3} y^\omega) \sqcup\!\sqcup_{t_4} z^\omega)$, where $t_3 = u(ur)^\omega$ and $t_2 = (ur^2)^\omega$.

Note that as a consequence of Theorem 16 and of Claim D, t_i, $1 \le i \le 4$ are in $\overline{T_2}$.

From the above Claim E, we deduce that there are the following formal descriptions:

$< x_1, x_2, \ldots, x_i, r^2(ru)^\omega >_{T_2}$ and $< y_1, y_2, \ldots, y_j, ur(ru)^\omega >_{T_2}$.

Using these notations we assert:

Claim F. If $< t_1, t_2, \ldots, t_n >_{T_2}$, then

$$< (ru)^\omega, x_1, ..., x_i, r^2(ru)^\omega, y_1, ..., y_j, ur(ru)^\omega, t_1, ..., t_n, rt_1, ut_1, ..., rt_n, ut_n >_{T_2} .$$

Proof of Claim F. We show that for all $1 \le k \le n$, the trajectories rt_k and ut_k are in $\overline{T_2}$. Assume $k = 1$ and note that $t_1 = (ru)^\omega$ or $t_1 = r(ru)^\omega$. Thus rt_1 and ut_1 are among the following trajectories: $r(ru)^\omega$, $u(ru)^\omega$, $r^2(ru)^\omega$, $ur(ru)^\omega$. Using Claims B and E we obtain that in each case they are in $\overline{T_2}$. Assume that $k > 1$. If t_k is in T_2, then the claim follows as in case $k = 1$. If t_k is obtained by using an operation $\diamond_i$, $1 \le i \le 4$, i.e., $t_k = \diamond_i(t_p, t_q)$, for some $1 \le i \le 4$,

$1 \le p, q < k$, then a similar proof as for Claim C shows that rt_k and ut_k are in $\overline{T_2}$.

Now we end the proof of Theorem 19.

Let $t = \alpha z^\omega$ be a ultimately periodic ω-word, $t \in V_+^\omega$. Let n be the length of α, i.e., $|\alpha|$. If $n = 0$, then t is a periodic ω word and from Theorem 16 we conclude that $t \in \overline{T_2}$. If $n \ge 1$, then using n times Claim F we obtain a formal description of t with respect to T_2. Hence, from Remark 3 it follows that $t \in \overline{T_2}$. Since, by Claim A, $\overline{T_2} = \overline{T_1}$, the proof is complete. □

Corollary 20. *Let t be an ω-word, $t \in V_+^*$. t is a ultimately periodic ω-word if and only if t can be obatined from the ω-word $r(ru)^\omega$ by finitely many applications of operations $\diamond_i$, $1 \le i \le 4$.* □

In the sequel we prove that the associative closure of any open set in V_+^ω is V_+^ω itself. We start by a preliminary result.

Proposition 21. *For every $1 \le i \le 4$ the mapping $\diamond_i$ is a surjective mapping, i.e., for any $t \in V_+^\omega$, there are $t_1, t_2 \in V_+^\omega$ such that $t = \diamond_i(t_1, t_2)$, $1 \le i \le 4$.*

Proof. Let t be in V_+^ω and consider an arbitrary trajectory $t' \in V_+^\omega$. Assume that $i = 1$ and define $t_1 = \diamond_3(t', t)$ and $t_2 = \diamond_4(t', t)$. It follows that $t = \diamond_1(t_1, t_2)$.

Analogously, for $i = 2$, define $t_1 = \diamond_3(t, t')$, $t_2 = \diamond_4(t, t')$ and observe that $t = \diamond_2(t_1, t_2)$.

If $i = 3$, then consider $t_1 = \diamond_2(t, t')$, $t_2 = \diamond_1(t, t')$ and notice that $t = \diamond_3(t_1, t_2)$.

Finally, if $i = 4$, then define $t_1 = \diamond_2(t', t)$, $t_2 = \diamond_1(t', t)$ and observe that $t = \diamond_4(t_1, t_2)$.

□

Notation. If w is a word from V^*, then wV_+^ω denotes the set $\{wt \mid t \in V_+^\omega\}$.

Proposition 22. *Let w be a nonempty word from V^*. There exists $w' \in V^*$ such that $|w'| < |w|$ and $w'V_+^\omega \subseteq \overline{wV_+^\omega}$.*

Proof. Let t be in V_+^ω and assume that $|w|_u = n$ and $|w|_r = m$. Let t_1 and t_2 be two trajectories such that $t = \diamond_1(t_1, t_2)$. Note that from Proposition 22 there exist such trajectories t_1 and t_2. We consider the two possible cases: $m \ne n$ and $m = n$.

Case 1. Assume that $m \ne n$ and consider that $n < m$. Now, observe that:

$$\diamond_1(wt_1, wr^n t_2) = \rho_1((x^\omega ⧢_{wt_1} y^\omega) ⧢_{wr^n t_2} z^\omega) =$$

$$= \rho_1((x^m ⧢_w y^n) ⧢_{wr^n} z^n)\rho_1((x^\omega ⧢_{t_1} y^\omega) ⧢_{t_2} z^\omega) =$$

$$= \rho_1((x^m ⧢_w y^n) ⧢_{wr^n} z^n)\diamond_1(t_1, t_2) = \rho_1((x^m ⧢_w y^n) ⧢_{wr^n} z^n)t = w't$$

Let α be the word $(x^m \sqcup\!\sqcup_w y^n) \sqcup\!\sqcup_{wr^n} z^n$. Notice that $|\alpha|_x = m$, $|\alpha|_y = |\alpha|_z = n$. Since ρ_1 mapps x in λ, y in r and z in u, it follows that $|w'| = |w'|_r + |w'|_u = n + n < n + m = |w|$.

Hence $w't = \diamond_1(wt_1, wr^n t_2) \in \overline{wV_+^\omega}$ and, moreover, $|w'| < |w|$.

Analogously, if $m < n$, then one may consider the value of $\diamond_3(wu^m t_1, wt_2)$.

Case 2. Now assume that $m = n$, i.e., $|w|_r = |w|_u = m$. Consider that $w = vr$ for some $v \in V^*$, $|v| = 2m - 1$. Let w'' be in V^* such that $|w''|_u = 1$. Notice that:

$$\diamond_3(wu^{m-1}w''t_1, wt_2) = \rho_3(x^\omega \sqcup\!\sqcup_{vru^{m-1}w''t_1} (y^\omega \sqcup\!\sqcup_{vrt_2} z^\omega)) =$$
$$= \rho_3(x^m \sqcup\!\sqcup_{vru^{m-1}} v(y,z)) \rho_3(x^k \sqcup\!\sqcup_{w''} y) \diamond_3(t_1, t_2) = w'w''t$$

where $v(y,z)$ is the prefix of length $2m-1$ of the ω-word $y^\omega \sqcup\!\sqcup_{vrt_2} z^\omega$, $k = |w''|_r$ and $w' = \rho_3(x^m \sqcup\!\sqcup_{vru^{m-1}} v(y,z))$.

One can easily verify that $|w'| = 2m - 1 < |w|$.
Hence $w'w''t = \diamond_3(wu^{m-1}w''t_1, wt_2) \in \overline{wV_+^\omega}$ and, moreover, $|w'| < |w|$.

Analogously, if $w = vu$, then let w'' be in V^* such that $|w''|_r = 1$ and one may compute the value of $\diamond_1(wt_1, wr^{m-1}w''t_2))$.

Hence, for each $w \in V^*$, $w \neq \lambda$, there exist w' and w'' in V^* such that $|w'| < |w|$ and $w'w''V_+^\omega \subseteq \overline{wV_+^\omega}$.

Since w'' is arbitrary, it follows that $w''V_+^\omega = V_+^\omega$.

Therefore, for each $w \in V^*$, $w \neq \lambda$, there exists $w' \in V^*$ such that $|w'| < |w|$ and $w'V_+^\omega \subseteq \overline{wV_+^\omega}$.

□

Corollary 23. *For each $w \in V^*$, $w \neq \lambda$, there exists $w' \in V^*$ such that $|w'| < |w|$ and $\overline{w'V_+^\omega} \subseteq \overline{wV_+^\omega}$.* □

Theorem 24. *For each $w \in V^*$, $\overline{wV_+^\omega} = V_+^\omega$.*

Proof. We apply Corollary 23 until $|w'| = 0$. □

Comment. The above result offers also a connection between associative closure of sets of ω-trajectories and the topological structure of V_+^ω. The most used topology on V^ω is defined taking as a base of neighbourhoods the sets wV^ω, where $w \in V^*$. For more details the reader may refer to [2] and [9]. Considering the induced topology on V_+^ω, a base of neighbourhoods consists of sets wV_+^ω. By using Theorem 24 we conclude that the associative closure of any open set in V_+^ω is V_+^ω itself.

Next theorem shows a remarkable property of the most famous infinite Sturmian word, known as the Fibonacci word. An infinite Sturmian word is an ω-word t over V such that the subword complexity of t is defined by the function $\varphi_t(n) = n + 1$, i.e., for any $n \geq 1$, the number of subwords of t of length n is exactly $n + 1$. See [1], [4] for other equivalent definitions of infinite Sturmian words as well as for a number of properties of these words.

The Fibonacci word f is defined as the limit of the sequence of words $(f_n)_{n\geq 0}$, where:

$$f_0 = r,\ f_1 = u,\ f_{n+1} = f_n f_{n-1},\ n > 0.$$

Note that f is an infinite word which has an initial prefix as follows:

$$f = uruururuuruururuururu\ldots$$

Remark. Note that for each n, $|f_n|$ is equal with the n-th term of the Fibonacci numerical sequence: $1, 1, 2, 3, 5, 8, 13\ldots$. Moreover, $|f_n|_r = |f_{n-1}|$ and $|f_n|_u = |f_{n-2}|$.

□

We recall some properties of the operation of shuffle on finite trajectories. For more details on this topic the reader is referred to [6].

Definition 25. Let W be the alphabet $W = \{x, y, z\}$ and consider the following four morphisms, ρ'_i, $1 \leq i \leq 4$, where $\rho'_i : W \longrightarrow V^*$, $1 \leq i \leq 4$, and

$$\rho'_1(x) = \lambda\ , \qquad \rho'_1(y) = r\ , \qquad \rho'_1(z) = u,$$

$$\rho'_2(x) = r\ , \qquad \rho'_2(y) = u\ , \qquad \rho'_2(z) = u,$$

$$\rho'_3(x) = r\ , \qquad \rho'_3(y) = u\ , \qquad \rho'_3(z) = \lambda,$$

$$\rho'_4(x) = r\ , \qquad \rho'_4(y) = r\ , \qquad \rho'_4(z) = u.$$

□

Definition 26. Let $\diamond'_i$, $1 \leq i \leq 4$ be the following partial operations on V^*.

$$\diamond'_i : V^* \times V^* \xrightarrow{o} V^*\ , \qquad 1 \leq i \leq 4,$$

Let t, t' be in V^* and assume that $|t| = n$, $|t|_r = p$, $|t|_u = q$, $|t'| = n'$, $|t'|_r = p'$, $|t'|_u = q'$,

1. if $n = p'$, then
$$\diamond'_1(t, t') = \rho_1((x^p ⧢_t y^q) ⧢_{t'} z^{q'})),$$
else, $\diamond'_1(t, t')$ is undefined.
2. if $n = p'$, then
$$\diamond'_2(t, t') = \rho_2((x^p ⧢_t y^q) ⧢_{t'} z^{q'})),$$
else, $\diamond'_2(t, t')$ is undefined.
3. if $n = q'$, then
$$\diamond'_3(t', t) = \rho_3(x^{p'} ⧢_{t'} (y^p (⧢_t z^q)),$$
else, $\diamond'_3(t, t')$ is undefined.

4. if $n = q'$, then
$$\diamond_4'(t', t) = \rho_4(x^{p'} ⧢_{t'} (y^p (⧢_t z^q)),$$
else, $\diamond_4'(t, t')$ is undefined.

□

Note that the morphisms ρ_i', $1 \leq i \leq 4$, as well as the operations $\diamond_i'$, $1 \leq i \leq 4$, are the versions of the morphisms ρ_i, $1 \leq i \leq 4$, and respectively of the operations $\diamond_i$, $1 \leq i \leq 4$, for the case of finite trajectories. In the sequel ρ_i' is denoted by ρ_i and $\diamond_i'$ is denoted by $\diamond_i$, $1 \leq i \leq 4$. Notice that this simplification does not produce any ambiguity.

Now we are in position to state our result:

Theorem 27. *The associative closure of the Fibonacci word, $\overline{f}$, contains all periodic trajectories and the containment is strict.*

Proof. We start by proving the following equality:
$$\diamond_1(f, \diamond_3(f, f)) = (ru)^\omega. \qquad (I)$$

Let $(e_n)_{n\geq 0}$ be the Fibonacci numerical sequence $1, 1, 2, 3, 5, 8, 13 \ldots$. Note that $e_n = |f_n|$ for every $n \geq 0$. Consider the notation $t = \diamond_1(f, \diamond_3(f, f))$. We show that each prefix of t of length $2e_n$ is of the form $(ru)^{e_n}$. The proof is by induction on n.

Let α_n be the value of $\diamond_3(f_{n+1}, f_n)$, i.e.,
$$\alpha_n = \rho_3(x^{e_{n-1}} ⧢_{f_{n+1}} (y^{e_{n-2}} ⧢_{f_n} z^{e_{n-1}})).$$

Therefore, we prove by induction on n that:
$$\diamond_1(f_{n-1}, \alpha_n) = (ru)^{e_{n-2}}. \qquad (II)$$

That is:
$$(x^{e_{n-3}} ⧢_{f_{n-1}} y^{e_{n-2}}) ⧢_{\alpha_n} z^{e_{n-2}} = (ru)^{e_{n-2}}.$$

The base of induction: $n = 3$.
Notice that:
$$\alpha_3 = \rho_3(x^2 ⧢_{uruur} (y ⧢_{uru} z^2)) = \rho_3(x^2 ⧢_{uruur} zyz) = \rho_3(zxyzx) = rur.$$

It follows that:
$$\diamond_1(f_2, \alpha_3) = \rho_1(ur, rur) = \rho_1((x ⧢_{ur} y) ⧢_{rur} z) =$$
$$= \rho_1(yx ⧢_{rur} z) = \rho_1(yzx) = ru = (ru)^{e_1}.$$

Hence, for $n = 3$ the equality (II) is true.
The inductive step: $n \vdash n + 1$
We start by computing the value of α_{n+1}:
$$\alpha_{n+1} = \rho_3(x^{e_n} ⧢_{f_{n+2}} (y^{e_{n-1}} ⧢_{f_{n+1}} z^{e_n})) =$$

$$= \rho_3(x^{e_n} ⧢_{f_{n+2}} (y^{e_{n-2}+e_{n-3}} ⧢_{f_n f_{n-1}} z^{e_{n_1}+e_{n-2}})) =$$

$$= \rho_3(x^{e_n} ⧢_{f_{n+2}} (y^{e_{n-2}} ⧢_{f_n} z^{e_{n_1}}) \cdot (y^{e_{n-3}} ⧢_{f_{n-1}} z^{e_{n_2}})) =$$

$$= \rho_3(x^{e_{n-1}+e_{n-2}} ⧢_{f_{n+1} f_n} (y^{e_{n-2}} ⧢_{f_n} z^{e_{n_1}}) \cdot (y^{e_{n-3}} ⧢_{f_{n-1}} z^{e_{n_2}})) =$$

$$= \rho_3((x^{e_{n-1}} ⧢_{f_{n+1}} (y^{e_{n-2}} ⧢_{f_n} z^{e_{n_1}})) \cdot (x^{e_{n-2}} ⧢_{f_n} (y^{e_{n-3}} ⧢_{f_{n-1}} z^{e_{n_2}})) =$$

$$= \alpha_n \alpha_{n-1}.$$

Therefore, we proved that:

$$\alpha_{n+1} = \alpha_n \alpha_{n-1}.$$

Using the above relation, the inductive hypothesis, the well-known properties of the numerical Fibonacci sequence and the properties of the Fibonacci sequence f, it follows that:

$$\diamond_1(f_n, \alpha_{n+1}) = \rho_1((x^{e_{n-2}} ⧢_{f_n} y^{e_{n-1}}) ⧢_{\alpha_{n+1}} z^{e_{n-1}}) =$$

$$= \rho_1((x^{e_{n-3}+e_{n-4}} ⧢_{f_{n-1} f_{n-2}} y^{e_{n-2}+e_{n-3}}) ⧢_{\alpha_{n+1}} z^{e_{n-1}}) =$$

$$= \rho_1((x^{e_{n-3}} ⧢_{f_{n-1}} y^{e_{n-2}}) \cdot (x^{e_{n-4}} ⧢_{f_{n-2}} y^{e_{n-3}}) ⧢_{\alpha_n \alpha_{n-1}} z^{e_{n-2}+e_{n-3}}) =$$

$$= \rho_1(((x^{e_{n-3}} ⧢_{f_{n-1}} y^{e_{n-2}}) ⧢_{\alpha_n} z^{e_{n-2}}) \cdot ((x^{e_{n-4}} ⧢_{f_{n-2}} y^{e_{n-3}}) ⧢_{\alpha_{n-1}} z^{e_{n-3}})) =$$

$$\diamond_1(f_{n-1}, \alpha_n) \diamond_1(f_{n-2}, \alpha_{n-1}) = (ru)^{e_{n-2}} (ru)^{e_{n-1}} = (ru)^{e_{n-1}+e_{n-2}} = (ru)^{e_{n-1}}.$$

Hence, equality (II), and consequently equality (I) are true.

Using Theorem 16, we deduce that all periodic ω-words from V_+^ω are contained in the associative closure of the Fibonacci sequence, $\overline{f}$.

Since the Fibonacci sequence is not a periodic ω-word, by Proposition 11, we conclude that the above containment is strict.

□

4 Conclusion

Many other problems remain to be investigated. For instance: does exist a proper ultimately periodic ω-word in the associative closure of the Fibonacci ω-word f? What can be said about the associative closure of some other Sturmian ω-words? The shuffle-like operations considered in this paper provide a new tool for studing properties of ω-words. Recently, a characterization of ω-words that are ultimately periodic has been obtained in [7]. This characterization is based on a different approach. Interrelations between this characterization and the characterization from the present paper are subject of further research.

References

[1] J. Berstel, *Recent Results on Sturmian Words*, in Developments in Language Theory, eds. J. Dassow, G. Rozenberg and A. Salomaa, World Scientific, 1996, 13-24.

[2] O. Carton, *Mots Infinis, ω-Semigroupes et Topologie*, These, Universite Paris 7, 1993.

[3] M. Lothaire, *Combinatorics on Words*, Addison Wesley, vol. 17 Enciclopedia of Mathematics and its Applications, 1983.

[4] A. de Luca, *Sturmian words: new combinatorial results* in Semigroups, Automata and Languages, eds. J. Almeida, G.M.S. Gomes and P.V. Silva, World Scientific, 1996, 67-84.

[5] A. Mateescu and G.D. Mateescu, "Associative closure and periodicity of ω-words", submitted.

[6] A. Mateescu, G. Rozenberg and A. Salomaa, "Shuffle on Trajectories: Syntactic Constraints", Technical Report 96-18, University of Leiden, 1996.

[7] F. Mignosi, A. Restivo and S. Salemi, "A Periodicity Theorem on Words and Applications", Proceedings of MFCS 1995, eds. J. Wiedermann and P. Hájek, LNCS 969, Springer-Verlag, 1995, 337-348.

[8] D. Park, "Concurrency and automata on infinite sequences", in *Theoretical Computer Science*, ed. P. Deussen, LNCS 104, Springer-Verlag, 1981, 167-183.

[9] D. Perrin and J. E. Pin, *Mots Infinis*, Report LITP 93.40, 1993.

[10] A. Salomaa, "Formal Languages", Academic Press, 1973.

[11] W. Thomas, "Automata on Infinite Objects", in *Handbook of Theoretical Computer Science*, Volume B, ed. J. van Leeuwen, Elsevier, 1990, 135-191.

Constructing Sequential Bijections

Christophe Prieur[1], Christian Choffrut[1] and Michel Latteux[2]

[1] Université Paris 7, LITP, 2 Pl. Jussieu,
72 251 Paris Cedex 05, France
[2] Université de Lille1, LIFL, bat. M3, Cité Scientifique,
59 655 Villeneuve d'Ascq Cedex, France

Abstract. We state a simple condition on a rational subset X of a free monoid B^* for the existence of a sequential function that is a one-to-one mapping of some free monoid A^* onto X. As a by-product we obtain new sequential bijections of a free monoid onto another.

1 Introduction

The starting point of the present paper is a simple question. Given two free monoid A^* (the "input" monoid) and B^* (the "output" monoid) respectively generated by m and n elements, design an effective function that is as elementary as possible and that maps bijectively A^* onto B^*. A natural solution uses the free monoid over m (resp. n) letters as a representation set of the integers in base m (resp. n); the problem of the leading zeros can be solved by resorting to the so-called p-adic representation instead of the standard one. But actually we use a less elaborate computation model.

Provide each transition of a finite deterministic automaton on A with an element in the free monoid B^*. The deterministic transducer thus obtained associates with every "input" string an "output" string, i. e., it defines a sequential function from A^* to B^*. This model relies on a memory that is only finite and does not allow afterthoughts. Indeed, it preserves prefixes so that any decision met for input string u envolves all strings having u as a prefix. The property of prefix-preservation can be viewed as follows. Consider the input monoid as a m-ary tree, and label each branch with a string from the output alphabet in such a way that all strings from the output monoids appear exactly once as the label of some path starting from the root. If we further require that the tree be produced by a finite deterministic transducer, then it has only finitely many different subtrees.

The condition on m and n for the existence of a solution was given in [3] by using a decomposition of the free monoid akin to the coset decomposition of groups. Here we address the more general issue of designing, when possible, a sequential function that performs a one-to-one mapping of a free monoid onto a given rational subset of $X \subseteq B^*$. The techniques utilized here rely on the completion of "partial" solutions. Indeed, the interpretation of the problem in terms of a labelled tree still makes sense in this more general setting. It should be intuitive that the input alphabet is immaterial, what counts is that every node has exactly m successors. The labels assigned to the different branches of a given node can be rearranged among the children without violating the bijectivity. Now consider a finite graph with labels in B^* and assume each string in X

is the label of exactly one path. Unroll this graph so as to obtain a (not necessarily m-ary) tree. The techniques developped in this paper allow, under certain conditions, to make the tree uniform without modifying the global set of labels, i. e., to obtain global solutions from partial ones.

In order to state the main result, we need some technical definitions. Denote by 1 the neutral element of B^*. For all $u, v \in B^*$, u is a *prefix* of v if there exists $x \in B^*$ such that $v = ux$. Given a subset $X \subseteq B^*$ and an element $u \in B^*$, set $u^{-1}X = \{x \in B^* \mid ux \in X\}$ and say $u \in X$ has *branching degree* d (relative to X) if the set $u^{-1}X - \{1\}$ has d minimal elements relative to the relation "being a prefix of".

Theorem 1. *Given a rational subset $X \subseteq B^*$ with $1 \in X$, and an integer $d > 1$, assume the following conditions are satisfied*

i) all elements $u \in X$ have branching degree smaller than or equal to d.

ii) for all $u \in X$ there exists $v \in X$, u a prefix of v, such that v has branching degree greater than 1.

Then there exists a sequential function that maps bijectively a free monoid with d generators onto X.

This result extends that of [3] in that it substitutes an arbitrary rational subset X for the free monoid B^*. It has a further interesting consequence taking us back to the initial question. If the complement of X (to which the neutral element 1 is added, to be more precise) satisfies the two conditions of the theorem too, then there exists a natural bijection from a free monoid onto B^* that is built by merging the two sequential functions for X and its complement in some natural way. In particular, this approach shows that suffix codes are not a necessary ingredient for sequential bijections of a free monoid onto another, as was implicitely assumed in [3]. But in spite of providing new and more elaborate solutions for the original problem, the present method fails, at least as exposed here, to recover all solutions given in [3].

Before tackling the problem, let us mention some related problems mainly concerned with the more powerful model of rational functions. Using slightly different definitions of sequential mappings (in fact introducing final states), algorithms for determining under which conditions there exists a sequential function mapping a given rational language onto another have been devised, cf., [7] and [9]. In [6] the same problem is considered where bijective rational transductions are used in place of sequential functions.

2 Preliminaries

We refer to [4], [2] and [8] for a more detailed exposition of the main notions briefly recalled here.

2.1 Free monoids

Given a set A (an *alphabet*) consisting of *letters*, A^* denotes the free monoid it generates. The elements of A^* are *words* or *strings*. The *length* of a word w is denoted by $|w|$. The *empty* word, denoted by 1, is the word of length 0 and A^+ denotes the set of non empty words.

2.2 Sequential functions

The notion of sequential functions of a free monoid into another is standard. We recall it for the sake of completeness. Intuitively, these mappings are realized by finite deterministic automata where all transitions are provided with an output.

Given two alphabets A and B, a *sequential transducer* $\mathcal{T}$ is a finite deterministic automaton whose states are all accepting, equipped with an *output function* from the set of transitions into B^*. Formally, we have $\mathcal{T} = (Q, i, \lambda)$ where Q is the set of states and i the initial state; the *next state* function of the *underlying* automaton associates with all pairs (q, a) a state $q.a$ and the *output* function λ associates with all pairs $(q, a) \in Q \times A$ a word in $\lambda(q, a) \in B^*$. These two functions are extended to A^* by induction on the length of the words:

i) for all $q \in Q$, $q.1 = q$
ii) for all $q \in Q, a \in A$, $u \in A^*$, $q.ua = (q.u).a$
iii) for all $q \in Q$, $\lambda(q, 1) = 1$
iv) for all $q \in Q$, $a \in A$, $u \in A^*$, $\lambda(q, ua) = \lambda(q, u)\lambda(q.u, a)$

The cardinality of Q is the *dimension* of the transducer. A function $f : A^* \to B^*$ is *sequential* if there exists a sequential transducer $\mathcal{T}$ such that $f(u) = \lambda(i, u)$ holds for all $u \in A^*$. Observe that all states are final and thus the function is total. We say the function f is *realized* by the transducer $\mathcal{T}$.

The present paper continues the study of the bijective sequential functions of a free monoid into another initiated in [3].

2.3 Rational series

The notion of power series is crucial, since the study of the sequential bijections can be reduced to it via a standard construction.

A $\mathbb{N}$ *-power series* in the *indeterminates* A is a function of A^* into the semiring of non negative integers $\mathbb{N}$. Such a series $\sigma : A^* \to \mathbb{N}$ is usually denoted as a formal sum

$$\sigma = \sum_{u \in A^*} \sigma(u)u$$

where $\sigma(u)$ is the *coefficient* of $u \in A^*$. The set $\mathbb{N}\langle\langle A^*\rangle\rangle$ of power series is provided with a structure of $*$-semiring by defining the addition, the product and the star operations. Indeed, if $\sigma_1, \sigma_2 \in \mathbb{N}\langle\langle A^*\rangle\rangle$ are two power series the sum and the product are defined by

$$(\sigma_1 + \sigma_2)(u) = \sigma_1(u) + \sigma_2(u),$$
$$(\sigma_1\sigma_2)(u) = \sum_{u = u_1 u_2} \sigma_1(u_1)\sigma_2(u_2),$$

for all $u \in A^*$. The star of a power series σ with $\sigma(1) = 0$ is defined by

$$\sigma^*(u) = 1 + \sum_{n>0, u = u_1 \ldots u_n} \sigma(u_1) \ldots \sigma(u_n)$$

for all $u \in A^*$. In particular a $\mathbb{N}$ -power series is *unambiguous* if its coefficients are 0 or 1.

The *support* of a power series $\sigma \in \mathbb{N}\langle\langle A^*\rangle\rangle$ is the subset consisting of all the elements $u \in A^*$ such that $\sigma(u) \neq 0$. In particular, an unambiguous series is the *characteristic* series of its support. The subsemiring of *polynomials* is the subset of all series having finite support.

The family of $\mathbb{N}$ *-rational* series, denoted by $\mathbb{N}$ -RatA^*, is the least family of series containing the polynomials and closed under addition, product and star.

Closely related to the notion of rational series is that of finite automata. For our purpose, we rely on a generalization of the notion of finite automaton. We call *generalized* automaton a triple $\mathcal{A} = (Q, i, D)$ where Q is the set of *states*, i is the *initial* state and $D \subseteq Q \times A^+ \times Q$ is a finite set of *transitions*, cf. [5]. Given a transition (q, u, q'), we say it *leaves* state q and *enters* state q'. The *degree* of state q, denoted by $\deg_{\mathcal{A}}(q)$ or simply $\deg(q)$ when the automaton $\mathcal{A}$ is understood, is the number of transitions leaving it. The automaton is *uniform* if all states have same degree. A *path* of *length* k in a generalized automaton is a sequence of k transitions of the form $(q_0, u_0, q_1), (q_1, u_1, q_2), \ldots, (q_{k-2}, u_{k-2}, q_{k-1})\ (q_{k-1}, u_{k-1}, q_k)$ where $(q_{i-1}, u_{i-1}, q_i) \in D$ for $i = 1, \ldots, k$. When $k = 1$ and $q_0 = q_k$, the path is a *loop*.

The *behaviour* of (or the series *recognized* by) $\mathcal{A}$ is the $\mathbb{N}$ -power series, denoted by $||\mathcal{A}||$, that assigns to every string $u \in A^*$ the number of paths starting in i and labelled with u. Observe that in the standard terminology we would say that all states are final. As follows from Kleene's Theorem (see e.g., [4], Theorem VII, 5. 1), $||\mathcal{A}||$ is a $\mathbb{N}$ -rational power series. A generalized automaton is *non ambiguous*, if $||\mathcal{A}||$ has coefficients 0 and 1. Two generalized automata are *equivalent* if they have the same behaviour.

3 Systems of equations

As said in the introduction, the present paper is concerned with sequential functions of a free monoid into another that are bijective. Let $\mathcal{T}$ be a sequential transducer realizing a sequential mapping of $f : A^* \to B^*$. Since there is a one to one correspondence between the words of A^* and the paths in the underlying automaton of $\mathcal{T}$, saying that f is a bijection is equivalent to saying that the behaviour of the generalized automaton $\mathcal{A}$ obtained from $\mathcal{T}$ by ripping off the input letter from every transition is the characteristic series of B^*.

From now on, we study sequential bijection in terms of generalized automata. Actually, it is customary to associate with such an automaton a system of equations in the variables $X_q \in \mathbb{N}\langle\langle B^*\rangle\rangle$, for each $q \in Q$, by setting

$$X_q = 1 + \sum_{q' \in Q} \alpha_{q,q'} X_{q'} \tag{1}$$

where $\alpha_{q,q'} = \sum\{u \in B^* \mid (q, u, q') \in D\}$.

It is well known that such a system has a solution and that this solution is unique if the support of the polynomials $\alpha_{q,q'}$ does not contain the empty element 1. In that case, the value of the component X_i (where i is the initial state) is the behaviour of the automaton $\mathcal{A}$, see [4] Prop. VII, 6.1.

Conversely, assume that the $\alpha_{q,q'}$'s in system (1) are polynomials with non negative integer coefficients and that for each $q \in Q$, the sum of the coefficients of $\sum_{q' \in Q} \alpha_{q,q'}$ equals the cardinality of A. If further the value of the X_i-component is the characteristic series of B^* then the system can be interpreted as defining a sequential bijection of A^* onto B^* by arbitrarily distributing the monomials of $\sum_{q' \in Q} \alpha_{q,q'}$ for each $q \in Q$, to the letters of A.

Example 1. With $B = \{x, y, z\}$ we have

$$X_0 = 1 + (y + z)X_0 + (x + xy)X_1$$
$$X_1 = 1 + zX_0 + (x + xy + yy)X_1$$

The sum of the coefficients of the 2 polynomials $y + z + x + xy$, and $z + x + xy + yy$ are equal to 4. Furthermore, the component X_0 is the characteristic series of $\{x, y, z\}^*$. Thus, the system gives rise to a bijective sequential mapping of the free monoid generated by $A = \{a, b, c, d\}$ onto B^*

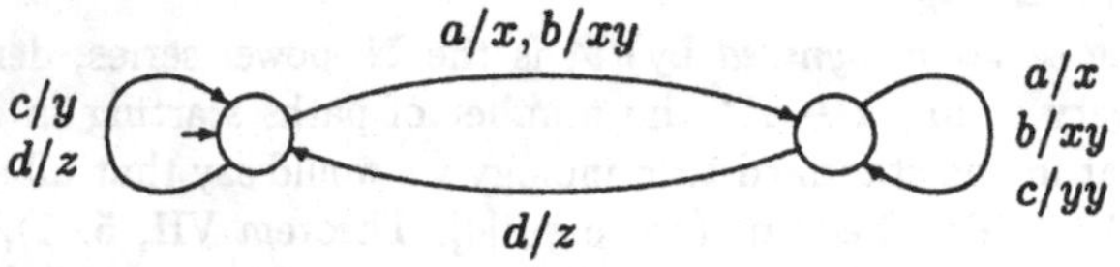

4 The coset construction

We recall here briefly a previous construction of a bijective mapping from a free monoid A^* into another B^*, [3]. Denoting by $\#A$ the cardinality of A, a necessary condition for the existence of such a mapping is that $\#A > \#B > 1$ or $\#A = \#B$ as can be readily seen. It turns out to be sufficient. The solution reminds the decomposition of a group relative to its cosets since it starts from the equality $B^* = SZ^*$ (as series) where Z is a finite complete suffix code and S the set of its proper suffixes, cf. [1].

Theorem 2. *Let A and B be two alphabets of cardinality m and n respectively. Then there exists a bijective sequential mapping of A^* onto B^* if and only if $m > n > 1$ or $m = n$ holds. Furthermore, if the condition is satsified, there exists a mapping of dimension 2.*

Proof. (Sketch) Let p be an integer such that $k(n - p - 1) = m - p - 1$ holds for some integer $k > 1$. Decompose $B = B_1 \cup B_2$ with $\#B_1 = p$. Let $Z \subseteq B_2^*$ be a complete suffix code with $m - p$ elements and let S be the set of its proper suffixes that are different from the empty string. Let b be an arbitrary letter of B_2. Then the pair (B^*, Z^*) is the solution of the following system.

$$X_0 = 1 + (B_1 + b)X_0 + (S + Z - b(1 + S))X_1$$
$$X_1 = 1 + B_1X_0 + ZX_1$$

Example 2. Example 1 corresponds to $B_1 = \{z\}$, $B_2 = \{x, y\}$, $b = y$ and $Z = \{x, xy, yy\}$

5 Transformation rules

Here we establish some transformation rules that preserve the behaviour of an unambiguous generalized automaton and that allow, under some hypotheses, to convert a given non uniform generalized automaton into a uniform one. The final degree however, depends on the degrees of the initial automaton. These rules are in essence simple transformations of systems of equations in the algebra of power series. We shall apply this technique to any generalized automaton recognizing the characteristic series of B^* in the last section in order to get novel sequential bijections.

5.1 By-pass rule

This operation makes a first state by-pass a second state. It increases the degree of the first state, which is all the purpose, and creates a new state of degree equal to the second one minus one.

By-pass rule: given an unambiguous generalized automaton $\mathcal{A} = (Q, i, D)$ and the non necessarily distinct transitions $(q_0, u, q_1), (q_1, v, q_2) \in D$, perform the following operations

1. create a new state $q_3 \notin Q$
2. delete transition (q_0, u, q_1)
3. add transitions (q_0, u, q_3), (q_0, uv, q_2)
4. for every $(q_1, w, p) \in D$ such that $w \neq v$, add the transition (q_3, w, p)

Proposition 3. *The by-pass rule tranforms an unambiguous generalized automaton $\mathcal{A}$ into an equivalent generalized automaton $\mathcal{A}'$.*

The degrees of the states are preserved except for the conditions $deg_{\mathcal{A}'}(q_3) = deg_{\mathcal{A}}(q_1) - 1$ and $deg_{\mathcal{A}'}(q_0) = deg_{\mathcal{A}}(q_0) + 1$.

Proof. Assign a system of equations with $\mathcal{A}$ as in section 3. If q_0 and q_1 are equal introduce a new variable and duplicate the equation associated with $q_0 = q_1$. Rename the indices of the variables X_j of the system as $\{0, 1, \ldots, n-1\}$ so that X_0, X_1 are associated with q_0 and q_1. Finally, let X_k, for some $0 \leq k < n$ be associated with q_2. This process ends up with a system of n equations in the n unknowns $X_j \in \mathbb{N}\langle\langle A^* \rangle\rangle$ where X_0, X_1 and X_k are associated with q_0 and q_1, q_2 respectively.

$$X_j = E_j \text{ for } 0 \leq j < n \tag{2}$$

By hypothesis, there exist $F_0, F_1 \in \mathbb{N}\langle\langle A^* \rangle\rangle$ such that

$$\begin{aligned} E_0 &= uX_1 + F_0 \\ E_1 &= vX_k + F_1 \end{aligned}$$

holds. The by-pass rule consists in introducing a new unknown X_n and the new system of equations

$$X_j = \begin{cases} uX_n + uvX_k + F_0 & \text{if } j = 0 \\ E_j & \text{if } j \neq 0, n \\ F_1 & \text{if } j = n \end{cases} \tag{3}$$

It suffices to prove that systems (2) and (3) are equivalent. Indeed, for every solution $(X_j)_{0\le j<n}$ of system (2), there is a X_n such that $(X_j)_{0\le j\le n}$ is a solution of system (3). The only unknown for which this requires a verification is X_0

$$\begin{aligned} X_0 = E_0 = uX_1 + F_0 &= u(vX_k + F_1) + F_0 \\ &= uvX_k + uX_n + F_0 \end{aligned}$$

Conversely, for every solution $(X_j)_{0\le j\le n}$ of system (3), $(X_j)_{0\le j<n}$ is a solution of system (2). Again, only X_0 needs verification

$$\begin{aligned} X_0 = uX_n + uvX_k + F_0 &= uF_1 + uvX_k + F_0 \\ &= u(F_1 + vX_k) + F_0 = uX_1 + F_0 = E_0 \end{aligned}$$

5.2 Pumping rule

The purpose of this operation is to use the existence of a loop in a first state in order to "pump" new transitions for the second state. It increases the degree of the first state by one.

Pumping rule: given an unambiguous generalized automaton $\mathcal{A} = (Q, D, i)$ and the non necessarily distinct transitions $(q_0, u, q_1), (q_1, s, q_1), (q_1, v, q_2) \in D$ with $(s, q_1) \neq (v, q_2)$, perform the following operations

1. create a new state $q_3 \notin Q$
2. delete transition (q_0, u, q_1)
3. add transitions (q_0, u, q_3), (q_0, uv, q_2)
4. add transitions (q_3, s, q_3) and (q_3, sv, q_2) and for every $(q_1, w, p) \in D$ such that $w \neq v$ and $w \neq s$ add the transition (q_3, w, p)

Proposition 4. *The pumping rule tranforms an unambiguous generalized automaton $\mathcal{A}$ into an equivalent generalized automaton $\mathcal{A}'$ The degrees of the states are preserved except for the conditions $deg_{\mathcal{A}'}(q_3) = deg_{\mathcal{A}}(q_1)$ and $deg_{\mathcal{A}'}(q_0) = deg_{\mathcal{A}}(q_0) + 1$.*

Proof. Proceeding as for the by-pass rule by assigning a system of equations to $\mathcal{A}$ we obtain a system of n equations in the n unknowns $X_j \in \mathbb{N}\langle\langle \mathcal{A}^* \rangle\rangle$ where X_0, X_1 and X_k are associated with q_0 and q_1, q_2 respectively.

$$X_j = E_j \text{ for } 0 \le j < n \tag{4}$$

By hypothesis, since $(s, q_1) \neq (v, q_2)$ holds, there exist $F_0, F_1 \in \mathbb{N}\langle\langle \mathcal{A}^* \rangle\rangle$ such that

$$\begin{aligned} E_0 &= uX_1 + F_0 \\ E_1 &= sX_1 + vX_k + F_1 \end{aligned}$$

holds. The pumping rule consists in introducing a new unknown X_n and the new system of equations

$$X_j = \begin{cases} uX_n + uvX_k + F_0 & \text{if } j = 0 \\ E_j & \text{if } j \neq 0, n \\ sX_n + svX_k + F_1 & \text{if } j = n \end{cases} \tag{5}$$

Every solution $(X_j)_{0\le j<n}$ of system (4), there is a X_n such that $(X_j)_{0\le j\le n}$ is a solution of system (5). Indeed, observe that unknown X_1 in system (4) satisfies

$$X_1 = s^*(vX_k + F_1)$$

and unknown X_n in system (5) satisfies

$$X_n = s^*(svX_k + F_1)$$

i. e., $sX_1 + F_1 = X_n$. Now compute X_0 in system (4)

$$\begin{aligned} X_0 = E_0 = uX_1 + F_0 &= u(sX_1 + vX_k + F_1) + F_0 \\ &= uX_n + uvX_k + F_0 \end{aligned}$$

Conversely, for every solution $(X_j)_{0\le j\le n}$ of system (5), $(X_j)_{0\le j<n}$ is a solution of system (4). Again, only X_0 needs verification

$$\begin{aligned} X_0 &= uX_n + uvX_k + F_0 = us^*(svX_k + F_1) + uvX_k + F_0 \\ &= us^*vX_k + us^*F_1 + F_0 = uX_1 + F_0 = E_0 \end{aligned}$$

6 Uniformizing an unambiguous generalized automaton

Uniformizing an unambiguous generalized automaton without modifying its behaviour, i.e., making all its nodes have the same degree, whatever the degree, is not always possible. We shall give sufficient conditions for achieving this goal. We first need a few definitions.

Given an unambiguous generalized automaton $\mathcal{A} = (Q, i, D)$ a subset $\emptyset \neq Q' \subseteq Q$ is ***stable*** if for any state $q \in Q'$ and any transition $(q, u, q') \in D$, $q' \in Q'$ holds. It is a ***minimal stable*** subset if it is stable and if it does not properly contain a stable subset. Clearly, for every state there exists a path from it to some minimal stable subset. Also, any minimal stable subset Q' is strongly connected in the sense that for all $q, q' \in Q'$ there exists a path from q to q'.

An unambiguous generalized automaton $\mathcal{A} = (Q, i, D)$ is ***uniform of degree d*** if all states q have degree d. It has ***limit degree*** $d \le 1$ if all states of some minimal stable subset have degree $d \le 1$. Otherwise it has limit degree $d = \max\{\deg(q) \mid q \in Q\}$.

The following procedure transforms an unambiguous generalized automaton $\mathcal{A}$ with limit degree $d > 1$ into an equivalent generalized automaton $\mathcal{A}'$ (i. e., $||\mathcal{A}|| = ||\mathcal{A}'||$) that is uniform of degree d' for some arbitrary $d' \ge d$.

1. Let Q_i, $i = 1, \ldots, r$ be the minimal stable subsets of the automaton. Identify in each Q_i a state ℓ_i as follows. If there exists $\ell \in Q_i$ with a loop, set $\ell_i = \ell$. Otherwise, by hypothesis on the limit degree, there exists an element $\ell \in Q_i$ and a cycle $\ell = q_0, q_1, \ldots, q_k = q_0$ with $\deg(q_{k-1}) > 1$. Apply repeatedly the by-pass rule backwards to the transitions $(q_{k-2}, u_{k-2}, q_{k-1})$, (q_{k-1}, u_{k-1}, q_k) for some $u_{k-2}, u_{k-1} \in B^*$, then to the transitions $(q_{k-3}, u_{k-3}, q_{k-2})$, $(q_{k-2}, u_{k-2}u_{k-1}, q_k)$ for some $u_{k-3} \in B^*$, ... then to the transitions (q_0, u_0, q_1), $(q_1, u_1 \ldots u_{k-2}u_{k-1}, q_k)$ for some $u_0 \in B^*$, cf. example 3. Then set $\ell_i = \ell$. In all cases we end up with a state ℓ_i having a transition (ℓ_i, u, ℓ_i).

2. If one of the above states ℓ_i does not have degree d', increase its degree by applying the by-pass rule to the transitions (ℓ_i, u, ℓ_i), (ℓ_i, u, ℓ_i), as many times as needed.
3. Whenever there exists a state with degree less than d', choose any such state q_0 with minimal distance to a state q with a loop and degree d'.
   i) if the distance is one, then apply the pumping rule to the states q_0, q.
   ii) if the distance is greater than one, let q_1 and q_2 be the first and second states along a path of minimal length from q_0 to q. Let (q_0, u, q_1), (q_1, w, q_2), $(q_1, v, q_3) \in D$ with $(q_1, v, q_3) \neq (q_1, w, q_2)$. Then apply the by-pass rule to the transitions $(q_0, u, q_1), (q_1, v, q_3)$.

Proposition 5. *The previous procedure transforms an unambiguous generalized automaton $\mathcal{A}$ with limit degree $d > 1$ into an equivalent generalized automaton $\mathcal{A}'$ that is uniform with degree d'.*

Proof. Fix a state $\ell = \ell_i$. Consider step 1 and use the same notations as in the above procedure. The by-pass rule applied to the transitions $(q_{k-i-1}, u_{k-i-1}, q_{k-i})$ and $(q_{k-i}, u_{k-i} \dots u_{k-1}, q_k)$ erases the only transition $(q_{k-i-1}, u_{k-i-1}, q_{k-i})$ and creates the transition $(q_{k-i-1}, u_{k-i-1}, q_k)$. Assume by contradiction that some state q that could reach ℓ before the application of the rule no longer reaches it afterwards. Then there existed a path from q to ℓ that used transition $(q_{k-i-1}, u_{k-i-1}, q_{k-i})$ exactly once. But now q_{k-i-1} is directedly connected to ℓ so that the former path can take the shortcut $(q_{k-i-1}, u_{k-i-1}, \ell)$. As for the new created state, say p, since q_{k-i} has degree at least 2, there is a transition (q_{k-i}, v, q') with $(v, q') \neq (u_{k-i} \dots u_{k-1}, q_k)$ so that (p, v, q') is created. Thus there exists a path from p to q' and from q' to ℓ. The same assertion holds for step 2 since no transition is erased.

Thus, before step 3, each state can reach a state of degree d' having a loop. Consider the quantity

$$E = \sum_{q \in Q} (d' - \deg(q)) \lambda_q$$

where $\deg(q)$ is the degree of $q \in Q$ and λ_q the minimal distance of q to a state with a loop and degree d'. When $\lambda_q = 1$, then each application of i) decrements the quantity E by one. When $\lambda_q > 1$, then ii) produces a new state of degree $d' - 1$ at distance $\lambda_q - 1$ of q and reduces the quantity $\deg(q)$ by one. Thus, E decreases in all cases by one.

Example 3 (Creating a loop in state 1).

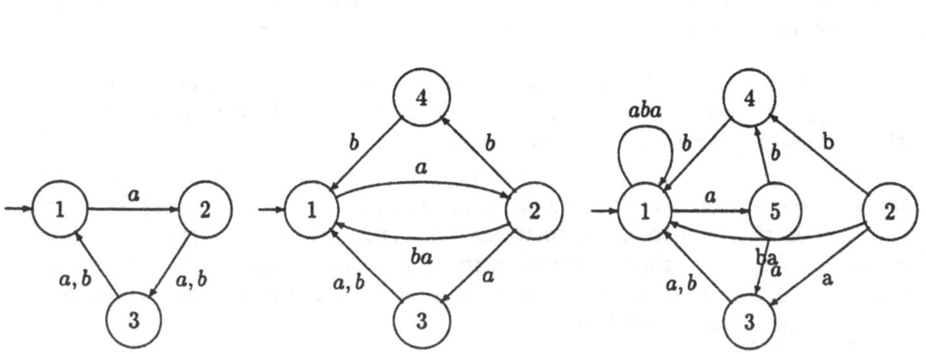

Example 4 (Application of the by-pass and the pumping rules). Starting with the generalized automaton reduced to one state 0 and having the only transitions $(0, a, 0)$ and $(0, b, 0)$ that recognizes $\{a, b\}^*$ we obtain in 3 steps a uniform generalized automaton of degree 3.

Step 1: by-pass of state 0 by itself, with transitions $(0, a, 0)$, $(0, a, 0)$

Step 2: pumping of state 0 from state 1, with transitions $(1, b, 0)$, $(0, b, 0)$, $(0, a^2, 0)$

Step 3: pumping of state 0 from state 1, with transitions $(1, ba^2, 0)$, $(0, b, 0)$, $(0, a^2, 0)$

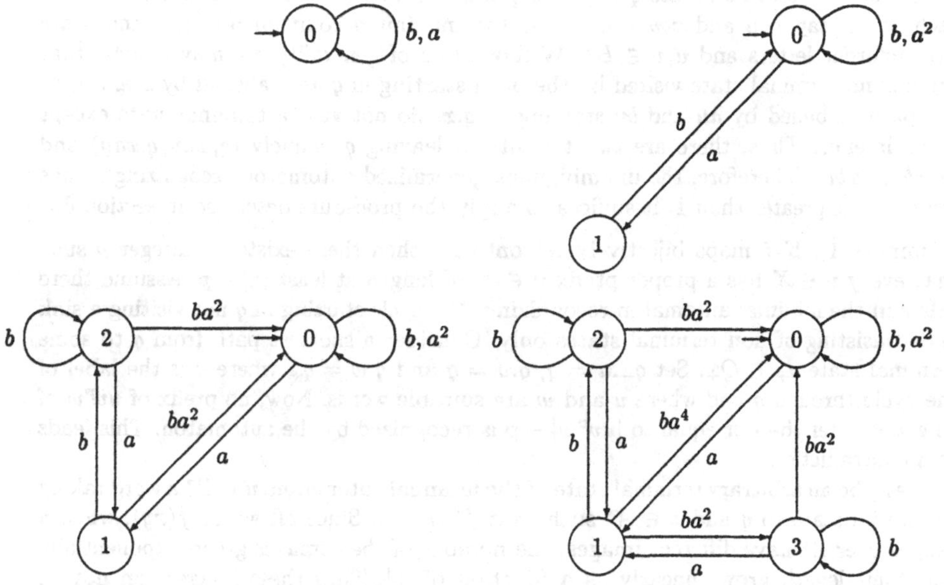

7 Applications to sequential bijections

In this last section we generalize the example of the previous section by showing for a fixed rational subset $X \subseteq B^*$ how to design, when possible, a bijective sequential mapping of some free monoid A^* onto X .

We recall that a state q of a finite deterministic automaton (in the usual sense), is a *sink* whenever $q = q.a$ holds for all letters a of the alphabet.

Theorem 6. *Given a rational subset $X \subseteq B^*$ with $1 \in X$, the following conditions are equivalent*

1. the minimal automaton of X satisfies the two conditions

i) *every cycle which does not visit a sink contains a terminal state*

ii) *for every terminal state there are two paths whose labels are not a prefix of one another that lead from this state to some terminal states.*

2. *there exists an alphabet A of cardinality greater than* 1 *and a sequential bijection of $f : A^* \to B^*$ that maps A^* onto X.*

Proof. 1) implies 2). Indeed, let $\mathcal{A} = (Q, q_-, Q_+)$ be the minimal automaton (in the usual sense) recognizing X where q_- stands for the initial state and Q_+ for the set of final states. The condition of the theorem is equivalent to saying that the maximal length of a path connecting two final states without visiting a final state inbetween, is bounded by the cardinality of the set of states. An unambiguous generalized automaton recognizing X is obtained as follows. The set of states is Q_+. The set of transitions is D where $D \subseteq Q_+ \times B^+ \times Q_+$ with $(q_0, u, q_1) \in D$ if and only if u is the label of a path from q_0 to q_1 in $\mathcal{A}$ that does not visit a terminal state between the two end states.

Consider a terminal state q and two paths labelled by two words, not a prefix of each other, say xau and xbv where x is the maximum common prefix, a and b are two different letters and $u, v \in B^*$. Without loss of generality, we may assume that there is no terminal state visited by the path starting in q and labelled by x and that the paths labelled by au and bv starting in $q.x$, do not visit a terminal state except at their ends. Thus, there are two transitions leaving q, namely $(q, xau, q.xau)$ and $(q, xbv, q.xbv)$. Therefore, the unambiguous generalized automaton recognizing X has limit degree greater than 1. It suffices to apply the procedure described in section 6.

2) implies 1). If f maps bijectively A^* onto X, then there exists an integer p such that every $u \in X$ has a proper prefix $v \in X$ of length at least $|u| - p$. Assume there exists in the minimal automaton recognizing X a cycle starting in q not visiting a sink and consisting of non terminal states only. Consider a shortest path from q to some terminal state $q_+ \in Q_+$. Set $q_-.u = q$, $q.v = q$ and $q.w = q_+$ where v is the label of the cycle through q and where u and w are suitable words. Now, no prefix of uv^pw of length greater than or equal to $|uv^pw| - p$ is recognized by the automaton. This leads to a contradiction.

Let q be an arbitrary terminal state of the minimal automaton, $u \in B^*$ a word taking the initial state to q and $x \in A^*$ such that $f(x) = u$. Since all words $f(xy)$, when y ranges over A^* have different images, the number of their images grows exponentially and their length grows linearly as a function of $|y|$. Thus these images can not be pairwise comparable relative to the relation "being a prefix of" which shows that there are among them two words that are not prefix of one another.

As a consequence, if a rational subset $X \subseteq B^*$ and $B^* - X \cup \{1\}$ satisfy the conditions of the previous theorem, then we proceed as follows. We add a new initial state with no entering transition to both unambiguous generalized automata and we make the union of two copies of them with the two initial states merged. Then we apply the uniformization procedure of section (6) in order to get a sequential bijection of a free monoid into B^*. The converse uses an analog argument as that of the previous theorem, thus we have

Theorem 7. *Given a rational subset $X \subseteq B^*$ with $1 \in X$, the following conditions are equivalent*

1. *the minimal automaton of X satisfies the two conditions*

   i) *every cycle which does not visit a sink contains a terminal and a non terminal state*

ii) *for every terminal (resp. non terminal) state there are two paths whose labels are not a prefix of one another that lead from this state to some terminal (resp. non terminal) states.*

2. *there exists an alphabet A of cardinality greater than* 1, *a decomposition $A = A_1 \cup A_2$ and a sequential bijection of $f : A^* \to B^*$ that maps $1 + A_1A^*$ onto X, and therefore A_2A^* onto $B^* - X$.*

Let us illustrate the theorem with the language $R = 1 + (xx + xy + yxx + yxy)^*$. We first construct the "canonical" generalized automaton recognizing R (the one obtained naturally from the minimal automaton as in the proof of theorem 6) then the canonical generalized automaton recognizing $A^* - R + 1$. Then we make the union of the two

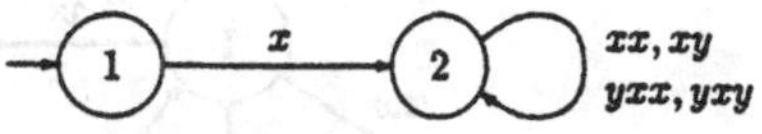

Generalized automaton of R

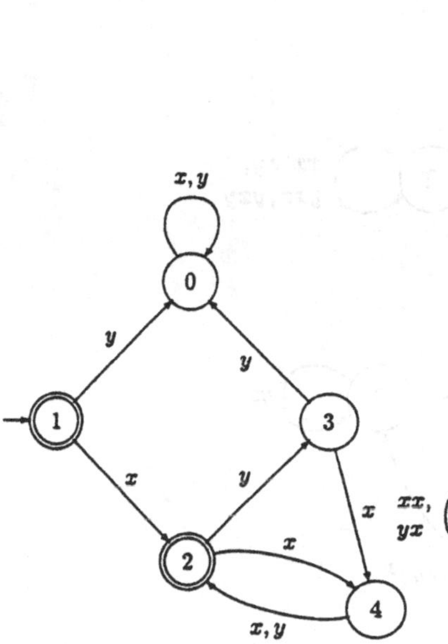

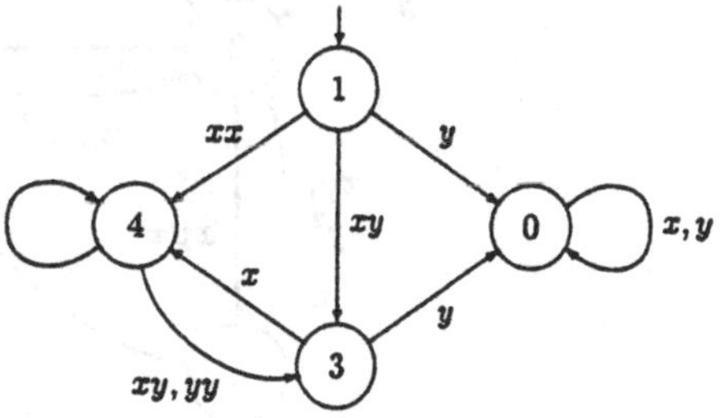

Deterministic complete automaton of R

Generalized automaton of $A^* - R + 1$

We obtain a new automaton by merging the two initial states and by identifying state 0 of $A^* - R + 1$ with the new initial state.

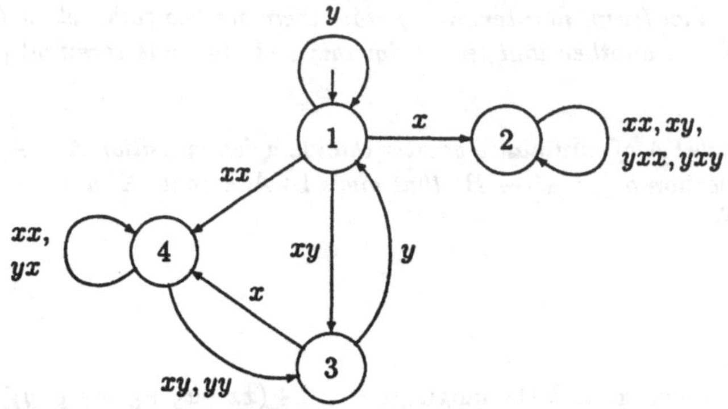

Applying the procedure of section 6 yields the final generalized wchich can be made into a bijection of a 4-generator onto a 2-generator free monoid.

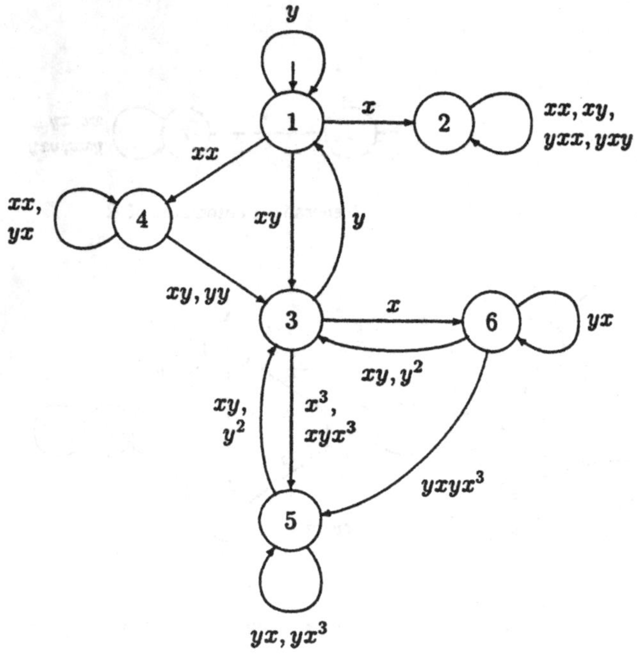

References

1. J. Berstel and D. Perrin. *Theory of Codes*. Academic Press, 1985.
2. J. Berstel and C. Reutenauer. *Rational Series and Their Languages*, volume 12 of *EATCS Monograph on Theoretical Computer Science*. Academic Press, 1988.
3. C. Choffrut. Bijective sequential mappings of a free monoid onto another. *RAIRO Informatique Théorique et Applications*, 28:265–276, 1994.
4. S. Eilenberg. *Automata, Languages and Machines*, volume A. Academic Press, 1974.
5. K. Hashiguchi. Algorithms for determining the number of non-terminals sufficient for generating a regular language. In B. Monien J. Leach Albert and M. Rodríguez Artalejo, editors, *ICALP 91*, number 510 in LNCS, pages 641–648. Springer Verlag, 1991.
6. H. A. Maurer and M. Nivat. Rational bijections of rational sets. *Acta Informatica*, 13:365–378, 1980.
7. R. Mac Naughton. A decision procedure for generalizd mappability-onto of regular sets. In *STOC Conference*, pages 206–218, 1971.
8. A. Salomaa and M. Soittola. *Automata-Theoretic Aspects of Formal Power Series*. Springer, 1978.
9. K. B. Salomon. The decidability of a mapping problem for generalized sequential machines with final states. *J. of Comput. and Sys. Sci.*, 10(2):200–218, 1975.

Rewriting Rules for Synchronization Languages

Kai Salomaa* and Sheng Yu**

Department of Computer Science
University of Western Ontario
London, Ontario, Canada N6A 5B7

Abstract. We study rewriting rules characterizing closure properties of synchronization languages. We introduce an extension of the syntactic definition of synchronization expressions, and an appropriate modification of their semantics. The extended definition has the advantage that it allows us to eliminate the less well motivated rewriting rules from the system under which the synchronization languages are closed. The modified system is shown to preserve regularity of the languages. We obtain a characterization of finite synchronization languages as the family consisting of languages satisfying the start-termination property and closed under three types of simple rewriting rules.

1 Introduction

Many theoretical models have been developed to formalize and analyse the concurrent behavior of distributed systems. Examples of such models are Petri nets, process algebra, and the theory of traces and its extensions. Among recent references we can list [1, 2, 4, 5, 7, 8, 9, 14, 18, 20]. Synchronization expressions were introduced in the ParC project [10, 11] as high level constructs for specifying synchronization constraints. Synchronization requests are specified as expressions of statement tags. The use of such specifications relieves programmers from the burden of imposing synchronization constraints with non-intuitive low-level constructs.

The synchronization languages [12, 13] define the semantics of synchronization expressions. Synchronization languages are a subfamily of regular languages that satisfy the *start-termination* property (st-property, for short), that is, the symbol denoting the start of a process b necessarily precedes the symbol denoting the termination of b which in turn precedes the symbol denoting the start of the next occurrence of b. However, clearly the st-property is not a sufficient condition for a regular language to be a synchronization language.

The synchronization languages over an alphabet Σ were shown to be closed under a set of rewriting rules $R(\Sigma)$. It was conjectured in [12, 13] that also the converse holds, namely that every regular st-language closed under $R(\Sigma)$ is a

* Research supported by the the Academy of Finland Grant 14018.
** Research supported by the Natural Sciences and Engineering Research Council of Canada Grant OGP0041630.

synchronization language. For two-letter alphabets the conjecture was proved by M. Clerbout, Y. Roos and I. Ryl [6]. Furthermore, in the same paper [6] they constructed a counter-example to the conjecture for general alphabets. It was observed in [13] that a drawback of the original formalism is that the rewriting system $R(\Sigma)$ needed for describing the closure properties of the synchronization language family does not even preserve the regularity of st-languages in general.

Here we extend the definition of synchronization expressions by allowing the operands of the join operation (shuffle) to have shared symbols. The semantic definition of the operation is then restricted to allow only shuffles of the operand words that satisfy the st-condition. Otherwise, the shuffle would not, in general, have any realistic interpretation. The new definition has the advantage that it allows us to eliminate from the system $R(\Sigma)$ rules of the form $a_t a_s b_t b_s \longrightarrow b_t b_s a_t a_s$ and their generalizations. The eliminated rules were, intuitively, the least well motivated rules of the system describing closure properties of the synchronization languages.

Using general properties of semi-commutations [5] we show that the restricted rewriting system R_Σ preserves regularity of arbitrary st-languages. The system $R(\Sigma)$ did not have this property. We show that a finite language L is a synchronization language if and only if L satisfies the st-property and L is closed under the rules of R_Σ. It is established that the new formalism has the projection property suggested in [6] as a necessary condition for obtaining a complete characterization of the synchronization languages. We describe a possible construction to extend the characterization result for the family of well-formed st-languages [12, 13].

2 Preliminaries

For more information on regular expressions and languages we refer the reader to [15, 19, 21]. A comprehensive presentation of string rewriting systems can be found in [3]. Here we briefly recall some definitions and results that will be needed in the following.

The cardinality of a finite set A is denoted $\#A$ and $\wp(A)$ is the power set of A. The set of positive integers is $\mathbb{N}$. Let Σ denote a finite alphabet. The set of finite words over Σ is Σ^* and λ denotes the empty word. Also, $\Sigma^+ = \Sigma^* - \{\lambda\}$. A Σ-language is any subset of Σ^*. The length of $w \in \Sigma^*$ is denoted $|w|$ and a singleton language $\{w\}$ is usually denoted simply by w. The set of symbols of Σ occurring in $w \in \Sigma^*$ is $\text{alph}(w)$ and the number of occurrences of $a \in \Sigma$ in a word $w \in \Sigma^*$ is denoted $\#_a(w)$. If A is a subset of Σ we denote by $\Pi_A : \Sigma^* \longrightarrow A^*$ the homomorphism (projection) defined by setting $\Pi_A(b) = b$ if $b \in A$ and $\Pi_A(b) = \lambda$ if $b \in \Sigma - A$.

The catenation of (Σ-)languages L_1 and L_2 is defined $L_1 \cdot L_2 = \{w \in \Sigma^* \mid (\exists v_i \in L_i, i = 1, 2)\ w = v_1 v_2\}$ and the catenation of n copies of $L \subseteq \Sigma^*$ is L^n ($n \geq 0$). Note that $L^0 = \{\lambda\}$. The Kleene star of the language L is $L^* = \cup_{i=0}^{\infty} L^i$. The family of regular (or rational) languages over Σ, $\text{REG}(\Sigma)$, is the smallest family of Σ-languages containing the empty language $\emptyset$, the elements of Σ and closed under (finite) union, catenation and the Kleene star operation. The set of *regular expressions* over the alphabet Σ, $\mathrm{reg}(\Sigma)$, is defined using the operations $+$, $\cdot$ and $*$ for, respectively, union, catenation and Kleene star. The language denoted by a regular expression $\alpha \in \mathrm{reg}(\Sigma)$ is $L(\alpha)$ and then $\mathrm{REG}(\Sigma) = \{L(\alpha) \mid \alpha \in \mathrm{reg}(\Sigma)\}$. When there is no danger of confusion we often denote also the language $L(\alpha)$ simply by α.

The *shuffle* of words $u, v \in \Sigma^*$ is the language $\omega(u,v) \subseteq \Sigma^*$ consisting of all words that can be written in the form $u_1v_1 \cdots u_nv_n$, $n \geq 0$, where $u = u_1 \cdots u_n$, $v = v_1 \cdots v_n$, $u_i, v_i \in \Sigma^*$, $i = 1, \ldots, n$. Note that without loss of generality we can assume that $u_i \neq \lambda$ when $i \neq 1$ and $v_i \neq \lambda$ when $i \neq n$. The shuffle operation is extended for languages in the natural way:

$$\omega(L_1, L_2) = [\bigcup_{w_1 \in L_1, w_2 \in L_2} \omega(w_1, w_2)\,], \quad L_1, L_2 \subseteq \Sigma^*.$$

A *string-rewriting system* (or Thue system) over Σ is a finite set R of rules $u \longrightarrow v$, $u, v \in \Sigma^*$. The rule $v \longrightarrow u$ is the *reversal* of a rule $u \longrightarrow v$, $u, v \in \Sigma^*$. The rules of R define the (single step) reduction relation $\longrightarrow_R \subseteq \Sigma^* \times \Sigma^*$ as follows. For $w_1, w_2 \in \Sigma^*$, $w_1 \longrightarrow_R w_2$ if and only if there exists a rule $u_1 \longrightarrow u_2 \in R$ and $r, s \in \Sigma^*$ such that $w_i = ru_is$, $i = 1, 2$. The *reduction relation* of R is the reflexive and transitive closure of $\longrightarrow_R$ and it is denoted $\longrightarrow_R^*$.

Let R be a rewriting system. For a language $L \subseteq \Sigma^*$ we denote $\Delta_R(L) = \{w \in \Sigma^* \mid (\exists w' \in L)\ w' \longrightarrow_R w\}$, and the *R-closure* of L is

$$\Delta_R^*(L) = \{w \in \Sigma^* \mid (\exists w' \in L)\ w' \longrightarrow_R^* w\}.$$

The language L is *closed under* R if $\Delta_R(L) \subseteq L$ which is equivalent to saying that $L = \Delta_R^*(L)$.

A *semi-commutation relation* on Σ (see, [5]) is an irreflexive subset $\theta \subset \Sigma \times \Sigma$, that is, θ is a subset of $\Sigma \times \Sigma - \{(a,a) \mid a \in \Sigma\}$. The semi-commutation system R_θ associated with θ is the set $\{ab \longrightarrow ba \mid (a,b) \in \theta\}$. Let θ be a semi-commutation on Σ and A a subset of Σ. The *non-commutation graph* of θ on A, $G(\theta, A)$, is the directed graph where the nodes are labeled by elements of A and there exists an edge from a to b if and only if $(a,b) \notin \theta$ $(a, b \in A,\ a \neq b)$. A word $w \in \Sigma^*$ is said to be *strongly connected* with respect to θ if the non-commutation graph of θ on $\mathrm{alph}(w)$ is strongly connected, that is, for any pair of distinct nodes n_1 and n_2 of $G(\theta, \mathrm{alph}(w))$ the graph has a directed path from n_1 to n_2.

We will use the following two results that are proved in section 12.6.3 of [5].

Proposition 1. [5] *Let θ be a semi-commutation on Σ. Assume that $L_1, L_2 \in \mathrm{REG}(\Sigma)$ are closed under R_θ. Then $\Delta_{R_\theta}^*(L_1 \cdot L_2)$ is regular.* □

Proposition 2. [5] *Let θ be a semi-commutation on Σ. We assume that $L \in \mathrm{REG}(\Sigma)$ is closed under R_θ and every word of L is strongly connected with respect to θ. Then $\Delta_{R_\theta}^*(L^*)$ is regular.* □

3 Synchronization Expressions and Start-Termination Languages

We define synchronization expressions following [12, 13]. However, we extend the definition by allowing arguments of the join operation to have shared symbols (corresponding to different occurrences of the same process). The semantics of the operation is then restricted in a suitable way in order to guarantee that the expression has a natural interpretation.

Definition 3. The set of *synchronization expressions* over Σ, $\mathrm{SE}(\Sigma)$, is the smallest subset of $(\Sigma \cup \{\phi, \rightarrow, \&, |, \|, *, (,)\})^*$ satisfying the following conditions.

(i) $\Sigma \cup \{\phi\} \subseteq \mathrm{SE}(\Sigma)$.
(ii) If $\alpha_1, \alpha_2 \in \mathrm{SE}(\Sigma)$ then $(\alpha_1 \diamond \alpha_2) \in \mathrm{SE}(\Sigma)$ for $\diamond \in \{\rightarrow, \&, |, \|\}$.
(iii) If $\alpha \in \mathrm{SE}(\Sigma)$ then $\alpha^* \in \mathrm{SE}(\Sigma)$.

The operators $\rightarrow$, $\|$, $|$, & and $*$ are called, respectively, the ***sequencing, join, selection, intersection,*** and ***repetition*** operators. The intuitive meaning will be clear from the semantic interpretation given below. For easier readability we usually omit the outermost parentheses of an expression. All the four binary operations will be associative, so when we are interested in the language denoted by the expression (as opposed to the expression itself) also other parentheses may be omitted.

For each symbol (or, process) $a \in \Sigma$ we associate symbols a_s and a_t to denote, respectively, the *start* and *termination* of the process. Also, we denote

$$\Sigma_s = \{a_s \mid a \in \Sigma\}, \quad \Sigma_t = \{a_t \mid a \in \Sigma\}.$$

For $a \in \Sigma$ we define a morphism $f_a : (\Sigma_s \cup \Sigma_t)^* \longrightarrow \{a_s, a_t\}^*$ by setting $f_a(a_s) = a_s$, $f_a(a_t) = a_t$ and $f_a(x) = \lambda$ if $x \notin \{a_s, a_t\}$. A word $w \in (\Sigma_s \cup \Sigma_t)^*$ is said to satisfy the *start-termination* condition (st-condition) if

$$(\forall a \in \Sigma)\ f_a(w) \in (a_s a_t)^*.$$

A word satisfying the above condition is said to be an *st-word*. A language L satisfies the st-condition if every word of L is an st-word, and then L is called an *st-language*. The set of all st-words over $\Sigma_s \cup \Sigma_t$ is denoted W_Σ^{st}.

Consecutive occurrences of symbols a_s and a_t in an st-word $w \in (\Sigma_s \cup \Sigma_t)^*$, $a \in \Sigma$, are called a *corresponding pair*. Thus in a word $u_1 a_s u_2 a_t u_3$, the denoted occurrences of a_s and a_t form a corresponding pair if and only if u_2 does not contain any symbols a_s or a_t.

For our definition of the semantics of synchronization expressions we need a restricted notion of shuffle where, for given st-words w_1, w_2, one allows only interleavings of w_1 and w_2 that produce an st-word. Similar notions of restricted shuffle are considered in [16] to model situations where it is necessary to ensure that some parts of distributed processes do not run concurrently. A very general method to define shuffle-like operations has recently been introduced in [17]. The below definition of st-shuffle could easily be presented within the formalism of trajectories considered there.

For $L_1, L_2 \subseteq (\Sigma_s \cup \Sigma_t)^*$ we define the st-*shuffle* of L_1 and L_2 by

$$\omega_{st}(L_1, L_2) = \omega(L_1, L_2) \cap W_{\Sigma}^{st}.$$

Note that if w_1, w_2 are st-words then $\omega_{st}(w_1, w_2)$ is always non-empty but otherwise the st-shuffle of given words can be empty. Furthermore, it is possible that $\omega_{st}(w_1, w_2) \neq \emptyset$ even if w_1 and/or w_2 is not an st-word.

Definition 4. For $\alpha \in \mathrm{SE}(\Sigma)$ we define the language denoted by α, $L(\alpha)$, inductively on the structure of α.

(i) $L(\phi) = \emptyset$ and $L(a) = \{a_s a_t\}$, $a \in \Sigma$.
(ii) $L(\alpha_1 \to \alpha_2) = L(\alpha_1) \cdot L(\alpha_2)$.
(iii) $L(\alpha_1 \parallel \alpha_2) = \omega_{st}(L(\alpha_1), L(\alpha_2))$.
(iv) $L(\alpha_1 \mid \alpha_2) = L(\alpha_1) \cup L(\alpha_2)$.
(v) $L(\alpha_1 \& \alpha_2) = L(\alpha_1) \cap L(\alpha_2)$.
(vi) $L(\alpha^*) = L(\alpha)^*$.

A language $L \subseteq (\Sigma_s \cup \Sigma_t)^*$ is a (Σ)-*synchronization language* if there exists $\alpha \in \mathrm{SE}(\Sigma)$ such that $L = L(\alpha)$. The family of synchronization languages is denoted

$$\mathcal{L}(\mathrm{SE}) = \{L(\alpha) \mid \alpha \in \mathrm{SE}(\Sigma) \text{ for some } \Sigma\}.$$

Note that $\mathcal{L}(\mathrm{SE})$ contains the empty word because $L(\phi^*) = \{\lambda\}$. The modifications to the definitions of [12, 13] are that in Definition 3 the arguments of the join operator need not be over disjoint alphabets and then, correspondingly, in Definition 4 (iii) the join operator is interpreted as the restricted st-shuffle. Allowing arbitrary shuffles of words having shared symbols would produce words that do not satisfy the st-condition. For instance, a word $a_s a_s a_t a_t$ does not have any meaningful interpretation: the process a cannot start again before the previous occurrence has terminated.

We can observe that if $\alpha \in \mathrm{SE}(\Sigma)$ is constructed so that the two arguments of any join operator do not contain shared symbols, then α is a synchronization expression as defined in [12, 13]. In this case, the language denoted by α according to Definition 4 is equal to the language denoted by α according to the definition of [12, 13]. Thus the new definition extends the family of synchronization languages. More importantly, the new definition has the advantage that it allows us to eliminate some unnatural closure properties of the language family.

When comparing Definitions 3 and 4 to the standard definition of regular expressions, we may ask whether the operator & is necessary, that is, whether the intersection operation could be represented using the other operations. Below we present a simple synchronization expression denoting a finite language that cannot be denoted by any expression not containing the intersection operator. Since it is natural to expect that a programmer may wish to use a conjunction of given synchronization conditions, in the definitions the explicit inclusion of the intersection operator is well motivated.

Lemma 5. *Let*

$$\alpha = ((a \to b) \parallel (c \to d)) \ \&\ ((a \to d) \parallel (b \parallel c)).$$

Then $L(\alpha)$ cannot be denoted by any synchronization expression that does not use the operator &.

Proof. Essentially the same example is given in [12]. The argument needed for the proof is now slightly different because with the new definition the join operators can have arbitrary arguments.

Consider the word $w = a_s c_s a_t b_s c_t d_s b_t d_t$. Assuming that $L(\alpha)$ is denoted by an expression β without the operator &, it follows that $w \in L(\beta') \subseteq L(\alpha)$ where β' is constructed using only the operators $\to$ and $\parallel$. (Since $L(\alpha)$ is finite, the expression β does not need to contain $*$, that is, the only possible $*$-subexpression in β is ϕ^* and these can be removed without changing the language denoted by the expression. Thus, β can be represented as a finite union of expressions β' as given above.) Since the word w is not the catenation of any non-empty st-words, it follows that β' cannot be of the form $\beta_1 \to \beta_2$, that is, the outermost operator cannot be sequencing. The expression β' contains exactly one occurrence of each of the symbols a, b, c and d. Using this observation, by a simple case analysis we conclude that β' also cannot be of the form $\beta_1 \parallel \beta_2$ except in the case where β_1 or β_2 equals to ϕ^*. □

The restriction in Definition 4 (iii) guarantees that we have, analogously with [12, 13], the following result. The straightforward proof uses structural induction on synchronization expressions.

Theorem 6. *Every synchronization language is an* st-*language.* □

Conversely, in the next section our aim is to present closure properties that guarantee that an st-language is a synchronization language.

4 Rewriting Rules

It is clear that if a Σ-synchronization language L contains a word $ua_s b_t v$ ($a, b \in \Sigma$, $a \neq b$, $u, v \in (\Sigma_s \cup \Sigma_t)^*$) then necessarily the denoted occurrences of a_s and b_t appear in different arguments of the join operator in a synchronization expression denoting L. This means that also the word $ub_t a_s v$ is in L. Similarly, it is observed that consecutive pairs of symbols (a_s, b_s) and (a_t, b_t) necessarily commute in any Σ-synchronization language, $a, b \in \Sigma$, $a \neq b$. On the other hand, it is clear that a subword of the form $a_t b_s$ cannot, in general, be replaced by $b_s a_t$. This is seen simply by considering the expression $a \to b$. By the above observations we see that every synchronization language is closed under a naturally defined semi-commutation system.

Definition 7. Let Σ be an alphabet. The semi-commutation relation $\theta_\Sigma \subseteq (\Sigma_s \cup \Sigma_t) \times (\Sigma_s \cup \Sigma_t)$ is defined to consist of the following pairs:

$$(a_s, b_t),\ (a_s, b_s),\ (a_t, b_t), \quad \text{where } a, b \in \Sigma, a \neq b.$$

We denote the semi-commutation system R_{θ_Σ} simply as R_Σ and it consists of the following rules, where $a, b \in \Sigma$, $a \neq b$:

(i) $a_s b_t \longrightarrow b_t a_s$,
(ii) $a_s b_s \longrightarrow b_s a_s$,
(iii) $a_t b_t \longrightarrow b_t a_t$.

Lemma 8. *If L is a Σ-synchronization language then L is closed under R_Σ.* □

When considering closure properties of synchronization languages our ultimate goal is to establish the converse of Lemma 8, namely that if a regular st-language is closed under R_Σ then it is a synchronization language. Such a characterization result would be very useful, for instance, for deciding whether the language accepted by a given finite automaton is a synchronization language. In many cases, it could be easy to verify whether a finite automaton accepts a language satisfying the simple semi-commutation properties of Definition 7.

The synchronization languages defined in [12, 13] were closed, in addition to the rules of R_Σ, under rules of the form

$$a_t a_s b_t b_s \longrightarrow b_t b_s a_t a_s, \quad a \neq b, \tag{1}$$

and their generalizations. We denote the extended set of rewriting rules by $R(\Sigma)$. It was shown in [6] that the rewriting system $R(\Sigma)$ did not give a characterization of the family of synchronization languages, except in the two-letter alphabet case. By proving a "switching property" for synchronization languages, [6] established that the $R(\Sigma)$-closure, $\Sigma = \{a, b, c\}$, of the language

$$b_s(a_s a_t)^* c_s b_t (a_s a_t)^* c_t \tag{2}$$

is not a synchronization language. The intuitive reason for the result is that in the original formalism of [12, 13] there is no way to differentiate between instances of the process a that occur, respectively, during b and during c.

It was briefly suggested in [6] that the difficulty could be fixed by further extending the rewriting system $R(\Sigma)$. We believe that such an approach would necessarily lead to very cumbersome definitions and the necessary rewriting system would probably be infinite. As an alternative solution to the problem we have here extended the syntactic definition of synchronization expressions which allows us to eliminate the more unnatural rules (1) from the rewriting system. Using our new definition, one may observe that the R_Σ-closure of the language (2) is denoted by the synchronization expression

$$(b \rightarrow a^*) \parallel (a^* \rightarrow c).$$

A major advantage of the new formalism, as opposed to the synchronization languages considered in [12, 13], is that the rewriting rules of R_Σ preserve the regularity of an arbitrary st-language.

Theorem 9. *Let $L \subseteq W_{\Sigma}^{st}$. If L is regular then $\Delta^*_{R_\Sigma}(L)$ is regular.*

Proof. We use induction on the structure of a regular expression α over $\Sigma_s \cup \Sigma_t$ denoting the language L. Since the subexpressions of α do not necessarily denote st-languages it is expedient to prove a somewhat stronger claim.

We say that $\alpha \in \text{reg}(\Sigma_s \cup \Sigma_t)$ has the *Σ-pair property* if always when β^* is a subexpression of α the following holds:

$$(\forall a \in \Sigma)(\forall w \in L(\beta)) \qquad \#_{a_s}(w) = \#_{a_t}(w) = 0 \\ \text{or } (\#_{a_s}(w) \neq 0 \text{ and } \#_{a_t}(w) \neq 0). \tag{3}$$

Clearly if α has the Σ-pair property then the same is true for all subexpressions of α. Note that if $L(\alpha)$ is an st-language then α necessarily has the Σ-pair property. This follows from the observation that by taking $n \geq 2$ iterations of a word violating the condition (3) we always obtain a word that cannot be a subword of any st-word.

Thus it is sufficient to show that the language $\Delta^*_{R_\Sigma}(L(\alpha))$ is regular always when $\alpha \in \text{reg}(\Sigma_s \cup \Sigma_t)$ has the Σ-pair property. The claim clearly holds for finite languages. Assume that for $L_i = L(\alpha_i)$, $\alpha_i \in \text{reg}(\Sigma_s \cup \Sigma_t)$, $i = 1, 2$, the languages $\Delta^*_{R_\Sigma}(L_i)$, $i = 1, 2$, are regular. We consider the cases where α is obtained from α_1 and α_2 by union, catenation or Kleene star. Firstly, $\Delta^*_{R_\Sigma}(L_1 \cup L_2) = \Delta^*_{R_\Sigma}(L_1) \cup \Delta^*_{R_\Sigma}(L_2)$ is regular. Secondly, $\Delta^*_{R_\Sigma}(L_1 \cdot L_2) = \Delta^*_{R_\Sigma}(\Delta^*_{R_\Sigma}(L_1) \cdot \Delta^*_{R_\Sigma}(L_2))$ is regular by Proposition 1.

Assume now that $\alpha = \alpha_1^*$. Since α has the Σ-pair property it follows that for any $a \in \Sigma$, an arbitrary word $w \in L(\alpha_1)$ either contains both symbols a_s and a_t or contains neither one of them. Thus, clearly an arbitrary word of $L(\alpha_1)$ (and of $\Delta^*_{R_\Sigma}(L(\alpha_1))$) is strongly connected with respect to θ_Σ. As an example, the non-commutation graph of θ_Σ on $\text{alph}(w) = \{a_s, a_t, b_s, b_t, c_s, c_t\}$ is given in Figure 1. Thus $\Delta^*_{R_\Sigma}(L(\alpha)) = \Delta^*_{R_\Sigma}((\Delta^*_{R_\Sigma}(L(\alpha_1)))^*)$ is regular by Proposition 2. □

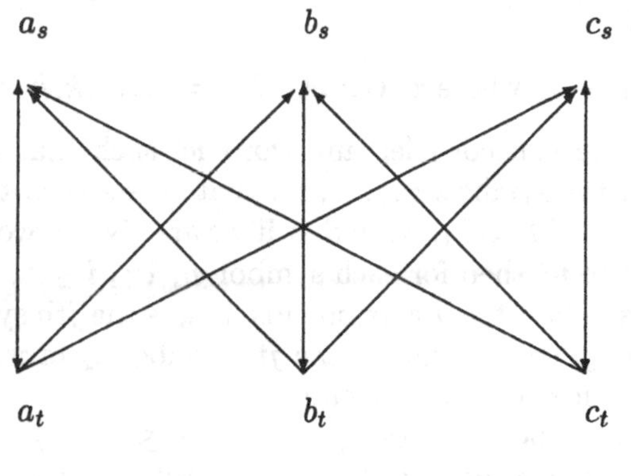

Figure 1: The non-commutation graph $G(\theta_\Sigma, \{a_s, a_t, b_s, b_t, c_s, c_t\})$.

Example 1. Let $\Sigma = \{a, b\}$ and consider the st-language

$$L = a_s(b_s a_t a_s b_t)^* a_t. \tag{4}$$

It can be verified that $\Delta^*_{R_\Sigma}(L)$ is denoted by the synchronization expression

$$((a \to b)^* \to a) \parallel (a \parallel b)^*$$

and thus $\Delta^*_{R_\Sigma}(L)$ is a synchronization language.

Let $R(\Sigma)$ be the rewriting system describing the closure properties of the synchronization language family considered in [12, 13], that is, in the two-letter alphabet case, $\Sigma = \{a, b\}$, $R(\Sigma)$ consists of the rules of R_Σ and the rule (1) and its reversal. It was shown in [13] that the closure under $R(\Sigma)$ of the regular st-language L as in (4) is not regular. The extended system $R(\Sigma)$ does not preserve regularity, that is, the analogy of Theorem 9 does not hold.

To conclude this section we state an obvious property of reductions on st-words that will be used in the next section.

Lemma 10. *Let $w_1, w_2 \in W^{\mathrm{st}}_\Sigma$. Then*

$$\Delta^*_{R_\Sigma}(w_1 w_2) = \Delta^*_{R_\Sigma}(w_1)\Delta^*_{R_\Sigma}(w_2).$$

□

5 Characterization Result

Here we will establish the characterization of finite synchronization languages in terms of closure under reductions of the system R_Σ and the st-condition. It is conjectured in [6] that if closure under a rewriting system R gives a characterization of the synchronization languages then R has to satisfy a certain type of projection property. The rewriting system of [12, 13] fails to satisfy the projection property. Below in Lemma 12 we show that the projection property holds for the system R_Σ and this fact will be used in the proof of the characterization result. First we need some definitions.

Consider a word

$$w = x_1 \cdots x_k \text{ where } x_i \in \Sigma_s \cup \Sigma_t, i = 1, \ldots, k, k \geq 1.$$

In the natural way we can consider any word w' such that $w \longrightarrow^*_{R_\Sigma} w'$ to be a permutation of the symbols $x_1, \ldots, x_k$, that is, we keep track of different occurrences of symbols of $\Sigma_s \cup \Sigma_t$. Note that if we are given a word $w' = y_1 \cdots y_k$ that is a descendant of w, then for each symbol y_i, $1 \leq i \leq k$, we can uniquely determine to which symbol of w it corresponds. If y_i is the jth symbol a_x ($a \in \Sigma$, $x \in \{s, t\}$) in w' then y_i corresponds to the jth symbol a_x in w. Below we often use this observation without further mention.

We say that symbol occurrences x_i and x_j, $1 \leq i < j \leq k$, are *parallel* in w, $\mathrm{Par}_w(x_i, x_j)$, if w can be rewritten to a word $u_1 x_j x_i u_2$ where $u_1 u_2$ is a permutation of the symbols x_h, $1 \leq h \leq k$, $h \notin \{i, j\}$. A rule of R_Σ that contains symbols belonging to $\{c_s, c_t, d_s, d_t\}$ is called a (c, d)-rule, $c, d \in \Sigma$.

Lemma 11. *Let $w = x_1 \cdots x_k \in (\Sigma_s \cup \Sigma_t)^*$ and $i, j \in \{1, \ldots, k\}$, $i < j$, be such that* $\mathrm{Par}_w(x_i, x_j)$. *Assume that $x_i \in \{a_s, a_t\}$ and $x_j \in \{b_s, b_t\}$, where $a, b \in \Sigma$, $a \neq b$. If a reduction*

$$w \longrightarrow^*_{R_\Sigma} w' \tag{5}$$

does not use any (a, b)-rule, then $\mathrm{Par}_{w'}(x_i, x_j)$, *that is, the corresponding occurrences of x_i and x_j are parallel also in w'.*

Proof. We have three possibilities to consider: (i) $(x_i, x_j) = (a_s, b_s)$, (ii) $(x_i, x_j) = (a_t, b_t)$ and (iii) $(x_i, x_j) = (a_s, b_t)$. Note that the possibility $(x_i, x_j) = (a_t, b_s)$ is excluded by the assumption that x_i and x_j are parallel in w and $i < j$. Since the reduction (5) does not use (a, b)-rules and $\mathrm{Par}_w(x_i, x_j)$, we can write $w' = u_1 x_i u_2 x_j u_3$ where u_2 contains the same symbols of $\{a_s, a_t, b_s, b_t\}$ that occur in w between x_i and x_j. Corresponding to the different cases we have: (i) $w' \longrightarrow^*_{R_\Sigma} u_1 u_2 x_j x_i u_3$, (ii) $w' \longrightarrow^*_{R_\Sigma} u_1 x_j x_i u_2 u_3$, and (iii) $w' \longrightarrow^*_{R_\Sigma} u_1 u_2 x_j x_i u_3$. For instance, in the first case the symbol $x_i = a_s$ can move right as far as it encounters a symbol a_t which cannot occur before $x_j = b_s$. Thus always $\mathrm{Par}_{w'}(x_i, x_j)$. □

The use of an (a, b)-rule may naturally transform parallel occurrences of symbols in, respectively, $\{a_s, a_t\}$ and $\{b_s, b_t\}$ to a non-parallel pair. The above lemma is an essential step for proving the projection property for R_Σ. The extended system $R(\Sigma)$ containing rules of the form (1) does not satisfy the statement of the lemma. In the word $w = b_s a_s c_s c_t \underline{a_t} a_s b_t \underline{b_s} a_t b_t$ the underlined occurrences of a_t and b_s can be transposed by rule (1). On the other hand, using only (a, c)-rules the word w can be rewritten to $b_s a_s \underline{a_t} c_s c_t a_s b_t \underline{b_s} a_t b_t$ where the same occurrences are not parallel.

Below we denote by $\Pi_{\{a,b\}}$ the projection $\Pi_{\{a_s, a_t, b_s, b_t\}} : (\Sigma_s \cup \Sigma_t)^* \longrightarrow \{a_s, a_t, b_s, b_t\}^*$, $a, b \in \Sigma$.

Lemma 12. *Let Σ be an alphabet and R_Σ be as in Definition 7. Then for any words $u, v \in (\Sigma_s \cup \Sigma_t)^*$,*

$$u \longrightarrow^*_{R_\Sigma} v \;\Leftrightarrow\; [\, (\forall a, b \in \Sigma)\; \Pi_{\{a,b\}}(u) \longrightarrow^*_{R_\Sigma} \Pi_{\{a,b\}}(v) \,].$$

Proof. The implication from left to right clearly holds. Assume that words u and v ($\in (\Sigma_s \cup \Sigma_t)^*$) satisfy the right side of the statement. Denote $u = x_1 \cdots x_k$ and $v = y_1 \cdots y_k$ and define the function $r : \{1, \ldots, k\} \longrightarrow \{1, \ldots, k\}$ by the condition $y_{r(i)} = x_i$, $i = 1, \ldots, k$. If $u = v$ we are done. Otherwise there exist

$$i < j \quad \text{such that} \quad r(j) < r(i), \quad i, j \in \{1, \ldots, k\}. \tag{6}$$

Furthermore, above we can assume that $j = i + 1$. Namely if $j > i + 1$ and $r(i + 1) < r(j)$ then in (6) we can replace j by $i + 1$; otherwise we inductively compare $i + 1$ and j. Assume that $x_i \in \{a_s, a_t\}$ and $x_{i+1} \in \{b_s, b_t\}$. Since $\Pi_{\{a,b\}}(u) \longrightarrow^*_{R_\Sigma} \Pi_{\{a,b\}}(v)$ it follows that $a \neq b$ and $(x_i, x_{i+1}) \neq (a_t, b_s)$.

Thus $u \longrightarrow_{R_\Sigma} u' = x_1 \cdots x_{i-1} x_{i+1} x_i x_{i+2} \cdots x_k$ and, by Lemma 11, the use of an (a, b)-rule does not change any pair of parallel occurrences of symbols $c, d \in \Sigma$, $\{c, d\} \neq \{a, b\}$, to be sequential in u'. Thus we have $\Pi_{\{c,d\}}(u') \longrightarrow^*_{R_\Sigma} \Pi_{\{c,d\}}(v)$ for all $c, d \in \Sigma$. Since v is obtained from u using a finite number of transpositions the claim follows. □

For proving the characterization result we make use of the following technical lemma. Intuitively, it states that if an st-word has two shuffle decompositions belonging to $\omega_{st}(W_1, a_s a_t)$ and $\omega_{st}(W_2, b_s b_t)$, $W_1, W_2 \subseteq (\Sigma_s \cup \Sigma_t)^*$, where in all pairs of words from, respectively, W_1 and W_2 in the suffixes of length n the numbers of occurrences of a's and b's differ by exactly one, then the "shuffled symbols" of $a_s a_t$ and $b_s b_t$ have to stay in the respective suffixes of length n.

Lemma 13. *Let* $\Sigma = \{a, b\}$ *and* $A_i, B_i \subseteq (\Sigma_s \cup \Sigma_t)^*$ *be* st-*languages,* $i = 1, 2$. *Assume that there exist* $h, k \in \mathbb{N}$ *such that*

$$(\forall u \in B_1)\ \#_{a_s}(u) = \#_{a_t}(u) = h,\ \#_{b_s}(u) = \#_{b_t}(u) = k + 1, \quad \textit{and}, \tag{7}$$

$$(\forall u \in B_2)\ \#_{a_s}(u) = \#_{a_t}(u) = h + 1,\ \#_{b_s}(u) = \#_{b_t}(u) = k. \tag{8}$$

Let $w \in \omega_{st}(A_1B_1, a_s a_t) \cap \omega_{st}(A_2B_2, b_s b_t)$ *be arbitrary. We claim that the suffix* w' *of* w *having length* $2(h + k + 2)$ *is an* st-*word such that* $\#_{a_s}(w') = h + 1$ *and* $\#_{b_s}(w') = k + 1$.

Proof. Let $u_i \in A_i$ and $v_i \in B_i$, $i = 1, 2$, be such that $w \in \omega_{st}(u_1 v_1, a_s a_t) \cap \omega_{st}(u_2 v_2, b_s b_t)$. It is sufficient to show that then necessarily

$$w \in u_1 \omega_{st}(v_1, a_s a_t) \text{ or } a_s a_t \text{ is a suffix of } u_1. \tag{9}$$

Intuitively, the condition (9) means that, in the shuffle of $u_1 v_1$ and $a_s a_t$ producing w, the symbols of $a_s a_t$ have to "stay in" the word v_1 except in the case where u_1 ends with $a_s a_t$ (and the suffix of u_1 can then be exchanged for the shuffled word $a_s a_t$).

For the sake of contradiction, assume that (9) does not hold. This means that in the shuffle of $u_1 v_1$ and $a_s a_t$ producing w, the symbol a_s has to bypass to the left of some occurrence of b_t in u_1, that is, if $u_1 = x_1 \cdots x_m$, $v_1 = y_1 \cdots y_n$, we can write

$$w = x_1 \cdots x_i a_s x_{i+1} \cdots x_m y_1 \cdots y_j a_t y_{j+1} \cdots y_n, \tag{10}$$

$(1 \leq i < m, 0 \leq j \leq n)$, or,

$$w = x_1 \cdots x_i a_s x_{i+1} \cdots x_{i+j} a_t x_{i+j+1} \cdots x_m y_1 \cdots y_n, \tag{11}$$

$(1 \leq i < m, 0 \leq j < m - i)$, and $b_t \in \{x_{i+1}, \ldots x_m\}$.

First consider the decomposition (10). Since $a_s a_t$ is not a suffix of $u_1 = x_1 \cdots x_m$ and w (as a result of an st-shuffle) is an st-word, without loss of generality we can assume that (i) $x_m = b_t$, or (ii) $x_{m-1} x_m = b_t a_t$ and $i < m - 1$.

In the case (i), we denote $w_1 = x_m y_1 \cdots y_j a_t y_{j+1} \cdots y_n$, and in the case (ii) let w_1 be the word $x_{m-1} x_m y_1 \cdots y_j a_t y_{j+1} \cdots y_n$. From (7) we have $\#_{a_s}(w_1) = h$ and $\#_{b_t}(w_1) = k + 2$. On the other hand, by (8) it follows that any suffix of $w \in \omega_{st}(u_2 v_2, b_s b_t)$ that has $k+2$ occurrences of b_t necessarily has to include the entire word v_2 and hence the suffix has to have at least $h+1$ occurrences of a_s. This is a contradiction. Exactly the same argument works for the decomposition (11). □

Now we can prove the main result of this section.

Theorem 14. *A finite language L over an alphabet $\Sigma_s \cup \Sigma_t$ is a synchronization language if and only if L is an* st*-language and closed under R_Σ.*

Proof. The "only if" part follows from Theorem 6 and Lemma 8. For the converse part it is sufficient to show that for an arbitrary st-word $w \in W_\Sigma^{st}$ there exists a synchronization expression α_w such that

$$L(\alpha_w) = \Delta^*_{R_\Sigma}(w). \tag{12}$$

We use induction on the length of w. For words $w = a_s a_t$, $a \in \Sigma$, we can choose $\alpha_w = a$. Assume then that $|w| = 2n$, $n > 1$, and that the claim holds for all st-words of length at most $2n - 2$. Let a_s be the rightmost start symbol occurring in w. We can write

$$w = u_1 a_s u_2 a_t,$$

where $u_1 \in (\Sigma_s \cup \Sigma_t)^*$ and $u_2 \in \Sigma_t^*$. Note that without loss of generality w can be assumed to end with a_t since the symbols of Σ_t commute in both directions with each other.

We say that a corresponding pair of occurrences of c_s and c_t is a ***dividing pair*** in w if c_s occurs in u_1 and c_t in u_2. Let

$$c_1, \ldots, c_k \in \Sigma, \ (k \geq 0), \tag{13}$$

be the sequence that contains $c \in \Sigma - \{a\}$ exactly when (c_s, c_t) is a dividing pair in w. The order of the sequence (13) can be arbitrary. Furthermore, we denote by u_1' the word obtained from u_1 by deleting every occurrence c_s belonging to a dividing pair of w.

We choose

$$\alpha_w = \alpha_1 \& \alpha_2 \tag{14}$$

where,

$$\alpha_1 = \alpha_{u_1 u_2} \parallel a \quad \text{and} \quad \alpha_2 = (\alpha_{u_1'} \to a) \parallel c_1 \parallel \ldots \parallel c_k.$$

Clearly, $\Delta^*_{R_\Sigma}(w) \subseteq L(\alpha_i)$, $i = 1, 2$, and hence in (12) the inclusion from right to left holds.

For the converse inclusion, consider an arbitrary word $v \in L(\alpha_w)$. Using Lemma 12 we see that it is sufficient to show that for all $b, c \in \Sigma$, $b \neq c$:

$$\Pi_{\{b,c\}}(w) \longrightarrow^*_{R_\Sigma} \Pi_{\{b,c\}}(v). \tag{15}$$

(i) Assume that in (15) $a \notin \{b, c\}$. Since $v \in L(\alpha_1)$ it follows that

$$\begin{aligned}\Pi_{\{b,c\}}(v) \in \Pi_{\{b,c\}}(L(\alpha_{u_1 u_2} \parallel a)) &= \Pi_{\{b,c\}}(L(\alpha_{u_1 u_2})) \\ &= \Pi_{\{b,c\}}(\Delta^*_{R_\Sigma}(u_1 u_2)) = \Delta^*_{R_\Sigma}(\Pi_{\{b,c\}}(w)).\end{aligned}$$

(ii) Consider then (15) where $c = a$ and $b \in \Sigma$ is arbitrary. First assume that (b_s, b_t) does not form a dividing pair in w. Then

$$\begin{aligned}\Pi_{\{a,b\}}(v) \in \Pi_{\{a,b\}}(L(\alpha_2)) &= \Pi_{\{a,b\}}(L(\alpha_{u_1'}))\{a_s a_t\} \\ &= \Pi_{\{a,b\}}(\Delta^*_{R_\Sigma}(u_1'))\{a_s a_t\} = \Delta^*_{R_\Sigma}(\Pi_{\{a,b\}}(w)).\end{aligned}$$

Secondly, assume that (b_s, b_t) occurs as a dividing pair in w. Then we can write

$$\begin{aligned}\Pi_{\{a,b\}}(w) &= w'w'' \\ &= w'b_s(a_sa_t)^{k_1}a_sb_t(b_sb_t)^{k_2}b_sa_t\cdots b_sa_t(a_sa_t)^{k_{2m-1}}a_sb_ta_t, \quad (16)\end{aligned}$$

where $m \geq 1$, $k_i \geq 0$, $i = 1, \ldots, m$, and w' is the maximal prefix of $\Pi_{\{a,b\}}(w)$ that is an st-word. (Strictly speaking, the suffix following w' may begin also with a_sb_s. However, the form (16) does not violate generality because a_s and b_s commute in both directions and we can choose $k_1 = 0$.) Furthermore, denote

$$w_1 = b_s(a_sa_t)^{k_1}a_sb_t(b_sb_t)^{k_2}b_sa_t\cdots b_sa_t(a_sa_t)^{k_{2m-1}}b_t, \text{ and,}$$

$$w_2 = b_s(a_sa_t)^{k_1}a_sb_t(b_sb_t)^{k_2}b_sa_t\cdots a_sb_t(b_sb_t)^{k_{2m-2}}a_t(a_sa_t)^{k_{2m-1}}a_sa_t.$$

Note that then $\Pi_{\{a,b\}}(u_1u_2) = w'w_1$ and $\Pi_{\{a,b\}}(u_1'a) = w'w_2$. By Lemma 10, we have

$$\begin{aligned}\Pi_{\{a,b\}}(L(\alpha_1)) &= \Pi_{\{a,b\}}(\omega_{\text{st}}(L(\alpha_{u_1u_2}), a_sa_t)) \\ &= \omega_{\text{st}}(\Delta^*_{R_\Sigma}(w')\Delta^*_{R_\Sigma}(w_1), a_sa_t), \quad (17)\end{aligned}$$

and,

$$\begin{aligned}\Pi_{\{a,b\}}(L(\alpha_2)) &= \Pi_{\{a,b\}}(\omega_{\text{st}}(L(\alpha_{u_1'} \to a), b_sb_t)) \\ &= \omega_{\text{st}}(\Delta^*_{R_\Sigma}(w')\Delta^*_{R_\Sigma}(w_2), b_sb_t). \quad (18)\end{aligned}$$

Furthermore,

$$\Pi_{\{a,b\}}(v) \in \Pi_{\{a,b\}}((L(\alpha_1) \cap L(\alpha_2)) \subseteq \Pi_{\{a,b\}}(L(\alpha_1)) \cap \Pi_{\{a,b\}}(L(\alpha_2)). \quad (19)$$

In the notations of Lemma 13 we choose $A_1 = A_2 = \Delta^*_{R_\Sigma}(w')$, $B_1 = \Delta^*_{R_\Sigma}(w_1)$ and $B_2 = \Delta^*_{R_\Sigma}(w_2)$. Observe that for all $z \in B_1$,

$$\#_{a_s}(z) = m - 1 + \sum_{i=1}^{m} k_{2i-1} \quad \text{and} \quad \#_{b_s}(z) = m + \sum_{i=1}^{m-1} k_{2i},$$

and for all $z \in B_2$,

$$\#_{a_s}(z) = m + \sum_{i=1}^{m} k_{2i-1} \quad \text{and} \quad \#_{b_s}(z) = m - 1 + \sum_{i=1}^{m-1} k_{2i}.$$

Thus combining (17), (18) and (19), by Lemma 13, it follows that $\Pi_{\{a,b\}}(v)$ has a decomposition v_1v_2 where $v_1 \in \Delta^*_{R_\Sigma}(w')$ and

$$v_2 \in \omega_{\text{st}}(\Delta^*_{R_\Sigma}(w_1), a_sa_t) \cap \omega_{\text{st}}(\Delta^*_{R_\Sigma}(w_2), b_sb_t). \quad (20)$$

By (16), to complete the proof it remains to show that $v_2 \in \Delta^*_{R_\Sigma}(w'')$. The words of $\Delta^*_{R_\Sigma}(w'')$ can be characterized as follows. They consist of sequential blocks $(a_sa_t)^{k_1}$, $(b_sb_t)^{k_2}$, ..., $(b_sb_t)^{k_{2m-2}}$, and $(a_sa_t)^{k_{2m-1}+1}$ where into each block $(a_sa_t)^{k_{2i-1}}$ one shuffles a string b_sb_t and into each block $(b_sb_t)^{k_{2i}}$ one shuffles a string $a_s a_t$. The situation is illustrated in Figure 2. In the figure, a symbol a_i (respectively, b_i) below an arrow indicates shuffle-insertion of the string $a_s a_t$ (respectively, $b_s b_t$) into the corresponding block. The subindices are used to distinguish between different occurrences of the strings. Note that the consecutive inserted occurrences of a_i and b_{i+1}, or, of b_i and a_{i+1} can overlap but any other overlaps are not possible.

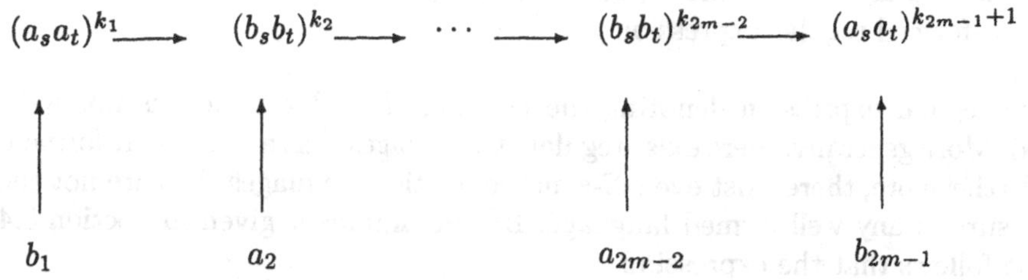

Figure 2: The words belonging to $\Delta^*_{R_\Sigma}(w'')$.

The words belonging to $\omega_{\rm st}(\Delta^*_{R_\Sigma}(w_1), a_s a_t)$ are as in Figure 2 except that the last block $(a_s a_t)^{k_{2m-1}+1}$ is replaced by $(a_s a_t)^{k_{2m-1}}$ and an additional string $a_s a_t$ is shuffled anywhere in the word. Similarly, the words of $\omega_{\rm st}(\Delta^*_{R_\Sigma}(w_2), b_s b_t)$ are obtained by removing b_{2m-1} from the figure and then shuffling $b_s b_t$ anywhere in the word. However, (20) implies that v_2 has both of the above decompositions and this turns out to be sufficient to guarantee that v_2 has a decomposition as in Figure 2.

To see this let $z_i \in \Delta^*_{R_\Sigma}(w_i)$, $i = 1, 2$, be such that $v_2 \in \omega_{\rm st}(z_1, a_s a_t)$ and $v_2 \in \omega_{\rm st}(z_2, b_s b_t)$. If in z_1 $a_s a_t$ is inserted into (or directly before or after) the last block, then clearly $v_2 \in \Delta^*_{R_\Sigma}(w'')$. Similarly, we are done if in z_2 the string $b_s b_t$ is inserted into the last block. For the sake of contradiction assume then that in z_1 the string $a_s a_t$ is inserted into the ith b-block, $(b_s b_t)^{k_{2i}}$, $1 \leq i < m$, (in the case $i = m - 1$ not after the block), and that in z_2 the string $b_s b_t$ is inserted into the jth a-block, $(a_s a_t)^{k_{2j-1}}$, $1 \leq j < m$. Assume that $j < i$, the other case is symmetric.

The shuffle of z_2 and $b_s b_t$ under consideration ends with $(a_s a_t)^{k_{2m-1}+1}$. This implies that in z_1 the string a_{2m-2} has to be inserted completely after b_{2m-1}, see Figure 2. Thus the last b-block in v_2 has length $k_{2m-2} + 1$ and then necessarily in the decomposition of z_2 the string a_{2m-2} precedes b_{2m-3}. Continuing in this way, we see that in z_1 b_{2r+1} precedes a_{2r} and that in z_2 a_{2r} precedes b_{2r-1} when $r > i$. When going to the left from the $2i$th block the strings a_r, b_r can be inserted freely in z_1 (according to Figure 2) but in z_2 b_{2r-1} strictly follows a_{2r} when $r > j$. Together these observations imply that v_2 has a decomposition as in Figure 2. □

6 Discussion and Open Problems

The next step would be to extend the characterization result of the previous section for the well-formed languages. Well-formed regular expressions and languages were defined in [12, 13].

Definition 15. Let $\alpha \in \text{reg}(\Sigma_s \cup \Sigma_t)$. We say that the regular expression α is *well-formed* if always when β^* is a subexpression of α, $L(\beta)$ is an st-language. A regular language L over the alphabet $\Sigma_s \cup \Sigma_t$ is *well-formed* if L is denoted by a well-formed regular expression.

The regular expression denoting the language L in Example 1 is not well-formed. More generally, there exist regular st-languages that are not well-formed, and, furthermore, there exist even Σ-synchronization languages that are not the R_Σ-closure of any well-formed language. By the argument given in Section 5.4 of [6] it follows that the expression

$$((a \parallel b) \to (a \parallel b)^*) \parallel (d \to (c \to d)^*) \tag{21}$$

denotes a language that is not the R_Σ-closure, $\Sigma = \{a, b, c, d\}$, of any well-formed language. The proof provided in [6] uses the extended rewriting system but the same argument works for R_Σ since it is verified that in reductions from words of the language (21) none of the extended rules can be applied. Note also that the semantics of the expression remains unchanged since in (21) the arguments of any join operator do not have symbols in common.

Below we describe how a construction similar to the one used in the proof of Theorem 14 could be used to show that the R_Σ-closure of an arbitrary well-formed regular language is a synchronization language. This would considerably strengthen the result of Theorem 14 although, by the above observations, it would still not give a complete characterization of all synchronization languages.

Following the method used in the proof of Theorem 14, we can inductively construct for a well-formed regular expression γ a synchronization expression α_γ that at least seems to denote the language $\Delta^*_{R_\Sigma}(L(\gamma))$. Using the distributivity of union and the fact that the synchronization expressions can employ the $*$-operation, it is seen that the only nontrivial case in the inductive construction is when γ is the catenation of smaller expressions, that is,

$$\gamma = \beta_1 a_s \beta_2 a_t,$$

where β_1 and β_2 are catenations of symbols of $\Sigma_s \cup \Sigma_t$ and $*$-expressions. (The case where γ ends with a $*$-expression can be handled simply using the sequencing operator.) Similarly as in the proof of Theorem 14, (c_s, c_t) is said to be a dividing pair of γ if c_s occurs in β_1 and c_t occurs in β_2 outside the $*$-expressions, and let $c_1, \ldots c_k \in \Sigma$ denote the sequence of symbols that occur as dividing pairs in γ. Furthermore, let β_1' (respectively, β_2') be the expression obtained from β_1 (respectively, β_2) by removing the occurrences of c_s (respectively, c_t) that belong to a dividing pair. Note that β'_i is well-formed although β_i, in general, is not. Then we conjecture that the expression

$$\alpha_\gamma = (\alpha_{\beta_1 \cdot \beta_2} \parallel a) \;\&\; [\,(\alpha_{\beta'_1} \to \alpha_{\beta'_2}) \parallel c_1 \parallel \ldots \parallel c_k\,]$$

denotes the language $\Delta^*_{R_\Sigma}(L(\gamma))$. However, we do not have a complete proof for the claim due to the fact that the analogy of Lemma 13 does not hold for sets of nonuniform length words and also the projection result of Lemma 12 is not directly applicable.

A construction as outlined above is not possible for arbitrary regular expressions because the subexpressions used for the $*$-operator need not in general be well-formed. A general characterization result may be more difficult, as witnessed by the fact that already for the very simple regular expression $\gamma = a_s(b_s a_t a_s b_t)^* a_t$ considered in Example 1, the synchronization expression denoting the language $\Delta^*_{R_\Sigma}(L(\gamma))$ seems to have little structural resemblance to γ.

References

1. Aalbersberg, I.J., Rozenberg, G.: Theory of traces. Theoret. Comput. Sci. **60** (1988) 1–82
2. Baeten, J.C.M., Weijland, W.P.: *Process algebra.* Cambridge University Press, Cambridge, 1990
3. Book, R.V., Otto, F.: *String-rewriting systems.* Texts and Monographs in Computer Science, Springer-Verlag, 1993
4. Bracho, F., Droste, M., Kuske, D.: Representations of computations in concurrent automata by dependence orders. Theoret. Comput. Sci. **174** (1997) 67–96
5. Clerbout, M., Latteux, M., Roos, Y.: Semi-commutations. In: *The Book of Traces.* (V. Diekert, G. Rozenberg, eds.) Chapter 12, pp. 487–552, World Scientific, Singapore, 1995
6. Clerbout, M., Roos, Y., Ryl, I.: Synchronization languages. To appear in Proc. of the 8th International Conference on Automata and Formal Languages, Salgótarján, Hungary, July 1996
7. De Nicola, R., Segala, R.: A process algebraic view of input/output automata. Theoret. Comput. Sci. **138** (1995) 391–423
8. Diekert, V., Gastin, P., Petit, A.: Rational and recognizable complex trace languages. Inform. Computation **116** (1995) 134–153
9. Droste, M.: Recognizable languages in concurrency monoids. Theoret. Comput. Sci. **150** (1995) 77–109
10. Govindarajan, R., Guo, L., Yu, S., Wang, P.: ParC Project: Practical constructs for parallel programming languages. Proc. of the 15th Annual IEEE International Computer Software & Applications Conference, 1991, pp. 183–189
11. Guo, L.: Synchronization expressions in parallel programming languages. PhD Thesis, University of Western Ontario, 1995
12. Guo, L., Salomaa, K., Yu, S.: Synchronization expressions and languages. Proc. of the 6th IEEE Symposium on Parallel and Distributed Processing, (Dallas, Texas). IEEE Computer Society Press, 1994, pp. 257–264

13. Guo, L., Salomaa, K., Yu, S.: On synchronization languages. Fundamenta Inform. **25** (1996) 423–436
14. Hennessy, M.: *Algebraic theory of processes*. The MIT Press, Cambridge, Mass., 1989
15. Hopcroft, J.E., Ullman, J.D.: *Introduction to Automata Theory, Languages, and Computation*. Addison-Wesley, Reading, MA, 1979
16. Mateescu, A.: On (left) partial shuffle. In: Results and Trends in Theoretical Computer Science. Lect. Notes in Comput. Sci. **812** (1994) 264–278
17. Mateescu, A., Rozenberg, G., Salomaa, A.: Shuffle on trajectories: Syntactic constraints. Turku Centre for Computer Science, TUCS Technical Report No 41, September 1996
18. Nielsen, M., Winskel, G.: Petri nets and bisimulation. Theoret. Comput. Sci. **153** (1995) 211–244
19. Salomaa, A.: *Formal Languages*. Academic Press, New York, 1973
20. Sassone, V., Nielsen, M., Winskel, G.: Models of concurrency: Towards a classification. Theoret. Comput. Sci. **170** (1996) 297–348
21. Yu, S.: Regular languages. In: *Handbook of Formal Languages, Vol. I.* (G. Rozenberg, A. Salomaa, eds.) pp. 41–110, Springer-Verlag, 1997

DNA Sequence Classification Using DAWGs

Samuel Levy and Gary D. Stormo

Molecular, Cellular and Developmental Biology,
University of Colorado, Boulder, CO 80309-0347, USA

Abstract. DNA sequence classification involves attributing sub-strings or words within a sequence to known distinct sequence classes. A query sequence was classified by comparing all of its words to words in databases representative of three classes of DNA, transcriptional promoters, exons and introns. The efficiency of this comparision was increased by constructing directed, acyclic word graphs (DAWGs) of all sequences and databases. The resulting *landscape* was scored to determine the preference of words in the query sequence for any one particular database class. Using this approach it was possible to detect 94% of a test set of individual promoter sequences, with only 4% incorrect detection of test exon sequences as promoters. Preliminary attempts were made to parse genomic DNA into promoter, exon and intron regions. Initial results indicate that a reasonably high degree of correlation exists between the predicted regions and known promoter-exon-intron domains.

1 Introduction

The ability to classify regions of DNA into functionally defined groups can provide an important first step towards an experimental demonstration of biological function. In order to identify the protein coding regions of genes, it is also useful to parse a sequence into known domains of non-coding. Thus the identification of transciptional activating promoter and intron domains are central to the identification of exon domains.

An initial step towards the complex problem of identifying the type and location of different sequence classes can be achieved by exact word matching of the query sequence with databases representative of a sequence class. Programs for identifying protein coding regions often use word frequencies, typically hexamers, as one type of evidence for classification (Uberbacher and Mural 1991, Snyder and Stormo 1993, 1995, Solovyev 1994). Word matching algorithms are also used for predicting promoter regions, where the words may be known binding sites for transcription factors, or simply words that are over-represented in known promoter regions (Prestridge 1995, Hutchinson 1996, Chen et al 1997). In this paper we describe an approach that uses the complete word frequencies, using all word lengths, to help determine the correct classification of regions of genomic DNA sequences. We start with databases of sequences known to belong to each class. For each database we build a directed, acyclic word graph (DAWG), which contains how many times each word occurs in the database (Blumer et al 1984, Blumer et al 1985). (In place of the DAWG we can also use a suffix array, which takes longer to construct but requires less memory (Manber and Myers 1993). A query sequence, which we wish to classify, is then compared to the DAWGs from the separate databases to determine which word frequencies it most closely matches.

The number of times each word occurs in a database can be conveniently displayed in a *landscape* above the query sequence (Clift et al 1986). Recent enhancements to the landscape program, including graphical display, data filtering and direct comparison of two landscapes on the same query sequence, will be described elsewhere (Levy *et al*, in preparation). In this communication the landscape approach will be employed as a basis for comparing word frequencies of a query sequence with databases of different sequence classes. Subsequent to obtaining the landscapes of the query sequence compared to promoter, exon and intron sequence databases, different scoring approaches are employed to identify regions of sequence belonging to any one sequence class.

2 Components of a Sequence Landscape

2.1 Single sequence, single database landscapes

Figure 1 provides a simple example of a landscape. In this case the database is the query sequence itself, so the landscape displays how many times each word in the sequence occurs. Every cell in the landscape, $L(j,k)$, indicates where the word begins, j, and its length, k. The value in the cell is the number of times that word occurs in the database. The word, *tcc*, occurs twice in the sequence as indicated by the frequency value of 2 in the cells found in row 3 at positions 3 and 8 of the sequence. The landscape also shows the frequency of two and one-letter words. For example, the word *tc* occurs three times starting at positions 3, 8 and 11, and the bases *t* and *c* each occur five times, *g* occurs three times and *a* occurs twice. The diagonal lines starting at positions 3 and 8 and ending at positions 4 and 9 respectively indicate that the words *cc* starting at positions 4 and 9 occur only within the context of the word *tcc* at positions 3 and 8.

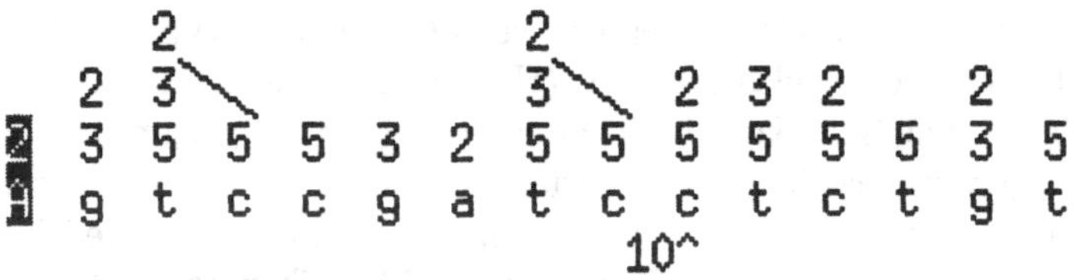

Fig. 1. Landscape of simple 15 base sequence compared to itself.

A small segment of the landscape of an authentic DNA sequence (Genbank locus HUMACTGA, bases 1125-1150) compared to a database of concatenated but distinct exon sequences shows the frequencies of up to 14 letter words (Fig.2). Frequency values greater than 99 are found but displayed as *. The vertical bar found at position 1131 extending from row 9 {1131,9} to 14 {1131,14}, indicates that the 9-14 letter words starting at position 1131 occur once in the exon database. The sequence row found at the base of the landscape shows the 14 letter word in upper case blocked letters (*CAGGGCGTCATGGT*) and a cursor is found in the landscape at {1131,14}. The boundary between intron B and exon 3 is found at position 1134 and cursory visual inspection of the landscape suggests that following position 1131 there is an increased occurrence of larger words found in the exon database. This suggests that from position 1131 the sequence exhibits word matching events that predict this region to be exon-like.

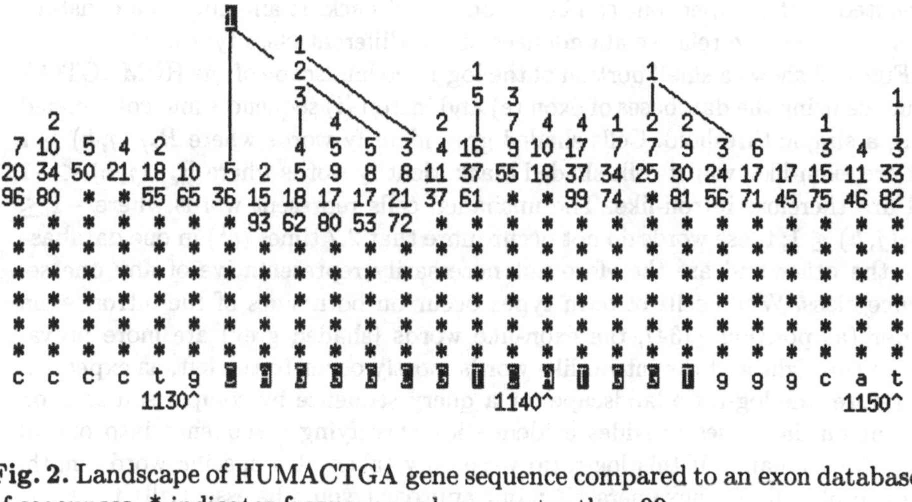

Fig. 2. Landscape of HUMACTGA gene sequence compared to an exon database of seqeunces. * indicates frequency values greater than 99.

2.2 Ratio landscapes

The goal of our research is, given a query sequence Q, and a set of DNA classes $\{C_1, C_2 \ldots\}$, identify which class Q belongs to, or determine its probability of belonging to each class. In practice, Q is often very long, and then we wish to identify the subsequences that belong to each class; i.e. we want to parse Q into sub-regions of the available classes. We do this by collecting databases of each class of sequence. D_1 is a collection of sequences, all of class C_1, D_2 sequences of class C_2, etc.

We base our classification on the words contained in Q and their relative frequencies in the different classes. If the databases for each class are representative of that class, then for a word w, occuring within Q, we have

$$P(C_i|w) \approx P(D_i|w) = P(w|D_i)\frac{P(D_i)}{P(w)} \qquad (1)$$

If we create a landscape of Q using the database D_i, and if w occurs at position j and is k-long, then $L_i(j,k)$ is the number of times w occurs in D_i, which is $P(w|D_i) \cdot |D_i|$. The analysis is facilitated by only considering relative probabilities, which we do with a log-ratio of different landscapes. From the landscapes using two different databases, i and h, we define the log-ratio for the word w that occurs at position j, of length k, as:

$$R_{h/i}(j,k) = \ln \frac{L_h(j,k)+1}{L_i(j,k)+1} + \ln \frac{|D_i|}{|D_h|} + \ln \frac{P(D_h)}{P(D_i)} = \ln \frac{P(D_h|w)}{P(D_i|w)} \tag{2}$$

The addition of 1 to the $L(j,k)$ terms serves as a small sample size correction and avoids any potential erroneous attempt to evaluate $\ln 0$. The last term, of the prior probabilities for the different classes, $\ln \frac{P(D_h)}{P(D_i)}$, is ignored in the work presented in this paper, but can easily be added back. It amounts to a constant offset based on the relative abundances of the different class types.

Figure 3 shows a small portion of the log-ratio landscape of the HUMACTGA sequence using the databases of exon (e) and intron (i) sequences and color coded using a simple threshold. Cells shaded grey identify words where $R_{e/i}(j,k) > 1$ and are exon-like, whilst cells shaded black identify words where $R_{e/i}(j,k) < -1$ and are therefore intron-like. The unshaded cells represent words where $-1 \leq R_{e/i}(j,h) \leq 1$; these words do not occur more that 2.7 times (e^1) in one database over the other and are therefore not necessarily representative of any one sequence class. While cells of both types occur on both sides of the intron/exon border (at position 1134), the exon-like words (shaded grey) are more prevalent to the right and the intron-like words mostly occur to the left, as expected. Therefore, the log-ratio landscape for a query sequence by comparison to exon and intron databases provides evidence for classifying a sequence into one of those two domains. If this log-ratio were only taken at a specific word length, for example only for hexamers, then our approach would be essentially the same as other methods that use relative hexamer frequencies in predicting exon and intron domains (Snyder and Stormo 1993, 1995, Fickett and Tung 1992). However, the advantage of our method is that useful information can be obtained from any length words, whichever ones are the most informative. The remainder of this paper describes approaches for combining the information from all of the words of the query sequence to aid in parsing it into the appropriate classes, and some preliminary evalutions of how well this method works.

3 Algorithms for Processing Landscapes

The log-ratio landscape contains the log-likelihoods of words belonging preferentially to one or other of the databases as outlined in (2). If a judicious choice of database were made then it would be possible to produce a log-ratio landscape of a query sequence that can be scored systematically for positive or negative values and ultimately it could be classified as belonging to one of the classes of database sequences.

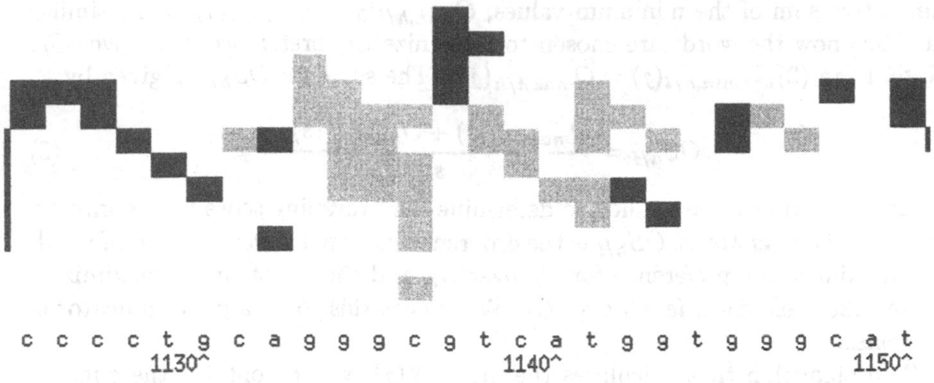

Fig. 3. Ratio landscape of HUMACTGA sequence. Grey shading represents exon-like words, black shading represents intron-like words. Note the higher frequency of exon-like words occurring to the right of the known intron-exon boundary (1134) and intron-like words occur mostly to the left of the boundary, as expected.

In order to optimize sequence recognition several different scoring algorithms were developed.

- Cumulative sum of all words
- Max/min cumulative sum of overlapping words
- Max/min cumulative sum of non-overlapping words

The cumulative sum of all words is represented by,

$$S_{h/i}(j) = S_{h/i}(j-1) + \sum_{k=1}^{c} R_{h/i}(j,k), \tag{3}$$

which produces the score, CS, such that

$$CS_{h/i} = \frac{S_{h/i}(s)}{s}, \tag{4}$$

where c is the maximum landscape height and s is the sequence length.

The algorithm that calculates the array $O(j)$, which contains the cumulative sum of either maximum (positive) or minimum (negative) values, is

$$O_{max,h/i}(j) = O_{max,h/i}(j-1) + \max_{m=1\ldots c} [R_{h/i}(j-m+1,m)]. \tag{5}$$

$O_{max,h/i}(j)$ is the sum of the $R_{h/i}$ values, using one word that ends at every position from 1 to j, with the words chosen so as to maximize the preference for class C_h over class C_i. A similar equation to (5) can be written that takes the cumulative sum of the minimum values, $O_{min,h/i}(j)$. $O_{min,h/i}(j)$ is the similar sum, but now the words are chosen to maximize the preference of C_i over C_h. (Note, from (2), $O_{min,h/i}(j) = O_{max,i/h}(j)$). The score for $OS_{h/i}$ is given by,

$$OS_{h/i} = \frac{O_{max,h/i}(s) + O_{min,h/i}(s)}{s} \tag{6}$$

Equation (6) can be applied to determine the prevailing score taking into account *overlapping* words. $OS_{h/i}$ is the difference in score between the set of words that maximize the preference for C_h over C_i, and the set of words maximizing the reverse preference, *ie.* C_i over C_h. We expect this to be a good indicator of true class.

The algorithm that calculates the array $N(j)$, which contains the cumulative sum of either maximum or minimum values is represented by the recursive dynamic programming algorithm,

$$N_{max,h/i}(j) = \max_{m=1\ldots c} [N_{max,h/i}(j-m) + R_{h/i}(j-m+1,m)] \tag{7}$$

N values are completely analogous to O values, the difference being the words are chosen to be *non-overlapping*. While more easily justified statistically, this may not give better performance empirically. Equation (7) can be applied to calculate the score, $NS_{h/i}$, of non-overlapping words in a sequence by,

$$NS_{h/i} = \frac{N_{max,h/i}(s) + N_{min,h/i}(s)}{s} \tag{8}$$

4 Promoter, Exon and Intron Sequence Recognition

The ratio landscape approach and the scoring algorithms outlined in the previous section in (3-8) were applied to the classification of individual test sequences known to belong to the three different classes of promoter, exon and intron DNA sequences.

In order to produce viable training (database) and test data it was necessary to obtain known promoter, exon and intron datasets. Promoter sequences were obtained from the Eukaryotic Promoter Database version 48 (Bucher, 1996) and included sequence data from -499 to +100 relative to the transcription initiation site. The 573 unique vertebrate promoter sequences were randomly assigned into a database of 430 concatenated but distinct sequences and 143 test sequences. In a similar manner, primate exon and intron sequences from GENBANK were randomly assigned into databases containing 724 (exons) and 709 (introns) sequences, leaving 241 and 236 sequences respectively as the test sets.

Subsquently, the following 12 ratio landscapes were obtained,

$$P_{p/e}, E_{p/e}, P_{p/i}, I_{p/i}$$
$$E_{e/p}, P_{e/p}, E_{e/i}, I_{e/i}$$
$$I_{i/p}, P_{i/p}, I_{i/e}, E_{i/e}$$

where the upper case letter represents the true class of the query (test) sequences and lower case represents the databases used. The first set of four log-ratio landscapes were used for promoter detection, the second set of four was used for exon detection and the last four were used for intron detection. Each log-ratio landscape was scored using the three scoring algorithms outlined in (3-8).

The scores, OS, NS, CS, obtained for each log-ratio landscape of a set of test sequences were plotted as a frequency distribution histogram with respect to the scores obtained. For example, the frequency distribution histograms for the scores obtained for the promoter test sequences (in $P_{p/e}$) and the exon test sequence (in $E_{p/e}$) are shown for the overlapping algorithm (Fig.4a), the non-overlapping algorithm (Fig.4c) and the landscape sum (Fig.4e). From these frequency distribution histograms, it is evident that the distributions have the least overlap for the pairs of tested sequences when either the overlapping or non-overlapping word scoring algorithm is used (Fig.4a-d). Thus, from the four landscapes, $P_{p/e}, E_{p/e}, P_{p/i}, I_{p/i}$ it is possible to determine a single score for each test sequence that can classify this sequence as either promoter or intron/exon. Similar comparisons can be used to distinguish between exons and introns (data not shown).

In order to determine the accuracy of this prediction method for true positives over false positives, a correlation coefficient (Mathews 1975),CC, was calculated across the range of bin values in each distribution histogram. The correlation coefficient is,

$$CC = \frac{(P^t N^t) - (N^f P^f)}{\sqrt{(N^t + N^f)(N^t + P^f)(P^t + N^f)(P^t + P^f)}} \tag{9}$$

where P^t are true positives, P^f are false positives, N^t and N^f are true and false negatives respectively. A value of 1 is the ideal situation of 100% true positives with 0% false positives. The optimal value for the correlation coefficient with the corresponding values for true and false positives shown for each set of test sequences are found in Table 1. The optimal true to false positive combination is indicated by the highest correlation coefficient value for each tested sequence class. Therefore, it becomes clear that promoter recognition occurs in 94% of test promoter sequences with 4% of test exon sequences detected as promoters (false positives). This level of recognition is very favorably high compared to other existing methods for eukaryotic promoter detection (Chen et al 1997, Prestridge 1995). Optimal exon and intron recognition is measured in 96% and 93% of test sequences with a 8% and 1% false positive detection, respectively.

5 Classification of Gene Sequence

The use of ratio landscapes coupled with scoring algorithm enables a high degree of recognition of known promoter, exon and intron sequences (see previous section). However, unannotated DNA sequences may contain all of these three

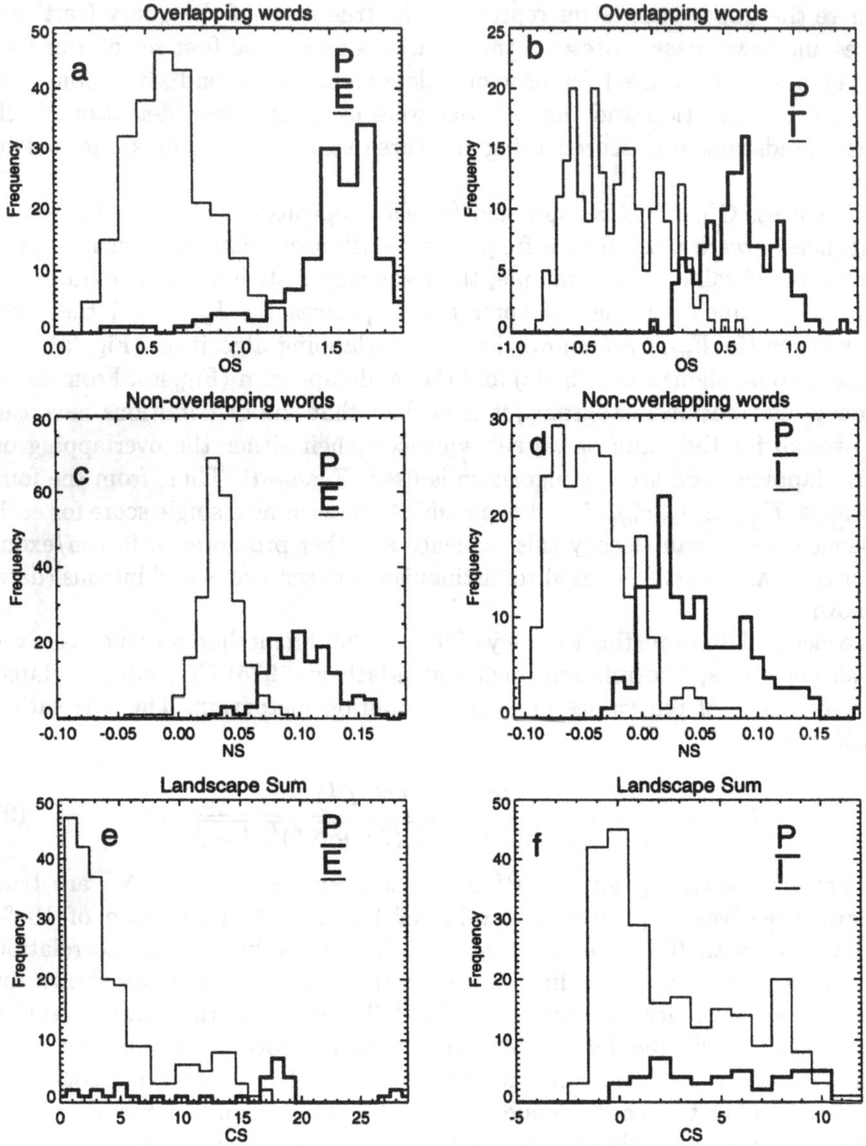

Fig. 4. Frequency distribution histograms of scores obtained from ratio landscapes using three different algorithms for the classification of promoter sequences. Note that the overlapping and non-overlapping word algorithms provide the best discrimination between promoter sequences and intron/exon sequences. P, test promoter sequences, E, test exon sequences, I, test intron sequences.

Table 1. Optimal levels of promoter, exon and intron sequence detection combining the log-ratio landscape method with scoring algorithms for either overlapping or non-overlapping words. *P*, promoter; *E*, exon; *I*, intron; ***ratio landscape types*** indicates which landscape were used for detection. ***true*** +, ***false*** +, true and false positives respectively, *CC*, correlation coefficient.

		P		E		I	
	Ratio landscape types	$P_{p/e}$, $E_{p/e}$	$P_{p/e}$, $I_{p/i}$	$E_{e/p}$, $P_{e/p}$	$E_{e/i}$, $I_{e/i}$	$I_{i/p}$, $P_{i/p}$	$I_{i/e}$, $E_{i/e}$
Overlapping	True + (%)	91	92	96	88	97	93
	False + (%)	1	5	8	2	9	1
	CC	0.89	0.87	0.88	0.86	0.88	0.92
Non-overlapping	True + (%)	94	94	90	88	88	93
	False + (%)	4	15	3	2	5	2
	CC	0.91	0.79	0.87	0.86	0.83	0.92

classes, amongst others, and the prevailing problem is to parse the sequence into domains of known class with known boundaries (Guigó et al 1992, Uberbacher and Mural, 1991, Fields and Soderlund 1990, Snyder and Stormo 1995, Burset and Guigó 1996). In this section, preliminary data is presented that attempts to use scored ratio landscapes to parse known DNA sequences in order to make three-state predictions.

In previous sections the scoring algorithms have been employed to produce a final value that indicates whether a sequence has a preference for C_h or C_i. The sequential sum of the arrays O and N can be represented as,

$$T_{O,h/i}(j) = O_{max,h/i}(j) + O_{min,h/i}(j)$$
$$T_{N,h/i}(j) = N_{max,h/i}(j) + N_{min,h/i}(j). \tag{10}$$

Therefore as the words in the query sequence display similarities to words in either D_h or D_i, the values in T will reflect these changes. For example, if a region of the query sequence contains words similar to those in D_h then T will increase (positive gradient), whereas if a local region contains more words found in D_i then T will decrease (negative gradient). The exact nature of these changes are more readily seen on a plot of $T_{O,e/i}$ taken from HUMACTGA log-ratio landscape (Fig. 5). A sliding average smoothing filter (Press et al 1992), with a window size of 50, was applied to $T_{O,e/i}$ (Fig. 5a) without significant loss of the overall trends. In Fig. 5a the inflection points represent boundaries between exon and introns, therefore positive gradients identify exons whilst negative gradients identify introns (Fig.5b). The exon-intron regions taken from the annotation of the HUMACTGA sequence are marked on Fig.5b and these correspond well with the positive and negative gradients.

The two-state prediction involving exons and introns only require one log-ratio landscape, $R_{e/i}$. In order to attempt a three-state prediction, two ad-

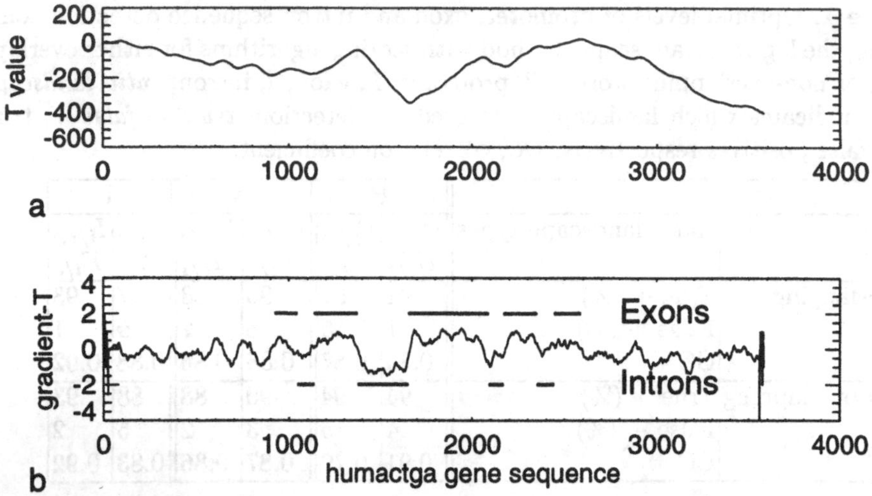

Fig. 5. Plots of $T_{O,e/i}$ for the HUMACTGA sequence taken from the log-ratio landscape of HUMACTGA. Smoothing performed with a sliding average filter (**a**). Gradient of smoothed $T_{O,e/i}$ (**b**) indicate exons (positive values) and introns (negative values). The exon-intron regions are shown, taken from the annotation of HUMACTGA sequence.

ditional log-ratio landscapes are needed, $R_{p/e}$ and $R_{p/i}$. Subsequently, gradients of the smoothed T array can be determined, which result in three arrays:- $\Delta T_{(O,e/i)}$, $\Delta T_{(O,p/e)}$, $\Delta T_{(O,p/i)}$, when the overlapping word algorithm is used to determine T. Using the resulting ΔT arrays from the three ratio landscapes after scoring using either the overlapping or non-overlapping method, it is possible to define predictor arrays for promoters (P), exons (E)and introns (I).

$$\begin{aligned} P_O(j) &= \Delta T_{(O,p/e)} + \Delta T_{(O,p/i)} \\ E_O(j) &= \Delta T_{(O,e/i)} - \Delta T_{(O,p/e)} \\ I_O(j) &= -\Delta T_{(O,e/i)} - \Delta T_{(O,p/i)} \end{aligned} \tag{11}$$

Equation (11) is written for the overlapping word algorithm but clearly three similar arrays exist for the non-overlapping word algorithm. The application of (11) to the HUMACTGA sequence shows the plots of the predictor array values along the sequence (Fig. 6a). The sequence can be parsed into different domains of promoter, exon and intron by choosing the predictor with the maximun value at each base. The maximum values found from 1 to approximately 1000 is promoter, which may be expected since promoter regions are generally, but not exclusively, found upstream from the first exon. However, it is clear that for the most part the promoter predictor maintains the maximum value of all three predictors through the sequence. A closer examination of the exon-intron predictors in Fig.6a does suggest that the exon-intron domains are reasonably correctly parsed if the promoter predictor did not exhibit such high values.

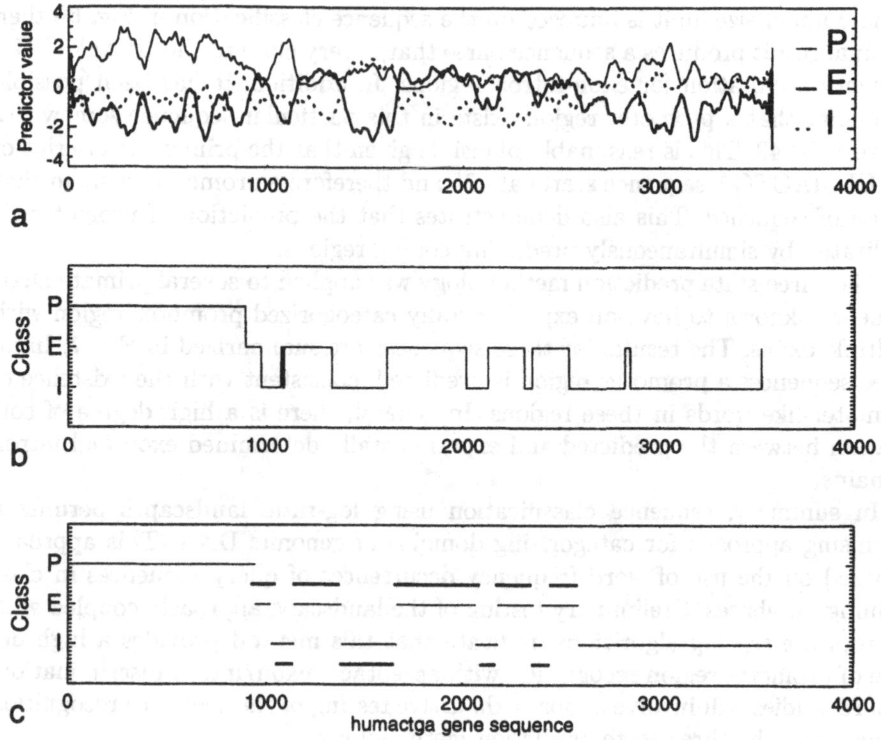

Fig. 6. Plots for three-state prediction of HUMACTGA sequence after overlapping word scoring.**a.** Plot of predictor arrays for promoter, exon and intron exhibit an over abundance of promoter regions. **b.** Three state prediction using the grammatical constraint $P \Rightarrow E \Leftrightarrow I$ **c.** Same as **b.** with the additional elimination of predicted domains < 40 bases in length. Exons and introns from the annotation of the HUMACTGA sequence are drawn in bold lines above and below the predicted exon and intron regions respectively. Note that the final sequence parse corresponds well with the known exon/intron boundaries.

One possible approach, given the apparent overprediction of promoter regions, is to apply a simple grammar scheme as had been employ in other genefinding approaches (Snyder and Stormo, 1995, Reese et al 1997, Stormo and Haussler, 1994). The sequence may be parsed such that the first transition event from promoter to exon essentially eliminates transition events from exon/intron back to promoter, *ie.* $P \Rightarrow E \Leftrightarrow I$. The application of this grammatical constraint clarifies the sequence classification considerably for the HUMACTGA sequence (Fig.6b). In the classification result in Fig.6b there is an $E \Rightarrow I \Rightarrow E$ transition between bases 1100-1200. In this particular case, the predicted intron domain is less than 40 bases in length and is probably too short to be an intron region. It is therefore necessary to impose a further constraint such that *any* predicted domain must be greater than 40 bases in length for the domain to be predicted. If the domain size limit is imposed on the sequence classification in Fig. 6b then the final result produces a sequence parse that is very close to the HUMACTGA sequence annotation for exon-intron regions. In addition, it has been possible to predict that a promoter region exists in this particular sequence somewhere between 1-943. This is reasonably plausible given that the primary transcript of the HUMACTGA sequence starts at 475 and therefore a promoter exists in that region of sequence. This also demonstrates that the prediction of promoters is facilitated by simultaneously predicting coding regions.

The three-state prediction methodology was applied to several primate DNA sequences known to have an experimentally cateogorized promoter region with multiple exons. The results for three sequences are summarized in Fig. 7. In all three sequences a promoter region is predicted, consistent with the existence of promoter-like words in these regions. In general, there is a high degree of correlation between the predicted and experimentally determined exon and intron domains.

In summary, sequence classification using log-ratio landscapes permits a promising approach for categorizing domains of genomic DNA. This approach is based on the use of word frequency occurrences of query sequences in class defining databases. Preliminary testing of the landscape approach, coupled with appropriate scoring algorithms, indicate that this method provides a high degree of promoter region recognition with acceptable exon/intron discrimination. Future studies will involve a more exhaustive testing of the sequence recognition scheme and the three-state prediction methodology.

6 Acknowledgments

This work was supported by the U.S. Department of Energy, Office of Energy Research, through grant ER61606. GDS would also like to acknowledge the long term contributions to our work from Andrzej Ehrenfeucht. He helped us get started on the research reported in this paper, and has been influential on several other projects. Much of our work would not have happened without the interactions we had with him.

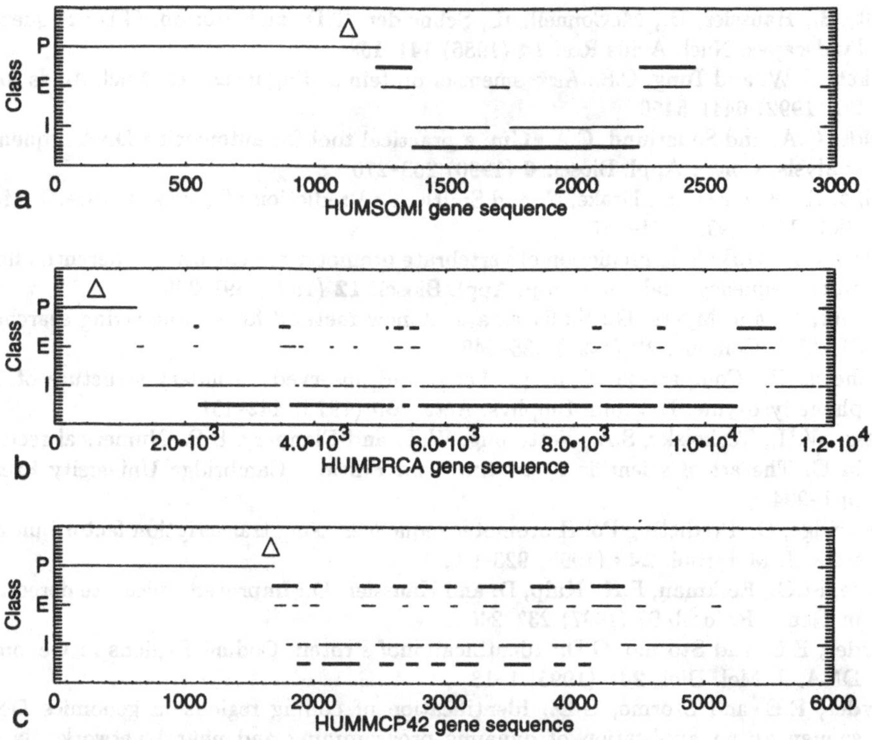

Fig. 7. Three-state prediction of promoter, exons and introns in the loci: HUMSOMI (a), HUMPRCA (b) and HUMMCP42 (c) . Bold lines drawn above exon and intron domains are taken from sequence annotation. Δ indicates the position of the transcription start site indicating that the predicted promoter regions occurrence correspond well with the known classification of these regions as containing promoter elements.

References

Blumer, A., Blumer, J., Ehrenfeucht, A., Haussler, D. and McConnell, R.: Building the minimal DFA for the set of all subwords of a word on-line in linear time. Lect. Notes Comp. Sci. **172** (1984) 109–118

Blumer, A., Blumer, J., Haussler, D., Ehrenfeucht, A., Chen, M.T. and Seiferas, J.: The smallest automaton recognizing the subwords of a text. Theoret. Comp. Sci. **40** (1985) 31–55

Bucher, P.: The eukaryotic promoter database EPD, EMBL nucleotide sequence data library, Release 48.

Burset, M. and Guigó, R.: Evaluation of gene structure prediction programs. Genomics, **34** (1996) 353–367

Chen, Q.K., Hertz, G.Z. and Stormo, G.D.: PromFD 1.0: a computer program that predicts eukaryotic pol II promoters using strings and IMD matrices. CABIOS (1997) In Press

Clift, B., Haussler, D., McConnell, R., Schneider, T.D. and Stormo, G.D.: Sequence landscapes. Nucl. Acids Res. **14** (1986) 141–158

Fickett, J.W. and Tung, C.S.: Assessment of protein coding measures. Nucl. Acids Res. **20** (1992) 6441–5450

Fields, C.A. and Soderlund, C.A.: Gm: a practical tool for automating DNA sequence analysis. Comp. Appl. Biosci. **6** (1990) 263–270

Guigó, R., Knudsen, S., Drake, N. and Smith, T.: Prediction of gene structure. J. Mol. Biol. **226** (1992) 141–157

Hutchinson, G.B.: The prediction of vertebrate promoter regions using differential hexamer frequency analysis. Comp. Appl. Biosci. **12** (1996) 391–398

Manber, U. and Myers, G.: Suffix arrays: A new method for on-line string searches. SIAM J. Comput. **22** (1993) 935–948

Mathews, B.: Comparison of the predicted and observed secondary structure of T4 phage lysozyme. Biochim. Biophys. Acta **405** (1975) 442–451

Press, W.H., Teukolsky, S.A., Vetterling, W.T. and Flannery, B.P.: Numerical recipes in C. The art of scientific computing. 2nd Edition. Cambridge University Press. pp1–994

Prestridge, D.: Predicting Pol II promoter sequences using transcription factor binding sites. J. Mol. Biol. **249** (1995) 923–932

Reese, M.G., Eeckman, F.H., Kulp, D. and Haussler, D.: Improved splice site detection in Genie. Recomb 97 (1997) 232–240

Snyder, E.E. and Stormo, G.D.: Identificationof Protein Coding Regions in Genomic DNA. J. Mol. Biol. **248** (1995) 1–18

Snyder, E.E. and Stormo, G.D.: Identification of coding regions in genomics DNA sequences: an application of dynamic programming and neural networks. Nucl. Acids Res. **21** (1993) 607–613

Solovyev, V.V., Salamov, A.A. and Lawrence, C.B.: Predicting internal exons by oligonucleotide composition and discriminant analysis of splicable open reading frames. Nucl. Acids Res. **22** (1994) 5156–5163

Stormo, G.D. and Haussler, D. : Optimally Parsing a Sequence into Different Classes Based on Multiple Types of Evidence. In: *Proceedings of the Second International Conference on Intelligent Systems in Molecular Biology*, pp. 369-375.

Uberbacher, E.C. and Mural, R.J.: Locating protein coding regions in human DNA sequences by a multiple sensor-neural network approach. Proc. Natl. Acad. Sci. **388** (1991) 11261–11265

DNA Computing: Distributed Splicing Systems*

Gheorghe Păun

Institute of Mathematics of the Romanian Academy
PO Box 1 - 764, 70700 Bucureşti, Romania
Email: gpaun@imar.ro

Abstract. Because splicing systems with a finite set of rules generate only regular languages, it is necessary to supplement such a system with a control mechanism on the use of rules. One fruitful idea is to use distributed architectures suggested by the grammar systems area. Three distributed computability (language generating) devices based on splicing are discussed here. First, we improve a result about the so-called communicating distributed H systems (systems with seven components are able to characterize the recursively enumerable languages - the best result known up to now is of ten components), then we introduce two new types of distributed H systems: the separated two-level H systems and the periodically time-varying H systems. In both cases we prove characterizations of recursively enumerable languages - which means that in all these cases we can design universal "DNA computers based on splicing".

1 Introduction

We continue here the study of *extended H systems*, language generating mechanisms introduced in [16], based on the splicing operation of [9]. This operation is a formal model of the recombinant behavior of DNA molecules (double stranded sequences) under the influence of restriction enzymes and ligases. Informally speaking, two DNA sequences are cut by two restriction enzymes and the fragments are recombined (by ligasion, provided that the ends produced by the enzymes match) such that possibly new sequences are produced. The sites where the enzymes can cut are encoded as pairs $(u_1, u_2), (u_3, u_4)$, and the fact that they produce matching ends is represented by the quadruple $((u_1, u_2), (u_3, u_4))$. We say that this is a *splicing rule*. In an H system, a set of axioms (initial strings) and a set of splicing rules are given. By an iterated application of these rules, starting from the axioms, we get a language. If also a terminal alphabet is provided and only strings on that alphabet are accepted, then we get the notion of an extended H system.

If only a finite set of rules are used, then even starting from a regular set of axioms, we can generate only regular languages. This has been proved in [4]; a simpler proof can be found in [17]. When using extended H systems, we obtain a characterization of regular languages, [16].

* Research supported by the Academy of Finland, Project 11281, and "NUFFIC: Cultural Cooperation between The Netherlands and Romania" Grant.

If the set of splicing rules is a regular language (each rule $((u_1, u_2), (u_3, u_4))$ is written as a string $u_1\#u_2\$u_3\#u_4$, hence the set of rules is a language), then extended H systems (with finite sets of axioms) characterize the recursively enumerable languages, that is they reach the full power of Turing machines/Chomsky type-0 grammars. This has been proved in [12]. Weaker results (for non-regular sets of rules) were previously given in [16].

However, working with infinite sets of rules, even regular, is not of much practical interest. Finite sets of rules give only regular languages, hence they stop at the level of finite automata/regular Chomsky grammars. It is therefore necessary to supplement the model with a feature able to increase its power. Many suggestions about how this can be done come both from the regulated rewriting area in formal language theory, see, e.g., [5], [7], and from the very proof in [12]. Several types of extended splicing systems with finite sets of axioms and of splicing rules were considered, with the application of splicing rules controlled in specific ways. We mention the control by ***permitting*** contexts (a rule is applied only to strings containing certain symbols associated to the rule), ***forbidding*** contexts (a rule is applied only to strings not containing certain symbols associated to the rule; the permitting contexts are a model of ***promoters***, the forbidding contexts correspond to ***inhibitors*** known in biochemistry), [2], [8], ***target*** languages (we accept the splicing only when the obtained strings belong to a given regular language), ***fitness*** mappings, as in the genetic algorithms area, [15], or working with ***multisets*** (keeping track of the number of copies of each string, starting with the axioms), [2], [8]. In all these cases, ***computational completeness*** is obtained, that is characterizations of recursively enumerable languages. Moreover, ***universal H systems*** of the mentioned types are obtained, in the usual sense: with all components fixed and able to simulate any given H system as soon as a code of it is introduced as an additional axiom in the universal system.

All these models have a common drawback (plus other specific shortcomings) when looking for implementing them: they use a large number of splicing rules, which means a large number of restriction enzymes. It is known that more restriction enzymes cannot work together, because each enzyme requires specific conditions (temperature, salinity, etc.). A possible idea to diminish this drawback is to use distributed architectures, as in grammar systems area, [1], [6], [11]. A variant of "distributed test tube systems" was introduced in [3]. Again, a characterization of recursively enumerable languages is obtained. The proof of this result from [3] does not give a bound on the number of components, but in [20] it is shown that distributed systems as in [3] with at most ten components can characterize the recursively enumerable languages. A stronger result is proved here: seven components suffice.

In search for models avoiding (some of) the shortcomings of the other models in this area, we propose here two new types of distributed H systems. The first one is a variant of a model discussed in [14]. Namely, we consider systems with the splicing occurring at two levels: at the local level, separate tubes exist which use internally their own splicing rules, at the global level there are splicing rules working on designated strings from each tube. The global splicing has priority on the local one. (In [14], also the global splicing rules are distributed to tubes, hence the global level is not separated from the local one.) We prove that systems of this form with three components characterize the recursively enumerable languages.

Finally, we consider systems with the sets of rules splitted in subsets which are used separately, one set at a time, in a periodical manner. This can be seen as a counterpart of the tables in an L system, modified according to environment changes. This fits very well with the DNA computing area, because the environment conditions have a crucial influence on biochemical reactions. As we have mentioned above, it is important to keep as small as possible the number of splicing rules used by a component of a system. This is realized in this last case: systems with at most three splicing rules in each component characterize the recursively enumerable languages.

Also these new models for DNA computing have non-realistic features from the point of view of the present day laboratory technology. However, *all* computationally complete models considered in this area are essentially based on such non-feasible yet assumptions, and all these assumptions are involved in controlling the splicing. Without such a control, we remain inside the borders of regular languages. Thus, we have to choose: either we are satisfied with the power of finite automata, or we must look for ways to implement the above mentioned assumptions, maybe also for other models, closer to the reality. This is a major challenge for the DNA computing area (which should be faced in an interdisciplinary manner).

2 Splicing systems

Let us consider an alphabet V and two special symbols, #, \$, not in V. By V^* we denote the set of all strings over V, including the empty one, denoted by λ. Two languages are considered equal if they differ by at most the empty string.

A *splicing rule* over V is a string $u_1\#u_2\$u_3\#u_4$, where $u_1, u_2, u_3, u_4 \in V^*$. For a splicing rule $r = u_1\#u_2\$u_3\#u_4$ and four strings $x, y, w, z \in V^*$ we write

$$\begin{aligned}(x, y) \vdash_r (w, z) \text{ iff } & x = x_1u_1u_2x_2, \; y = y_1u_3u_4y_2, \\ & w = x_1u_1u_4y_2, \; z = y_1u_3u_2x_2, \\ & \text{for some } x_1, x_2, y_1, y_2 \in V^*.\end{aligned}$$

We say that we *splice* the strings x, y at the *sites* u_1u_2, u_3u_4, respectively.

A pair $\sigma = (V, R)$, where V is an alphabet and R is a set of splicing rules over V is called an *H scheme*. With respect to a splicing scheme $\sigma = (V, R)$ and a language $L \subseteq V^*$ we define

$$\begin{aligned}&\sigma(L) = \{w \in V^* \mid (x, y) \vdash_r (w, z) \text{ or } (x, y) \vdash_r (z, w), \text{ for some } x, y \in L, r \in R\},\\ &\sigma^0(L) = L,\\ &\sigma^{i+1}(L) = \sigma^i(L) \cup \sigma(\sigma^i(L)), \; i \geq 0,\\ &\sigma^*(L) = \bigcup_{i\geq 0} \sigma^i(L).\end{aligned}$$

The fundamental notion for what follows if that of an H system.

An *extended H system* is a construct

$$\gamma = (V, T, A, R),$$

where V is an alphabet, $T \subseteq V, A \subseteq V^*$, and $R \subseteq V^*\#V^*\$V^*\#V^*$. ($T$ is the *terminal* alphabet, A is the set of *axioms*, and R is the set of *splicing rules*.) When $T = V$, the system is said to be non-extended. The pair $\sigma = (V, R)$ is the *underlying H scheme* of γ.

The language generated by γ is defined by

$$L(\gamma) = \sigma^*(A) \cap T^*.$$

(We iterate the splicing operation according to rules in R, starting from strings in A, and we keep only the strings composed of terminal symbols.)

We denote by $EH(F_1, F_2)$ the family of languages generated by extended H systems $\gamma = (V, T, A, R)$, with $A \in F_1, R \in F_2$, where F_1, F_2 are two given families of languages. (Note that R is a language, hence the definition makes sense.)

By *FIN, REG, CF, CS, RE* we denote the families of finite, regular, context-free, context-sensitive, recursively enumerable languages, respectively. For further elements of formal language theory we refer to [18], [19].

Two basic results concerning the power of extended H systems are the following ones.

Theorem 1. $EH(FIN, FIN) = EH(REG, FIN) = REG$.

Theorem 2. $EH(FIN, REG) = RE$.

The inclusion $EH(REG, FIN) \subseteq REG$ follows from the results in [4], [17], the inclusion $REG \subseteq EH(FIN, FIN)$ is proved in [16]. Theorem 2 is proved in [12].

3 Communicating distributed H systems

A *communicating distributed H system* (of degree $n, n \geq 1$) is a construct

$$\Gamma = (V, (A_1, R_1, V_1), \ldots, (A_n, R_n, V_n)),$$

where V is an alphabet, A_i is a finite subset of V^*, R_i is a finite subset of $V^*\#V^*\$V^*\#V^*$, and $V_i \subseteq V$, $1 \leq i \leq n$.

Each triple (A_i, R_i, V_i) is called a *component* of the system, or a *test tube*; A_i is the set of axioms of the tube i, R_i is the set of splicing rules of the tube i, V_i is the *selector* of the tube i.

We denote

$$B = V^* - \bigcup_{i=1}^{n} V_i^*.$$

The pair $\sigma_i = (V, R_i)$ is the underlying H scheme associated to the component i of the system.

An n-tuple $(L_1, \ldots, L_n), L_i \subseteq V^*, 1 \leq i \leq n$, is called a *configuration* of the system; L_i is also called the *contents* of the ith tube.

For two configurations $(L_1, \ldots, L_n), (L'_1, \ldots, L'_n)$, we define

$$(L_1, \ldots, L_n) \Longrightarrow (L'_1, \ldots, L'_n) \text{ iff, for each } i, 1 \leq i \leq n,$$

$$L'_i = \bigcup_{j=1}^{n} (\sigma_j^*(L_j) \cap V_i^*) \cup (\sigma_i^*(L_i) \cap B).$$

In words, the contents of each tube is spliced according to the associated set of rules (we pass from L_i to $\sigma_i^*(L_i), 1 \leq i \leq n$), and the result is redistributed among the n components according to the selectors $V_1, \ldots, V_n$; the part which cannot be redistributed (does not belong to some $V_k^*, 1 \leq k \leq n$) remains in the tube. Because we have imposed no restrictions over the alphabets V_i, for example, we did not suppose that they are pairwise disjoint, when a string in $\sigma_j^*(L_j)$ belongs to several languages V_i^*, then copies of this string will be distributed to all tubes i with this property.

The *language generated* by Γ is

$$L(\Gamma) = \{w \in V^* \mid w \in L_1 \text{ for } (A_1, \ldots, A_n) \Longrightarrow^* (L_1, \ldots, L_n), t \geq 0\},$$

where $\Longrightarrow^*$ is the reflexive and transitive closure of the relation $\Longrightarrow$.

By $CDH_n, n \geq 1$, we denote the family of languages generated by communicating distributed H systems with at most n components. When n is not specified, we write CDH_*.

In [3] it is proved that $RE = CDH_*$, then the result is improved in [20] to

Theorem 3. $RE = CDH_{10}$.

A strenghtening of this result is proved here, thus (partially) answering the question formulated in [20], whether or not this result is optimal: the result is not optimal, but still we do not know which is the best one

Theorem 4. $RE = CDH_7$.

Proof. We only have to prove the inclusion $\subseteq$.

Consider a type-0 grammar $G = (N, T, S, P)$, take a new symbol, B, and denote, for an easy reference, $N \cup T \cup \{B\} = \{X_1, X_2, \ldots, X_n\}$. Because $N \neq \emptyset, T \neq \emptyset$, we have $n \geq 3$. We construct the communicating distributed H system

$$\Gamma = (V, (A_1, R_1, V_1), \ldots, (A_7, R_7, V_7)),$$

with

$$V = N \cup T \cup \{X, X', X'', Y, Y', Y'', Z, B, C, E\},$$
$$A_1 = \emptyset,$$

$$R_1 = \emptyset,$$
$$V_1 = \emptyset,$$
$$A_2 = \{XBSY, ZE\} \cup \{ZvY \mid u \to v \in P\}$$
$$\cup \{ZC^iY \mid 1 \le i \le n\},$$
$$R_2 = \{\#uY\$\#vY \mid u \to v \in P\}$$
$$\cup \{\#X_iY\$Z\#C^iY' \mid 1 \le i \le n\}$$
$$\cup \{\#Y\$Z\#E\},$$
$$V_2 = N \cup T \cup \{X, Y, B\},$$
$$A_3 = \{ZY''\} \cup \{X''X_iZ \mid 1 \le i \le n\},$$
$$R_3 = \{\#CY'\$Z\#Y''\} \cup \{XC^i\#\$X''X_i\#Z \mid 1 \le i \le n\},$$
$$V_3 = N \cup T \cup \{X, Y', B, C\},$$
$$A_4 = \{X'CZ, EZZ\},$$
$$R_4 = \{X\#\$X'C\#Z,\ E\#ZZ\$XB\#\},$$
$$V_4 = N \cup T \cup \{X, Y'', B, C, E\},$$
$$A_5 = \{ZY'\},$$
$$R_5 = \{\#Y''\$Z\#Y'\},$$
$$V_5 = N \cup T \cup \{X', Y'', B, C\},$$
$$A_6 = \{XZ\},$$
$$R_6 = \{X'\#\$X\#Z\},$$
$$V_6 = N \cup T \cup \{X', Y', B, C\},$$
$$A_7 = \{XZ, ZY, ZZ\},$$
$$R_7 = \{X\#Z\$X''\#,\ \#Y'\$Z\#Y,\ \#E\$ZZ\#,\ \#ZZ\$E\#\},$$
$$V_7 = N \cup T \cup \{X'', Y', B, E\}.$$

Let us examine the work of Γ. The first component only selects the strings produced by other components which are terminal with respect to G. No terminal string (that is, a string in T^*) can enter a splicing, because all splicing rules in $R_2 - R_7$ contain control symbols $X, X', X'', Y, Y', Y'', Z$.

In the initial configuration $(A_1, \ldots, A_7)$ only the second component can execute a splicing. There are three possibilities: to use a rule $\#uY\$Z\#vY$, for some $u \to v \in P$, a rule $\#X_iY\$Z\#C^iY'$, for some $1 \le i \le n$, or the rule $\#Y\$Z\#E$.

Consider the general case, of having in component 2 a string XwY with $w \in (N \cup T \cup \{B\})^*$; initially, $w = BS$. We have three possible types of splicings, as mentioned above:

1. $(Xw_1|uY, Z|vY) \vdash (Xw_1vY, ZuY)$, for $u \to v \in P$, providing that $w = w_1u$;
2. $(Xw_1|X_iY, Z|C^iY') \vdash (Xw_1C^iY', ZX_iY)$, for $1 \le i \le n$, providing that $w = w_1X_i$;
3. $(Xw|Y, Z|E) \vdash (XwE, ZY)$.

No string containing a symbol Z can be communicated from a component to another one. The string Xw_1vY will remain in the first component (no selector

$V_i, 3 \leq i \leq 7$, contains both X and Y), and new splicings of types 1, 2, 3 can be applied to it. The string ZuY will enter new splicings only if a rule $x \rightarrow u$ exists in P, which means that ZuY is already in A_2. The string ZX_iY can enter only splicings with other axioms in A_2, hence always producing two strings containing the symbol Z. The same happens for ZY, which can be spliced only with ZE.

Let us examine first the string XwE obtained in the third case above. No further splicings can be done on it in component 2 and the only selector containing both X and E is V_4. If we communicate this string to the fourth component and we apply to it the rule $X\#\$X'C\#Z$, then we get

$$(X|wE, X'C|Z) \vdash (XZ, X'CwE).$$

The string $X'CwE$ cannot be communicated to another component and no further splicing can involve it. The string XZ can be spliced only with other axioms in A_4, hence no terminal string can be produced in this way. Therefore, XwE must be spliced in the fourth component by using the rule $E\#ZZ\$XB\#$, providing that $w = Bw'$:

$$(E|ZZ, XB|w'E) \vdash (Ew'E, XBZZ).$$

No further splicing can be done here on $Ew'E$, the string must be communicated to component 7. Two possible splicings in this component are

$$(|ZZ, E|w'E) \vdash (w'E, EZZ),$$
$$(w'|E, ZZ|) \vdash (w', ZZE).$$

Their order can be interchanged and the result is the same.

The string w' contains no control symbol X, X', X'', Y, Y', Y'', hence it cannot enter new splicings; it can be communicated to the first component only when $w' \in T^*$.

Splicings of the form

$$(|ZZ, Ew'E|) \vdash (\lambda, Ew'EZZ),$$

must be followed by splicings as above:

$$(Ew'|EZZ, ZZ|) \vdash (Ew', ZZEZZ),$$
$$(|ZZ, E|w') \vdash (w', EZZ).$$

In all cases, the same string w' is obtained after removing the two occurrences of E.

Consider now the string Xw_1C^iY' produced in the second component by a splicing of type 2 above. No further splicing in component 2 can involve this string, it will be communicated to the third component. There are two possibilities here:

$$(i)\ (Xw_1C^{i-1}|CY', Z|Y'') \vdash (Xw_1C^{i-1}Y'', ZCY'),$$
$$(ii)\ (XC^j|w_1'C^iY', X''X_j|Z) \vdash (XC^jZ, X''X_jw_1'C^iY'),$$
$$\text{providing that } w_1 = C^jw_1', 1 \leq j \leq n.$$

Let us again examine first the second case. The string XC^jZ can be spliced only with axioms, hence the symbol Z will be present in both the resulting strings. If the string $X''X_jw_1'C^iY'$ is spliced again in component 3, then one produces a string containing both the symbols X'', Y''; there is no component accepting such a string. Therefore, the string $X''X_jw_1'C^iY'$ must be communicated to component 7, providing that w_1' contains no occurrence of C and $i = 0$. This means that the block C^j in the left hand end of w is maximal and that no occurrence of C still exists to the left hand of w_1 above. In component 7, the string $X''X_jw_1'Y'$ can enter two splicings:

$$(X|Z, X''|X_jw_1'Y') \vdash (XX_jw_1'Y', X''Z),$$
$$(X''X_jw_1'|Y', Z|Y) \vdash (X''X_jw_1'Y, ZY').$$

The strings $X''Z, ZY'$ cannot produce terminal strings. The string $XX_jw_1'Y'$ can be communicated only to component 3, but none of the rules in R_3 can be applied, because each such rule contains an occurrence of C, whereas $XX_jw_1'Y'$ contains no occurrence of C. The string $X''X_jw_1'Y$ cannot be communicated.

Therefore, both the strings must enter new splicings in component 7 in order to lead to terminal strings:

$$(XX_jw_1'|Y', Z|Y) \vdash (XX_jw_1'Y, ZY'),$$
$$(X|Z, X''|X_jw_1'Y) \vdash (XX_jw_1'Y, X''Z).$$

Besides the "garbage" strings $ZY', X''Z$, we obtain the same string, $XX_jw_1'Y$. Moreover, we have started from Xw_1X_iY and we have obtained $XX_jw_1'Y$, passing through Xw_1C^iY'. The continuation of case (i) above moves occurrences of the symbol C from the right hand end of the string bounded by X, Y' to its left hand end (see below). All these occurrences were moved and all of them were replaced by X_j, otherwise, we have seen, we cannot reach the string $XX_jw_1'Y$. This implies $i = j$ and $w_1 = w_1'$, hence the symbol X_j has been moved from the right hand end to the left hand end of the string.

Let us continue now the case (i) above. The string $Xw_1C^{i-1}Y''$ should be communicated unmodified to component 4. (If we splice it again in component 3, then we produce a string containing both the symbols X'', Y'', and no component accepts it.) If we apply to it the rule $E\#ZZ\$XB\#$ in R_4, then we get

$$(E|ZZ, XB|w_2C^{i-1}Y'') \vdash (Ew_2C^{i-1}Y'', XBZZ),$$

providing that $w_1 = Bw_2$. No further splicing in this component can involve the string $Ew_2C^{i-1}Y''$ and no selector accepts this string. Therefore, we have to continue with

$$(X|w_1C^{i-1}Y'', X'C|Z) \vdash (XZ, X'Cw_1C^{i-1}Y'').$$

Consequently, the occurrence of C removed from the right-hand end of the string has been reintroduced in the left hand end. The string $X'Cw_1C^{i-1}Y''$ will be communicated to component 5, where we perform

$$(X'Cw_1C^{i-1}|Y'', Z|Y') \vdash (X'Cw_1C^{i-1}Y', ZY'').$$

The string $X'Cw_1C^{i-1}Y'$ should be communicated to component 6, where we perform

$$(X'|Cw_1C^{i-1}Y', X|Z) \vdash (X'Z, XCw_1C^{i-1}Y').$$

The string $XCw_1C^{i-1}Y'$ will be communicated to component 3, hence again splicings of types (i), (ii) are possible. The process can be iterated.

Consequently, component 2 simulates the rules in P, starts rotating the string by replacing X_i by C^i, or starts finishing the process by introducing E instead of Y. Components 3, 4, 5, 6 move occurrences of symbol C from the right hand end to the left hand end. Component 3 also reintroduces symbols X_i in the left hand end, replacing substrings C^i. Only when all symbols C were moved from the right hand end to the left hand end and all the moved symbols are replaced by the corresponding symbol X_i, the process can continue. In this way, the rotate-and-simulate procedure is correctly implemented, the system Γ will produce, as strings collected by the first component, exactly the strings generated by G. Thus, $L(G) = L(\Gamma)$. □

If we also provide a terminal alphabet, then component 1 can be removed, we can characterize RE by (extended) systems with only six components.

It is an *open problem* whether the "magic number" seven above can be replaced by a smaller one. Anyway, systems with three components are able to generate non-context-free languages, [3].

A variant of communicating distributed H systems can be the following one: after producing a string x in a component i we immediately check whether or not x can be communicated to another component, j; if this is the case, the string x enters no further splicing in component i; however, it is not immediately communicated to component j, but only after completing the σ_i^* splicing.

More formally, for $\sigma_i = (V, R_i)$ as above and

$$B_i = V^* - \bigcup_{j \neq i} V_j^*,$$

we define

$$\bar{\sigma}_i(L) = \sigma_i(L) \cap B_i.$$

Then $\bar{\sigma}_i^k, k \geq 0, \bar{\sigma}_i^*$ and, based on them, the language generated by a communicating distributed H system are defined in the same way as above. We denote by CDH_n', $n \geq 1$, and CDH_* the families of languages obtained in this way.

Examining the proof of Theorem 4, we can see that it works in the sense of $\bar{\sigma}_i$ (after splicing a string used in the simulation of a derivation in G, no further splicings can be done on the same string before communicating it to another component), with one exception: after communicating a string $X''wY'$ from component 3 to component 7, if in component 7 we first produce $X''wY$, this string cannot be communicated, hence one further splicing is done here, by which we get XwY; this string cannot enter new splicings in component 7. Therefore, the process is done according to the operation $\bar{\sigma}_7^*$ (no splicing is performed to a string which can be communicated). If we first splice such that we obtain XwY', and this string is communicated to component 3, which is not what we intend, and what is needed in order to simulate a derivation in G, then it cannot enter other splicings here. Consequently, working in the mode $\bar{\sigma}_i^*$, we generate the same language as working in the mode σ_i^*. Thus, we obtain

Theorem 5. $RE = CDH_7'$.

We close this section with the remark that the proof of Theorems 4 and 5 is constructive and, when starting from a universal type-0 grammar, it provides a universal communicating distributed H system, in the usual sense of the universality. The "program" of the particular H system to be simulated by the universal one is introduced in the axiom set of the second component, instead of the axiom $XBSY$ (the axiom S should be replaced by the *code* of the simulated grammar).

4 Two-level distributed H systems

The distributed H systems in the previous section involve a feature which can raise serious difficulties from a practical point of view: the communication between components observing the filters restriction. The models below try to avoid communication (in the form above).

A *two-level distributed H system* (of degree $n, n \geq 1$), is a construct, [14],

$$\Gamma = (V, T, C_1, \ldots, C_n),$$

where V is an alphabet, $T \subseteq V$, and $C_i = (w_i, A_i, I_i, E_i)$, $1 \leq i \leq n$, where $w_i \in V^*, A_i \subseteq V^*, I_i, E_i \subseteq V^*\#V^*\$V^*\#V^*$, for $\#, \$$ symbols not in V.

The meaning of these elements is as follows: V is the *total alphabet*, T is the *terminal alphabet*, and $C_i, 1 \leq i \leq n$, are the *components* of the system (we also call them *test tubes*); for each component, w_i is the *active axiom*, A_i is the set of *passive axioms*, I_i is the set of *internal splicing rules*, and E_i is the set of *external splicing rules*.

The *contents* of a tube $C_i, 1 \leq i \leq n$, is described by a pair (x_i, M_i), where $x_i \in V^*$ is the *active* string and $M_i \subseteq V^*$ is the set of *passive* strings. An n-tuple $\pi = [(x_1, M_1), \ldots, (x_n, M_n)]$ is called a *configuration* of the system. For $1 \leq i \leq n$ and a given configuration π as above, we define

$$\mu(x_i, \pi) = \begin{cases} \textit{external}, & \text{if there are } r \in E_i \text{ and } x_j, j \neq i, \\ & \text{such that } (x_i, x_j) \vdash_r (u, v), u, v \in V^*, \\ \textit{internal}, & \text{otherwise.} \end{cases}$$

Then, for two configurations $\pi = [(x_1, M_1), \ldots, (x_n, M_n)]$ and $\pi' = [(x_1', M_1'), \ldots, (x_n', M_n')]$, we write $\pi \Longrightarrow_{ext} \pi'$ if the following conditions hold:

1. there is $i, 1 \leq i \leq n$, such that $\mu(x_i, \pi) = \textit{external}$,
2. for each $i, 1 \leq i \leq n$, with $\mu(x_i, \pi) = \textit{external}$, we have $(x_i, x_j) \vdash_r (x_i', z_i)$, for some $j, 1 \leq j \leq n, j \neq i$, and $r \in E_i$; moreover, $M_i' = M_i \cup \{z_i\}$,

3. for each $i, 1 \leq i \leq n$, with $\mu(x_i, \pi) = internal$, we have $(x'_i, M'_i) = (x_i, M_i)$.

For two configurations π and π' as above, we write $\pi \Longrightarrow_{int} \pi'$ if the following conditions hold:

1. for all $i, 1 \leq i \leq n$, we have $\mu(x_i, \pi) = internal$,
2. for each $i, 1 \leq i \leq n$, either $(x_i, z) \vdash_r (x'_i, z')$, for some $z \in M_i, z' \in V^*, r \in I_i$, and $M'_i = M_i \cup \{z'\}$, or
3. no rule $r \in I_i$ can be applied to (x_i, z), for any $z \in M_i$, and then $(x'_i, M'_i) = (x_i, M_i)$.

The relation $\Longrightarrow_{ext}$ defines an external splicing, $\Longrightarrow_{int}$ defines an internal splicing. Note that in both cases all the splicing operations are performed in parallel and the components not able to use a splicing rule do not change their contents. We stress the fact that the external splicing has priority over the internal one and that all operations have as the first term an active string; the first string obtained by splicing becomes the new active string of the corresponding component, the second string becomes an element of the set of passive strings of that component.

We write $\Longrightarrow$ for both $\Longrightarrow_{ext}$ and $\Longrightarrow_{int}$, and $\Longrightarrow^*$ for the reflexive and transitive closure of $\Longrightarrow$. The *language generated* by a two-level distributed H system Γ is defined by

$$L(\Gamma) = \{w \in T^* | [(w_1, A_1), \ldots, (w_n, A_n)] \Longrightarrow^* [(x_1, M_1), \ldots, (x_n, M_n)], \text{ for } w = x_1, x_i \in V^*, 2 \leq i \leq n, \text{ and } M_i \subseteq V^*, 1 \leq i \leq n\}.$$

We denote by LDH_n the family of languages generated by two-level H systems with at most n components, $n \geq 1$, all of them having *finite sets of axioms and finite sets of splicing rules.* When no restriction is imposed on the number of components, we write LDH_*.

A more natural variant of the model above are the *separated two-level distributed H systems*, which are two-level H systems with the sets $E_i, 1 \leq i \leq n$, of external splicing rules, identical for all components. Then this common set of rules can be considered at the level of the system, and the system can be written in the more suggestive way

$$\Gamma = (V, T, (w_1, A_1, I_1), \ldots, (w_n, A_n, I_n), E).$$

The language generated is equal to the language generated by the two-level system $\Gamma' = (V, T, (w_1, A_1, I_1, E), \ldots, (w_n, A_n, I_n, E))$. We denote by $SLDH_n, n \geq 1$, the family of languages generated by separated two-level H systems of degree at most n; when n is not specified, we write $SLDH_*$.

From the definitions we have

Lemma 6. $SLDH_n \subseteq LDH_n, n \geq 1$.

Also in this case the hierarchy on the number of components collapses:

Theorem 7. $RE = SLDH_3$.

Proof. Consider a type-0 grammar $G = (N, T, S, P)$. We construct the separated two-level distributed H system

$$\Gamma = (V, T, (w_1, A_1, I_1), (w_2, A_2, I_2), (w_3, A_3, I_3), E),$$

with

$$\begin{aligned}
V &= N \cup T \cup \{X, Z, Z_s, Z_l, Z_r, Y, C_1, C_2, C_3\},\\
w_1 &= SXXC_1,\\
A_1 &= \{ZvXZ_s \mid u \to v \in P\}\\
&\cup \{ZXX\alpha Z_l, Z\alpha XXZ_r \mid \alpha \in N \cup T\},\\
I_1 &= \{\#uXZ\$Z\#vXZ_s \mid u \to v \in P\}\\
&\cup \{\#\alpha XZ\$Z\#XX\alpha Z_l,\ \#XZ\$Z\#\alpha XXZ_r \mid \alpha \in N \cup T\},\\
w_2 &= C_2Z,\\
A_2 &= \{C_2Y\},\\
I_2 &= \{C_2\#Y\$C_2\#Z\},\\
w_3 &= C_3Z,\\
A_3 &= \{C_3Y\},\\
I_3 &= \{C_3\#Z\$C_3\#Y,\ C_3\#Y\$C_3\#Z\},\\
E &= \{C_2\#Z\$X\#X,\ X\#X\$C_3\#Z,\ X\#Z_s\$C_2\#X,\\
&\quad C_2\#X\$C_3\#Y,\ \#XXC_1\$C_2Z\#\}\\
&\cup \{XX\alpha\#Z_l\$C_2X\#,\ \alpha XX\#Z_r\$C_2X\alpha\# \mid \alpha \in N \cup T\}.
\end{aligned}$$

The intuition behind this construction is the following. We simulate the work of G in the first two components, on their active strings. Namely, the substring XX of the active string of the first component shows the place where the simulation is done: immediately to the left of XX. To this aim, the active string of the first component is cut between the two occurrences of X, the prefix remains in the first component and the suffix is saved in the second component. This is done by external splicings performed simultaneously in the first and the second components. The simulation is performed by an internal splicing in the first component. Then the two strings are concatenated again by an external splicing, in the presence of the symbol X_s (the subscript s stands for "simulation"). The substring XX can be moved to the left or to the right, over one symbol, in a similar way for both directions; the symbols X_l, X_r control the operation (l = left, r = right). After removing the substring XX, together with the symbol C_1, no further splicing is possible.

Let us examine in some details the work of Γ.

Consider a configuration

$$[(w_1XXw_2C_1, M_1), (C_2Z, M_2), (C_3Z, M_0)] \qquad (*)$$

Initially we have $w_1 = S, w_2 = \lambda, M_1 = A_1, M_2 = A_2, M_3 = A_3$. We have to splice the active strings according to the rules $C_2\#Z\$X\#X, X\#X\$C_3\#Z$ in E and we obtain the configuration:

$$[(w_1XZ, M_1' = M_1 \cup \{C_3Xw_2C_1\}), (C_2Xw_2C_1, M_2' = M_2 \cup \{w_1XZ\}), (C_3Z, M_3)].$$

No external splicing is possible, we perform internal splicings in the first and the third components. In the first component there are three possibilities:

1) If $w_1 = w_1'u$, for some $u \to v \in P$, then we can use the rule $u\#XZ\$Z\#vXZ_s$ in component 1 and $C_3\#Z\$C_3\#Y$ in component 3 and we get:

$$[(w_1'vXZ_s, M_1'' = M_1' \cup \{ZuXZ\}), (C_2Xw_2C_1, M_2'), (C_3Y, M_3' = M_3 \cup \{C_3Z\})].$$

External splicings are possible, using the rules $X\#Z_s\$C_2\#X, C_2\#X\$C_3\#Y$, and leading to the configuration

$$[(w_1'vXXw_2C_1, M_1''' = M_1'' \cup \{C_2Z_s\}), (C_2Y, M_2'' = M_2' \cup \{C_3Xw_2C_1\}), (C_3Y, M_3')].$$

No external splicing is possible and no internal splicing in the first component, but we can perform internal splicings in components 2 and 3, leading to:

$$[(w_1'vXXw_2C_1, M_1'''), (C_2Z, M_2''' = M_2'' \cup \{C_2Y\}), (C_3Z, M_3'' = M_3' \cup \{C_3Y\})].$$

We have returned to a configuration of the form we have started with, $(*)$.

The new passive strings, produced during these operations, will enter no new splicing in components 1 and 2, and they are always C_3Y, C_3Z in the third component.

2) If $w_1 = w_1'\alpha$, for some $\alpha \in N \cup T$, then we can use the rule $\#\alpha XZ\$Z\#XX\alpha Z_l$ in component 1 and the rule $C_3\#Z\$C_3\#Y$ in component 3, and we get:

$$[(w_1'XX\alpha Z_l, M_1'' = M_1' \cup \{Z\alpha XZ\}), (C_2Xw_2C_1, M_2'), (C_3Y, M_3' = M_3 \cup \{C_3Z\})].$$

External splicing must be done, using the rules $XX\alpha\#Z_l\$C_2X\#$ and $C_2\#X\$C_3\#Y$, leading to:

$$[(w_1'XX\alpha w_2C_1, M_1''' = M_1'' \cup \{C_2XZ_l\}), (C_2Y, M_2'' = M_2' \cup \{C_3Xw_2C_1\}), (C_3Y, M_3')].$$

No external splicing is possible and no internal splicing in the first component, but we can perform internal splicings in the other components; we obtain the configuration

$$[(w_1'XX\alpha w_2C_1, M_1'''), (C_2Z, M_2''' = M_2' \cup \{C_2Y\}), (C_3Z, M_3'' = M_3' \cup \{C_3Y\})].$$

We have also returned to a configuration of type $(*)$.

3) If in configuration $(*)$ we have $w_2 = \alpha w_2'$, for some $\alpha \in N \cup T$, then we can proceed as follows; by using the rule $\#XZ\$Z\#\alpha XXZ_r$ in the first component and $C_3\#Z\$C_3\#Y$ in the third one, we first produce

$$[(w_1\alpha XXZ_r, M_1'' = M_1' \cup \{ZXZ\}), (C_2Xw_2C_1, M_2'), (C_3Y, M_3' = M_3 \cup \{C_3Z\})],$$

then we have to perform external splicings, according to the rules $\alpha XX\#Z_r\$C_2X\alpha\#$, $C_2\#X\$C_3\#Y$, leading to

$$[(w_1\alpha XXw_2'C_1, M_1''' = M_1''\cup\{C_2X\alpha Z_r\}), (C_2Y, M_2'' = M_2'\cup\{C_3Xw_2C_1\}), (C_3Y, M_3')].$$

No external splicing is possible and no internal splicing in the first component; splicing internally in the other components, we get

$$[(w_1\alpha XXw_2'C_1, M_1'''), (C_2Z, M_2''' = M_2''\cup\{C_2Y\}), (C_3Z, M_3'' = M_3'\cup\{C_3Y\})].$$

We have again returned to a configuration of type $(*)$.

In case 1 we have simulated the rule $u \to v$ in the presence of XX, in case 2 we have moved XX across a symbol to the left, in case 3 we have moved XX across a symbol to the right. The operations above can be repeated an arbitrary number of times. Changing in this way the place of XX, we can simulate the rules of G in any place of the word.

When we have a configuration of the form

$$[(wXXC_1, M_1), (C_2Z, M_2), (C_3Z, M_3)],$$

then also the external splicing is possible using the rules $\#XXC_1\$C_2Z\#$, $C_2\#Z\$X\#X$, leading to

$$[(w, M_1' = M_1\cup\{C_2ZXXC_1\}), (C_2XC_1, M_2' = M_2\cup\{wXZ\}), (C_3Z, M_3)].$$

No further splicings, internal or external, are possible in the first component. If w is a terminal string, then it is accepted in the language generated by Γ, if not, then it will never lead to a terminal string.

Consequently, we have $L(G) = L(\Gamma)$. □

We do not know whether the threshold 3 in this theorem is optimal or not. Again, starting from a universal type-0 grammar, we obtain a universal two-level distributed H system.

5 Time-varying distributed H systems

The model introduced in this section, reminding periodically time-varying grammars in regulated rewriting area and tabled L systems, starts from the observation that the splicing rules are based on enzymes whose work essentially depends on the environment conditions. Hence, in any moment, only a subset of the set of all available rules are active.

A *time-varying distributed H system* (of degree $n, n \geq 1$) is a construct

$$\Gamma = (V, T, A, R_1, R_2, \ldots, R_n),$$

where V is an alphabet, $T \subseteq V$ (terminal alphabet), A is a finite subset of V^* (axioms), and R_i are finite sets of splicing rules over $V, 1 \leq i \leq n$.

At each moment $k = n \cdot j + i, j \geq 0, 1 \leq i \leq n$, the component R_i is used for splicing the currently available strings. Specifically, we define

$$L_0 = A,$$
$$L_k = \sigma_i(L_{k-1}), \text{ for } i \equiv k (mod\ n), k \geq 1,$$

where $\sigma_i = (V, R_i), 1 \leq i \leq n$.

Therefore, from a step k to the next step, $k+1$, one passes only the result of splicing the strings in L_k according to the rules in R_k; the strings in L_k which cannot enter a splicing are removed.

The language generated by Γ is defined by

$$L(\Gamma) = (\bigcup_{k \geq 0} L_k) \cap T^*.$$

We denote by $VDH_n, n \geq 1$, the family of languages generated by time-varying distributed H systems of degree at most n, and by VDH_* the family of all languages of this type.

In contrast to the results in the previous sections, we were not able to characterize RE by systems with a bounded number of components, but, and this is more attractive from a practical point of view, we have bounded the size of each component: components consisting of only three splicing rules are enough.

Theorem 8. $RE = VDH_*$.

Proof. Consider a type-0 grammar $G = (N, T, S, P)$ with $N \cup T = \{\alpha_1, \ldots, \alpha_n\}$ and $P = \{u_i \to v_i \mid 1 \leq i \leq m\}$. Let $\alpha_{n+1} = B$ be a new symbol. We construct the time-varying distributed H system

$$\Gamma = (V, T, A, R_1, \ldots, R_{2n+m+2}),$$

with

$$V = N \cup T \cup \{X, Y, Y', Z, B\},$$
$$A = \{XBSY, ZY, ZY', ZZ\}$$
$$\cup \{ZvY \mid u \to v \in P\}$$
$$\cup \{X\alpha_i Z \mid 1 \leq i \leq n\},$$

and the following sets of splicing rules

$$R_i = \{\#u_iY\$Z\#v_iY,\ \#Y\$Z\#Y,\ Z\#\$Z\#\},\ 1 \leq i \leq m,$$
$$R_{m+2j-1} = \{\#\alpha_jY\$Z\#Y,\ \#Y\$Z\#Y',\ Z\#\$Z\#\},\ 1 \leq j \leq n,$$
$$R_{m+2j} = \{X\alpha_j\#Z\$X\#,\ \#Y'\$Z\#Y,\ Z\#\$Z\#\},\ 1 \leq j \leq n,$$
$$R_{m+2n+1} = \{XB\#\$\#ZZ,\ \#Y\$Z\#Y',\ Z\#\$Z\#\},$$
$$R_{m+2n+2} = \{\#Y\$ZZ\#,\ \#Y'\$Z\#Y,\ Z\#\$Z\#\}.$$

The idea behind this construction is again the simulate-and-rotate one, used also in the proof of Theorem 4. The components $R_i, 1 \leq i \leq m$, simulate the rules in P, in the right hand end of the strings of the form XwY produced by Γ (starting with $XBSY$). The components $R_{m+2j-1}, 1 \leq j \leq n$, remove one occurrence of the corresponding symbol α_j from the right hand end of the strings, whereas the pair components $R_{m+2j}, 1 \leq j \leq n$, reintroduce an occurrence of α_j in the left hand end of the strings. Thus, these components $R_i, m+1 \leq i \leq m+2n$, circularly permute the strings, making possible the simulation of rules in P in any desired position. The components R_{m+2n+1}, R_{m+2n+2} remove the end markers X (only in the presence of B, hence in the right permutation) and Y. All components contain rules used just for passing the strings unmodified to the next step. These rules are $\#Y\$Z\#Y$ in $R_i, 1 \leq i \leq m$, $\#Y\$Z\#Y'$ and $\#Y'\$Z\#Y$ alternating in $R_{m+2j-1}, R_{m+2j}, 1 \leq j \leq n$, and in R_{m+2n+1}, R_{m+2n+2}, as well as the rules $Z\#\$Z\#$ present in all components, and used for passing from a step to the next one the axioms. The role of Y' is to prevent wrong splicing, by introducing a symbol α_j in a string from which a symbol α_j has not been removed. This is the essential point of this construction, hence we shall examine its implementation in some detail.

Consider a string XwY and assume that the component R_1 works on it. The rule $\#Y\$Z\#Y$ changes nothing, it only passes the string to the next component. If we reach a component $R_i, 1 \leq i \leq n$, and a rule $\#u_iY\$Z\#v_iY$ can be used, then again a string bounded by X, Y is obtained.

When reaching R_{m+1}, a string of the form XwY can enter the splicing

$$(Xw|Y, Z|Y) \vdash (XwY', ZY)$$

or, when $w = w'\alpha_1$, the splicing

$$(Xw'|\alpha_1Y, Z|Y) \vdash (Xw'Y, Z\alpha_1Y).$$

In the first case, XwY' can also enter two different splicings in the next component, R_{m+2}:

$$(Xw|Y', Z|Y) \vdash (XwY, ZY'),$$
$$(X\alpha_1|Z, X|wY) \vdash (X\alpha_1wY', XZ).$$

By the first splicing, we have returned to XwY, which is passed to the third component. Also the string $X\alpha_1wY'$ obtained in the second case is passed to the third component, but there is no splicing rule here which can be applied to this string. Thus $X\alpha_1wY'$ is no more present at the next steps, preventing the production of "wrong" strings: α_1 is introduced to the left of w without first removing an occurrence of α_1 from the right hand of w.

In the second case, that when α_1 has been removed from $w'\alpha_1$, the string $Xw'Y$ reaches R_{m+2} where only one splicing is possible:

$$(X\alpha_1|Z, X|w'Y) \vdash (X\alpha_1w'Y, XZ).$$

The string $X\alpha_1w'Y$ is a correct one-step circular permutation of $Xw'\alpha_1Y$, and it is again bounded by X, Y.

Therefore, we pass to R_{m+3} a string of the form XzY. The case of $j = 1$ is similar to the case of arbitrary j, $1 \leq j \leq n$, hence the components R_{m+2j-1}, R_{m+2j}, perform the desired rotations, or they pass the strings unmodified to R_{m+2n+1}.

A string of the form XwY can enter in R_{m+2n+1} two splicings:

$$(Xw|Y, Z|Y') \vdash (XwY', ZY),$$
$$(XB|w'Y, |ZZ) \vdash (XBZZ, w'Y), \text{ if } w = Bw.$$

In the first case, there is only one possible splicing in R_{m+2n+2}:

$$(Xw|Y', Z|Y) \vdash (XwY, ZY'),$$

hence we pass the string XwY unchanged to R_1, resuming the cycle. In the second case, we also have only one possibility of splicing in R_{m+2n+2}:

$$(w'|Y, ZZ|) \vdash (w', ZZY).$$

The string w' is not marked by X, Y, hence it cannot enter new splicings if it is passed to R_1. If it is not terminal, then it is lost. Consequently, any derivation in G can be simulated in Γ, following the usual simulate-and-rotate procedure. Also like in the proofs in the previous sections, the "by-products" of the splicings, strings which are not of the forms XwY, XwY', are never producing terminal strings outside $L(G)$. Thus, $L(G) = L(\Gamma)$, completing the proof. □

6 Conclusions

We have solved a problem formulated as open in [20] (seven test tubes in a communicating distributed H systems are enough in order to characterize the recursively enumerable languages – and the same result holds true also for a variant of these systems), then we have introduced separated two-level distributed H systems and time-varying distributed H systems. Also these two types of distributed H systems characterize RE (three components are sufficient in the first case, components consisting of three rules only suffice in the second case). On the basis of the proofs, *universal* H systems of the mentioned types are obtained. This theoretically proves that "universal programmable DNA computers based on splicing" can be designed (and, moreover, proves that language theory is a suitable framework for investigating, at the conceptual level, the power of DNA computing).

Acknowledgments. Some of the variants of the systems above have occcurred during discussions (of course, distributed...) with Jürgen Dassow, Rudolf Freund, Grzegorz Rozenberg, Arto Salomaa.

References

1. E. Csuhaj-Varju, J. Dassow, J. Kelemen, Gh. Păun, *Grammar Systems. A Grammatical Approach to Distribution and Cooperation*, Gordon and Breach, London, 1994.
2. E. Csuhaj-Varju, L. Freund, L. Kari, Gh. Păun, DNA computing based on splicing: universality results, *First Annual Pacific Symp. on Biocomputing*, Hawaii, Jan. 1996.
3. E. Csuhaj-Varju, L. Kari, Gh. Păun, Test tube distributed systems based on splicing, *Computers and AI*, 15, 2-3 (1996), 211 - 231.
4. K. Culik II, T. Harju, Splicing semigroups of dominoes and DNA, *Discrete Appl. Math.*, 31 (1991), 261 - 277.
5. J. Dassow, Gh. Păun, *Regulated Rewriting in Formal Language Theory*, Springer-Verlag, Berlin, Heidelberg, 1989.
6. J. Dassow, Gh. Păun, G. Rozenberg, Grammars systems, a chapter in vol. 2 of *Handbook of Formal Languages* (G. Rozenberg, A. Salomaa, eds.), Springer-Verlag, Heidelberg, 1997.
7. J. Dassow, Gh. Păun, A. Salomaa, Grammars with controlled derivations, a chapter in vol. 2 of *Handbook of Formal Languages* (G. Rozenberg, A. Salomaa, eds.), Springer-Verlag, Heidelberg, 1997.
8. R. Freund, L. Kari, Gh. Păun, DNA computing based on splicing: The existence of universal computers, *Technical Report 185-2/FR-2/95*, TU Wien, 1995.
9. T. Head, Formal language theory and DNA: an analysis of the generative capacity of specific recombinant behaviors, *Bull. Math. Biology*, 49 (1987), 737 - 759.
10. T. Head, Gh. Păun, D. Pixton, Language theory and molecular genetics. Generative mechanisms suggested by DNA recombination, a chapter in vol. 2 of *Handbook of Formal Languages* (G. Rozenberg, A. Salomaa, eds.), Springer-Verlag, Heidelberg, 1996.
11. Gh. Păun, Grammar systems. A grammatical approach to distribution and cooperation, ICALP 1995, *LNCS 944* (Z. Fulop, F. Gecseg, eds.), Springer-Verlag, 1995, 429 - 443.
12. Gh. Păun, Regular extended H systems are computationally universal, *J. Automata, Languages, Combinatorics*, 1, 1 (1996), 27 - 36.
13. Gh. Păun, Computing by splicing: How simple rules ?, *Bulletin of the EATCS*, 60 (1996), 144 - 150.
14. Gh. Păun, Five (plus two) universal DNA computing models based on the splicing operation, *Second DNA Based Computers Workshop*, Princeton, 1996, 67 - 86.
15. Gh. Păun, Splicing systems with targets are computationally complete, *Inform. Processing Letters, Inform. Processing Letters*, 59 (1996), 129 - 133.
16. Gh. Păun, G. Rozenberg, A. Salomaa, Computing by splicing, *Theoretical Computer Sci.*, 168, 2 (1996), 321 - 336.
17. D. Pixton, Regularity of splicing languages, *Discrete Appl. Math.*, 69 (1996), 101 - 124.
18. G. Rozenberg, A. Salomaa, eds., *Handbook of Formal Languages*, 3 volumes, Springer-Verlag, Heidelberg, 1997.
19. A. Salomaa, *Formal Languages*, Academic Press, New York, 1973.
20. C. Zandron, C. Ferretti, G. Mauri, A reduced distributed splicing system for RE languages, in vol. *Control, Coooperation, Combinatorics. New Trends in Formal Languages* (Gh. Păun, A. Salomaa, eds.), *LNCS* 1218, Springer-Verlag, 1997

Author Index

Lecture Notes in Computer Science

For information about Vols. 1–1192

please contact your bookseller or Springer-Verlag

Vol. 1193: J.P. Müller, M.J. Wooldridge, N.R. Jennings (Eds.), Intelligent Agents III. XV, 401 pages. 1997. (Subseries LNAI).

Vol. 1194: M. Sipper, Evolution of Parallel Cellular Machines. XIII, 199 pages. 1997.

Vol. 1195: R. Trappl, P. Petta (Eds.), Creating Personalities for Synthetic Actors. VII, 251 pages. 1997. (Subseries LNAI).

Vol. 1196: L. Vulkov, J. Waśniewski, P. Yalamov (Eds.), Numerical Analysis and Its Applications. Proceedings, 1996. XIII, 608 pages. 1997.

Vol. 1197: F. d'Amore, P.G. Franciosa, A. Marchetti-Spaccamela (Eds.), Graph-Theoretic Concepts in Computer Science. Proceedings, 1996. XI, 410 pages. 1997.

Vol. 1198: H.S. Nwana, N. Azarmi (Eds.), Software Agents and Soft Computing: Towards Enhancing Machine Intelligence. XIV, 298 pages. 1997. (Subseries LNAI).

Vol. 1199: D.K. Panda, C.B. Stunkel (Eds.), Communication and Architectural Support for Network-Based Parallel Computing. Proceedings, 1997. X, 269 pages. 1997.

Vol. 1200: R. Reischuk, M. Morvan (Eds.), STACS 97. Proceedings, 1997. XIII, 614 pages. 1997.

Vol. 1201: O. Maler (Ed.), Hybrid and Real-Time Systems. Proceedings, 1997. IX, 417 pages. 1997.

Vol. 1203: G. Bongiovanni, D.P. Bovet, G. Di Battista (Eds.), Algorithms and Complexity. Proceedings, 1997. VIII, 311 pages. 1997.

Vol. 1204: H. Mössenböck (Ed.), Modular Programming Languages. Proceedings, 1997. X, 379 pages. 1997.

Vol. 1205: J. Troccaz, E. Grimson, R. Mösges (Eds.), CVRMed-MRCAS'97. Proceedings, 1997. XIX, 834 pages. 1997.

Vol. 1206: J. Bigün, G. Chollet, G. Borgefors (Eds.), Audio- and Video-based Biometric Person Authentication. Proceedings, 1997. XII, 450 pages. 1997.

Vol. 1207: J. Gallagher (Ed.), Logic Program Synthesis and Transformation. Proceedings, 1996. VII, 325 pages. 1997.

Vol. 1208: S. Ben-David (Ed.), Computational Learning Theory. Proceedings, 1997. VIII, 331 pages. 1997. (Subseries LNAI).

Vol. 1209: L. Cavedon, A. Rao, W. Wobcke (Eds.), Intelligent Agent Systems. Proceedings, 1996. IX, 188 pages. 1997. (Subseries LNAI).

Vol. 1210: P. de Groote, J.R. Hindley (Eds.), Typed Lambda Calculi and Applications. Proceedings, 1997. VIII, 405 pages. 1997.

Vol. 1211: E. Keravnou, C. Garbay, R. Baud, J. Wyatt (Eds.), Artificial Intelligence in Medicine. Proceedings, 1997. XIII, 526 pages. 1997. (Subseries LNAI).

Vol. 1212: J. P. Bowen, M.G. Hinchey, D. Till (Eds.), ZUM '97: The Z Formal Specification Notation. Proceedings, 1997. X, 435 pages. 1997.

Vol. 1213: P. J. Angeline, R. G. Reynolds, J. R. McDonnell, R. Eberhart (Eds.), Evolutionary Programming VI. Proceedings, 1997. X, 457 pages. 1997.

Vol. 1214: M. Bidoit, M. Dauchet (Eds.), TAPSOFT '97: Theory and Practice of Software Development. Proceedings, 1997. XV, 884 pages. 1997.

Vol. 1215: J. M. L. M. Palma, J. Dongarra (Eds.), Vector and Parallel Processing – VECPAR'96. Proceedings, 1996. XI, 471 pages. 1997.

Vol. 1216: J. Dix, L. Moniz Pereira, T.C. Przymusinski (Eds.), Non-Monotonic Extensions of Logic Programming. Proceedings, 1996. XI, 224 pages. 1997. (Subseries LNAI).

Vol. 1217: E. Brinksma (Ed.), Tools and Algorithms for the Construction and Analysis of Systems. Proceedings, 1997. X, 433 pages. 1997.

Vol. 1218: G. Păun, A. Salomaa (Eds.), New Trends in Formal Languages. IX, 465 pages. 1997.

Vol. 1219: K. Rothermel, R. Popescu-Zeletin (Eds.), Mobile Agents. Proceedings, 1997. VIII, 223 pages. 1997.

Vol. 1220: P. Brezany, Input/Output Intensive Massively Parallel Computing. XIV, 288 pages. 1997.

Vol. 1221: G. Weiß (Ed.), Distributed Artificial Intelligence Meets Machine Learning. Proceedings, 1996. X, 294 pages. 1997. (Subseries LNAI).

Vol. 1222: J. Vitek, C. Tschudin (Eds.), Mobile Object Systems. Proceedings, 1996. X, 319 pages. 1997.

Vol. 1223: M. Pelillo, E.R. Hancock (Eds.), Energy Minimization Methods in Computer Vision and Pattern Recognition. Proceedings, 1997. XII, 549 pages. 1997.

Vol. 1224: M. van Someren, G. Widmer (Eds.), Machine Learning: ECML-97. Proceedings, 1997. XI, 361 pages. 1997. (Subseries LNAI).

Vol. 1225: B. Hertzberger, P. Sloot (Eds.), High-Performance Computing and Networking. Proceedings, 1997. XXI, 1066 pages. 1997.

Vol. 1226: B. Reusch (Ed.), Computational Intelligence. Proceedings, 1997. XIII, 609 pages. 1997.

Vol. 1227: D. Galmiche (Ed.), Automated Reasoning with Analytic Tableaux and Related Methods. Proceedings, 1997. XI, 373 pages. 1997. (Subseries LNAI).

Vol. 1228: S.-H. Nienhuys-Cheng, R. de Wolf, Foundations of Inductive Logic Programming. XVII, 404 pages. 1997. (Subseries LNAI).

Vol. 1230: J. Duncan, G. Gindi (Eds.), Information Processing in Medical Imaging. Proceedings, 1997. XVI, 557 pages. 1997.

Vol. 1231: M. Bertran, T. Rus (Eds.), Transformation-Based Reactive Systems Development. Proceedings, 1997. XI, 431 pages. 1997.

Vol. 1232: H. Comon (Ed.), Rewriting Techniques and Applications. Proceedings, 1997. XI, 339 pages. 1997.

Vol. 1233: W. Fumy (Ed.), Advances in Cryptology — EUROCRYPT '97. Proceedings, 1997. XI, 509 pages. 1997.

Vol 1234: S. Adian, A. Nerode (Eds.), Logical Foundations of Computer Science. Proceedings, 1997. IX, 431 pages. 1997.

Vol. 1235: R. Conradi (Ed.), Software Configuration Management. Proceedings, 1997. VIII, 234 pages. 1997.

Vol. 1236: E. Maier, M. Mast, S. LuperFoy (Eds.), Dialogue Processing in Spoken Language Systems. Proceedings, 1996. VIII, 220 pages. 1997. (Subseries LNAI).

Vol. 1238: A. Mullery, M. Besson, M. Campolargo, R. Gobbi, R. Reed (Eds.), Intelligence in Services and Networks: Technology for Cooperative Competition. Proceedings, 1997. XII, 480 pages. 1997.

Vol. 1239: D. Sehr, U. Banerjee, D. Gelernter, A. Nicolau, D. Padua (Eds.), Languages and Compilers for Parallel Computing. Proceedings, 1996. XIII, 612 pages. 1997.

Vol. 1240: J. Mira, R. Moreno-Díaz, J. Cabestany (Eds.), Biological and Artificial Computation: From Neuroscience to Technology. Proceedings, 1997. XXI, 1401 pages. 1997.

Vol. 1241: M. Akşit, S. Matsuoka (Eds.), ECOOP'97 – Object-Oriented Programming. Proceedings, 1997. XI, 531 pages. 1997.

Vol. 1242: S. Fdida, M. Morganti (Eds.), Multimedia Applications, Services and Techniques – ECMAST '97. Proceedings, 1997. XIV, 772 pages. 1997.

Vol. 1243: A. Mazurkiewicz, J. Winkowski (Eds.), CONCUR'97: Concurrency Theory. Proceedings, 1997. VIII, 421 pages. 1997.

Vol. 1244: D. M. Gabbay, R. Kruse, A. Nonnengart, H.J. Ohlbach (Eds.), Qualitative and Quantitative Practical Reasoning. Proceedings, 1997. X, 621 pages. 1997. (Subseries LNAI).

Vol. 1245: M. Calzarossa, R. Marie, B. Plateau, G. Rubino (Eds.), Computer Performance Evaluation. Proceedings, 1997. VIII, 231 pages. 1997.

Vol. 1246: S. Tucker Taft, R. A. Duff (Eds.), Ada 95 Reference Manual. XXII, 526 pages. 1997.

Vol. 1247: J. Barnes (Ed.), Ada 95 Rationale. XVI, 458 pages. 1997.

Vol. 1248: P. Azéma, G. Balbo (Eds.), Application and Theory of Petri Nets 1997. Proceedings, 1997. VIII, 467 pages. 1997.

Vol. 1249: W. McCune (Ed.), Automated Deduction – CADE-14. Proceedings, 1997. XIV, 462 pages. 1997. (Subseries LNAI).

Vol. 1250: A. Olivé, J.A. Pastor (Eds.), Advanced Information Systems Engineering. Proceedings, 1997. XI, 451 pages. 1997.

Vol. 1251: K. Hardy, J. Briggs (Eds.), Reliable Software Technologies – Ada-Europe '97. Proceedings, 1997. VIII, 293 pages. 1997.

Vol. 1252: B. ter Haar Romeny, L. Florack, J. Koenderink, M. Viergever (Eds.), Scale-Space Theory in Computer Vision. Proceedings, 1997. IX, 365 pages. 1997.

Vol. 1253: G. Bilardi, A. Ferreira, R. Lüling, J. Rolim (Eds.), Solving Irregularly Structured Problems in Parallel. Proceedings, 1997. X, 287 pages. 1997.

Vol. 1254: O. Grumberg (Ed.), Computer Aided Verification. Proceedings, 1997. XI, 486 pages. 1997.

Vol. 1255: T. Mora, H. Mattson (Eds.), Applied Algebra, Algebraic Algorithms and Error-Correcting Codes. Proceedings, 1997. X, 353 pages. 1997.

Vol. 1256: P. Degano, R. Gorrieri, A. Marchetti-Spaccamela (Eds.), Automata, Languages and Programming. Proceedings, 1997. XVI, 862 pages. 1997.

Vol. 1258: D. van Dalen, M. Bezem (Eds.), Computer Science Logic. Proceedings, 1996. VIII, 473 pages. 1997.

Vol. 1259: T. Higuchi, M. Iwata, W. Liu (Eds.), Evolvable Systems: From Biology to Hardware. Proceedings, 1996. XI, 484 pages. 1997.

Vol. 1260: D. Raymond, D. Wood, S. Yu (Eds.), Automata Implementation. Proceedings, 1996. VIII, 189 pages. 1997.

Vol. 1261: J. Mycielski, G. Rozenberg, A. Salomaa (Eds.), Structures in Logic and Computer Science. X, 371 pages. 1997.

Vol. 1262: M. Scholl, A. Voisard (Eds.), Advances in Spatial Databases. Proceedings, 1997. XI, 379 pages. 1997.

Vol. 1263: J. Komorowski, J. Zytkow (Eds.), Principles of Data Mining and Knowledge Discovery. Proceedings, 1997. IX, 397 pages. 1997. (Subseries LNAI).

Vol. 1264: A. Apostolico, J. Hein (Eds.), Combinatorial Pattern Matching. Proceedings, 1997. VIII, 277 pages. 1997.

Vol. 1265: J. Dix, U. Furbach, A. Nerode (Eds.), Logic Programming and Nonmonotonic Reasoning. Proceedings, 1997. X, 453 pages. 1997. (Subseries LNAI).

Vol. 1266: D.B. Leake, E. Plaza (Eds.), Case-Based Reasoning Research and Development. Proceedings, 1997. XIII, 648 pages. 1997 (Subseries LNAI).

Vol. 1267: E. Biham (Ed.), Fast Software Encryption. Proceedings, 1997. VIII, 289 pages. 1997.

Vol. 1268: W. Kluge (Ed.), Implementation of Functional Languages. Proceedings, 1996. XI, 284 pages. 1997.

Vol. 1269: J. Rolim (Ed.), Randomization and Approximation Techniques in Computer Science. Proceedings, 1997. VIII, 227 pages. 1997.

Vol. 1270: V. Varadharajan, J. Pieprzyk, Y. Mu (Eds.), Information Security and Privacy. Proceedings, 1997. XI, 337 pages. 1997.

Vol. 1271: C. Small, P. Douglas, R. Johnson, P. King, N. Martin (Eds.), Advances in Databases. Proceedings, 1997. XI, 233 pages. 1997.

Vol. 1272: F. Dehne, A. Rau-Chaplin, J.-R. Sack, R. Tamassia (Eds.), Algorithms and Data Structures. Proceedings, 1997. X, 476 pages. 1997.